CITY COLLEGE OF SAN FRANCISCO
ROSENBERG LIBRARY
50 PHELAN AVENUE
SAN FRANCISCO, CA 94112

Ref T 185 .A67 2012 v.1

Applied science

# FOR REFERENCE

**Do Not Take From This Room**

# Applied Science

# Applied Science

## Volume 1

Editor

**Donald R. Franceschetti, Ph.D.**

*The University of Memphis*

CITY COLLEGE OF SAN FRANCISCO
ROSENBERG LIBRARY
50 PHELAN AVENUE
SAN FRANCISCO, CA 94112

**SALEM PRESS**
A Division of EBSCO Publishing

Ipswich, Massachusetts     Hackensack, New Jersey

Ref T 185 .A67 2012 v.1

Applied science

MAR 11 2015

LA

*Cover Photo:* Aerobraking Simulation on a Supercomputer (© Roger Ressmeyer/CORBIS)

Copyright © 2012, by SALEM PRESS, a Division of EBSCO Publishing, Inc.

All rights reserved. No part of this work may be used or reproduced in any manner whatsoever or transmitted in any form or by any means, electronic or mechanical, including photocopy, recording, or any information storage and retrieval system, without written permission from the copyright owner except in the case of brief quotations embodied in critical articles and reviews or in the copying of images deemed to be freely licensed or in the public domain. For permissions requests, contact proprietarypublishing@ebscohost.com.

∞ The paper used in these volumes conforms to the American National Standard for Permanence of Paper for Printed Library Materials, Z39.48-1992 (R1997).

**Library of Congress Cataloging-in-Publication Data**

Applied science / editor, Donald R. Franceschetti.
    p. cm.
 ISBN 978-1-58765-781-8 (set) — ISBN 978-1-58765-782-5 (v. 1) — ISBN 978-1-58765-783-2 (v. 2) — ISBN 978-1-58765-784-9 (v. 3) — ISBN 978-1-58765-785-6 (v. 4) — ISBN 978-1-58765-786-3 (v. 5)
 1. Engineering. 2. Technology. I. Franceschetti, Donald R., 1947-
  600--dc23

                                                                              2012002375

FIRST PRINTING
PRINTED IN THE UNITED STATES OF AMERICA

# CONTENTS

# PUBLISHER'S NOTE

*Applied Science* is a comprehensive five-volume reference set that examines the relationship between science and technology, providing insight into the many ways in which science affects daily life. Understanding the interconnectedness of the different and varied branches of science and technology is important for anyone preparing for a career or endeavor in science or technology. Toward that end, essays look beyond basic principles to examine a wide range of topics, including industrial and business applications, historical and social contexts, and the impact a particular field of science or technology will have on future jobs and careers. A career-oriented feature details core jobs within the field, with a focus on fundamental and recommended coursework.

*Applied Science* is specifically designed for a high-school audience and is edited to align with secondary or high-school curriculum standards. The content is readily accessible, as well, to patrons of both academic and university libraries. Librarians and general readers alike will also turn to this reference work for both basic information and current developments, from stem cell engineering to renewable energy production, presented in accessible language with copious reference aids. Pedagogical tools and elements, including a bibliographical directory of scientists, a timeline of major scientific milestones, and a glossary of key terms and concepts, round out this unprecedented and unique resource.

## SCOPE OF COVERAGE

Comprising over 300 lengthy and alphabetically arranged essays on a broad range of subject fields, from long-established engineering fields to cutting-edge fields such as micro- and nanotechnologies, this excellent reference work addresses applied sciences in areas as diverse as aerospace, communications, energy, information, medical, military, transportation, forensic, and food technologies. In particular, a multitude of entries fall under the emerging fields of allied health, environmental science, and electronics. *Applied Science* is also richly illustrated with photos, charts, and tables, as well as "Fascinating Facts" related to each applied-science field.

## ESSAY LENGTH AND FORMAT

Essays in the encyclopedic set range in length from 3,000 to 4,000 words. All the entries begin with ready-reference top matter, including an indication of the relevant discipline or field of study and a summary statement in which the writer explains why the topic is important to the study of the applied sciences. A selection of key terms and concepts within that major field or technology is presented, with basic principles examined. Essays then place the subject field in historical or technological perspective and examine its development and implication. Also discussed are applications and applicable products and the field's impact on industry. Cross-references direct readers to related topics, and further reading suggestions accompany all articles.

## SPECIAL FEATURES

Several features continue to distinguish this reference series from other works on applied science. The front matter includes a table detailing common units of measure, with equivalent units of measure provided for the user's convenience. Additional features are provided to assist in the retrieval of information. An index illustrates the breadth of the reference work's coverage, directing the reader to appearances of important names, terms, and topics throughout the text, and is followed by the category list, which groups article titles by general area of interest. Each of these reader aids is cumulative, referencing the entire five volumes of the complete set.

The back matter includes several appendixes and indexes, including a biographical dictionary of scientists, a compendium of the people most influential in shaping the discoveries and technologies of applied science. A time line provides a chronology of all major scientific events across significant fields and technologies, from agriculture to computers to engineering to medical to space sciences. Additionally, a complete glossary covers all technical and other specialized terms used throughout the set, while a general bibliography offers a comprehensive list of works on applied science for students seeking out more information on a particular field or subject.

## ACKNOWLEDGMENTS

Many hands went into the creation of this work. Special mention must be made of editor Donald R. Franceschetti, who played a principal role in shaping the reference work and its contents. Thanks are also due to the many academicians and professionals who worked to communicate their expert understanding of the applied sciences to the general reader; a list of these individuals and their affiliations appears at the beginning of the volume. The contributions of all are gratefully acknowledged.

# EDITOR'S INTRODUCTION

It has been traditional to describe areas of scientific endeavor as either fundamental or applied. Work in the fundamental areas is said to be curiosity driven, while work in the applied areas is motivated by human needs, or at least the desire to bring a more appealing product to market. It is not, however, possible to draw a rigid dividing line between fundamental and applied science. Knowledge of no identifiable value today may be urgently needed for application tomorrow. Take the chemistry of uranium compounds as an example. Once an obscure academic backwater, the chemistry of uranium compounds became a national concern during World War II, when it was realized that a new class of weapons based on uranium fission could be developed. Similarly, the essential genetic material, deoxyribonucleic acid (DNA), was identified in 1868, but its role in genetics was not suspected for decades and its applications to genetic engineering and DNA fingerprinting not for another century. On the other hand, advances in applied science such as the ability to produce a good vacuum, which was needed for electron tubes and used in the first computers, radio, and television, were also essential to the discovery of the electron and the development of the particle accelerators used to gain a better understanding of the forces of nature. This introductory essay presents an overview of the history of applied science and its interplay with the development of fundamental scientific knowledge and the larger culture.

Something should be said at the outset about the relationship between science and technology. Terence Kealey states in his book *The Economic Laws of Scientific Research* (1996) that technology is the activity of manipulating nature, while science is the activity of learning about nature. Technology has been a feature of human culture since the Stone Age, when humans first learned to fabricate tools. At first, technology was advanced by trial and error. As scientific knowledge developed, it was occasionally put into the service of technology, particularly in the nineteenth and twentieth centuries. A clear distinction between science and technology is not possible either. In modern times, advances in scientific knowledge have often stimulated entirely new technologies. When American inventor Thomas Alva Edison sought a long-lasting filament for the first electric light, he proceeded mainly by trial and error. His notebooks report numerous attempts with different filament materials. However, Edison did proceed with a background of fundamental science. He understood that most materials when heated in air would react with the oxygen to form an oxide, and therefore he focused on filaments carrying current in either a vacuum or an inert atmosphere. In 1883, Edison stumbled on the Edison effect—the fact that electric current would flow from a heated metal filament across an evacuated space to a positively charged conductor. Edison recorded the effect in his notebooks, but seeing no application, did not pursue it. In 1897, the English physicist Joseph John Thomson discovered the electron, and it soon became apparent that the Edison effect was simply the emission of electrons from a hot metal surface. By 1904, English physicist John Ambrose Fleming had invented the vacuum tube diode, and in 1906, the American inventor Lee de Forest showed how incorporating a third electrode would allow a very small voltage to control a much larger current, the effect that made radio and television possible.

## COMMON THEMES

As we consider the evolution of pure and fundamental science, a number of recurrent themes can be noted. One is the growth of the human population. With improvements in agriculture and medical knowledge, the human population increased, resulting in a competition for scarce resources and a greater coordination of human activities. At times, warfare resulted, with many scientific advances coming from military necessity. The science of physics has been valued particularly for providing an understanding of projectiles, ranging from rocks to arrows to cannonballs to ballistic missiles. In the middle of the twentieth century, the science of computing proved of great military value as an aid in locating the enemy, cracking its codes, and designing the most advanced weapon systems.

A second theme concerns the increasing investment that has been made in scientific research and development. To the ancient Greeks, scientific thinking was a luxury proper to the leisure class. The notion of systematic scientific research, to be done by a skilled professional class paid for its scientific work and involving possibly lengthy investigation, and the development of instruments such as the telescope, the microscope, and the voltaic pile, had to wait for the scientific revolution of the seventeenth century. At first, scientists sought patrons among the wealthy and the nobility, much as composers or painters did. Later, corporations would be formed to exploit scientific discoveries. Corporate investment was stimulated by the development of patent law, which gave inventors salable rights to their inventions for a limited time in exchange for public disclosure of how they worked. When large fortunes began to be amassed by families, such as the Rockefellers and the Du Ponts, the introduction of an inheritance tax, and later the income tax, led to the creation of great foundations to support medical and astronomical research. A further change in the scale and organization of scientific research would occur when in 1942 the United States government committed itself to the Manhattan Project, the development of a weapon of unprecedented power.

At the close of World War II, there was considerable debate as to the proper relation between science and government in peacetime. Some conservatives wanted to impose strong military control over scientific research. In 1945, Vannevar Bush, who had been director of the wartime Office of Scientific Research and Development, published the report "Science: The Endless Frontier," which painted a very rosy picture of the gains to be derived by public investment in basic scientific research. Among other things, Bush emphasized the role that government could play in supporting research in universities, research to fill in the middle ground between purely academic research and research directed toward immediate objectives. Further, universities had a natural role to play in training the next generation of scientists. Bush's arguments led to establishment of the National Science Foundation and expanded research funding within and by the National Institutes of Health in the United States. This trend was accelerated by the Soviet Union's launch of Sputnik 1, the first Earth satellite, in 1957, an event that shook the American public's faith in the inevitable superiority of American technology. In the twenty-first century, science and technology are supported by many sources, public and private, and the pace of

development is perhaps even greater than at any time in the past.

The third theme is the ultimate interconnectedness of the different branches of science and technology. This will be particularly important for anyone preparing for a career in science or technology. It is no longer wise to concentrate on just one field of endeavor, such as solid-state physics or computer engineering. One never knows when one's one field will be revolutionized by developments originating in another. One must keep an eye on technology as a whole.

## THE PRELITERATE WORLD

According to archaeologists and anthropologists, the development of written language occurred relatively recently in human history, perhaps about 3000 B.C.E. Many of the basic components of applied science date to those preliterate times when humans struggled to secure the basic necessities— food, clothing, and shelter—against a background of growing population and changing climate. When food collection was limited to hunting and gathering, knowledge of the seasons and animal behavior was important for survival. The development of primitive stone tools and weapons greatly facilitated both hunting and obtaining meat from animal carcasses and also the preservation of the hides for clothing and shelter. Sometime in the middle Stone Age, humans obtained control over fire, making it possible to soften food by cooking and to separate some metals from their ores. Control over fire also made it possible to harden earthenware pottery and keep predatory animals away at night.

With gradually improved living conditions, human fertility and longevity both increased, as did competition for necessities of life. Spoken language, music, magical thinking, and myth developed as a means of coordinating activity. Warfare, along with more peaceful approaches, was adopted as a means of settling disputes, while society was reorganized to ensure access to the necessities of life, including protection from military attack.

## THE ANCIENT WORLD

With the invention of written language it became possible to enlarge and coordinate human activity on an unprecedented scale. Several new areas of applied science and engineering were needed. Cities were established so that skilled workers could be freed from direct involvement in food production. Logistics and management become functions of the scribe class, which could read and write. Libraries were built and manuscripts collected. The beginnings of mathematics can be seen in building, surveying, and tabulating wealth. Engineers built roads so that a ruler could oversee the enlarged domain and troops could move rapidly to where they were needed. Taxes were imposed to support the central government, and accounting methods were introduced. Aqueducts were needed to bring fresh water to the cities.

The applied science of materials became critical at this juncture. Clay could be made into bricks under the tropic sun, and the biblical book of Exodus describes the use of straw fibers to increase tensile strength. While terra-cotta vessels continued to be made, kilns could reach high enough temperatures to produce glazed vessels. Bronze became the material of choice both for weapons and heavy duty pottery.

About 500 B.C.E., the ancient Greeks begin speculating on why substances behave as they do. Empedocles advanced the theory that all matter is composed of four elements: air, earth, fire, and water. This basic theory was embraced by the great medical thinker Hippocrates in his humoral theory of disease. Democritus and his teacher Leucippus argued that matter is composed of indestructible atoms. Plato proposed that the atoms were shaped like the regular solids and added a fifth substance, or quintessence, to account for heavenly bodies. His student Aristotle accepted the four-element theory but rejected the existence of atoms. Aristotle would hold great sway over medieval thinkers, and the existence of atoms, now assumed by all scientists, would remain in doubt until the beginning of the twentieth century. One can only speculate on whether acceptance of the atomic theory (which could explain, for example, why alloys such as bronze were stronger than the metals of which they were composed) would have made for a much more rapid advance of technology.

While the Babylonians and Egyptians proved themselves as master builders, the science of building proceeded largely by trial and error. Centuries before Euclid, these societies discovered the 3:4:5 right triangle and used it to prepare right angles where needed in building. The Greeks excelled at deductive logic and developed the notion of mathematical proof. Ironically, though, Greek thinkers thought that the terrestrial world was too imperfect to be described by mathematics and so missed the opportunity to develop physics along modern lines.

## THE SO-CALLED DARK AGES AND THE COMMERCIAL REVOLUTION

In 313 B.C.E., the Roman emperor Constantine converted to Christianity. While this meant better living conditions for the growing Christian population, it also meant the destruction of the products of pagan culture. The famous library at Alexandria was burned in 391 on the orders of the emperor. The fall of the Roman Empire and the destruction of the library at Alexandria in the third century B.C.E. marked a decline in learning that would continue for more than five hundred years. At the same time, Christianity, which had become the official religion of the empire with the conversion of the emperor, emphasized the world to come over the present one. Nonetheless, progress in technology continued, new agricultural techniques were introduced, and the horse became a potent contributor to both warfare and agriculture. Scientific ideas eventually made their way back into European culture.

The Crusades of 1095 to 1291 made Europeans more aware of the larger world and of the possibilities provided by trade with the Middle East and Far East. Trade required the development of durable forms of money and a system of banking to finance expeditions that could be lost or could return great profits. Notions of risk had to be quantified and a reliable system of bookkeeping introduced. The leaders in this undertaking were the republics of the Italian peninsula and the Dutch republic. With accumulations of capital, inventors could obtain financing to market their inventions and the notion of a patent— the grant of a legal monopoly to an inventor for a limited time—developed. The first patent appears to have been awarded by the Republic of Florence in 1421 to the architect Filippo Brunelleschi. Provisions for the granting of patents were written into the Statute of Monopolies in England in 1623 and into Article 1 of the United States Constitution of 1787.

## ASTRONOMY, THE CALENDAR, AND LONGITUDE

One might think that astronomy could serve as a paradigmatic example of fundamental science, but astronomy, and later space science, provides an example of a scientific activity often pursued for practical benefit. Fixing the calendar, in particular, provides an interesting illustration of the interaction between pure astronomy, practicality, and religion.

The calendar provided a scheme of dates for planting and harvesting. Ancient stone monuments in Europe and the Americas may have functioned in part as astronomical observatories, to keep track of the solstices and equinoxes. The calendar also provided a means for keeping religious feasts such as Easter in synchrony with celestial events such as the vernal equinox. Roman emperor Julius Caesar introduced a calendar based on a year of 365 days. This calendar, called the Julian, would endure for more than fifteen hundred years. Eventually, however, it was realized that the Julian year was about six hours shorter than Earth's orbital period and so was out of sync with astronomical events. In 1583, Pope Gregory XIII introduced the system of leap-year days. The Gregorian calendar is out of sync with Earth's orbit by only one day in thirty-three hundred years.

Modern astronomy begins with the work of Nicolaus Copernicus, Galileo Galilei, and Sir Isaac Newton. Copernicus was the first to advance the heliocentric model of the solar system, suggesting that it would be simpler to view Earth as a planet orbiting a stationary sun. Copernicus, however, was a churchman, and fear of opposition from church leaders led him to postpone publication of his ideas until the year of his death. Galileo was the first to use a technological advance, the invention of the telescope, to record numerous observations that called into question the geocentric model of the known universe. Among these was the discovery of four moons that orbited the planet Jupiter.

At the same time, advances in shipbuilding and navigation served to bring the problem of longitude to prominence. Out of sight of land, a sailing ship could easily determine its latitude on a clear night by noting the elevation of the pole star above the horizon. To determine longitude, however, required an accurate measurement of time, which was difficult to do onboard a moving ship at sea. Galileo was quick to propose that occultations by Jupiter of its moons could be used as a universal clock. The longitude problem also drew the attention of the great Newton. Longitude would eventually be solved by John Harrison's invention of a chronometer that could be

used on ship, and the deciphering of it also fostered an improved understanding of celestial mechanics.

Since the launching of the first artificial satellites, astronomy, or rather, space and planetary science, has assumed an even greater role in applied science. The safety of astronauts working in space requires understanding the dynamics of solar flares. A deeper understanding of the solar atmosphere and its dynamics could also have important consequences for long-range weather prediction.

### SCIENTIFIC REVOLUTION

The Renaissance and the Protestant Reformation marked something of a rebirth of scientific thinking. With wealthy patrons, natural philosophers felt secure in challenging the authority of Aristotle. Galileo published arguments in favor of the Copernican solar system. In the *Novum Organum* (*New Instrument*), Sir Francis Bacon formalized the inductive method, in which generalizations could be made from observations, which then could be tested by further observation or experiment. With the nominal support of the British Crown, the Royal Society was formed to serve as a forum for the exchange of scientific ideas and the support and publication of research results. Need for larger scale studies brought craftsmen into the sciences, culminating in the recognition of the professional scientist. Bacon also proposed that the government undertake the support of scientific investigation for the common good. Bacon himself tried his hand at applied science. He conceived the idea that low temperatures could preserve meat while on a coach trip. He stopped the coach, purchased a chicken from a farmer's wife, and stuffed it with snow. Unfortunately he contracted pneumonia while doing this experiment and died forthwith.

The Industrial Revolution followed on the heels of the scientific revolution in England. Key to the Industrial Revolution was the technology of the steam engine, which was the first portable source of motive power not dependent on human or animal muscle. The steam engine powered factories, ships, and later locomotives. In the case of the steam engine, technological advance preceded the development of the pertinent science—thermodynamics and present-day understanding of heat as a random molecular form of energy.

To do justice to the full scale of applied science would indeed take a multivolume encyclopedia such as this one. In the remainder of this introduction, consideration will be made of only a few representative fields, highlighting the evolution of each area and its interconnectedness with fundamental and applied science as a whole.

## APPLIED CHEMISTRY

In 1792, Scottish inventor William Murdock discovered a way to produce illuminating gas by the destructive distillation of coal, producing a cleaner and more dependable source of light than previously available and bringing about the gaslight era. The production of illuminating gas left behind a nasty residue called coal tar. A search was launched to find an application for this major industrial waste. An early use, the waterproofing of cloth, was discovered by Scottish chemist Charles Macintosh, resulting in the raincoat that now carries his name. In 1856, English chemist William Henry Perkin discovered the first of the coal tar dyes, mauve. The color mauve, a deep purple, had been obtained from plant sources and had become something of a fashion fad in Paris in 1857. The fad spread to London in 1858, when Queen Victoria chose to wear a mauve velvet dress to her daughter's wedding. The demand for mauve outstripped the supply of vegetable sources, and the discovery of several other dyes followed.

The possibility of dyeing living tissue was rapidly seized on and applied to the tissues of the human body and the microorganisms that afflict it. German bacteriologist Paul Ehrlich proposed that the selective adsorption of dyes could serve as the basis for a chemically based therapy to kill infectious disease-bearing organisms.

## OPTICS, THE MICROSCOPE, AND MICROBIOLOGY

The use of lenses as an aid to vision may date to China in 500 B.C.E. Marco Polo, in his journeys more than seventeen hundred years later, reported seeing many Chinese wearing eyeglasses. English physicist (and curator of experiments for the Royal Society) Robert Hooke published his book *Micrographia* (*Tiny Handwriting*) in 1665, which included many illustrations of living tissue. Antoni van Leeuwenhoek was influenced by Hooke and reported many observations of microbial life to the Royal Society. The simple microscopes of Hooke and Leeuwenhoek suffered from many forms of aberration or distortion. Subsequent investigators introduced combinations of lenses to reduce the aberration, and good compound microscopes became available for the study of microscopic life around 1830.

An accomplished physical chemist, Louis Pasteur is best known as the father of microbiology. Pasteur's work on microbes began with the study of the problems of the fermentation industry. While Leeuwenhoek had reported the existence of microorganisms, the notion that they might be responsible for disease or agricultural problems met considerable resistance.

Pasteur was drawn into applied research by problems arising in the fermentation industry. In 1857, he announced that fermentation was the result of microbial action. He also showed that the souring of milk resulted from microorganisms, leading to the development of pasteurization as a technique for preserving milk. As a sequel to his work on fermentation, Pasteur brought into question the commonly held idea that living organisms could generate spontaneously. Through carefully designed experiments, he demonstrated that broth could be maintained indefinitely, even when exposed to the air, provided that bacteria-carrying dust was excluded.

Pasteur's further research included investigating the diseases that plagued the French silk industry. He developed a means of vaccinating sheep against infection by *Bacillus anthracis* and a vaccine to protect chickens against cholera. Pasteur's most impressive achievement may have been the development of a treatment effective against the rabies virus for people bitten by rabid dogs or wolves.

Pasteur's scientific achievement illustrates the close interplay of fundamental and applied advances that occur in many scientific fields. Political scientist Donald Stokes has termed this arena of application-driven scientific research as Pasteur's quadrant, to distinguish it from purely curiosity-driven research (as in modern particle physics), advance by trial and error (for example, Edison's early work on the electric light), and the simple cataloging of properties and behaviors (as in classical botany and zoology). The study of applied science is a detailed examination of Pasteur's quadrant.

## ELECTROMAGNETIC TECHNOLOGY

The history of electromagnetic devices provides an excellent example of the complex interplay of fundamental and applied science. The phenomena of static electricity and natural magnetism were described by Thales of Miletus but remained curiosities through much of history. The magnetic compass was developed by Chinese explorers in about 1100 B.C.E., and the nature of Earth's magnetic field was explored by William Gilbert, physician to Queen Elizabeth I, around 1600. By the late eighteenth century, a number of devices for producing and storing static electricity were being used in popular demonstrations, and the lightning rod invented by Benjamin Franklin greatly reduced the damage due to lightning strikes on tall buildings. In 1800, Italian physicist Alessandro Volta developed the first electrical battery. Equipped with a source of continuous electric current, electrical and electromagnetic discoveries, practical and fundamental, accumulated at a breakneck pace.

The voltaic pile, or battery, was employed by British scientist Sir Humphry Davy to isolate a number of chemical elements for the first time. In 1820, Danish physicist Hans Christian Ørsted discovered that any current-carrying wire was surrounded by an electric field. In 1831, English physicist Michael Faraday discovered that a changing magnetic field would induce an electric current in a loop of wire, thus paving the way for the electric generator and the transformer. In Albany, New York, schoolteacher Joseph Henry set his students the challenge of building the strongest possible electromagnet. Henry would move on to be professor of natural philosophy at Princeton University, where he invented a primitive telegraph.

The basic laws of electromagnetism were summarized in 1865 by Scottish physicist James Clerk Maxwell in a set of four differential equations that yielded a number of practical results almost immediately. For free space, these equations had wavelike solutions that traveled at the speed of light, which was immediately seen to be a form of electromagnetic radiation. Further, it turned out that visible light covered only a small frequency range. Applied scientists soon discovered how to transmit messages by radio waves: electromagnetic waves of much lower frequency.

## THE COMPUTER

One of the most clearly useful of modern artifacts, the digital electronic computer, as it has come to be known, has a lineage that includes the most abstract

of mathematics, the automated loom, the vacuum tubes of the early twentieth century, and the modern sciences of semiconductor physics and photochemistry. Although computing devices such as the abacus and slide rule themselves have a long history, the programmable digital computer has advanced computational power by many orders of magnitude. However, the basic logic of the computer and the computer program arose from a mathematical logician's attempt to answer a problem arising in the foundations of mathematics.

From the time of the ancient Greeks to the end of the nineteenth century, mathematicians had assumed that their subject was essentially a study of the real world, the part amenable to purely deductive reasoning. This included the structure of space and the basic rules of counting, which lead to the rules of arithmetic and algebra. With the discovery of non-Euclidean geometries and the paradoxes of set theory, mathematicians felt the need for a closer study of the foundations of mathematics, to make sure that the objects that might exist only in their minds could be studied and talked about without risking inconsistency.

David Hilbert, a professor of mathematics at the University of Göttingen, was the recognized leader of German mathematics. At a mathematics conference in 1928, Hilbert identified three questions about the foundations of mathematics that he hoped would be resolved in short order. The third of these was the so-called decidability problem: Was there was a foolproof procedure to determine whether a mathematical statement was true or false? Essentially, if one had the statement in symbolic form, was there a procedure for manipulating the symbols in such a way that one could determine whether the statement was true in a finite number of steps?

British mathematician Alan Turing presented an analysis of the problem by showing that any sort of mathematical symbol manipulation was in essence a computation and thus a manipulation of symbols not unlike the addition or multiplication one learns in elementary school. Any such symbolic manipulation could be emulated by an abstract machine that worked with a finite set of symbols that would store a simple set of instructions and process a one dimensional array of symbols, replacing it with a second array of symbols. Turing showed that there was no solution in general to Hilbert's decision problem but in the process also showed how to construct a machine (now called a Turing machine) that could execute any possible calculation. The machine would operate on a string of symbols recorded on a tape and would output the result of the same calculation on the same tape. Further, Turing showed the existence of machines that could read instructions given in symbolic form and then perform any desired computation on a one-dimensional array of numbers that followed. The universal Turing machine was a programmable digital computer. The instructions could be read from a one-dimensional tape, a magnetically stored memory, or a card punched with holes, as used for mechanized weaving of fabric.

The earliest electronic computers were developed at the time of World War II and involved numerous vacuum tubes. Since vacuum tubes are based on thermionic emission, the Edison effect mentioned above, they produced immense amounts of heat and involved the possibility that the heating element in one of the tubes might well burn out during the computation. In fact, it was standard procedure to run a program, one that required proper function of all the vacuum tubes, both before and after the program of interest. If the results of the first and last computations did not vary, one could assume that no tubes had burned out in the mean time.

World War II ended in 1945. In addition to the critical role computing machines played in the design of the first atomic bombs, computational science played an important role in predicting the behavior of targets. The capabilities of computing machines would grow rapidly following the invention of the transistor by John Bardeen, Walter Brattain, and William Shockley in 1947. In this case, fundamental science led to tremendous advances in applied science.

The story of semiconductor science is worth telling. Silicon was unusual in displaying an increase in electrical conductivity as the temperature was raised. In general, when one finds an interesting property of a material, one tries to purify and refine the material. However, purified silicon lost most of its conductivity. On further investigation, it was found that tiny concentrations of impurities could vastly change both the amount of electrical conductivity and the mechanism by which it occurs. Because the useful properties of semiconductors depend critically on the impurities or "dirt" in the material,

solid-state (and other) physicists sometimes refer to the field as dirt physics. Adding a small amount of phosphorus to pure silicon resulted in n-type conductivity, the type due to electrons moving in response to an electric field. Adding an impurity such as boron produced p-type conductivity, in which electron vacancies (in chemical bonds) moved through the material. Creating a p-type region next to an n-type produced a junction that let current flow in one direction and not the other, just as in a vacuum tube diode. Placing a p-type region between two n-types produced the equivalent of Lee de Forest's diode—a transistor. The transistor, however, did not require a heater and could be miniaturized.

In the 1960's, the production of integrated circuits—many transistors and other circuit elements on a single silicon wafer or chip—began. Currently hundreds of thousands of circuit elements are available on a single chip, and anyone who buys a laptop computer will command more computational power than any government could control in 1950.

*Donald R. Franceschetti*

### FURTHER READING

Bell Telephone Laboratories. *A History of Engineering and Science in the Bell System: Electronics Technology, 1925-1975.* Edited by M.D. Fagen. 7 vols. New York: Bell Laboratories, 1975. Provides detailed information on the development of the transistor and the integrated circuit.

Bodanis, David. *Electric Universe: How Electricity Switched on the Modern World.* New York: Three Rivers Press, 2005. Popular exposition of the applications of electronics and electromagnetism from the time of Joseph Henry to the microprocessor age.

Burke, James. *Connections.* New York: Simon & Schuster, 2007. Describes linkages between inventions throughout history.

Cobb, Cathy, and Harold Goldwhite. *Creations of Fire: Chemistry's Lively History from Alchemy to the Atomic Age.* Cambridge, Mass.: Perseus, 1995. History of pure and applied chemistry from the beginning through the late twentieth century.

Garfield, Simon. *Mauve: How One Man Invented a Color That Changed the World.* New York: W. W. Norton, 2000. Focuses on how the single and partly accidental discovery of coal tar dyes led to several new areas of chemical industry.

Kealey, Terence. *The Economic Laws of Scientific Research.* New York: St. Martin's Press, 1996. Makes the case that government funding of scientific research is relatively inefficient and emphasizes the role of private investment and hobbyist scientists.

Schlager, Neil, ed. *Science and Its Times: Understanding the Social Significance of Scientific Discovery.* 8 vols. Detroit: Gale Group, 2000. Massive reference work on the impact of scientific and technological developments from the earliest times to the present.

Sobel, Dava. *Longitude.* New York: Walker & Company, 2007. Story of the competition among scientists and inventors to develop a reliable means of determining longitude at sea.

Stokes, Donald E. *Pasteur's Quadrant: Basic Science and Technological Innovation.* Washington, D.C.: Brookings Institution Press, 1997. Presents an extended argument that many fundamental scientific discoveries originate in application-driven research, and that the distinction between pure and applied science is not, of itself, very useful.

# CONTRIBUTORS

Richard Adler
*University of Michigan-Dearborn*

Jeongmin Ahn
*Syracuse University*

Ezinne Amaonwu
*Rockville, Maryland*

Michael Auerbach
*Marblehead, Massachusetts*

Mihaela Avramut
*Verlan Medical Communications*

Dana K. Bagwell
*Memory Health and Fitness Institute*

Craig Belanger
*Journal of Advancing Technology*

Raymond D. Benge
*Tarrant County College*

Harlan H. Bengtson
*Southern Illinois University, Edwardsville*

Lakhdar Boukerrou
*Florida Atlantic University*

Victoria M. Breting-García
*Houston, Texas*

Joseph Brownstein
*Atlanta, Georgia*

Michael A. Buratovich
*Spring Arbor University*

Byron D. Cannon
*University of Utah*

Christina Capriccioso
*University of Michigan College of Engineering*

Richard P. Capriccioso
*University of Phoenix*

Christine M. Carroll
*American Medical Writers Association*

Michael J. Caulfield
*Gannon University*

Martin Chetlen
*Moorpark College*

Edward N. Clarke
*Worcester Polytechnic Institute*

Robert L. Cullers
*Kansas State University*

Christopher Dean
*Curtin University of Technology, Perth, Australia*

Joseph Dewey
*University of Pittsburgh*

Thomas Drucker
*University of Wisconsin-Whitewater*

Jeremy Dugosh
*American Board of Internal Medicine*

Elvira R. Eivazova
*Vanderbilt University School of Medicine*

David Elliott
*Northern Arizona University*

Renée Euchner
*American Medical Writers Association*

Jack Ewing
*Boise, Idaho*

Ronald J. Ferrara
*Middle Tennessee State University, Murfreesboro*

June Gastón
*Borough of Manhattan Community College, City University of New York*

Jennifer L. Gibson
*Marietta, Georgia*

James S. Godde
*Monmouth College, Illinois*

Dalton R. Gossett
*College of Arts and Sciences at Louisiana State University, Shreveport*

Glenda Griffin
*Newton Gresham Library, Sam Houston State University*

Gina Hagler
*Washington, D.C.*

Wendy C. Hamblet
*North Carolina Agricultural and Technical State University*

Howard V. Hendrix
*California State University, Fresno*

Robert M. Hordon
*Rutgers University*

Ngaio Hotte
*Vancouver, British Columbia*

Carly L. Huth
*Publication Services, Inc.*

April D. Ingram
*Kelowna, British Columbia*

Micah L. Issitt
*Philadelphia, Pennsylvania*

Jerome A. Jackson
*Florida Gulf Coast University*

Domingo M. Jariel, Jr.
*Louisiana State University, Eunice*

Bruce E. Johansen
*University of Nebraska at Omaha*

Cheryl Pokalo Jones
*Townsend, Delaware*

Vincent Jorgensen
*Sunnyvale, California*

Bassam Kassab
*Santa Clara Valley Water District*

Marylane Wade Koch
*Loewenberg School of Nursing,
University of Memphis*

Narayanan M. Komerath
*Georgia Institute of Technology*

Jeanne L. Kuhler
*Benedictine University*

Lisa LaGoo
*Medtronic*

Dawn A. Laney
*Atlanta, Georgia*

Jeffrey Larson
*Tioga Medical Center, Tioga,
North Dakota*

M. Lee
*Independent Scholar*

Donald W. Lovejoy
*Palm Beach Atlantic University*

Arthur J. Lurigio
*Loyola University Chicago*

R. C. Lutz
*CII Group*

Marianne M. Madsen
*University of Utah*

Mary E. Markland
*Argosy University*

Sergei A. Markov
*Austin Peay State University*

Jordan M. Marshall
*Indiana University-Purdue
University*

Amber M. Mathiesen
*University of Utah*

Laurence W. Mazzeno
*Alvernia University*

Roman Meinhold
*Assumption University, Bangkok,
Thailand*

Julia M. Meyers
*Duquesne University*

Randall L. Milstein
*Oregon State University*

Mary R. Muslow
*Louisiana Tech University*

M. Mustoe
*Eastern Oregon University*

Terrence R. Nathan
*University of California, Davis*

Holly Nyple
*San Jose, California*

David Olle
*Eastshire Communications*

Ayman Oweida
*University of Western Ontario*

Robert J. Paradowski
*Rochester Institute of
Technology*

Ellen E. Anderson Penno
*Western Laser Eye Associates*

John R. Phillips
*Purdue University Calumet*

George R. Plitnik
*Frostburg State University*

Michael L. Qualls
*Fort Valley State University*

Cynthia F. Racer
*American Medical Writers
Association*

Corie Ralston
*Lawrence Berkeley National
Laboratory*

Steven J. Ramold
*Eastern Michigan University*

Diane C. Rein
*University at Buffalo*

Richard M. J. Renneboog
*Independent Scholar*

Barbara J. Rich
*Bend, Oregon*

Joseph Di Rienzi
*College of Notre Dame of
Maryland*

James L. Robinson
*University of Illinois*

Charles W. Rogers
*Southwestern Oklahoma State
University*

Carol A. Rolf
*Rivier College*

Lars Rose
*Author, The Nature of Matter*

Charles Rosenberg
*Milwaukee, Wisconsin*

Julia A. Rosenthal
*Chicago, Illinois*

Joseph R. Rudolph
*Towson University*

Elizabeth D. Schafer
*Loachapoka, Alaska*

Sibani Sengupta
*American Medical Writers
Association*

Martha A. Sherwood
*Kent Anderson Law Office*

Linda R. Shoaf
*American Dietetic Association*

Paul P. Sipiera
*Harper College, Palatine, Illinois*

Billy R. Smith, Jr.
*Anne Arundel Community College*

Dwight G. Smith
*Southern Connecticut State
University*

Roger Smith
*Portland, Oregon*

Ruth Waddell Smith
*Michigan State University*

Max Statman
*Eastman Chemical Company*

Polly D. Steenhagen
*Delaware State University*

Judith L. Steininger
*Milwaukee School of Engineering*

Martin V. Stewart
*Middle Tennessee State
University*

Robert E. Stoffels
*St. Petersburg, Florida*

Rena Christina Tabata
*University of British Columbia*

John M. Theilmann
*Converse College*

Bethany Thivierge
*Technicality Resources*

Anh Tran
*Wichita State University, Kansas*

Christine Watts
*University of Sydney*

Shawncey Jay Webb
*Taylor University*

Judith Weinblatt
*New York, New York*

George M. Whitson III
*University of Texas at Tyler*

Edwin G. Wiggins
*Webb Institute, Glen Cove, New York*

Thomas A. Wikle
*Oklahoma State University*

Bradley R. A. Wilson
*University of Cincinnati*

Barbara Woldin
*American Medical Writers
Association*

Jessica C. Y. Wong
*Environment Canada*

Robin L. Wulffson
*Faculty, American College of
Obstetrics and Gynecology*

Susan M. Zneimer
*U.S. Labs, Irvine, California*

# COMMON UNITS OF MEASURE

Common prefixes for metric units—which may apply in more cases than shown below—include *giga-* (1 billion times the unit), *mega-* (one million times), *kilo-* (1,000 times), *hecto-* (100 times), *deka-* (10 times), *deci-* (0.1 times, or one tenth), *centi-* (0.01, or one hundredth), *milli-* (0.001, or one thousandth), and *micro-* (0.0001, or one millionth).

| Unit | Quantity | Symbol | Equivalents |
|------|----------|--------|-------------|
| Acre | Area | ac | 43,560 square feet<br>4,840 square yards<br>0.405 hectare |
| Ampere | Electric current | A *or* amp | 1.00016502722949 international ampere<br>0.1 biot *or* abampere |
| Angstrom | Length | Å | 0.1 nanometer<br>0.0000001 millimeter<br>0.000000004 inch |
| Astronomical unit | Length | AU | 92,955,807 miles<br>149,597,871 kilometers<br>(mean Earth-Sun distance) |
| Barn | Area | b | $10^{-28}$ meters squared<br>(approx. cross-sectional area of 1 uranium nucleus) |
| Barrel<br>(dry, for most produce) | Volume/capacity | bbl | 7,056 cubic inches; 105 dry quarts; 3.281 bushels, struck measure |
| Barrel<br>(liquid) | Volume/capacity | bbl | 31 to 42 gallons |
| British thermal unit | Energy | Btu | 1055.05585262 joule |
| Bushel<br>(U.S., heaped) | Volume/capacity | bsh *or* bu | 2,747.715 cubic inches<br>1.278 bushels, struck measure |
| Bushel<br>(U.S., struck measure) | Volume/capacity | bsh *or* bu | 2,150.42 cubic inches<br>35.238 liters |
| Candela | Luminous intensity | cd | 1.09 hefner candle |
| Celsius | Temperature | C | 1° centigrade |
| Centigram | Mass/weight | cg | 0.15 grain |
| Centimeter | Length | cm | 0.3937 inch |
| Centimeter, cubic | Volume/capacity | cm³ | 0.061 cubic inch |
| Centimeter, square | Area | cm² | 0.155 square inch |
| Coulomb | Electric charge | C | 1 ampere second |
| Cup | Volume/capacity | C | 250 milliliters<br>8 fluid ounces<br>0.5 liquid pint |

| Unit | Quantity | Symbol | Equivalents |
|------|----------|--------|-------------|
| Deciliter | Volume/capacity | dl | 0.21 pint |
| Decimeter | Length | dm | 3.937 inches |
| Decimeter, cubic | Volume/capacity | dm³ | 61.024 cubic inches |
| Decimeter, square | Area | dm² | 15.5 square inches |
| Dekaliter | Volume/capacity | dal | 2.642 gallons<br>1.135 pecks |
| Dekameter | Length | dam | 32.808 feet |
| Dram | Mass/weight | dr *or* dr avdp | 0.0625 ounce<br>27.344 grains<br>1.772 grams |
| Electron volt | Energy | eV | $1.5185847232839 \times 10^{-22}$ Btus<br>$1.6021917 \times 10^{-19}$ joules |
| Fermi | Length | fm | 1 femtometer<br>$1.0 \times 10^{-15}$ meters |
| Foot | Length | ft *or* ' | 12 inches<br>0.3048 meter<br>30.48 centimeters |
| Foot, square | Area | ft² | 929.030 square centimeters |
| Foot, cubic | Volume/capacity | ft³ | 0.028 cubic meter<br>0.0370 cubic yard<br>1,728 cubic inches |
| Gallon (British Imperial) | Volume/capacity | gal | 277.42 cubic inches<br>1.201 U.S. gallons<br>4.546 liters<br>160 British fluid ounces |
| Gallon (U.S.) | Volume/capacity | gal | 231 cubic inches<br>3.785 liters<br>0.833 British gallon<br>128 U.S. fluid ounces |
| Giga-electron volt | Energy | GeV | $1.6021917 \times 10^{-10}$ joule |
| Gigahertz | Frequency | GHz | — |
| Gill | Volume/capacity | gi | 7.219 cubic inches<br>4 fluid ounces<br>0.118 liter |
| Grain | Mass/weight | gr | 0.037 dram<br>0.002083 ounce<br>0.0648 gram |
| Gram | Mass/weight | g | 15.432 grains<br>0.035 avoirdupois ounce |

| Unit | Quantity | Symbol | Equivalents |
|------|----------|--------|-------------|
| Hectare | Area | ha | 2.471 acres |
| Hectoliter | Volume/capacity | hl | 26.418 gallons<br>2.838 bushels |
| Hertz | Frequency | Hz | $1.08782775707767 \times 10^{-10}$ cesium atom frequency |
| Hour | Time | h | 60 minutes<br>3,600 seconds |
| Inch | Length | in *or* " | 2.54 centimeters |
| Inch, cubic | Volume/capacity | in³ | 0.554 fluid ounce<br>4.433 fluid drams<br>16.387 cubic centimeters |
| Inch, square | Area | in² | 6.4516 square centimeters |
| Joule | Energy | J | $6.2414503832469 \times 10^{18}$ electron volt |
| Joule per kelvin | Heat capacity | J/K | $7.24311216248908 \times 10^{22}$ Boltzmann constant |
| Joule per second | Power | J/s | 1 watt |
| Kelvin | Temperature | K | -272.15° Celsius |
| Kilo-electron volt | Energy | keV | $1.5185847232839 \times 10^{-19}$ joule |
| Kilogram | Mass/weight | kg | 2.205 pounds |
| Kilogram per cubic meter | Mass/weight density | kg/m³ | $5.78036672001339 \times 10^{-4}$ ounces per cubic inch |
| Kilohertz | Frequency | kHz | — |
| Kiloliter | Volume/capacity | kl | — |
| Kilometer | Length | km | 0.621 mile |
| Kilometer, square | Area | km² | 0.386 square mile<br>247.105 acres |
| Light-year<br>(distance traveled by light in one Earth year) | Length/distance | lt-yr | 5,878,499,814,275.88 miles<br>$9.46 \times 10^{12}$ kilometers |
| Liter | Volume/capacity | L | 1.057 liquid quarts<br>0.908 dry quart<br>61.024 cubic inches |
| Mega-electron volt | Energy | MeV | — |
| Megahertz | Frequency | MHz | — |
| Meter | Length | m | 39.37 inches |
| Meter, cubic | Volume/capacity | m³ | 1.308 cubic yards |

| Unit | Quantity | Symbol | Equivalents |
|------|----------|--------|-------------|
| Meter per second | Velocity | m/s | 2.24 miles per hour<br>3.60 kilometers per hour |
| Meter per second per second | Acceleration | m/s$^2$ | 12,960.00 kilometers per hour per hour<br>8,052.97 miles per hour per hour |
| Meter, square | Area | m$^2$ | 1.196 square yards<br>10.764 square feet |
| Metric. *See* unit name | | | |
| Microgram | Mass/weight | mcg *or* μg | 0.000001 gram |
| Microliter | Volume/capacity | μl | 0.00027 fluid ounce |
| Micrometer | Length | μm | 0.001 millimeter<br>0.00003937 inch |
| Mile<br>(nautical international) | Length | mi | 1.852 kilometers<br>1.151 statute miles<br>0.999 U.S. nautical miles |
| Mile<br>(statute or land) | Length | mi | 5,280 feet<br>1.609 kilometers |
| Mile, square | Area | mi$^2$ | 258.999 hectares |
| Milligram | Mass/weight | mg | 0.015 grain |
| Milliliter | Volume/capacity | ml | 0.271 fluid dram<br>16.231 minims<br>0.061 cubic inch |
| Millimeter | Length | mm | 0.03937 inch |
| Millimeter, square | Area | mm$^2$ | 0.002 square inch |
| Minute | Time | m | 60 seconds |
| Mole | Amount of substance | mol | $6.02 \times 10^{23}$ atoms or molecules of a given substance |
| Nanometer | Length | nm | 1,000,000 fermis<br>10 angstroms<br>0.001 micrometer<br>0.00000003937 inch |
| Newton | Force | N | 0.224808943099711 pound force<br>0.101971621297793 kilogram force<br>100,000 dynes |
| Newton meter | Torque | N·m | 0.7375621 foot-pound |
| Ounce<br>(avoirdupois) | Mass/weight | oz | 28.350 grams<br>437.5 grains<br>0.911 troy or apothecaries' ounce |

| Unit | Quantity | Symbol | Equivalents |
|------|----------|--------|-------------|
| Ounce (troy) | Mass/weight | oz | 31.103 grams<br>480 grains<br>1.097 avoirdupois ounces |
| Ounce (U.S., fluid or liquid) | Mass/weight | oz | 1.805 cubic inch<br>29.574 milliliters<br>1.041 British fluid ounces |
| Parsec | Length | pc | 30,856,775,876,793 kilometers<br>19,173,511,615,163 miles |
| Peck | Volume/capacity | pk | 8.810 liters |
| Pint (dry) | Volume/capacity | pt | 33.600 cubic inches<br>0.551 liter |
| Pint (liquid) | Volume/capacity | pt | 28.875 cubic inches<br>0.473 liter |
| Pound (avoirdupois) | Mass/weight | lb | 7,000 grains<br>1.215 troy or apothecaries' pounds<br>453.59237 grams |
| Pound (troy) | Mass/weight | lb | 5,760 grains<br>0.823 avoirdupois pound<br>373.242 grams |
| Quart (British) | Volume/capacity | qt | 69.354 cubic inches<br>1.032 U.S. dry quarts<br>1.201 U.S. liquid quarts |
| Quart (U.S., dry) | Volume/capacity | qt | 67.201 cubic inches<br>1.101 liters<br>0.969 British quart |
| Quart (U.S., liquid) | Volume/capacity | qt | 57.75 cubic inches<br>0.946 liter<br>0.833 British quart |
| Rod | Length | rd | 5.029 meters<br>5.50 yards |
| Rod, square | Area | $rd^2$ | 25.293 square meters<br>30.25 square yards<br>0.00625 acre |
| Second | Time | s or sec | $1/60$ minute<br>$1/3600$ hour |
| Tablespoon | Volume/capacity | T or tb | 3 teaspoons<br>4 fluid drams |
| Teaspoon | Volume/capacity | t or tsp | 0.33 tablespoon<br>1.33 fluid drams |

| Unit | Quantity | Symbol | Equivalents |
|------|----------|--------|-------------|
| Ton (gross or long) | Mass/weight | t | 2,240 pounds<br>1.12 net tons<br>1.016 metric tons |
| Ton (metric) | Mass/weight | t | 1,000 kilograms<br>2,204.62 pounds<br>0.984 gross ton<br>1.102 net tons |
| Ton (net or short) | Mass/weight | t | 2,000 pounds<br>0.893 gross ton<br>0.907 metric ton |
| Volt | Electric potential | V | 1 joule per coulomb |
| Watt | Power | W | 1 joule per second<br>0.001 kilowatt<br>$2.84345136093995 \times 10^{-4}$ ton of refrigeration |
| Yard | Length | yd | 0.9144 meter |
| Yard, cubic | Volume/capacity | $yd^3$ | 0.765 cubic meter |
| Yard, square | Area | $yd^2$ | 0.836 square meter |

# COMPLETE LIST OF CONTENTS

## Volume 1

## Volume 2

## Volume 3

Contents . . . . . . . . . . . . . . . . . . . . . . . . . . . . . . . . v

**Volume 4**

Contents . . . . . . . . . . . . . . . . . . . . . . . . . . . . . . . v
Common Units of Measure . . . . . . . . . . . . . . . . vii
Complete List of Contents . . . . . . . . . . . . . . . . xiii

## Volume 5

# Applied Science

# ACOUSTICS

## FIELDS OF STUDY

Electrical, chemical, and mechanical engineering; architecture; music; speech; psychology; physiology; medicine; atmospheric physics; geology; oceanography.

## SUMMARY

Acoustics is the science dealing with the production, transmission, and effects of vibration in material media. If the medium is air and the vibration frequency is between 18 and 18,000 hertz (Hz), the vibration is termed "sound." Sound is used in a broader context to describe sounds in solids and underwater and structure-borne sounds. Because mechanical vibrations, whether natural or human induced, have accompanied humans through the long course of human evolution, acoustics is the most interdisciplinary science. For humans, hearing is a very important sense, and the ability to vocalize greatly facilitates communication and social interaction. Sound can have profound psychological effects; music may soothe or relax a troubled mind, and noise can induce anxiety and hypertension.

## KEY TERMS AND CONCEPTS

- **Cochlea:** Inner ear, which converts pressure waves of sound into electric impulses that are transmitted to the brain via the auditory nerves.
- **Decibel (dB):** Unit of sound intensity used to quantify the loudness of a vibration.
- **Destructive Interference:** Interference that occurs when two waves having the same amplitude in opposite directions come together and cancel each other.
- **Doppler Effect:** Apparent change in frequency of a wave because of the relative motions of the source and an observer. Wavelengths of approaching objects are shortened, and those of receding objects are lengthened.
- **Hertz (Hz):** Unit of frequency; the number of vibrations per second of an oscillation.
- **Infrasound:** Air vibration below 20 hertz; perceived as vibration.
- **Physical Acoustics:** Theoretical area concerned with the fundamental physics of wave propagation and the use of acoustics to probe the physical properties of matter.
- **Resonance:** Large amplitude of vibration that occurs when an oscillator is driven at its natural frequency.
- **Sound:** Vibrations in air having frequencies between 20 and 20,000 hertz and intensities between 0 and 135 decibels and therefore perceptible to humans.
- **Sound Spectrum:** Representation of a sound in terms of the amount of vibration at a each individual frequency. Usually presented as a graph of amplitude (plotted vertically) versus frequency (plotted horizontally).
- **Spectrogram:** Graph used in speech research that plots frequency (vertical axis) versus the time of the utterance (horizontal axis). The amplitude of each frequency component is represented by its darkness.
- **Transducer:** Device that transmutes one form of energy into another. Acoustic examples include microphones and loudspeakers.
- **Ultrasound:** Frequencies above 20,000 hertz used by bats for navigation and by humans for industrial applications and nonradiative ultrasonic imaging.

### DEFINITION AND BASIC PRINCIPLES
The words "acoustics," and "phonics" evolved from ancient Greek roots for hearing and speaking, respectively. Thus, acoustics began with human communication, making it one of the oldest if not the most basic of sciences. Because acoustics is ubiquitous in human endeavors, it is the broadest and most interdisciplinary of sciences; its most profound contributions have occurred when it is commingled with

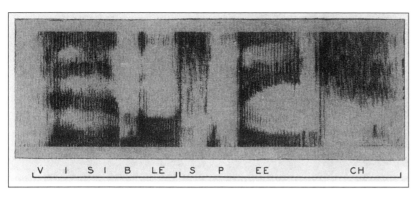

*A spectrogram of the words "visible speech" being spoken.* (Science Source).

Acoustics arguably originated with human communication and music. The caves in which the prehistoric Cro-Magnons displayed their most elaborate paintings have resonances easily excited by the human voice, and stalactites emit musical tones when struck or rubbed with a stick. Paleolithic societies constructed flutes of bird bone, used animal horns to produce drones, and employed rattles and scrapers to provide rhythm.

In the sixth century B.C.E., Pythagoras was the first to correlate musical sounds and mathematics by relating consonant musical intervals to simple ratios of integers. In the fourth century B.C.E., Aristotle deduced that the medium that carries a sound must be compressed by the sounding body, and the third century B.C.E. philosopher Chrysippus correctly depicted the propagation of sound waves with an expanding spherical pattern. In the first century B.C.E., the Roman architect and engineer Marcus Vitruvius Pollio explained the acoustical characteristics of Greek theaters, but when the Roman civilization declined in the fourth century, scientific inquiry in the West ceased for the next millennium.

In the seventeenth century, modern experimental acoustics originated when the Italian mathematician Galileo explained resonance as well as musical consonance and dissonance, and theoretical acoustics got its start with Sir Isaac Newton's derivation of an expression for the velocity of sound. Although this yielded a value considerably lower than the experimental result, a more rigorous derivation by Pierre-Simon Laplace in 1816 obtained an equation yielding values in complete agreement with experimental results.

During the eighteenth century, many famous mathematicians studied vibration. In 1700, French mathematician Joseph Sauveur observed that strings vibrate in sections consisting of stationary nodes located between aggressively vibrating antinodes and that these vibrations have integer multiple frequencies, or harmonics, of the lowest frequency. He also noted that a vibrating string could simultaneously produce the sounds of several harmonics. In 1755,

an independent field. The interdisciplinary nature of acoustics has often consigned it to a subsidiary role as an minor subdivision of mechanics, hydrodynamics, or electrical engineering. Certainly, the various technical aspects of acoustics could be parceled out to larger and better established divisions of science, but then acoustics would lose its unique strengths and its source of dynamic creativity. The main difference between acoustics and more self-sufficient branches of science is that acoustics depends on physical laws developed in and borrowed from other fields. Therefore, the primary task of acoustics is to take these divergent principles and integrate them into a coherent whole in order to understand, measure, and control vibration phenomena.

The Acoustical Society of America subdivides acoustics into fifteen main areas, the most important of which are ultrasonics, which examines high-frequency waves not audible to humans; psychological acoustics, which studies how sound is perceived in the brain; physiological acoustics, which looks at human and animal hearing mechanisms; speech acoustics, which focuses on the human vocal apparatus and oral communication; musical acoustics, which involves the physics of musical instruments; underwater sound, which examines the production and propagation of sound in liquids; and noise, which concentrates on the control and suppression of unwanted sound. Two other important areas of applied acoustics are architectural acoustics (the acoustical design of concert halls and sound reinforcement systems) and audio engineering (recording and reproducing sound).

Daniel Bernoulli proved that this resultant vibration was the independent algebraic sum of the various harmonics. In 1750, Jean le Rond d'Alembert used calculus to obtain the wave equation for a vibrating string. By the end of the eighteenth century, the basic experimental results and theoretical underpinnings of acoustics were extant and in reasonable agreement, but it was not until the following century that theory and a concomitant advance of technology led to the evolution of the major divisions of acoustics.

Although mathematical theory is central to all acoustics, the two major divisions, physical and applied acoustics, evolved from the central theoretical core. In the late nineteenth century, Hermann von Helmholtz and Lord Rayleigh (John William Strutt), two polymaths, developed the theoretical aspects. Helmholtz's contributions to acoustics were primarily in explaining the physiological aspects of the ear. Rayleigh, a well-educated wealthy English baron, synthesized virtually all previous knowledge of acoustics and also formulated an appreciable corpus of experiment and theory.

Experiments by Georg Simon Ohm indicated that all musical tones arise from simple harmonic vibrations of definite frequency, with the constituent components determining the sound quality. This gave birth to the field of musical acoustics. Helmholtz's studies of instruments and Rayleigh's work contributed to the nascent area of musical acoustics. Helmholtz's knowledge of ear physiology shaped the field that was to become physiological acoustics.

Underwater acoustics commenced with theories developed by the nineteenth-century mathematician Siméon-Denis Poisson, but further development had to await the invention of underwater transducers in the next century.

Two important nineteenth-century inventions, the telephone (patented 1876) and the mechanical phonograph (invented 1877), commingled and evolved into twentieth-century audio acoustics when united with electronics. Some products in which sound production and reception are combined are microphones, loudspeakers, radios, talking motion pictures, high-fidelity stereo systems, and public sound-reinforcement systems. Improved instrumentation for the study of speech and hearing has stimulated the areas of physiological and psychology acoustics, and ultrasonic devices are routinely used for medical diagnosis and therapy, as well as for burglar alarms and rodent repellants. Underwater transducers are employed to detect and measure moving objects in the water, while audio engineering technology has transformed music performance as well as sound reproduction. Virtually no area of human activity has remained unaffected by continually evolving technology based on acoustics.

## HOW IT WORKS

**Ultrasonics.** Dog whistles, which can be heard by dogs but not by humans, can generate ultrasonic frequencies of about 25 kilohertz (kHz). Two types of transducers, magnetostrictive and piezoelectric, are used to generate higher frequencies and greater power. Magnetostrictive devices convert magnetic energy into ultrasound by subjecting ferric material (iron or nickel) to a strong oscillating magnetic field. The field causes the material to alternately expand and contract, thus creating sound waves of the same frequency as that of the field. The resulting sound waves have frequencies between 20 Hz and 50 kHz and several thousand watts of power. Such transducers operate at the mechanical resonance frequency where the energy transfer is most efficient.

Piezoelectric transducers convert electric energy into ultrasound by applying an oscillating electric field to a piezoelectric crystal (such as quartz). These transducers, which work in liquids or air, can generate frequencies in the megahertz region with considerable power. In addition to natural crystals, ceramic piezoelectric materials, which can be fabricated into any desired shape, have been developed.

**Physiological and Psychological Acoustics.** Physiological acoustics studies auditory responses of the ear and its associated neural pathways, and psychological acoustics is the subjective perception of sounds through human auditory physiology. Mechanical, electrical, optical, radiological, or biochemical techniques are used to study neural responses to various aural stimuli. Because these techniques are typically invasive, experiments are performed on animals with auditory systems that are similar to the human system. In contrast, psychological acoustic studies are noninvasive and typically use human subjects.

A primary objective of psychological acoustics is to define the psychological correlates to the physical parameters of sound waves. Sound waves in air may be characterized by three physical parameters: frequency, intensity, and their spectrum. When a sound wave impinges on the ear, the pressure variations in the air are transformed by the middle ear to mechanical vibrations in the inner ear. The cochlea then decomposes the sound into its constituent frequencies and transforms these into neural action potentials, which travel to the brain where the sound is evidenced. Frequency is perceived as pitch, the intensity level as loudness, and the spectrum determines the timbre, or tone quality, of a note.

Another psychoacoustic effect is masking. When a person listens to a noisy version of recorded music, the noise virtually disappears if the music is being enjoyed. This ability of the brain to selectively listen has had important applications in digitally recorded music. When the sounds are digitally compressed, such as in MP3 (MPEG-1 audio layer 3) systems, the brain compensates for the loss of information; thus one experiences higher fidelity sound than the stored content would imply. Also, the brain creates information when the incoming signal is masked or nonexistent, producing a psychoacoustic phantom effect. This phantom effect is particularly prevalent when heightened perceptions are imperative, as when danger is lurking.

Psychoacoustic studies have determined that the frequency range of hearing is from 20 to about 20,000 Hz for young people, and the upper limit progressively decreases with age. The rate at which hearing acuity declines depends on several factors, not the least of which is lifetime exposure to loud sounds, which progressively deteriorate the hair cells of the cochlea. Moderate hearing loss can be compensated for by a hearing aid; severe loss requires a cochlear implant.

**Speech Acoustics.** Also known as acoustic phonetics, speech acoustics deals with speech production and recognition. The scientific study of speech began with Thomas Alva Edison's phonograph, which allowed a speech signal to be recorded and stored for later analysis. Replaying the same short speech segment several times using consecutive filters passing through a limited range of frequencies creates a spectrogram, which visualizes the spectral properties of vowels and consonants. During the first half of the twentieth century, Bell Telephone Laboratories invested considerable time and resources to the systematic understanding of all aspects of speech, including vocal tract resonances, voice quality, and prosodic features of speech. For the first time, electric circuit theory was applied to speech acoustics, and analogue electric circuits were used to investigate synthetic speech.

**Musical Acoustics.** A conjunction of music, craftsmanship, auditory science, and vibration physics, musical acoustics analyzes musical instruments to better understand how the instruments are crafted, the physical principles of their tone production, and why each instrument has a unique timbre. Musical instruments are studied by analyzing their tones and then creating computer models to synthesize these sounds. When the sounds can be recreated with minimal software complications, a synthesizer featuring realistic orchestral tones may be constructed. The second method of study is to assemble an instrument or modify an existing instrument to perform nondestructive (or on occasion destructive) testing so that the effects of various modifications may be gauged.

**Underwater Sound.** Also know as hydroacoustics, this field uses frequencies between 10 Hz and 1 megahertz (MHz). Although the origin of hydroacoustics can be traced back to Rayleigh, the deployment of submarines in World War I provided the impetus for the rapid development of underwater listening devices (hydrophones) and sonar (sound navigation ranging), the acoustic equivalent of radar. Pulses of sound are emitted and the echoes are processed to extract information about submerged objects. When the speed of underwater sound is known, the reflection time for a pulse determines the distance to an object. If the object is moving, its speed of approach or recession is deduced from the frequency shift of the reflection, or the Doppler effect. Returning pulses have a higher frequency when the object approaches and lower frequency when it moves away.

**Noise.** Physically, noise may be defined as an intermittent or random oscillation with multiple frequency components, but psychologically, noise is any unwanted sound. Noise can adversely affect human health and well-being by inducing stress, interfering with sleep, increasing heart rate, raising blood pressure, modifying hormone secretion, and even inducing depression. The physical effects of noise are no less severe. The vibrations in irregular road surfaces caused by large rapid vehicles can cause adjacent buildings to vibrate to an extent that is intolerable to the buildings' inhabitants, even without structural damage. Machinery noise in industry is a serious problem because continuous exposure to loud sounds will induce hearing loss. In apartment buildings, noise transmitted through walls is always problematic; the goal is to obtain adequate sound insulation using lightweight construction materials.

Traffic noise, both external and internal, is ubiquitous in modern life. The first line of defense is to reduce noise at its source by improving engine enclosures, mufflers, and tires. The next method, used primarily when interstate highways are adjacent to residential areas, is to block the noise by the construction of concrete barriers or the planting of sound-absorbing vegetation. Internal automobile noise has been greatly abated by designing more aerodynamically efficient vehicles to reduce air

---

## Fascinating Facts About Acoustics

- Scientists have created an acoustic refrigerator, which uses a standing sound wave in a resonator to provide the motive power for operation. Oscillating gas particles increase the local temperature, causing heat to be transferred to the container walls, where it is expelled to the environment, cooling the interior.
- A cochlear implant, an electronic device surgically implanted in the inner ear, provides some hearing ability to those with damaged cochlea or those with congenital deafness. Because the implants use only about two dozen electrodes to replace 16,000 hair cells, speech sounds, although intelligible, have a robotic quality.
- MP3 files contain audio that is digitally encoded using an algorithm that compresses the data by a factor of about eleven but yields a reasonably faithful reproduction. The quality of sound reproduced depends on the data sampling rate, the quality of the encoder, and the complexity of the signal.
- Sound cannot travel through a vacuum, but it can travel four times faster through water than through air.
- The cocktail party effect refers to a person's ability to direct attention to one conversation at a time despite the many conversations taking place in the room.
- Continued exposure to noise over 85 decibels will gradually cause hearing loss. The noise level on a quiet residential street is 40 decibels, a vacuum cleaner 60-85, a leafblower 110, an ambulance siren 120, a rifle 163, and a rocket launching from its pad 180.

---

turbulence, using better sound isolation materials, and improving vibration isolation.

Aircraft noise, particularly in the vicinity of airports, is a serious problem exacerbated by the fact that as modern airplanes have become more powerful, the noise they generate has risen concomitantly. The noise radiated by jet engines is reduced by two structural modifications. Acoustic linings are placed around the moving parts to absorb the high frequencies caused by jet whine and turbulence, but this modification is limited by size and weight constraints. The second modification is to reduce the number of rotor blades and stator vanes, but this is somewhat inhibited by the desired power output. Special noise problems occur when aircraft travel at supersonic speeds (faster than the speed of sound), as this propagates a large pressure wave toward the ground that is experienced as an explosion. The unexpected sonic boom startles people, breaks windows, and damages houses. Sonic booms have been known to destroy rock structures in national parks. Because of these concerns, commercial aircraft are prohibited from flying at supersonic speeds over land areas.

Construction equipment (such as earthmoving machines) creates high noise levels both internally and externally. When the cabs of these machines are not closed, the only feasible manner of protecting operators' hearing is by using ear plugs. By carefully designing an enclosed cabin, structural vibration can be reduced and sound leaks made less significant, thus quieting the operator's environment. Although manufacturers are attempting to reduce the external noise, it is a daunting task because the rubber tractor treads occasionally used to replace metal are not as durable.

### APPLICATIONS AND PRODUCTS

**Ultrasonics.** High-intensity ultrasonic applications include ultrasonic cleaning, mixing, welding, drilling, and various chemical processes. Ultrasonic cleaners use waves in the 150 to 400 kHz range on items (such as jewelry, watches, lenses, and surgical instruments) placed in an appropriate solution. Ultrasonic cleaners have proven to be particularly effective in cleaning surgical devices because they loosen contaminants by aggressive agitation irrespective of an instrument's size or shape, and disassembly is not required. Ultrasonic waves are effective in cleaning most metals and alloys, as well as wood, plastic, rubber, and cloth.

Ultrasonic waves are used to emulsify two nonmiscible liquids, such as oil and water, by forming the liquids into finely dispersed particles that then remain in homogeneous suspension. Many paints, cosmetics, and foods are emulsions formed by this process.

Although aluminum cannot be soldered by conventional means, two surfaces subjected to intense ultrasonic vibration will bond—without the application of heat—in a strong and precise weld. Ultrasonic drilling is effective where conventional drilling is problematic, for instance, drilling square holes in glass. The drill bit, a transducer having the required shape and size, is used with an abrasive slurry that chips away the material when the suspended powder oscillates. Some of the chemical applications of ultrasonics are in the atomization of liquids, in electroplating, and as a catalyst in chemical reactions.

Low-intensity ultrasonic waves are used for nondestructive probing to locate flaws in materials for which complete reliability is mandatory, such as those used in spacecraft components and nuclear reactor vessels. When an ultrasonic transducer emits a pulse of energy into the test object, flaws reflect the wave and are detected. Because objects subjected to stress emit ultrasonic waves, these signals may be used to interpret the condition of the material as it is increasingly stressed. Another application is ultrasonic emission testing, which records the ultrasound emitted by porous rock when natural gas is pumped into cavities formed by the rock to determine the maximum pressure these natural holding tanks can withstand.

Low-intensity ultrasonics is used for medical diagnostics in two different applications. First, ultrasonic waves penetrate body tissues but are reflected by moving internal organs, such as the heart. The frequency of waves reflected from a moving structure is Doppler-shifted, thus causing beats with the original wave, which can be heard. This procedure is particularly useful for performing fetal examinations on a pregnant woman; because sound waves are not electromagnetic, they will not harm the fetus. The second application is to create a sonogram image of the body's interior. A complete cross-sectional image may be produced by superimposing the images scanned by successive ultrasonic waves passing through different regions. This procedure, unlike an X ray, displays all the tissues in the cross section and also avoids any danger posed by the radiation involved in X-ray imaging.

**Physiological and Psychological Acoustics.** Because the ear is a nonlinear system, it produces beat tones that are the sum and difference of two frequencies. For example, if two sinusoidal frequencies of 100 and 150 Hz simultaneously arrive at the ear, the brain will, in addition to these two tones, create tones of 250 and 50 Hz (sum and difference, respectively). Thus, although a small speaker cannot reproduce the fundamental frequencies of bass tones, the difference between the harmonics of that pitch will re-create the missing fundamental in the listener's brain.

Another psychoacoustic effect is masking. When a person listens to a noisy version of recorded music, the noise virtually disappears if the individual is enjoying the music. This ability of the brain to selectively listen has had important applications in digitally recorded music. When sounds are digitally compressed, as in MP3 systems, the brain compensates for the loss of information, thus creating a higher fidelity sound than that conveyed by the stored content alone.

As twentieth-century technology evolved, environmental noise increased concomitantly; lifetime exposure to loud sounds, commercial and recreational, has created an epidemic of hearing loss, most noticeable in the elderly because the effects are cumulative. Wearing a hearing aid, fitted adjacent to or inside the ear canal, is an effectual means of counteracting this handicap. The device consists of one or several microphones, which create electric signals that are amplified and transduced into sound waves redirected back into the ear. More sophisticated hearing aids incorporate an integrated circuit to control volume, either manually or automatically, or to switch to volume contours designed for various listening environments, such conversations on the telephone or where excessive background noise is present.

**Speech Acoustics.** With the advent of the computer age, speech synthesis moved to digital processing, either by bandwidth compression of stored speech or by using a speech synthesizer. The synthesizer reads a text and then produces the appropriate phonemes on demand from their basic acoustic parameters, such as the vibration frequency of the vocal cords and the frequencies and amplitudes of the vowel formants. This method of generating speech is considerably more efficient in terms of data storage than archiving a dictionary of prerecorded phrases.

Another important, and probably the most difficult, area of speech acoustics is the machine recognition of spoken language. When machine recognition programs are sufficiently advanced, the computer will be able to listen to a sentence in any reasonable dialect and produce a printed text of the utterance. Two basic recognition strategies exist, one dealing with words spoken in isolation and the other with continuous speech. In both cases, it is desirable to teach the computer to recognize the speech of different people through a training program. Because recognition of continuous speech is considerably more difficult than the identification of isolated words, very sophisticated pattern-matching models must be employed. One example of a machine recognition system is a word-driven dictation system that uses sophisticated software to process input speech. This system is somewhat adaptable to different voices and is able to recognize 30,000 words at a rate of 30 words per minute. The ideal machine recognition system would translate a spoken input language into another language in real time with correct grammar. Although some progress is being made, such a device has remained in the realm of speculative fantasy.

**Musical Acoustics.** The importance of musical acoustics to manufacturers of quality instruments is apparent. During the last decades of the twentieth century, fundamental research led, for example, to vastly improved French horns, organ pipes, orchestral strings, and the creation of an entirely new family of violins.

**Underwater Sound.** Applications for underwater acoustics include devices for underwater communication by acoustic means, remote control devices, underwater navigation and positioning systems, acoustic thermometers to measure ocean temperature, and echo sounders to locate schools of fish or other biota. Low-frequency devices can be used to explore the seabed for seismic research.

Although primitive measuring devices were developed in the 1920's, it was during the 1930's that sonar systems began incorporating piezoelectric transducers to increase their accuracy. These improved systems and their increasingly more sophisticated progeny became essential for the submarine warfare of World War II. After the war, theoretical advances in underwater acoustics coupled with computer technology have raised sonar systems to ever more sophisticated levels.

**Noise.** One system for abating unwanted sound is active noise control. The first successful application of active noise control was noise-canceling headphones, which reduce unwanted sound by using microphones placed in proximity to the ear to record the incoming noise. Electronic circuitry then generates a signal, exactly opposite to the incoming sound, which is reproduced in the earphones, thus canceling the noise by destructive interference. This system enables listeners to enjoy music without having to use excessive volume levels to mask outside noise and allows people to sleep in noisy vehicles such as airplanes. Because active noise suppression is more effective with low frequencies, most commercial systems rely on soundproofing the earphone to attenuate high frequencies. To effectively cancel high frequencies, the microphone and emitter would have to be situated adjacent to the user's eardrum, but this is not technically feasible. Active noise control is also being considered as a means of controlling low-frequency airport noise, but because of its complexity and expense, this is not yet commercially feasible.

## IMPACT ON INDUSTRY

Acoustics is the focus of research at numerous governmental agencies and academic institutions, as well as some private industries. Acoustics also plays an important role in many industries, often as part of product design (hearing aids and musical instruments) or as an element in a service (noise control consulting).

**Government Research.** Acoustics is studied in many government laboratories in the United States, including the U.S. Naval Research Laboratory (NRL), the Air Force Research Laboratory (AFRL), the Los Alamos National Laboratory, and the Lawrence Livermore National Laboratory. Research at the NRL and the AFRL is primarily in the applied acoustics area, and Los Alamos and Lawrence Livermore are oriented toward physical acoustics. The NRL emphasizes fundamental multidisciplinary research focused on creating and applying new materials and technologies to maritime applications. In particular, the applied acoustics division, using ongoing basic scientific research, develops improved signal processing systems for detecting and tracking underwater targets. The AFRL is heavily invested in research on auditory localization (spatial hearing), virtual auditory display technologies, and speech intelligibility in noisy environments. The effects of high-intensity noise on humans, as well as methods of attenuation, constitute a significant area of investigation at this facility. Another important area of research is the problem of providing intelligible voice communication in extremely noisy situations, such as those encountered by military or emergency personnel using low data rate narrowband radios, which compromise signal quality.

**Academic Research.** Research in acoustics is conducted at many colleges and universities in the United States, usually through physics or engineering departments, but, in the case of physiological and psychological acoustics, in groups that draw from multiple departments, including psychology, neurology, and linguistics. The Speech Research Laboratory at Indiana University investigates speech perception and processing through a broad interdisciplinary research program. The Speech Research Lab, a collaboration between the University of Delaware and the A. I. duPont Hospital for Children, creates speech synthesizers for the vocally impaired. A human speaker records a data bank of words and phrases that can be concatenated on demand to produce natural-sounding speech.

Academic research in acoustics is also being conducted in laboratories in Europe and other parts of the world. The Laboratoire d'Acoustique at the Université de Maine in Le Mans, France, specializes in research in vibration in materials, transducers, and musical instruments. The Andreyev Acoustics Institute of the Russian Acoustical Society brings together researchers from Russian universities, agencies, and businesses to conduct fundamental and applied research in ocean acoustics, ultrasonics, signal processing, noise and vibration, electroacoustics, and bioacoustics. The Speech and Acoustics Laboratory at the Nara Institute of Science and Technology in Nara, Japan, studies diverse aspects of human-machine communication through speech-oriented multimodal interaction. The Acoustics Research Centre, part of the National Institute of Creative Arts and Industries in New Zealand, is concerned with the impact of noise on humans. A section of this group, Acoustic Testing Service, provides commercial testing of building materials for their noise attenuation properties.

**Industry and Business.** Many businesses (such as the manufacturers of hearing aids, ultrasound medical devices, and musical instruments) use acoustics in their products or services and therefore employ experts in acoustics. Businesses also are involved in many aspects of acoustic research, particularly controlling

noise and facilitating communication. Raytheon BBN technologies (Cambridge, Massachusetts) has developed low data rate Noise Robust Vocoders (electronic speech synthesizers) that generate comprehensible speech at data rates considerably below other state-of-the-art devices. Acoustic Research Laboratories in Sydney, Australia, designs and manufactures specialized equipment for measuring environmental noise and vibration, in addition to providing contract research and development services.

## CAREERS AND COURSE WORK

**Career opportunities occur in academia (teaching and research), industry, and national laboratories.** Academic positions dedicated to acoustics are few, as are the numbers of qualified applicants. Most graduates of acoustics programs find employment in research-based industries in which acoustical aspects of products are important, and others work for government laboratories.

Although the subfields of acoustics are integrated into multiple disciplines, most aspects of acoustics can be learned by obtaining a broad background in a scientific or technological field, such as physics, engineering, meteorology, geology, or oceanography. Physics probably provides the best training for almost any area of acoustics. An electrical engineering major is useful for signal processing and synthetic speech research, and a mechanical engineering background is requisite for comprehending vibration. Training in biology is expedient for physiological acoustic research, and psychology course work provides essential background for psychological acoustics. Architects often employ acoustical consultants to advise on the proper acoustical design of concert halls, auditoriums, or conference rooms. Acoustical consultants also assist with noise reduction problems and help design soundproofing structures for rooms. Although background in architecture is not a prerequisite for becoming this type of acoustical consultant, engineering or physics is.

Acoustics is not a university major; therefore, specialized knowledge is best acquired at the graduate level. Many electrical engineering departments have at least one undergraduate course in acoustics, but most physics departments do not. Nevertheless, a firm foundation in classical mechanics (through physics programs) or a mechanical engineering vibration course will provide, along with numerous courses in mathematics, sufficient underpinning for successful graduate study in acoustics.

## SOCIAL CONTEXT AND FUTURE PROSPECTS

Acoustics affects virtually every aspect of modern life; its contributions to societal needs are incalculable. Ultrasonic waves clean objects, are routinely employed to probe matter, and are used in medical diagnosis. Cochlear implants restore people's ability to hear, and active noise control helps provide quieter listening environments. New concert halls are routinely designed with excellent acoustical properties, and vastly improved or entirely new musical instruments have made their debut. Infrasound from earthquakes is used to study the composition of Earth's mantle, and sonar is essential to locate submarines and aquatic life. Sound waves are used to explore the effects of structural vibrations. Automatic speech recognition devices and hearing aid technology are constantly improving.

Many societal problems related to acoustics remain to be tackled. The technological advances that made modern life possible have also resulted in more people with hearing loss. Environmental noise is ubiquitous and increasing despite efforts to design quieter machinery and pains taken to contain unwanted sound or to isolate it from people. Also, although medical technology has been able to help many hearing- and speech-impaired people, other individuals still lack appropriate treatments. For example, although voice generators exist, there is considerable room for improvement.

*George R. Plitnik, B.A., B.S., M.A., Ph.D.*

## FURTHER READING

Bass, Henry E., and William J. Cavanaugh, eds. *ASA at Seventy-five.* Melville, N.Y.: Acoustical Society of America, 2004. An overview of the history, progress, and future possibilities for each of the fifteen major subdivisions of acoustics as defined by the Acoustical Society of America.

Beyer, Robert T. *Sounds of Our Times: Two Hundred Years of Acoustics.* New York: Springer-Verlag, 1999. A history of the development of all areas of acoustics. Organized into chapters covering twenty-five to fifty years. Virtually all subfields of acoustics are covered.

Crocker, Malcolm J., ed. *The Encyclopedia of Acoustics.* 4 vols. New York: Wiley, 1997. A comprehensive work detailing virtually all aspects of acoustics.

Everest, F. Alton, and Ken C. Pohlmann. *Master Handbook of Acoustics.* 5th ed., New York: McGraw-Hill, 2009. A revision of a classic reference work designed for those who desire accurate information on a level accessible to the layperson with limited technical ability.

Rossing, Thomas, and Neville Fletcher. *Principles of Vibration and Sound.* 2d ed. New York: Springer-Verlag, 2004. A basic introduction to the physics of sound and vibration.

Rumsey, Francis, and Tim McCormick. *Sound and Recording: An Introduction.* 5th ed. Boston: Elsevier/Focal Press, 2004. Presents basic information on the principles of sound, sound perception, and audio technology and systems.

Strong, William J., and George R. Plitnik. *Music, Speech, Audio.* 3d ed. Provo, Utah: Brigham Young University Academic Publishing, 2007. A comprehensive text, written for the layperson, which covers vibration, the ear and hearing, noise, architectural acoustics, speech, musical instruments, and sound recording and reproduction.

Swift, Gregory. "Thermoacoustic Engines and Refrigerators." *Physics Today* (July, 1995): 22-28. Explains how sound waves may be used to create more efficient refrigerators with no moving parts.

## WEB SITES

*Acoustical Society of America*
http://asa.aip.org

*Institute of Noise Control Engineering*
http://www.inceusa.org

*International Commission for Acoustics*
http://www.icacommission.org

*National Council of Acoustical Consultants*
http://www.ncac.com

**See also:** Applied Physics; Communication; Earthquake Prediction; Music Technology; Noise Control; Pattern Recognition; Speech Therapy and Phoniatrics; Telecommunications; Ultrasonic Imaging.

# AERONAUTICS AND AVIATION

## FIELDS OF STUDY

Algebra; calculus; inorganic chemistry; organic chemistry; physical chemistry; optics; modern physics; statics; aerodynamics; thermodynamics; strength of materials; propulsion; propeller and rotor theory; vehicle performance; aircraft design; avionics; orbital mechanics; spacecraft design.

## SUMMARY

Aeronautics is the science of atmospheric flight. Aviation is the design, development, production, and operation of flight vehicles. Aerospace engineering extends these fields to space vehicles. Transonic airliners, airships, space launch vehicles, satellites, helicopters, interplanetary probes, and fighter planes are all applications of aerospace engineering.

## KEY TERMS AND CONCEPTS

- **Airfoil:** Structure, such as a wing or a propeller, designed to interact, in motion, with the surrounding airflow in a manner that optimizes the desired reaction, whether that be to minimize air resistance or to maximize lift.
- **Boundary Layer:** Thin region near a surface of an aircraft where the flow slows down because of viscous friction.
- **Bypass Ratio:** Ratio of turbofan engine mass flow rate bypassing the hot core, to that through the core.
- **Delta V:** Speed difference corresponding to the difference in energies between two orbital states.
- **Fuselage:** Body of an aircraft, other than engines, wings, tails, or control surfaces.
- **Lift To Drag Ratio:** Ratio of the lift to drag in cruise; the aerodynamic efficiency metric for transport aircraft and gliders.
- **Oblique Shock:** Thin wave in a supersonic flow through which flow turns and decelerates sharply.
- **Prandtl-Meyer Expansion:** Ideal model of a supersonic flow accelerating through a turn.
- **Stall:** Condition in which flow separates from most of a lifting surface, sharply lowering lift and raising drag.
- **Takeoff Gross Weight:** Mass or weight of an aircraft at takeoff with full payload and fuel load; the highest design weight for liftoff.
- **Wind Tunnel:** Facility where a smooth, uniform flow helps simulate flow around an object in flight.
- **Wing:** Object that generates lift with low drag and supports the weight of the aircraft in flight.

### DEFINITION AND BASIC PRINCIPLES

Aeronautics is the science of atmospheric flight. The term ("aero" referring to flight and "nautics" referring to ships or sailing) originated from the activities of pioneers who aspired to navigate the sky. These early engineers designed, tested, and flew their own creations, many of which were lighter-than-air balloons. Modern aeronautics encompasses the science and engineering of designing and analyzing all areas associated with flying machines.

Aviation (based on the Latin word for "bird") originated with the idea of flying like the birds using heavier-than-air vehicles. "Aviation" refers to the field of operating aircraft, while the term "aeronautics" has been superseded by "aerospace engineering," which specifically includes the science and engineering of spacecraft in the design, development, production, and operation of flight vehicles.

A fundamental tenet of aerospace engineering is to deal with uncertainty by tying analyses closely to what is definitely known, for example, the laws of physics and mathematical proofs. Lighter-than-air airships are based on the principle of buoyancy, which derives from the law of gravitation. An object that weighs less than the equivalent volume of air experiences a net upward force as the air sinks around it.

Two basic principles that enable the design of heavier-than-air flight vehicles are those of aerodynamic lift and propulsion. Both arise from Sir Isaac Newton's second and third laws of motion. Aerodynamic lift is a force perpendicular to the direction of motion, generated from the turning of flowing air around an object. In propulsion, the reaction to the acceleration of a fluid generates a force that propels an object, whether in air or in the vacuum of space. Understanding these principles allowed aeronauts to design vehicles that could fly steadily despite being much heavier than the air they displaced and allowed rocket scientists to develop vehicles that could accelerate in space. Spaceflight occurs at speeds so high that the vehicle's kinetic energy is comparable to the potential energy due to gravitation. Here the principles of orbital mechanics derive from the laws of dynamics and gravitation and extend to the regime of relativistic phenomena. The engineering sciences of building vehicles that can fly, keeping them stable, controlling their flight, navigating, communicating, and ensuring the survival, health, and comfort of occupants, draw on every field of science.

## BACKGROUND AND HISTORY

The intrepid balloonists of the nineteenth century were followed by aeronauts who used the principles of aerodynamics to fly unpowered gliders. The Wright brothers demonstrated sustained, controlled, powered aerodynamic flight of a heavier-than-air aircraft in 1903. The increasing altitude, payload, and speed capabilities of airplanes made them powerful weapons in World War I. Such advances improved flying skills, designs, and performance, though at a terrible cost in lives.

The monoplane design superseded the fabric-and-wire biplane and triplane designs of World War I. The helicopter was developed during World War II and quickly became an indispensable tool for medical evacuation and search and rescue. The jet engine, developed in the 1940's and used on the Messerschmitt 262 and Junkers aircraft by the Luftwaffe and the Gloster Meteor by the British, quickly enabled flight in the stratosphere at speeds sufficient to generate enough lift to climb in the thin air. Such innovations led to smooth, long-range flights in pressurized cabins and shirtsleeve comfort. Fatal crashes of the de Havilland Comet airliner in 1953 and 1954 focused attention on the science of metal fatigue.

The Boeing 707 opened up intercontinental air travel, followed by the Boeing 747, the supersonic Concorde, and the EADS Airbus A380. A series of manned research aircraft designated X-planes since the 1930's investigated various flight regimes and also drove the development of better wind tunnels and high-altitude simulation chambers. German ballistic missiles led to U.S. and Soviet missile programs that grew into a space race, culminating in the first humans landing on the Moon in 1969. Combat-aircraft development enabled advances that resulted in safer and more efficient airliners.

## HOW IT WORKS

**Force Balance in Flight.** Five basic forces acting on a flight vehicle are aerodynamic lift, gravity, thrust, drag, and centrifugal force. For a vehicle in steady level flight in the atmosphere, lift and thrust balance gravity (weight) and aerodynamic drag. Centrifugal force due to moving steadily around the Earth is too weak at most airplane flight speeds but is strong for a maneuvering aircraft. Aircraft turn by rolling the lift vector toward the center of curvature of the desired flight path, balancing the centrifugal reaction due to inertia. In the case of a vehicle in space beyond the atmosphere, centrifugal force and thrust counter gravitational force.

**Aerodynamic Lift.** Aerodynamics deals with the forces due to the motion of air and other gaseous fluids relative to bodies. Aerodynamic lift is generated perpendicular to the direction of the free stream as the reaction to the rate of change of momentum of air turning around an object, and, at high speeds, to compression of air by the object. Flow turning is accomplished by changing the angle of attack of the surface, by using the camber of the surface in subsonic flight, or by generating vortices along the leading edges of swept wings.

**Propulsion.** Propulsive force is generated as a reaction to the rate of change of momentum of a fluid moving through and out of the vehicle. Rockets carry all of the propellant onboard and accelerate it out through a nozzle using chemical heat release, other heat sources, or electromagnetic fields. Jet engines "breathe" air and accelerate it after reaction with fuel. Rotors, propellers, and fans exert lift force on the air and generate thrust from the reaction to this force. Solar sails use the pressure of solar radiation to push large, ultralight surfaces.

**Static Stability.** An aircraft is statically stable if a small perturbation in its attitude causes a restoring aerodynamic moment that erases the perturbation. Typically, the aircraft center of gravity must be ahead of the center of pressure for longitudinal stability. The tails or canards help provide stability about the different axes. Rocket engines are said to be stable if the rate of generation of gases in the combustion chamber does not depend on pressure stronger than by a direct proportionality, such as a pressure exponent of 1.

**Flight Dynamics and Controls.** Static stability is not the whole story, as every pilot discovers when the

## Fascinating Facts About Aeronautics and Aviation

- The X-29 test vehicle demonstrated that an aircraft could be built to be statically unstable and yet maintain stable flight. The flight control computer reliably provides rapid updates of control surface actuators to compensate for disturbances before they amplify. Modern fighter aircraft are marginally unstable but use control systems to augment stability, thereby increasing maneuver performance.

- Future space travelers may fly up in hypersonic air-breathing vehicles that take off and land like airplanes. High-pressure intake air is liquefied in ducts cooled by the liquid-hydrogen fuel, enabling oxygen to be separated out from nitrogen and stored for use with hydrogen fuel beyond the atmosphere. This decreases takeoff weight.

- A long conductor trailed from a spacecraft generates an electric current when moving through the Earth's magnetic field. Conversely, a current passed through a tether between two objects in different orbits generates a force. This is a proposed electrodynamic space broom, which will drag debris into orbits where they will quickly burn up in the atmosphere.

- A conventional helicopter requires a tail rotor to provide anti-torque and keep the craft from spinning in reaction to the torque put in to turn the rotor. The rotor must also change its pitch angle through every cycle in order to keep the lift moment equal between the advancing side rotor blade and the retreating side blade. The counter-rotating rotor helicopter has two rotors on the same axis but rotating in opposite directions. This eliminates the need for a tail rotor and for large cyclic pitch variation. This has allowed helicopters to fly substantially faster. Counter-rotating compressor and turbine stages in jet engines eliminate the need for stator stages that do not contribute useful work, thereby reducing the mass and number of stages needed in modern aircraft engines.

- When aerodynamic lift is generated, the flow around the body is pushed down (against the direction of lift), while the flow outside the body's span gets lifted up. The energy in this updraft is wasted unless another aircraft benefits from following in close proximity off to one side. Birds use this feature routinely and so have air forces since World War I to save fuel. Swarms of small unmanned aerial vehicles, or micro spacecraft, can generate the same resolution and efficiency as a very large antenna by maintaining relative position in flight.

airplane drifts periodically up and down instead of holding a steady altitude and speed. Flight dynamics studies the phenomena associated with aerodynamic loads and the response of the vehicle to control surface deflections and engine-thrust changes. The study begins with writing the equations of motion of the aircraft resolved along the six degrees of freedom: linear movement along the longitudinal, vertical and sideways axes, and roll, yaw, and pitch rotations about them. Maneuvering aircraft must deal with coupling between the different degrees of freedom, so that roll accompanies yaw, and so on.

The autopilot system was an early flight-control achievement. Terrain-following systems combine information about the terrain with rapid updates, enabling military aircraft to fly close to the ground, much faster than a human pilot could do safely. Modern flight-control systems achieve such feats as reconfiguring control surfaces and fuel to compensate for damage and engine failures; or enabling autonomous helicopters to detect, hover over, and pick up small objects and return; or sending a space probe at thousands of kilometers per hour close to a planetary moon or landing it on an asteroid and returning it to Earth. This field makes heavy use of ordinary differential equations and transform techniques, along with simulation software.

**Orbital Missions.** The rocket equation attributed to Russian scientist Konstantin Tsiolkovsky related the speed that a rocket-powered vehicle gains to the amount and speed of the mass that it ejects. A vehicle launched from Earth's surface goes into a trajectory where its kinetic energy is exchanged for gravitational potential energy. At low speeds, the resulting trajectory intersects the Earth, so that the vehicle falls to the surface. At high enough speeds, the vehicle goes so far so fast that its trajectory remains in space and takes the shape of a continuous ellipse around Earth. At even higher kinetic energy levels, the vehicle goes into a hyperbolic trajectory, escaping Earth's orbit into the solar system. The key is thus to achieve enough tangential speed relative to Earth.

Most rockets rise rapidly through the atmosphere so that the acceleration to high tangential speed occurs well above the atmosphere, thus minimizing air-drag losses.

**Hohmann Transfer.** Theoretically, the most efficient way to impart kinetic energy to a vehicle is impulsive launch, expending all the propellant instantly so that no energy is wasted lifting or accelerating propellant with the vehicle. Of course, this would destroy any vehicle other than a cannonball, so large rockets use gentle accelerations of no more than 1.4 to 3 times the acceleration due to gravity. The advantage of impulsive thrust is used in the Hohmann transfer maneuver between different orbits in space. A rocket is launched into a highly eccentric elliptical trajectory. At its highest point, more thrust is added quickly. This sends the vehicle into a circular orbit at the desired height or into a new orbit that takes it close to another heavenly body. Reaching the same final orbit using continuous, gradual thrust would require roughly twice as much expenditure of energy. However, continuous thrust is still an attractive option for long missions in space, because a small amount of thrust can be generated using electric propulsion engines that accelerate propellant to extremely high speeds compared with the chemical engines used for the initial ascent from Earth.

## APPLICATIONS AND PRODUCTS

**Aerospace Structures.** Aerospace engineers always seek to minimize the mass required to build the vehicle but still ensure its safety and durability. Unlike buildings, bridges, or even (to some degree) automobiles, aircraft cannot be made safer merely by making them more massive, because they must also be able to overcome Earth's gravity. This exigency has driven development of new materials and detailed, accurate methods of analysis, measurement, and construction. The first aircraft were built mostly from wood frames and fabric skins. These were superseded by all-metal craft, constructed using the monocoque concept (in which the outer skin bears most of the stresses). The Mosquito high-speed bomber in World War II reverted to wood construction for better performance. Woodworkers learned to align the grain (fiber direction) along the principal stress axes. Metal offers the same strength in all directions for the same thickness. Composite structures allow fibers with high tensile strength to be

*Replica of the Wright Flyer undergoing aerodynamic tests in a wind tunnel at NASA's Langley research center, in Virginia.* (Chuck Thomas/Old Dominion University/Photo Researchers, Inc.)

placed along the directions where strength is needed, bonding different layers together.

**Aeroelasticity.** Aeroelasticity is the study of the response of structurally elastic bodies to aerodynamic loads. Early in the history of aviation, several mysterious and fatal accidents occurred wherein pieces of wings or tails failed in flight, under conditions where the steady loads should have been well below the strength limits of the structure. The intense research to address these disasters showed that beyond some flight speed, small perturbations in lift, such as those due to a gust or a maneuver, would cause the structure to respond in a resonant bending-twisting oscillation, the perturbation amplitude rapidly rising in a "flutter" mode until structural failure occurred. Predicting such aeroelastic instabilities demanded a highly mathematical approach to understand and apply the theories of unsteady aerodynamics and structural dynamics. Modern aircraft are designed so that the flutter speed is well above any possible speed achieved. In the case of helicopter rotor blades and gas turbine engine blades, the problems of ensuring aeroelastic stability are still the focus of leading-edge research. Related advances in structural dynamics have enabled development of composite structures and of highly efficient turbo machines that use counter-rotating stages, such as those in the F135 engines used in the F-35 Joint Strike Fighter. Such advances also made it possible for earthquake-surviving high-rise buildings to be built in cities such as San

Francisco, Tokyo, and Los Angeles, where a number of sensors, structural-dynamics-analysis software, and actuators allow the correct response to dampen the effects of earth movements even on the upper floors.

**Smart Materials.** Various composite materials such as carbon fiber and metal matrix composites have come to find application even in primary aircraft structures. The Boeing 787 is the first to use a composite main spar in its wings. Research on nano materials promises the development of materials with hundreds of times as much strength per unit mass as steel. Another leading edge of research in materials is in developing high-temperature or very low-temperature (cryogenic) materials for use inside jet and rocket engines, the spinning blades of turbines, and the impeller blades of liquid hydrogen pumps in rocket engines. Single crystal turbine blades enabled the development of jet engines with very high turbine inlet temperatures and, thus, high thermodynamic efficiency. Ceramic designs that are not brittle are pushing turbine inlet temperatures even higher. Other materials are "smart," meaning they respond actively in some way to inputs. Examples include piezoelectric materials.

**Wind Tunnels and Other Physical Test Facilities.** Wind tunnels, used by the Wright brothers to develop airfoil shapes with desirable characteristics, are still used heavily in developing concepts and proving the performance of new designs, investigating causes of problems, and developing solutions and data to validate computational prediction techniques. Generally, a wind tunnel has a fan or a high-pressure reservoir to add work to the air and raise its stagnation pressure. The air then flows through means of reducing turbulence and is accelerated to the maximum speed in the test section, where models and measurement systems operate.

The power required to operate a wind tunnel is proportional to the mass flow rate through the tunnel and to the cube of the flow speed achieved. Low-speed wind tunnels have relatively large test sections and can operate continuously for several minutes at a time. Supersonic tunnels generally operate with air blown from a high-pressure reservoir for short durations. Transonic tunnels are designed with ventilating slots to operate in the difficult regime where there may be both supersonic waves and subsonic flow over the test configuration. Hypersonic tunnels require heaters to avoid liquefying the air and to simulate the

high stagnation temperatures of hypersonic flight and operate for millisecond durations. Shock tubes generate a shock from the rupture of a diaphragm, allowing high-energy air to expand into stationary air in the tube. They are used to simulate the extreme conditions across shocks in hypersonic flight. Many other specialized test facilities are used in structural and materials testing, developing jet and rocket engines, and designing control systems.

**Avionics and Navigation.** Condensed from the term "aviation electronics," the term "avionics" has come to include the generation of intelligent software systems and sensors to control unmanned aerial vehicles (UAVs), which may operate autonomously. Avionics also deals with various subsystems such as radar and communications, as well as navigation equipment, and is closely linked to the disciplines of flight dynamics, controls, and navigation.

During World War II, pilots on long-range night missions would navigate celestially. The gyroscopes in their aircrafts would spin at high speed so that their inertia allowed them to maintain a reference position as the aircraft changed altitude or accelerated. Most modern aircraft use the Global Positioning System (GPS), Galileo, or GLONASS satellite constellations to obtain accurate updates of position, altitude, and velocity. The ordinary GPS signal determines position and speed with fair accuracy. Much greater precision and higher rates of updates are available to authorized vehicle systems through the differential GPS signal and military frequencies.

**Gravity Assist Maneuver.** Yuri Kondratyuk, the Ukrainian scientist whose work paved the way for the first manned mission to the moon, suggested in 1918 that a spacecraft could use the gravitational attraction of the moons of planets to accelerate and decelerate at the two ends of a journey between planets. The Soviet Luna 3 probe used the gravity of the Moon when photographing the far side of it in 1959. American mathematician Michael Minovitch pointed out that the gravitational pull of planets along the trajectory of a spacecraft could be used to accelerate the craft toward other planets. The Mariner 10 probe used this "gravitational slingshot" maneuver around Venus to reach Mercury at a speed small enough to go into orbit around Mercury. The Voyager missions used the rare alignment of the outer planets to receive gravitational assists from Jupiter and Saturn to go on to Uranus and Neptune, before doing another slingshot around Jupiter and Saturn to escape the solar system. Gravity assist has become part of the mission planning for all exploration missions and even for missions near Earth, where the gravity of the Moon is used.

## IMPACT ON INDUSTRY

Aeronautics and aviation have had an immeasurable impact on industry and society. Millions of people fly long distances on aircraft every day, going about their business and visiting friends and relatives, at a cost that is far lower in relative terms than the cost of travel a century ago.

Every technical innovation developed for aeronautics and aviation finds its way into improved industrial products. Composite structural materials are found in everything from tennis rackets to industrial machinery. Bridges, stadium domes, and skyscrapers are designed with aerospace structural-element technology and structural-dynamics instrumentation and testing techniques. Electric power is generated in utility power plants using steam generators sharing jet engine turbo machine origins.

Satellite antennae are found everywhere. Much digital signal processing, central to digital music and cell phone communications, came from research projects driven by the need to extract low-level signatures buried in noise. Similarly, image-processing algorithms that enable computed tomography (CT) scans of the human body, eye and cardiac diagnostics, image and video compression, and laser printing came from aerospace image-processing projects. The field of geoinformatics has advanced immensely, with most mapping, navigation, and remote-sensing enterprises assuming the use of space satellites. The GPS has spawned numerous products for terrestrial drivers on land and navigators on the ocean. Aerospace medicine research has developed advances in diagnosing and monitoring the human body and its responses to acceleration, bone marrow loss, muscular degeneration, and their prevention through exercise, hypoxia, radiation protection, heart monitoring, isolation from microorganisms, and drug delivery. Teflon coatings developed for aerospace products are also used in cookware.

## CAREERS AND COURSE WORK

Aerospace engineers work on problems that push the frontiers of technology. Typical employers in this industry are manufacturers of aircraft or their parts and subsystems, airlines, government agencies and laboratories, and the defense services. Many aerospace engineers are also sought by financial services and other industries seeking those with excellent quantitative (mathematical and scientific) skills and talents.

University curriculum generally starts with a year of mathematics, physics, chemistry, computer graphics, computer science, language courses, and an introduction to aerospace engineering, followed by sophomore-year courses in basic statics, dynamics, materials, and electrical engineering. Core courses include low-speed and high-speed aerodynamics, linear systems analysis, thermodynamics, propulsion, structural analysis, composite materials, vehicle performance, stability, control theory, avionics, orbital mechanics, aeroelasticity and structural dynamics, and a two-semester sequence on capstone design of flight vehicles. High school students aiming for such careers should take courses in mathematics, physics, chemistry and natural sciences, and computer graphics. Aerospace engineers are frequently required to write clear reports and present complex issues to skeptical audiences, which demands excellent communication skills. Taking flying lessons or getting a private pilot license is less important to aerospace engineering, as exhilarating as it is, and should be considered only if one desires a career as a pilot or astronaut.

The defense industry is moving toward using aircraft that do not need a human crew and can perform beyond the limits of what a human can survive, so the glamorous occupation of combat jet pilot may be heading for extinction. Airline pilot salaries are also coming down from levels that compared with surgeons toward those more comparable to bus drivers. Aircraft approach, landing, traffic management, emergency response, and collision avoidance systems may soon become fully automated and will require maneuvering responses that are beyond what a human pilot can provide in time and accuracy.

Opportunities for spaceflight may also be minimal unless commercial and military spaceflight picks up to fill the void left by the end of civilian programs discussed below. This is not a unique situation in aviation history. Early pilots, even much later than the intrepid "aeronauts," also worked much more for love of the unparalleled experience of flying, rather than for looming prospects of high-profile careers or the salaries paid by struggling startup airline companies. The only reliable prediction that can be made about aerospace careers is that they hold many surprises.

## SOCIAL CONTEXT AND FUTURE PROSPECTS

Airline travel is under severe stress in the first part of the twenty-first century. This is variously attributed to airport congestion, security issues, rising fuel prices, predatory competition, reduction of route monopolies, and leadership that appears to offer little vision beyond cost cutting. Meanwhile, the demand for air travel is rising all over the world. Global demand for commercial airliners is estimated at nearly 30,000 aircraft through 2030 and is valued at more than $3.2 trillion—in addition to 17,000 business jets valued at more than $300 billion.

Detailed design and manufacturing of individual aircraft are distributed between suppliers worldwide, with the wings, tails, and engines of a given aircraft often designed and built in different parts of the world. Japan and China are expected to increase their aircraft manufacturing, while major U.S. companies appear to be moving more toward becoming system integrators and away from manufacturing.

The human venture in space is also under stress as the U.S. space shuttle program ends without another human-carrying vehicle to replace it. The future of the one remaining space station is in doubt, and there are no plans to build another.

On the other hand, just over one century into powered flight, the human venture into the air and beyond is just beginning. Aircraft still depend on long runways and can fly only in a very limited range of conditions. Weather delays are still common because of uncertainty about how to deal with fluctuating winds or icing conditions. Most airplanes still consist of long tubes attached to thin wings, because designing blended wing bodies is difficult with the uncertainties in modeling composite structures. The aerospace and aviation industry is a major generator of atmospheric carbon releases. This will change only when the industry switches to renewable hydrogen fuel, which may occur faster than most people anticipate.

The human ability to access, live, and work in space or on extraterrestrial locations is extremely limited, and this prevents development of a large space-based economy. This situation may be expected to change over time, with the advent of commercial space launches. New infrastructure will encourage commercial enterprises beyond Earth.

The advancements in the past century are truly breathtaking and bode well for the breakthroughs that one may hope to see. Hurricanes and cyclonic storms are no longer surprise killers; they are tracked from formation in the far reaches of the oceans, and their paths are accurately predicted, giving people plenty of warning. Crop yields and other resources are accurately tracked by spacecraft, and ground-penetrating radar from Earth-sensing satellites has discovered much about humankind's buried ancient heritage and origins. Even in featureless oceans and deserts, GPS satellites provide accurate, reliable navigation information. The discovery of ever-smaller distant planets by orbiting space telescopes, and of unexpected forms of life on Earth, hint at the possible discovery of life beyond Earth.

*Narayanan M. Komerath, Ph.D.*

## FURTHER READING

Anderson, John D., Jr. *Introduction to Flight.* 5th ed. New York: McGraw-Hill, 2005. This popular textbook, setting developments in a historical context, is derived from the author's tenure at the Smithsonian Air and Space Museum.

Bekey, Ivan. *Advanced Space System Concepts and Technologies, 2010-2030+.* Reston, Va.: American Institute of Aeronautics and Astronautics, 2003. Summaries of various advanced concepts and logical arguments used to explore their feasibility.

Design Engineering Technical Committee. *AIAA Aerospace Design Engineers Guide.* 5th ed. Reston, Va.: American Institute of Aeronautics and Astronautics, 2003. A concise book of formulae and numbers that aerospace engineers use frequently or need for reference.

Gann, Ernest K. *Fate Is the Hunter.* 1961. Reprint. New York: Simon and Schuster, 1986. Describes an incident that was the basis for a 1964 film of the same name. Autobiography of a pilot, describing the early days of commercial aviation and coming close to the age of jet travel.

Hill, Philip, and Carl Peterson. *Mechanics and Thermodynamics of Propulsion.* 2d ed. Upper Saddle River, N.J.: Prentice Hall, 1991. A classic textbook on propulsion that covers the basic science and engineering of jet and rocket engines and their components. Also gives excellent sets of problems with answers.

Jenkins, Dennis R. *X-15: Extending the Frontiers of Flight.* NASA SP-2007-562. Washington, D.C.: U.S. Government Printing Office, 2007. Contains various copies of original data sheets, memos, and pictures from the days when the X-15 research vehicle was developed and flown.

Lewis, John S. *Mining the Sky: Untold Riches from the Asteroids, Comets and Planets.* New York: Basic Books, 1997. The most readable answer to the question "What resources are there beyond Earth to make exploration worthwhile?" Written from a strong scientific background, it sets out the reasoning to estimate the presence and accessibility of extraterrestrial water, gases, minerals, and other resources that would enable an immense space-based economy.

Liepmann, H. W., and A. Roshko. *Elements of Gas Dynamics.* Mineola, N.Y.: Dover Publications, 2001. A textbook on the discipline of gas dynamics as applied to high-speed flow phenomena. Contains several photographs of shocks, expansions, and boundary layer phenomena.

O'Neill, Gerard K. *The High Frontier: Human Colonies in Space.* 3d ed. New York: William Morrow & Company, 1977. Reprint. Burlington, Ontario, Canada: Apogee Books, 2000. Sets out the logic, motivations, and general parameters for human settlements in space. This formed the basis for NASA/ASEE (American Society for Engineering Education) studies in 1977-1978 and beyond, to investigate the design of space stations for permanent habitation.

Fascinating exposition of how ambitious concepts are systematically analyzed and engineering decisions are made on how to achieve them, or why they cannot yet be achieved.

Peebles, Curtis. *Road to Mach 10: Lessons Learned from the X-43A Flight Research Program.* Reston, Va.: American Institute of Aeronautics and Astronautics, 2008. A contemporary experimental flight-test program description.

**WEB SITES**

*Aerospace Digital Library*
http://www.adl.gatech.edu

*American Institute of Aeronautics and Astronautics*
http://www.aiaa.org

*Jet Propulsion Laboratory*
A Gravity Assist Primer
http://www2.jpl.nasa.gov/basics/grav/primer.php

*National Aeronautics and Space Administration*
Born of Dreams, Inspired by Freedom. U.S. Centennial of Flight Commission
http://www.centennialofflight.gov

**See also:** Applied Mathematics; Applied Physics; Atmospheric Sciences; Avionics and Aircraft Instrumentation; Communications; Computer Science; Jet Engine Technology; Propulsion Technologies; Space Science; Space Stations; Spacecraft Engineering.

# AGRICULTURAL SCIENCE

Animal breeding and husbandry; plant breeding and propagation; agroforestry; agronomy; horticulture; soil science.

## SUMMARY

Agriculture, the practice of producing foods, fibers, and other useful products from domesticated plants and animals, has developed to its present state through the application of scientific discoveries that have been made throughout history. Agricultural science (the study of agriculture) gathers and analyzes information from basic research conducted at federal, state, and private facilities and applies it to agricultural settings. Agriculture encompasses a wide spectrum of applications; therefore, agricultural science is extremely diverse, covering almost every aspect of plant and animal science as well as the humans involved in the process.

## KEY TERMS AND CONCEPTS

- **Agribusiness:** Corporation that controls the entire agricultural process from production to marketing.
- **Agricultural Chemical:** Chemical such as a fertilizer or pesticide that is used to enhance agricultural production.
- **Agricultural Scientist:** Person who studies some aspect of agriculture to understand and improve it.
- **Agriculturist:** Person such as a farmer or rancher who is directly involved in agricultural production.
- **Animal Husbandry:** Practice of breeding or caring for farm animals.
- **Biotechnology:** Manipulation of the genetic material of plants and animals in order to change the expressed characteristics.
- **Monoculture:** Practice of devoting large acreages to the production of only one crop.
- **Sustainable Agriculture:** Agriculture that is diversified, ecologically sound, economically viable, socially just, and culturally appropriate.

## DEFINITION AND BASIC PRINCIPLES

Agricultural science is the study of agriculture (also called farming), the practice of producing foods, fibers, and other useful products from domesticated plants and animals. The production of food is one of the oldest professions in the history of humankind, and even in the twenty-first century, agriculture remains the most common occupation worldwide. The term "agriculture" covers a wide spectrum of practices, beginning with sustenance agriculture, in which a farmer produces a variety of crops sufficient to feed the farmer and his or her family as well as a small excess, which can be sold for cash or bartered for other goods or services. At the other end of the scale, commercialized industrial agriculture involves many acres of land and large numbers of livestock; a considerable input of resources such as fuel, pesticides, and fertilizers; and a high level of mechanization to maximize profit from the enterprise.

Agricultural science uses information gained from basic research and applies it to agricultural settings to increase the yield and quality of agricultural products. The overall goal is to produce sufficient food and fibers to feed and clothe the population and increase the profit margin for those involved in the production process. Agricultural science covers almost every aspect of plant and animal science, and the application of scientifically sound agricultural principles can enhance production regardless of the level of agriculture being practiced.

## BACKGROUND AND HISTORY

The beginnings of agriculture predate recorded history. No one knows when the first crop was cultivated, but beginning around 8500 B.C.E., humans began the gradual transition from a hunter-gatherer to an agricultural society. By 4000 B.C.E., agriculture had been firmly established in Asia, India, Mesopotamia, Egypt, Mexico, Central America, and South America. The transition to agriculture occurred because humans discovered that seeds from certain wild grasses could be collected and planted on land that could be controlled, and that the resulting crop could be harvested for food. This process represents the earliest stage of agricultural science. Through continued observation and selection of preferred plant and animal characteristics, agriculture continued to develop, and by the start of the Bronze Age (around 3000 B.C.E. in the Middle East), humankind had become fully dependent on domestic crops.

Agriculture developed slowly, and even in the nineteenth century, most agricultural enterprises practiced sustenance agriculture. The Industrial Revolution changed the agriculture industry with the invention and production of agricultural machinery such as the cotton gin (by Eli Whitney in 1793), mechanical reaper (by Cyrus Hall McCormick in 1833) and steel plow (by John Deere in 1837). Mechanization led to the development of commercial agriculture and an increase in interest in all areas of agricultural science.

In the twentieth century, the combined action of many agricultural scientists in the Green Revolution led to the development of high-yielding varieties of numerous crops. As a result of numerous other developments in agricultural science, a modern agricultural unit requires relatively few employees, is highly mechanized, devotes large amounts of land to the production of only one crop, and is highly reliant on agricultural chemicals.

## HOW IT WORKS

**Research.** Agricultural science can be divided into two broad categories, research and extension. Agricultural scientists who engage in research are involved in designing experiments, gathering and analyzing data, drawing conclusions, and publishing their results. In the United States, formal training in agricultural science can be traced back to 1862, with the establishment of the Department of Agriculture and the passage of the Morrill Act, which provided the means for the establishment of land-grant colleges in each state. One of the major roles of the land-grant colleges was to provide training in agriculture. The Hatch Act of 1887 provided the means to fund the establishment of agricultural research stations in each state. In most states, the agricultural research station is connected with a land-grant college or university, and the major function of the research station is to conduct agricultural research.

The research being conducted at land-grant colleges and universities and state and federal agricultural research stations is as varied as agriculture itself. Some of the disciplines involved in the study of agriculture include agronomy and horticulture, weed science, forestry, animal and poultry science, dairy and food science, agricultural engineering, soil chemistry and physics, rural sociology, agricultural economics, and plant and animal physiology, pathology, and genetics.

**Extension.** Basic research usually is conducted at land-grant institutions, and applied research typically takes place at agricultural research stations. Basic research is designed to study some fundamental principle and generally leads to a better understanding of how nature works. Applied research usually involves taking information discovered in basic research and using it in real-world situations. In agricultural science, this means applying the knowledge in an agricultural setting. When applied research shows that a particular technique or development is an improvement over the existing practice (for example, resulting in higher yield or quality), the information is passed on to those who can actually use it. This is most often done through pamphlets and other materials published by the state's Cooperative Extension Service. The Cooperative Extension Service was established by the Smith-Lever Act of 1914. In most states, the Cooperative Extension Service is a component of the agricultural research station, and its primary function is to transfer technology from those who discover it (agricultural scientists) to those who produce agricultural commodities. In many areas, agricultural extension agents will visit individually with the producers to ensure that the technology is transferred.

## APPLICATIONS AND PRODUCTS

**Food and Fiber Production.** In a sense, humans have been applying some aspects of agricultural science for more than 10,000 years. Even though they were not aware of it, people were acting in accordance with basic scientific principles every time they selected seeds for planting and animals for breeding in hopes of reproducing a desired trait. In the modern world, essentially no food, fiber, or other agricultural product has not been affected by the application of agricultural science. Many of these applications resulted in higher yields. For example, in the later part of the nineteenth century, soil scientists learned that soil erosion could be dramatically reduced by planting windbreaks and plowing perpendicular to the slope rather than in the direction of the slope.

This helped conserve topsoil and increased yields. Another practice that has increased yields has been the use of fertilizers. In the nineteenth century, plant scientists discovered which nutrients are required by plants, and agriculturalists used this knowledge to return fertility to depleted soil and increase yields.

The use of agricultural machinery led to tremendous increases in crop yields. To use their machinery more efficiently, agriculturalists turned to monoculture. Although this practice raised yields, it also made crops more susceptible to damage by pests. Scientists have developed many different agricultural chemicals to combat these pests and maintain high yields.

Animal husbandry has also benefitted from the efforts of agricultural scientists. In the poultry industry, scientists discovered that by controlling the temperature and lighting in the poultry house, the number of consecutive days on which hens lay eggs can be increased from around fifty to more than three hundred. This has resulted in tremendous yields in egg production.

Discoveries in genetics have also contributed significantly to increases in agricultural yields. Plant breeders have been able to use knowledge of genetics to produce numerous varieties of high-yielding crops, and animal breeders have produced breeds that grow much larger than the parental stock. In combination with genetics, biotechnology is being used to enhance the quality of many agricultural products. For example, genetic engineering has produced a variety of tomatoes that maintains its flavor during storage, thereby increasing its shelf life. In the future, the quality of many other agricultural products is likely to be improved through genetic engineering.

**Agricultural Products.** The products of agricultural science are the foods, fibers, and other commodities that are produced by agricultural endeavors around the world and that affect the lives of everyone on a daily basis. Agricultural products are generally divided into those that come from animals and those that are derived from plants. Each of these products exists in its present state as a result of the application of knowledge gained from agricultural science.

The major animal-derived products can be divided into edible and nonedible red meat products, milk and milk products, poultry and egg products, and wool and mohair. Edible red meat products primarily come from cattle, swine, sheep, goats, and animals such as horses and Asian or African buffalo. The major nonedible red meat products include rendered fat, which is used to make soap and formula animal feeds; bone meal, which is used in fertilizer and animal feeds; manures used as fertilizers; and hides and skins, which are tanned and used to make leather products. Milk and milk products (also known as dairy products) are produced primarily by dairy cattle and include whole milk, evaporated and condensed milk, cultured milk products, cream products, butter, cheese, and ice cream. Poultry (chickens, turkeys, ducks, geese, pigeons, and guinea hens) and egg products are nutritious, relatively inexpensive, and used by humans throughout the world. The hair covering the skin of some farm animals (wool and mohair) is also considered an agricultural product.

Plant-derived agricultural products are very diverse. Timber products include those materials derived from the trees of renewable forests. Grain crops such as corn, rice, and wheat are grasses that produce edible seed. Cotton, flax, and hemp

*A USDA researcher manipulating DNA of wheat insect pests.* (Richard T. Nowitz/Photo Researchers, Inc.)

are the principal fiber plants grown in the United States, although less important crops such as ramie, jute, and sisal are also grown. Fruit crops refer to the fruit from a variety of perennial plants that are harvested for their refreshing flavors and nourishment. Nut crops such as pecan, walnut, and almond refer to those woody plants that produce seed with firm shells and an inner kernel. Vegetable crops are extremely diverse and range from starchy calorie sources (such as potatoes) to foods that supply mainly vitamins and minerals (such as lettuce). The world's three most popular nonalcoholic beverages are derived from coffee, tea, and cocoa plants. Spice crops are plants grown for their strong aroma or flavor, and drug crops are plants that have a medicinal property. Ornamental crops are grown for aesthetic purposes and are divided into florist crops (flower and foliage plants) and landscape crops (nursery plants). Forage crops include clover, alfalfa, and other small grain grasses that are grown to feed livestock and a variety of straw crops for haymaking. Specialty crops, grown for their high cash value, include sugarcane, tobacco, artichokes, and rubber.

## IMPACT ON INDUSTRY

Throughout the world, agricultural scientists can be found working for governments, colleges and universities, international organizations, and industry.

**Government Regulation and Research.** Most developed countries have an agency responsible for overseeing the nation's food and fiber production. Often this agency is a federally funded agricultural agency similar to the United States Department of Agriculture (USDA) that is responsible for conducting agricultural research and ensuring that the latest developments are made available to the agricultural industry. Agencies such as the USDA generally have research facilities located in various parts of the country. These facilities may stand alone or be located on a university campus.

In most developed countries, each state or province will have a state agency that oversees regional agricultural research stations and extension specialists. The type of agriculture practiced in a given area is influenced by a number of factors, including water availability, soil, and climate; therefore, these facilities are usually in different geographical areas, and the agricultural scientists who are employed at these facilities generally focus their research on the agricultural commodities produced in that area. In the United States, for example, a research station in the dairy region of Wisconsin may conduct research related to the dairy industry; in the corn belt of Iowa, the focus may be on corn research; and cotton research may receive the major research emphasis in the high plains of Texas.

**Academic Research.** Throughout the world, a significant amount of agricultural research takes place at colleges and universities. Much of this research is conducted in institutions such as the land-grant universities, which were partially founded on the premise that training in agricultural science would be provided; however, research projects related to agriculture can also be found at other educational institutions. Scientists at these institutions often conduct basic research, but discoveries made at this level often lead to practical developments in agricultural science. Regardless of the institution, research is expensive and in most cases is supported by grant funds primarily provided by government agencies and to a lesser extent by private industry.

**International Organizations.** In underdeveloped countries, agricultural development through the application of scientific agricultural practices is often accomplished through the actions of international organizations. One such organization, the International Fund for Agricultural Development, was established in 1977 as a specialized agency of the United Nations. Its primary function is to provide financial support for agricultural development projects, particularly those involving food production in developing countries. Although the agency may fund only agricultural development, other groups such as the Consultative Group on International Agricultural Research may actually conduct agricultural research. This group's mission is to use scientific research and research-related activities in the fields of agriculture, forestry, fisheries, and the environment to create sustainable food sources and reduce poverty in developing countries. Through the activities of organizations such as these, agricultural science is being applied to help reduce poverty and hunger and improve human health and nutrition throughout the world.

**Industry.** Agricultural science plays a major role in the financial health of many businesses, ranging from small feed, seed, and fertilizer retail outlets to major corporations. At the same time, industry has contributed to the advancement of agricultural science. Sometimes this contribution is in the form of grants to finance agricultural research by federal or state scientists or university faculty, in the hope that the research will be useful in developing an agricultural product. However, many companies employ agricultural scientists to conduct research specifically for the company.

Agricultural chemicals and machinery are two of the areas in which industry has played a major role. Because of the low availability and high cost of labor, mechanization is a must; therefore, agricultural engineers are constantly developing new cost- and time-saving machinery. Two prominent companies producing agricultural equipment are Case IH and Deere & Company.

Modern agricultural is highly reliant on agricultural chemicals such as fertilizers and pesticides, and large-scale agriculture would be impossible without the use of these chemicals; therefore, agricultural chemical companies are constantly conducting research on new, environmentally safe, and effective chemicals. Leaders in the production of agricultural chemicals include Monsanto and Bayer CropScience.

## CAREERS AND COURSE WORK

Careers in agricultural science are as varied as the field of agriculture itself. A large number of agricultural scientists work in basic or applied research and development for federal or state governmental agencies, colleges and universities, or private industry. Some people trained in agricultural science advance to administrative positions, where they manage research and development programs or marketing or production operations in companies that produce food products or agricultural chemicals, supplies, and machinery. Other agricultural scientists work as consultants to agricultural enterprises. The actual nature of the work being performed by agricultural scientists can be broadly divided into food science and technology, plant science, soil science, and animal science. Of the roughly 31,000 people employed in agricultural science in the United States in 2008, about 43 percent were employed as food scientists or technologists, 45 percent were working as plant and soil scientists, and 12 percent were animal scientists. These numbers do not include those college and university faculty members who were trained in one of the areas of agricultural science. As of the 2010's,

### Fascinating Facts About Agricultural Science

- In 1810, 90 percent of the people in the United States lived on farms. In 1860, only about 60 percent of American people were involved in agriculture, and by 1972, only 4.6 percent of the U.S. population was involved in farming. In 2010, less than 2 percent of Americans were actively engaged in agricultural production.
- In 1800, a single American farmer produced enough food to feed five people; in 2010, each farmer produced enough food to sustain ninety-seven people.
- In the United States, more than $9 billion per year is awarded in grants to fund research in agricultural science.
- As a result of research in agricultural science, modern American farmers can produce 76 percent more crops than their parents did on the same amount of land.
- American supermarkets offer 6,000 to 8,000 agricultural products, more than half of which were not available in 1980.
- More than 40 percent of the world's laborers are employed in some aspect of food production. This means that more than 2 billion people depend on agricultural science for their livelihood.

employment opportunities in agricultural science were expected to be higher than the average for all occupations.

The course work required for majors in agricultural science varies considerably. The specific curriculum depends on the area of agricultural science in which the student is interested, but a strong math and science background is generally needed. Most entry-level positions in the farming or food-processing industry require a bachelor's degree; however, to acquire a research position in most universities, state or federal agencies, or private industry, a master's or doctoral degree in agricultural science, in an engineering specialty, or in a related science such as biology, chemistry, or physics is usually mandatory.

## SOCIAL CONTEXT AND FUTURE PROSPECTS

Agriculture has been and will continue to be extremely important to the development of human culture. Modern civilization would not have developed without agriculture, and agriculture has been successful primarily because of myriad discoveries made in various disciplines associated with agricultural science. The continued advancement of civilization depends on the ability to produce sufficient food and fiber to feed and clothe the world's populations; therefore, agricultural science will continue to be of paramount importance in the future as scientists try to determine the best methods to produce enough food to feed an ever-growing number of people.

Biotechnology is likely to have a tremendous impact on agricultural science. Agricultural scientists have begun to use biotechnology in an attempt to improve a wide variety of characteristics in various plants and animals. Some genetically modified plants are already commercially available. As genetically modified foods are created, scientists must deal with the social issues and scientific questions that they create.

Although agricultural science has increased the yields of agricultural products, some of these gains have come at a cost to the environment. Agricultural scientists are studying ways to practice sustainable agriculture and to raise crop yields and quality without causing damage to the environment. Environmental concerns have also sparked interest in the production of biofuels, fuels manufactured from agricultural derivatives. This raises the question of how much agricultural land should be devoted to food production and how much to growing crops as energy sources.

*Dalton R. Gossett, Ph.D.*

## Further Reading

Brown, Robert C. *Biorenewable Resources: Engineering New Products from Agriculture.* Hoboken, N.J.: Wiley-Blackwell, 2003. Describes the interface between agricultural science and process engineering.

Conkin, Paul K. *A Revolution Down on the Farm: The Transformation of American Agriculture Since 1929.* Lexington: University Press of Kentucky, 2009. Discusses changes in agriculture during the twentieth century and presents a concise introduction to agriculture in the United States.

Gardner, Bruce L. *American Agriculture in the Twentieth Century: How It Flourished and What It Cost.* Cambridge, Mass.: Harvard University Press, 2006. Describes how both mechanical and biotechnical inventions and innovations have contributed to increases in productivity in American agriculture.

Hurt, R. Douglas. *American Agriculture: A Brief History.* Rev. ed. West Lafayette, Ind.: Purdue University Press, 2002. An introductory work that describes the beginning of agriculture in the United States.

Janick, Jules. *Horticulture Science.* 4th ed. San Francisco: W. H. Freeman, 1986. Describes vegetable crop production and contains sections on the major horticultural crops.

Metcalfe, Darrel S., and D. M. Elkins. *Crop Production: Principles and Practices.* 4th ed. New York: Macmillan, 1980. One of the most valuable sources available on the practical aspects of crop production.

Rasmussen, R. Kent, ed. *Agriculture in History.* Pasadena, Calif.: Salem Press, 2010. A collection of essays on events in agricultural history, arranged in chronological order to demonstrate trends.

Smith, Bruce D. *Emergence of Agriculture.* San Francisco: W. H. Freeman, 1999. Examines the development of agriculture as it occurred in the Middle East, Europe, China, Africa, and the Americas.

Taylor Robert E., and Thomas G. Field. *Scientific Farm Animal Production.* 9th ed. New York: Prentice Hall, 2007. An excellent overview of animal production and animal products.

## Web Sites

*Consultative Group on International Agricultural Research*
http://www.cgiar.org

*International Fund for Agricultural Development*
http://www.ifad.org

*U.S. Department of Agriculture*
http://www.usda.gov

**See also:** Agroforestry; Animal Breeding and Husbandry; Bioengineering; Biofuels and Synthetic Fuels; Egg Production; Erosion Control; Fisheries Science; Food Science; Genetically Modified Food Production; Genetic Engineering; Horticulture; Plant Breeding and Propagation; Soil Science.

# AGROFORESTRY

## FIELDS OF STUDY

Forestry; silviculture; agriculture; botany; soil science; horticulture; animal husbandry; hydrology; sustainability.

## SUMMARY

Agroforestry is an interdisciplinary scientific field focused on combining and applying forestry and agricultural principles and techniques to create a sustainable land-use system. Agroforestry provides a wide range of ecological benefits, including improvements to water quality and wildlife habitat and reductions in soil erosion. By combining trees and shrubs with crops and livestock, agroforestry systems attempt to optimize the benefits of short-term crop and livestock rotations and long-term forest rotations. Blending of production practices, which has occurred for centuries, allows a landowner to reap the economic benefits of annual crops while waiting for forest products. In addition to economic benefits, agroforestry results in ecological benefits that may not occur in a traditional agricultural system, including the formation of windbreaks, wildlife corridors, and riparian buffers.

## KEY TERMS AND CONCEPTS

- **Agriculture:** Practice of establishing, growing, managing, and harvesting crops and livestock.
- **Agronomy:** Breeding and raising of crops.
- **Carrying Capacity:** Number of individuals of a plant or animal species that a given piece of land can support.
- **Competition:** In plants, the interaction between two individuals in the attempt to use the same resource, resulting in negative impacts on both.
- **Husbandry:** Breeding and raising of livestock.
- **Monocultures:** Single crops grown on large farms, typical of industrialized agriculture since the early twentieth century.
- **Mycorrhiza:** Mutualistic relationship between a fungus and a plant (tree or crop) wherein the plant gains access to limited soil nutrients from the fungus and the fungus gains energy from the plant.

- **Silviculture:** Practice of establishing, growing, managing, and harvesting trees in a forest.
- **Taungya:** Burmese word for an agroforestry technique of growing crops alongside trees during tree-plantation establishment.

### DEFINITION AND BASIC PRINCIPLES

In many situations, land is managed using a single-approach (or monoculture) system, such as a farm devoted solely to raising crops or livestock or a forest that is cultivated for pulp or lumber production. These separate approaches can be managed successfully to produce a sustainable product. However, through the application of key forestry and agricultural techniques, agroforestry practitioners attempt to develop a sustainable land-use practice producing both short-term and long-term benefits. Although many combinations of forest- and crop-livestock-management techniques exist, agroforestry typically takes the forms categorized as tree-crop, tree-animal, shelterbelt, riparian forest, and natural-specialty crop systems.

Agroforestry provides the framework for a systems approach to land management, allowing for understanding how the physical structure, ecological influences, and economic outputs of trees and forests are interconnected with those same characteristics in crops and livestock. In many climatic regions, from tropical to semiarid zones, constant cropping can place a drain on soil nutrients eventually creating infertile land. By incorporating fallow periods in the crop rotation, farmers can replenish these soil nutrients. However, during the fallow period no crops are produced and, depending on local policies and food-production demand, the fallow period may be excessively shortened and no longer serve its purpose. The goal of agroforestry is to incorporate deeper-rooted perennials (trees and shrubs) in order to maintain nutrient cycling and soil fertility so agriculture can continue for longer periods of time.

### BACKGROUND AND HISTORY

Agroforestry as a technique for land management has been used for centuries, but it was not until the mid-1970's that the term was coined and defined. Early adoptions of what would later be defined as agroforestry probably incorporated the production of crops, animals, and building materials. Although there was a subsequent shift to farming and forestry with a focus on monocultures, agroforestry continued to exist. In later decades, agroforestry has become an important approach to land management in developing regions of the world. Focusing on sustainable land-management practices that allow for increases in economic and ecological diversity has facilitated survival and development of regions where heavy forestry or heavy agriculture could have destroyed resources. Because of the large economic and ecological impact on local communities, government and nongovernmental organizations have expanded research since the 1990's to refine agroforestry practices and disseminate information to the public.

## HOW IT WORKS

Practitioners cannot assume that all forms of agroforestry will succeed in all environments. The surrounding ecological and economic environment, previous land-use practices, and goals of the manager, owner, and community all play into the success of any one of the numerous and overlapping agroforestry strategies. Although the techniques used may be structurally different, the outputs are the same: Establish an ecologically and economically sustainable system that provides short- and long-term products in the forms of agricultural crops and livestock, as well as tree and non-tree forestry products. The five common agroforestry systems (tree-crop, tree-animal, shelterbelt, riparian forests, and natural-specialty crops) are broad, and there is overlap between each one. The application of a specific system is typically related to the ecological and economic environment surrounding an individual owner and piece of land, which includes the selection of particular crop and tree species. Because of the broad nature of these systems, the individual goals defined by the practitioner may separate or merge the agroforestry practice being conducted, resulting in several agroforestry systems existing on a single piece of land.

**Adding Trees to Agriculture.** Two important decisions a practitioner must make regarding the incorporation of trees into an agricultural operation concern are spacing and timing of tree growth. First, trees can be grown in zones or scattered across the land. Zones can be in lines, plots, blocks, or any other systematic arrangement on the land. Spacing between trees is a powerful management tool, in both zoned and scattered arrangements, because it determines the level and intensity of competition and facilitates management. If the trees are growing too close together, growth may be reduced as a result of competition for moisture and nutrients, as well as issues related to crop planting, fertilizing, and harvesting by machinery.

Second, timing can influence competition between trees and crops, as well as labor costs for planting and harvesting. If trees and crops are grown sequentially, direct competition between the two groups of plants can be greatly reduced. However, there may be increases to required labor and associated costs by staggering the land-management techniques. If trees and crops are grown simultaneously, then direct competition between the two groups of plants may be increased, but labor costs may be reduced because of the combined management.

**Adding Crops and Livestock to Silviculture.** As with adding trees to establish agriculture, the same issues of spacing and timing exist when adding crops to a forest. With application of agroforestry techniques to land under more forest-focused management, the overall goals shift from long-term forest management to include short-term crop rotations. One different strategy of adding crops to silviculture is called *taungya*, which is a Burmese term for growing crops during the first few years of tree-plantation establishment. The goal of taungya is to gain short-term crop benefits during the early stages of tree establishment. After one to three years, the crops are no longer sown and the focus is turned to the management of the trees. Whether cropping ends after a few years or is continued indefinitely, there is a need to space trees effectively so that the future plantation can successfully reach the long-term goals defined at the outset.

## APPLICATIONS AND PRODUCTS

**Tree-Crop Systems.** Planting trees in rows, or any spacing configuration that allows for sowing of crops between the trees, is known as intercropping, alley-cropping, and agrosilviculture. Often, the spacing is arranged to allow modern agricultural machinery use and limit the need for physical labor. Two main approaches exist in tree-crop systems, which include planting trees on cropped land and planting crops on forested land. The former typically results in planting straight rows of trees while the latter may result in establishing tree plantations or utilizing natural forests, which may add difficulty in properly spacing trees. In either approach, there is a need for practitioners to make management decisions related to the pattern of trees on the land (zoned or scattered), spacing (between zones, between individual trees), and diversity of tree species. Crops may be temporarily sown during the establishment phase of a forest, which can have an added complication of multiple owners between trees and crops. Tree-crop systems are common in both temperate and tropical zones and have been used for fruit, nut, olive, and grape production for centuries in Europe and North and South America.

**Tree-Animal Systems.** Silvopasture is the management of livestock, their forage, and trees. If a pasture has trees, but those trees are not part of the active land management, then it is not agroforestry. The application of tree-animal agroforestry may occur as an agriculture-dominated, integrated, or forestry-dominated approach. In an agriculture-dominated approach, the focus is placed on maximizing the livestock and food-product outputs. This approach may occur when trees are planted in a pasture to provide shelter for livestock or when livestock are allowed to graze in an orchard. With an integrated approach, the focus is balanced between the management of livestock outputs and tree-product outputs. In temperate areas, such as the southeastern United States, this approach is used for the management of pine plantations with grazing for cattle. Forestry-dominated approaches place management emphasis on tree outputs and utilizing livestock to maximize those tree products. In tropical areas where tree growth can be rapid, introduction of livestock may be beneficial to the trees. However, in temperate areas, tree growth may be slower. If livestock are introduced too early, there may be grazing and damage to the trees, reducing survival and success of the tree-animal system.

**Shelterbelt Systems.** When planted upwind from croplands, windbreaks provide linear shelter that reduces soil erosion. These windbreaks may include mixtures of trees and shrubs. A single row of trees does not provide a solid wall of protection, so it is necessary for multiple rows to be planted in order to create a stratified, continuous canopy to block wind. To reduce excess runoff and soil erosion related to water flow, contour hedgerows are planted along contours of moderate to steep slopes and can slow water movement. Contour hedgerows can be a less costly and less time-consuming strategy for reducing water erosion than the construction of terraces along a slope. Depending on strategies and goals for the trees, shelterbelts can function merely as filter strips, reducing the volume of soil erosion, or they can function as alley crops, providing mulch or tree crops mixed in with the annual row crops.

**Riparian Forest Systems.** Riparian forests are stands of trees adjacent to a linear, flowing body of water (a river or stream). The soils of these forests are either continuously saturated or repeatedly inundated with water. Hydrology and soil types play very important roles in determining which trees can survive and be productive. Riparian forest systems can benefit more than just the immediate community of landowners, as compared with tree-crop or tree-animal systems. Forested zones adjacent to rivers and streams can maintain water quality by filtering agricultural chemicals in runoff through infiltration and immobilization by plants and soil microbes; stabilize banks and reduce erosion by creating a soil-retaining structure with roots and coarse woody debris, thereby improving water infiltration, which reduces the quantity of runoff and the erosive energy; and reduce average and fluctuations of water temperature, maintaining fish and other aquatic-organism populations. Riparian forest systems benefit both the land adjacent to the river and adjacent land downstream.

**Natural-Specialty Crop Systems.** Forest farming takes advantage of the shade produced by a forest canopy to produce special, high-value non-timber forest products. In temperate agroforestry, these products include fruits, nuts, and mushrooms as food crops; ginseng, catnip, and echinacea as herbal medicines; and ferns and flowers for decorative and ornamental uses. Often, these specialty crops are added as products to forests where high-value timber is being produced in order to diversify the economic outputs.

## IMPACT ON INDUSTRY

Incorporation of agroforestry techniques into an agricultural or silvicultural market can have significant impacts on the market value of products, as well as the economic development of a village. With external support from both governmental and nongovernmental organizations, such growth and development can improve standards of living; increase and improve infrastructure related to product storage, transportation, and market stability; and improve food quality and quantity. Although the agricultural industry being affected by agroforestry initially is localized to a village or community, expansion of agroforestry techniques into a region, state, or nation, can impact the industry at those scales, greatly improving the economic and ecological aspects.

**Government and University Research.** Research conducted by government agencies, universities, and nongovernmental organizations is essential to the improvement of agroforestry strategies. These studies provide a basis for understanding competition levels, improvements to growth and production, and ecological and economic benefits. Since not every agroforestry practice is effective in all environments, it is important to understand what components of a particular practice will work and what components fail within an environment. Modifying strategies may mean the difference between marginal or no gains in crops or tree products, and complete success of a farm incorporating agroforestry. Questions related to the operational side of agroforestry (species selection and combinations, spacing, timing), as well as the social side (market fluctuations, lifestyle improvements, economic development), are addressed by research supported by different organizations, both governmental and nongovernmental.

**Industry and Business Sectors.** The commercial aspect of agroforestry is most typically related to local and regional economic growth and development, with the exception of some extremely high-value specialty crops. The addition of a crop to a tree system, or vice versa, can be quite lucrative for local farmers. However, a lack of knowledge or inadequate infrastructure may pose obstacles to getting those products to a suitable market. To overcome barriers related to market knowledge, timing of sales, production processing, or product transport, local farmers may need to form cooperative associations. In some cases, this effort requires external support in the form of micro loans to individual farms from banks or government agencies in order to meet production needs while products are warehoused until markets improve.

## CAREERS AND COURSE WORK

Because of the combination of techniques and principles used in agroforestry, practitioners benefit from a strong educational background in both agriculture and silviculture. Specialized degree programs in agroforestry, both undergraduate and graduate, are more likely found at universities in tropical regions where agroforestry is more commonly practiced. Courses in plant and animal biology, soil and water resources, crop science, and forestry are necessary to understand the application of agroforestry techniques. In addition to the actual production side of agroforestry, a social aspect related to the communities surrounds the practice, so courses in social and economic issues related to natural resources and rural areas are beneficial.

Careers in agroforestry will most likely not have the job title of "agroforester." Typically, a professional in the field will either be a landowner actually practicing agroforestry or an outside consultant assisting landowners in the application of techniques. The latter may be an individual working for a government agency as a local, state, or regional forester or extension specialist, or for a nongovernmental organization with a mission to assist local landowners. Within the United States Department of Agriculture there is the National Agroforestry Center and in Canada there is the Agroforestry Development Centre within the country's Agriculture and Agri-Food department. Outside governmental employment, opportunities

---

### Fascinating Facts About Agroforestry

- More than 1 billion hectares of the world's agricultural land (about 46 percent) has greater than 10 percent tree cover.

- Researchers estimate that 17 percent of agricultural land managed using agroforestry methods involves tree cover great then 30 percent. It is also estimated that 46 percent of agricultural land has tree cover of 10 percent or more.

- Agroforestry is a major land-management strategy in South America and Southeast Asia, where it accounts for more than 80 percent of the agricultural land in each of these two regions.

- Planting of leguminous trees with maize provides necessary fertilization of soils in sub-Saharan Africa resulting in tripling of yields. The trees fix atmospheric nitrogen, making it available in the soil for maize. The trees' limbs are pruned creating mulch for soil protection, and the wood is used as fuel. Without the use of trees, typically a third of the area planted in maize is abandoned.

do exist with organizations such as the World Agroforestry Centre, which is headquartered in Nairobi, Kenya. Because of the scale and scope of such an organization, as well as with national-level agencies, professionals usually work as part of interdisciplinary groups. Individuals with specialized education and experience in tree, crop, animal, or social sciences can be involved with agroforestry research and applications, adding to the growth and development of the field.

## SOCIAL CONTEXT AND FUTURE PROSPECTS

Implementation of agroforestry techniques has improved the monetary, social, and ecological economies of local villages and regions in many tropical and temperate areas. However, in order to garner those benefits at the national level there is a need to scale up the techniques and practices used in agroforestry. To expand agroforestry and its benefits, national governmental agencies, as well as multinational nongovernment organizations, must take an active role in the dissemination of information and support to farmers. This has been demonstrated in several countries, including Cameroon, Indonesia, and Uganda, where increasing involvement of farmers in agroforestry programs has significantly improved the quality of products and the return on farmer investments. In each case, cooperative efforts by governmental agencies, as well as international relief organizations and other nongovernmental organizations, has expanded the application of agroforestry techniques to magnify the geographic scope of monetary, social, and ecological benefits.

*Jordan M. Marshall, Ph.D*

## FURTHER READING

Gordon, Andrew M., and Steven M. Newman, eds. *Temperate Agroforestry Systems*. Wallingford, England: CAB International, 1997. Includes chapters on the application of agroforestry approaches in several countries in North and South America, Europe, and Asia.

Huxley, Peter A. *Tropical Agroforestry*. Hoboken, N.J.: Wiley-Blackwell, 1999. Discusses general agroforestry topics, strategies, and concerns with strong focus on tree-specific aspects.

Jose, Shibu. "Agroforestry for Ecosystem Services and Environmental Benefits: An Overview." *Agroforestry Systems* 76, no. 1 (May, 2009): 1-10. Reviews the environmental benefits of agroforestry in terms of carbon sequestration, biodiversity conservation, soil enrichment, and air and water quality, which are often more difficult to quantify compared to the economic benefits of agroforestry.

Smith, David M., et al. *The Practice of Silviculture: Applied Forest Ecology*. 9th ed. New York: John Wiley & Sons, 1997. Plantation silviculture as well as broader forestry theories are covered in easy-to-understand language.

## WEB SITES

*Association for Temperate Agroforestry*
Agroforestry: An Integrated Land-Use Management System for Production and Farmland Conservation
http://www.aftaweb.org/resources1.php?page=36

*National Association of University Forest Resources Programs*
http://www.naufrp.org

*Society of American Foresters*
http://www.safnet.org

*U.S. Forest Service*
http://www.fs.fed.us

*World Agroforestry Centre*
http://www.worldagroforestrycentre.org

**See also:** Agricultural Science; Agronomy; Animal Breeding and Husbandry; Erosion Control; Forestry; Horticulture; Land-Use Management; Plant Breeding and Propagation; Silviculture; Soil Science.

# AGRONOMY

## FIELDS OF STUDY

Biology; chemistry; earth science; biotechnology; mineralogy; ecology; field crop production; soil management; horticulture; meteorology; climatology; entomology; plant physiology; plant genetics; turf science.

## SUMMARY

Agronomy is the interdisciplinary field in which plant and soil sciences are applied to the production of crops. Agronomists develop ways in which crop yields can be increased and their quality improved. Some agronomists specialize in soil management and land use, which seeks to protect existing farmland and reclaim land for future use in growing crops. Other specialties cover areas such as weed and pest management, meteorology, and the impact of climate change on crop production. The growing importance of biofuels such as ethanol has increased interest in agronomy as a scientific and professional field.

## KEY TERMS AND CONCEPTS

- **Biomass:** Plant or animal matter, particularly when used as an energy source.
- **Crop:** Plant product grown for use as food, animal feed, fiber, or fuel.
- **Forage:** Crop category that includes grasses and is used primarily for feeding animals.
- **Herbicide:** Product that kills or controls weeds and other plants that reduce crop yields.
- **Input:** Product added to soil to increase or improve crop yield, such as fertilizer.
- **Irrigation:** Watering of fields to supplement rainfall.
- **Rotation:** System under which the types of crops grown in a field are changed from one season to the next to improve yields.
- **Yield:** Amount of a crop produced within a defined geographic area, such as a field, during one growing season.

## DEFINITION AND BASIC PRINCIPLES

Agronomy is the study of plants grown as crops for food, animal feed, and nonfood uses such as energy.

In the United States, these crops include wheat, corn, soybeans, grasses, cotton, and a wide variety of fruits and vegetables. Leading crops in other countries vary widely, depending on the nature of the local soil, geography, and growing season.

Plant science is a major component of agronomy. Many agronomists look for ways to grow stronger, hardier plants with higher yields. New types of plants are bred by agronomists to contain specific improvements, such as increased nutrient levels or resistance to pests, over earlier breeds. An area of strong interest is the development of plant types that require fewer inputs such as fertilizers and insecticides to perform well.

The field of agronomy also covers the many factors in the environment that play a role in whether a crop succeeds or fails. The chemical makeup and water balance within a crop's soil are leading factors. Weather and climate patterns, both within a single season and over many years, affect the quantity and quality of crop yields. Technology and economics influence demand for certain types of crops, which in turn pushes market prices up and down. Agronomists help producers respond to these factors.

## BACKGROUND AND HISTORY

Agronomy is nearly as old as human civilization. According to archaeological findings, people have been growing plants for food for more than 10,000 years, starting in the western Asian regions of what was Mesopotamia and the Levant.

Many historians believe that plant cultivation, the earliest form of farming, led to a major change in human culture. The growing season required people to live in one place for long periods of time. Permanent settlements near fields most likely evolved into some of the first villages. These settlements were often near water sources such as rivers, which were needed to irrigate field crops. Some of the first developments in agronomy involved the design and building of water-delivery systems.

The industrial revolution brought widespread change to the field of agronomy. Steam-powered farming equipment replaced draft animals such as horses and mules. Plant scientists developed and standardized new breeds of field crops, which increased yields. By the mid-twentieth century, nearly all corn grown in the United States was from hybrid stock.

The use of inputs such as fertilizers and pesticides also increased, but in some cases caused significant environmental harm. Since the 1990's, agronomists have focused more closely on ways to improve crops without damaging local ecosystems.

### HOW IT WORKS

Field crops require the right plant type and breed, healthy soil, adequate water and nutrients, appropriate growing temperatures and rainfall, and the control of disease and pests in order to succeed.

**Plant Breeding and Genetics.** When choosing a type of field crop to plant, farmers and growers consider factors such as the hardiness of certain breeds and their expected yields at the end of the harvest season. Buyers of agricultural products such as food-manufacturing companies look for products that are high in quality and contain specific nutritional or chemical properties. Agronomists who specialize in plant breeding and genetics support the needs of both farms and buyers.

Multiple methods are used to create hybrid plants. Some hybrid strains are created by planting one breed next to another and allowing the two breeds to cross-pollinate. Plant scientists also use in vitro techniques, in which plant tissues are combined in a laboratory setting to create strains that would not occur in nature. One technique that has received significant media attention is genetic modification. Genetically modified plants contain genes introduced directly from other sources that create changes in the plant much more quickly than could be generated through traditional breeding.

**Soil Health.** To support a crop with the highest possible yields, the soil in which the seeds are planted must be in good condition and must match the needs of the particular plant breed. The health of soil can be measured on the basis of its physical properties, its chemical makeup, and the biological material it contains. These qualities are tracked by soil surveys that are conducted and published regularly. Many farmers switch the types of crops they grow in a particular field every few years in order to keep certain soil nutrients from being depleted. This practice is known as crop rotation.

**Hydration and Irrigation.** Field crops require vast amounts of water. When sources such as rainfall do

*A USDA soil researcher checks ground porosity and collects samples.* (Richard T. Nowitz/Photo Researchers, Inc.)

not provide enough water for healthy plant growth, hydration must be supplemented by irrigation systems. These systems are often based on networks of pipes or hoses connected to sprinklers or drip mechanisms that can supply water to an entire field.

**Weather and Climate.** Even when a hardy breed of plant is grown in healthy soil and receives enough water, an entire crop can be damaged or destroyed by unexpected weather patterns. Many farmers and growers protect themselves against weather-related risks by purchasing crop insurance, which covers losses in situations such as storms or early freezes. While climate has less variance for individual farmers on a season-by-season basis, changes in climate over the long term can affect the types of plants that grow successfully in a given area. In some cases, climate change can increase or decrease the amount of land suited for growing crops.

**Pathology and Pest Control.** Like any living organism, field crops are susceptible to natural threats such as disease, predators, and competition from other plants. Agronomists specializing in plant pathology look for ways to fight disease through direct treatment as well as the breeding of new, hardier strains for future crops. Pests such as insects are controlled through the application of inputs such as insecticides. Pest control also influences plant genetics, as in the case of cotton bred to contain a natural compound toxic to boll weevils. Weeds are managed through a combination of herbicides, adjustments to soil properties such as adding or removing water, and the development of plant breeds resistant to weeds.

## APPLICATIONS AND PRODUCTS

Agronomic crops can be broken down into categories in a number of different ways, such as plant type or climate in which the crop is most likely to be found. One of the most common ways in which field crops are grouped is by the end use of the raw material. Nearly all crops in the world can be considered a form of food, animal feed, fiber, energy, or tools for environmental preservation. Some major crops such as corn can be classified in multiple ways, as corn is used for feeding both people and animals and is also refined into ethanol.

**Food.** Food represents one of the most diverse categories of agronomic crops in the United States and worldwide. When people think of field crops and food, the types of plants that come to mind first are grains such as corn, wheat, and rice. When measured by acres of land planted, grains make up the largest share of agronomic crops grown throughout the world. This category also includes fruits, tree nuts, vegetables, plants grown to be refined into sugar and sweeteners (such as beets and sugarcane), and plants from which oil is made (such as soybeans). While tobacco is not considered a food product, tobacco crops are often included in this category because cigarettes and other items made from tobacco leaves can be consumed only once.

The demand for food crops worldwide grows only as fast as the global population. Individual types of food crops may face sharp increases and drops in demand, however. These changes are influenced by factors such as weather patterns and crop failures, prices set by the commodities markets (which, in turn, make the prices of consumer items rise and fall), and the changing tastes of food buyers. The spike in popular interest in low-carbohydrate diets in the late 1990's had an impact on crops used to make products such as flours and sugars. Similarly, populations in developing countries where incomes are rising often change their food-buying habits and choose items more prominent in North American and European diets than in local ones. This pattern can push up prices for crops such as corn and lower demand for locally grown fruits and vegetables. Interestingly, this trend also works in reverse. Consumers in affluent economies such as the United States have become more interested in buying produce from crops grown locally in an effort to reduce the overall impact on the environment. These kinds of changes can lead to rapid shifts in demand for individual types of agronomic crops.

**Animal Feed.** As with food, the category of agronomic crops grown for animal feed is dominated by grains. In the United States, the leading feed grains tracked by the U.S. Department of Agriculture (USDA) are corn, sorghum, barley, and oats. Corn makes up the largest share of this category by volume. The USDA estimates that the 2010-2011 growing season will yield a total of nearly 12.5 billion bushels of corn, much of which will be used for animal feed. The combined yield for sorghum, barley, and oats is estimated at less than 1 billion bushels.

A second type of animal-feed crop is hay. Much of the hay grown and harvested in the United States comes from alfalfa plants or a mixture of grass types. Demand for hay is influenced in part by the weather. Farmers feed more hay to their livestock—particularly cattle—in drier conditions. Hay is often grouped by the USDA with a type of crop known as silage. Silage is not a unique plant, but rather the plant stalks and leaves left after the harvesting and processing of grains such as corn and sorghum. Hay and silage together belong to a category of crops known as forage. While forage was once defined as plant matter eaten by livestock grazing in fields, it has been expanded to include plants that are cut, dried, and brought to the animals.

**Fiber.** Plants grown for nonfood use often fall into the category of fiber. Fiber crops are processed for use in making cloth, rope, paper and packaging, and composite materials such as insulation for homes. In the United States, most of the yearly agronomic fiber crop is made up of cotton plants. The United States is the third-largest grower of cotton in the world. Other plants in this category are jute, sisal, and flax. Jute and sisal are frequently used to make rope, burlap, and rugs. Flax is refined into linen and used in a wide variety of applications ranging from fine clothing and home-decorating products to high-grade papers. Flax fiber is also used in making rope and burlap.

**Energy and Environmental Preservation.** Most of the agronomic crops in these categories also appear in one of the three categories above. Of the crops grown in the United States as sources of bioenergy, corn tops the list as a source of ethanol. Ethanol sources in other countries include crops such as sugarcane and grasses. Vegetable-based oils made from soybeans are blended into diesel fuel to make a composite known as biodiesel. The refining processes for turning crops into bioenergy sources are not yet as cost-effective as traditional sources of fuel such as petroleum and coal, but this situation is likely to change as technologies improve.

Environmental preservation from the standpoint of agronomic crops is a broad and developing category. It includes the strategic planting and rotating of crops to return depleted nutrients to the soil. It also includes the growth of plants near fields to minimize soil runoff and to protect endangered areas such as wetlands.

### IMPACT ON INDUSTRY

Government agencies, academic institutions, and the private sector all play an important role in the field of agronomy. Funding from government sources provides much of the financial support needed for technological development. Agencies also lead many of the research initiatives, which are supplemented by the work of scholars at colleges and universities. Private corporations help to spread innovation from one country to another and devote a portion of their profits to research.

**Government Agencies.** Nearly every country in the world has a national-level government department or agency devoted to agriculture. The departments frequently oversee agencies at the state or local level. The missions of these departments and agencies are to manage each country's agricultural policies and to ensure that public funding is spent on projects that improve the performance of the farming sector. Information on the country's agricultural practices and yields is gathered, published, and used to support policy decisions. The USDA is one of the largest agencies of this type in the world, employing a significant number of agronomists. The USDA also holds regulatory powers over farms and agricultural businesses by setting standards and ensuring that approved practices are followed.

**Academic Institutions.** A significant amount of the research and development conducted in agronomy takes place at colleges and universities. Academic institutions offer advanced programs of study in subfields of agricultural science such as crop science, environmental management and land use, and soil science. In the United States, state universities established as land-grant institutions are required to fund departments in agricultural science. Many of

the members of professional associations such as the American Society of Agronomy and the European Society for Agronomy are faculty members at academic institutions.

**Corporations and the Private Sector.** Because agriculture is a part of every country's economy, the business of agronomy is one of the most global in scope of any industry. Some of the largest private-sector corporations in the world, such as Cargill, Monsanto and ADM, are focused on agricultural science. These firms develop new seed hybrids to meet goals such as higher yields per growing season and greater resistance to pests such as insects and weeds. Private-sector firms also design and manufacture soil inputs and manage the processing and shipping of raw materials from crops.

Because many of the largest firms in this industry have operations throughout the world, they are well positioned to spread new technological developments quickly from one region to another. However, these firms also receive negative attention from the media and from consumers because of concerns about issues such as the environmental impact of new technologies. The development of genetically modified plants has been a topic of heightened interest in modern times, as the plants do not occur in nature and their long-term environmental effects are not fully documented. Other advances have been less controversial. These include the creation of new crop breeds that require fewer inputs or offer a higher concentration of nutrients such as vitamins, minerals, and proteins.

### CAREERS AND COURSE WORK

A career in agronomy requires a solid background in agricultural science. Most agronomists hold bachelor's degrees, while many specialists—particularly those in research or teaching positions—have master's degrees or doctorates. All state colleges and universities established as land-grant institutions offer programs in agronomy or agricultural science as part of their educational mission. These programs allow students to specialize in fields such as plant genetics and breeding, soil science, meteorology and climatology, and agronomic finance and business management.

Students majoring in agronomy take courses in a wide variety of areas. Common fields are mathematics (particularly calculus and geometry), physics, and

## Fascinating Facts About Agronomy

- The Weed Science Society of America tracks nearly 3,500 types of weeds in its database. Many have colorful names such as kangaroo thorn and sneezewort yarrow.
- Some historians say that people in the first farming villages appeared not only to tend field crops but also to brew beer and other alcoholic beverages.
- More food is going straight from the farm to the fork. From 1997 to 2007, the amount of food sold by farmers directly to consumers more than doubled, helped by farmers' markets and community-supported agriculture (CSA) programs.
- One of the leading sources of sweeteners in U.S. food products is corn. One bushel of corn provides enough syrup to sweeten more than 400 cans of soda.
- A combine, one of the most common types of farm equipment in the United States, can harvest enough wheat in nine seconds to make seventy loaves of bread.
- Soybeans are used in many nonfood products. Soy-based wax is used to make crayons, which have brighter colors and are easier to use than those made from petroleum-based wax.
- The top three growers of fiber crops in the world are China, India, and the United States. Cotton makes up the largest share in all three countries. In the United States, more than half of all cotton goes into clothing production, which is led by jeans.

mechanics. Depending on the field a student pursues, advanced course work in biology, botany and plant science, and organic chemistry may be needed. Many courses are highly focused in scope, such as plant pathology or the physical properties of soil.

Demand for professionals with agronomy degrees is rising, according to the U.S. Department of Labor. Job growth in the field of agricultural and food science is projected to be 16 percent between 2008 and 2018. The need for reliable, efficient, environmentally sound sources of plant-based food is a major factor. The increasing use of biomass as an energy source is also contributing to the need for agronomists with an up-to-date knowledge of science and technology.

Many agronomists work for companies that serve farms. These companies manufacture inputs such as fertilizers, create new breeds of field crops, and process raw materials such as grains and fibers. Other agronomists work for government agencies, primarily within the USDA, or teach and conduct research at colleges and universities. An estimated 12 percent of agricultural scientists are independent consultants.

### SOCIAL CONTEXT AND FUTURE PROSPECTS

One of the turning points in public consciousness about agronomy was the release of Rachel Carson's book *Silent Spring* in 1962. Carson's book linked a number of ecological problems, particularly the deaths of wild plants and birds, to the widespread use of pesticides such as DDT. Many of these pesticides were used on field crops. The book led to the United States ban on DDT in 1972 and increased public awareness of the potential environmental harm in certain farming practices.

Consumer interest in the quality of food sources has been on the rise since the 1990's. While there has been demand for products such as organic foods for much of the twentieth century, the category has grown most quickly at the beginning of the twenty-first century. Some consumers participate in community-supported agriculture (CSA) programs in which fruits, vegetables, dairy, and other items are delivered directly from local farmers. Gourmet and chain restaurants are more likely to advertise their use of locally grown and environmentally sound food ingredients. This demand extends to nonfood items ranging from organic cotton fibers in clothing and linens to plant-based, biodegradable home products. It has also influenced the growth of plant-based, non-petroleum energy sources such as ethanol.

Agronomists are well positioned to benefit professionally from these trends. Upcoming issues of interest for agronomists are likely to include the impact of changing weather and climate patterns and the ways in which crop yields can be increased without causing environmental harm.

*Julia A. Rosenthal, B.A.*

### FURTHER READING

Carson, Rachel. *Silent Spring*. 1962. Reprint. New York: Houghton Mifflin, 2002. A landmark examination of the impact of pesticides in agriculture, particularly DDT, and on the environment that sharply increased consumer awareness when it was first published—and continues to do so.

Fageria, Nand Kumar, Virupax C. Baligar, and Charles Allan Jones. 3d ed. *Growth and Mineral Nutrition of Field Crops*. Boca Raton, Fla.: CRC Press, 2011. Covers the biology of crops and the factors that affect soil quality.

Kingsbury, Noel. *Hybrid: The History and Science of Plant Breeding*. Chicago: University of Chicago Press, 2009. An engaging overview of the history of plants and their cultivation for human use throughout the world.

Miller, Fred P. "After 10,000 Years of Agriculture, Whither Agronomy?" *Agronomy Journal* 100, No. 1 (2007): 22-34. Available at https://www.agronomy.org/files/about-agronomy/future-of-agronomy.pdf.

Reed, Matthew. *Rebels for the Soil: The Rise of the Global Organic Food and Farming Movement*. London: Earthscan, 2010. An extensively researched history of organic farming and its political implications.

Vandermeer, John H. *The Ecology of Agroecosystems*. Sudbury, Mass.: Jones and Bartlett, 2011. A reference on the relationship between agronomy and environmental issues, supported by case studies on historical crises.

**WEB SITES**
*Agricultural Council of America*
http://www.agday.org/index.php

*American Society of Agronomy*
http://www.agronomy.org

*Crop Science Society of America*
http://www.crops.org

*Soil Science Society of America*
https://www.soils.org

*United States Department of Agriculture*
http://www.usda.gov

*Weed Science Society of America*
http://www.wssa.net

**See also:** Agricultural Science; Climatology; Erosion Control; Food Science; Genetically Modified Food Production; Horticulture; Meteorology; Soil Science.

# AIR-QUALITY MONITORING

Meteorology; electronics; engineering; environmental planning; environmental studies; statistics; physics; chemistry; mathematics.

## SUMMARY

Air-quality monitoring involves the systematic sampling of ambient air, analysis of samples for pollutants injurious to human health or ecosystem function, and integration of the data to inform public policy decision making. Air-quality monitoring in the United States is governed by the federal Clean Air Act of 1963 and subsequent amendments and by individual state implementation plans. Data from air monitoring are used to propose and track remediation strategies that have been credited with dramatically lowering levels of some pollutants in recent years.

## KEY TERMS AND CONCEPTS

- **Aerosol:** Suspension of solid particles or very fine droplets of liquid, one of the major components of smog.
- **Air Pollutants:** Foreign or natural substances occurring in the atmosphere that may result in adverse effects to humans, animals, vegetation, or materials.
- **Air Quality Index (AQI):** Numerical index used for reporting severity of air pollution levels to the public. The AQI incorporates five criteria pollutants — ozone, particulate matter, carbon monoxide, sulfur dioxide, and nitrogen dioxide —into a single index, which is used for weather reporting and public advisories.
- **Carbon Monoxide (CO):** Colorless, odorless gas resulting from the incomplete combustion of hydrocarbon fuels. CO interferes with the blood's ability to carry oxygen to the body's tissues and results in numerous adverse health effects. More than 80 percent of the CO emitted in urban areas is contributed by motor vehicles.
- **Continuous Emission Monitoring (CEM):** Used for determining compliance of stationary sources with their emission limitations on a continuous basis by installing a system to operate continuously inside of a smokestack or other emission source.
- **Criteria Air Pollutant:** Air pollutant for which acceptable levels of exposure can be determined and for which an ambient air-quality standard has been set. Examples include: ozone, carbon monoxide, nitrogen dioxide, sulfur dioxide, PM10, and PM2.5.
- **Ozone:** Reactive toxic chemical gas consisting of three oxygen atoms, a product of the photochemical process involving the Sun's energy and ozone precursors, such as hydrocarbons and oxides of nitrogen. Ozone in the troposphere causes numerous adverse health effects and is a major component of smog.
- **PM 10 and PM 2.5:** Particulate matter with maximum diameters of 10 and 2.5 micrometers (μm), respectively.
- **Regional Haze:** Haze produced by a multitude of sources and activities that emit fine particles and their precursors across a broad geographic area. National regulations require states to develop plans to reduce the regional haze that impairs visibility in national parks and wilderness areas.
- **Volatile Organic Compounds (VOCs):** Carbon-containing compounds that evaporate into the air. VOCs contribute to the formation of smog and may themselves be toxic.

## DEFINITION AND BASIC PRINCIPLES

Air-quality monitoring aims to track atmospheric levels of chemical compounds and particulate matter that are injurious to human health or cause some form of environmental degradation. The scope varies, from single rooms to the entire globe. There has been a tendency since the 1990's to shift focus away from local exterior air pollution to air quality inside buildings, both industrial and residential, to regional patterns including air-quality issues that cross international boundaries, and to worldwide trends including global warming and depletion of the ozone layer.

Data from air-quality monitoring are used to alert the public to hazardous conditions, track remediation efforts, and suggest legislative approaches to

*An air quality sampling station at the Upper Ringwood Superfund site in New Jersey.* (Martin Shields/Photo Researchers, Inc.)

environmental policy. Agencies conducting local and regional exterior monitoring include national and state environmental protection agencies, the National Oceanic and Atmospheric Administration (NOAA, popularly known as the Weather Bureau), and the National Aeronautics and Space Administration (NASA). Industries conduct their own interior monitoring with input from state agencies and the Occupational Safety and Health Administration (OSHA). There is an increasing market for inexpensive devices homeowners can install themselves to detect household health hazards.

The Clean Air Act of 1963 and amendments of 1970 and 1990 mandate monitoring of six criteria air pollutants—carbon monoxide, nitrogen dioxide, sulfur dioxide, ozone, particulate matter, and lead—with the aim of reducing release into the

environment and minimizing human exposure. With the exception of particulate matter, atmospheric concentrations of these pollutants have decreased dramatically since the late 1980's.

A key concept in implementing monitoring programs is that of probable real-time exposure, with monitoring protocols matched to real human experience. Increasingly compact and automated equipment that can be left permanently at a site and collect data at intervals over a period of weeks and months has been a real boon to establishing realistic tolerance levels.

## BACKGROUND AND HISTORY

The adverse effects of air pollution from burning coal were first noticed in England in the Middle Ages and were the reason for the ban on coal in London in 1306. By the mid-seventeenth century, when John Evelyn wrote *Fumifugium or, the Inconvenience of the Aer and the Smoake of London Dissipated* (1661) the problem of air pollution in London had become acute, and various measures, including banning certain industries from the city, were proposed to combat it. Quantifying or even identifying the chemicals responsible exceeded contemporary scientific knowledge, but English statistician John Graunt, correlating deaths from lung disease recorded in the London bills of mortality with the incidence of "great stinking fogs," obtained objective evidence of health risks, and scientist Robert Boyle proposed using the rate of fading of strips of dyed cloth to monitor air quality.

Chemical methods for detecting nitrogen and sulfur oxides existed in the mid-eighteenth century. By the late nineteenth century, scientists in Great Britain were undertaking chemical analyses of rainwater and measuring rates of soot deposition with the aim of encouraging industries either to adopt cleaner technologies or to move to less populated areas. Early examples of environmental legislation based on scientific evidence are the Alkali Acts, the first of which was enacted in 1862, requiring industry to mitigate dramatic environmental and human health effects by reducing hydrogen chloride (HCl) emissions by 95 percent.

Until the passage of the first national Clean Air Act in 1963, monitoring and abatement of air pollution in the United States was mainly a local matter, and areas in which the residents were financially dependent on a single polluting industry were reluctant to take any steps toward abatement. Consequently, although air quality in most large metropolitan areas improved in the first half of the twentieth century, grave health hazards remained in some industries. This discrepancy was highlighted by the 1948 tragedy in Donora, Pennsylvania, when twenty people died and almost 6,000 (about half the population) became acutely ill during a prolonged temperature inversion that trapped effluents from a zinc smelter, including highly toxic fluorides. Although autopsies and blood tests revealed a pattern of chronic and acute fluoride poisoning, no specific legislation regulating the industry followed the investigation. Blood tests and autopsies have also been used to track incidences of lead and mercury poisoning, some of it from atmospheric pollution.

## HOW IT WORKS

**Structure of Air-Monitoring Programs.** In the United States, state environmental protection agencies, overseen by the federal government, conduct the largest share of air-quality monitoring out of doors, while indoor monitoring is usually the responsibility of the owner or operator.

The Clean Air Act and its amendments require each state to file a State Implementation Plan (SIP) for air-pollution monitoring, prevention, and remediation. States are responsible for most of the costs of implementing regulations, which may be more stringent than federal guidelines but cannot fall below them. Detailed regulations specify which pollutants must be monitored, frequency and procedures for monitoring, and acceptable equipment. The regulations change constantly in response to developing technology, shifting patterns of pollution, and political considerations. There is a tendency to respond rapidly and excessively to new threats while grandfathering in older programs, such as those aimed at asbestos and lead, which address hazards that are far less acute than they once were.

The United States Environmental Protection Agency (EPA) collects and analyzes data from state monitoring stations to track national and long-term trends, make recommendations for expanded programs, and ensure compliance. Some pollutants can be tracked using remote sensing from satellites, a function performed by NASA. The World Health Organization (WHO), a branch of the United Nations, integrates data from national programs and conducts monitoring of its own. Air pollution is an international problem, and the lack of controls in developing nations spills over into the entire biosphere.

Workplace air monitoring falls under the auspices of state occupational safety and health administrations. In addition to protecting workers against the by-products of manufacturing processes, monitoring also identifies allergens, ventilation problems, and secondhand tobacco smoke.

**Monitoring Methods and Instrumentation.** Methods of monitoring are specific to the pollutant. A generalized monitoring device consists of an air pump capable of collecting samples of defined volume, a means of concentrating and fixing the pollutants of interest, and either an internal sensor that registers and records the level of the pollutant or a removable collector.

An ozone monitor is an example of a continuous emission sampler based on absorption spectroscopy. A drop in beam intensity is proportional to ozone concentration in the chamber. Absorption spectrometers exist for sulfur and nitrogen oxides. This type of technology is portable and relatively inexpensive to run and can be used under field conditions, for example monitoring in-use emissions of motor vehicles. Absorption spectroscopy is also used in satellite remote sensing and has been adapted to remote sensing devices deployed on the ground to measure vehicular emissions.

There are a number of methods for measuring particulate matter, the simplest of which, found in some home smoke detectors, involves a photoelectric cell sensitive to the amount that a light beam is obscured. More sophisticated mass monitors measure scattering of a laser light beam, with the degree of scattering proportional to particle size and density. Forcing air through a filter traps particles for further analysis. X-ray fluorescence, in which a sample is bombarded with X rays and emitted light is measured, will detect lead, mercury, and cadmium at very low concentrations. Asbestos fibers present will turn up either by X-ray fluorescence or visual inspection of filters. Pollen and mold spores, important as allergens, are detected by visual inspection. A drawback of filters is cost and the skilled labor required to process them.

A total hydrocarbon analyzer used by the auto industry uses the flame ionization detection principle to identify specific hydrocarbons in auto exhaust.

Volatile organic compounds (VOCs) present a challenge because total concentration is low outside of enclosed spaces and certain industrial sites, and rapid efficient methodology is not available for distinguishing between different classes of organic compounds.

## APPLICATIONS AND PRODUCTS

**Reporting and Predicting Air Quality.** Media report air-quality indices along with other weather data, and the general public has become accustomed to using this information to plan activities such as outdoor recreation. Projections are also used to schedule unavoidable industrial-emissions release to coincide with favorable weather patterns and minimize public inconvenience. Predicting air quality is an evolving and inexact science involving predicting and tracking weather patterns but also integrating myriad human activities.

**Vehicular Emissions.** Federal law mandates that urban areas with unhealthy levels of vehicular pollution require testing of automobile emissions. Laws concerning testing vary considerably from state to state and jurisdictions within states. Additionally, new vehicles manufactured or sold in the United States must undergo factory testing, both of the model and of the individual units, to ensure the vehicle meets federal standards.

Typical vehicle-inspection protocol requires motorists to bring the vehicle to a garage where automatic equipment samples exhaust and analyzes it for CO, aggregate hydrocarbons, nitrogen and sulfur oxides, and particulate matter. Some state departments of motor vehicles operate their own inspection stations, while others license private garages. There are a number of compact units on the market that provide the required information with little operator input.

With improving air quality and a decreasing proportion of older cars that lack pollution-control equipment, some jurisdictions are withdrawing from vehicle testing.

Public-health effects of air-pollution monitoring and mitigation on public health deserve mentioning. Adverse effects on human health were the principal rationale for instituting laws curbing air pollution, and the decline in certain health problems associated with atmospheric pollution since those laws were enacted is testimony to their effectiveness.

Rates of lung cancer and chronic obstructive pulmonary disease (COPD) have declined, and ages of onset have increased substantially since the mid-twentieth century. Although the bulk of this is due to declining tobacco use, some of it is related to pollution control.

## IMPACT ON INDUSTRY

Impact on industry is twofold, including the effects, both positive and negative, of complying with emissions regulations, and the creation and marketing of new products to meet the needs of monitoring.

Costs of installing pollution-mitigation equipment and monitoring emissions are nontrivial, and the benefits to a community of cleaning up an industry do not necessarily compensate the manufacturer's bottom line. Earlier fears that stringent environmental regulations would force plants to close or relocate elsewhere have to some extent been realized, although the relative contributions of environmental legislation, labor costs, tax structure, and other government policies to the exodus of heavy industry from the United States are debated. When existing facilities are exempted from regulations but new facilities must comply, it discourages growth and investment. Few people doubt that the benefits of clean air outweigh the costs.

Air-pollution monitoring and abatement is a major industry in its own right, employing people in state and federal environmental agencies, university- and industry-based research laboratories, and in the manufacture, sale, and servicing of pollution-control devices and monitoring equipment. Because devices and their deployment must comply with constantly shifting state and federal regulations, this is an industry in which the manufacturing end for industrial and research equipment remains based in the United States.

The consumer market for pollution-detection devices and services is growing rapidly, particularly for interior air. Smoke detectors have become a standard household fixture and CO monitors are recommended for fuel-oil- and gas-heated homes. Testing for formaldehyde, mold, and in some localities radon is often part of building inspections during real estate sales.

Although the field is not expanding at the rate it once was, identification of additional areas of concern, the rapid expansion of industry in other parts of the world, and the continuing need for trained technicians make this one of the better employment prospects for would-be research scientists.

## CAREERS AND COURSE WORK

Air-pollution monitoring is a field with solid career prospects, in government and private industry. The majority of openings call for field technicians who supervise monitoring facilities, conduct a varying amount of chemical and physical analyses (now mostly automated), and collect and analyze data. For this type of position a bachelor's degree in a field that includes substantial grounding in chemistry, mathematics, and data management, and training on the types of equipment used in monitoring, is usually required. A number of state colleges offer undergraduate degrees in environmental engineering that provide a solid background. Online programs offered by for-profit institutions lack the rigor and hands-on experience necessary for this demanding occupation.

For research positions in government laboratories and educational institutions, an advanced degree in meteorology, environmental science, or environmental engineering is generally required.

This is a rapidly evolving field for which knowledge of the latest techniques and regulations is essential. Degree programs that offer a solid internship program, integrating students into working government or industrial laboratories, offer a tremendous advantage in a job market where actual experience is essential.

The development, sale, and servicing of monitoring equipment offers other employment options. While course work can provide the general level of knowledge necessary to sell, adjust, and repair sophisticated automated electronic equipment, such a career objective will also require extensive on-the-job training. Some manufacturers offer factory training for service people.

## SOCIAL CONTEXT AND FUTURE PROSPECTS

The regulations and remediation efforts that monitoring informs and supports have clearly had a positive impact on the health and well-being of Americans in the nearly half-century since the passage of the Clean Air Act of 1963. Heavy-metal exposure from atmospheric sources has dropped dramatically. Mandatory pollution-control devices on new passenger vehicles have curbed emissions in states where annual vehicle emission tests are not even required. Older diesel trucks, farm vehicles, and stationary engines remain a concern in the United States.

Air pollution remains a significant problem in the developing world, especially in China, where the rapid growth of coal-fired industries has created

---

### Fascinating Facts About Air-Quality Monitoring

- In 1987, the Environmental Protection Agency (EPA) ranked indoor air pollution from cigarette smoke, poorly functioning heating systems, and formaldehyde from building materials as fourth in thirteen environmental health problems analyzed.
- Ninety percent of Californians were exposed to unhealthful air conditions at some point in 2008.
- The earliest known complaint about noxious air pollution is one registered in Nottingham, England, in 1257 by Queen Eleanor of Provence.
- Medieval and early modern physicians were quick to associate the smell of sulfur in coal smoke as a health hazard because of the prevailing belief that disease was caused by miasmas, or vaporous exhalations from swamps and rotting material including sewage and garbage.
- In 1955 the city of Los Angeles instituted the first systematic air-monitoring program to track levels of the city's notorious smog, caused mainly by motor vehicles. The first monitors used the rate of degradation of thin rubber strips to determine when smog levels were unhealthful and vulnerable individuals should stay indoors.
- Tetraethyl lead, a gasoline additive and significant pollutant, was removed from gasoline not because of human health issues but because it poisoned catalytic converters, needed for fuel efficiency.
- Haze from coal-fired electric generating plants is adversely affecting tourism in Great Smoky Mountains National Park and other Southeastern vacation destinations noted for splendid vistas.
- Ironically, children who engage in vigorous exercise out of doors are vulnerable to respiratory damage from smog, making real-time monitoring of sulfur, nitrogen oxides, and ozone and prompt reporting to the public an important public-health measure.

conditions in urban areas reminiscent of Europe in the nineteenth century. Addressing these problems is a matter of international concern, because air pollution is no respecter of national boundaries.

Low-end environmental monitoring devices for the consumer market are a growth industry. With respect to genuine hazards such as smoke and carbon monoxide, this is a positive development, but there is concern that overzealous salespeople and environmental-consulting firms will exaggerate risks and push for costly solutions in order to enhance their bottom line, as occurred with asbestos abatement in the 1970's and 1980's.

Although often criticized, an integrated approach to environmental policy that includes "cap and trade"—allowing industries to use credits for exceeding standards in one area to offset lagging performance in other areas, or to sell these credits to other industries so long as an industry-wide target is met—helps ease the nontrivial burdens of complying with constantly evolving environmental standards.

Much of the information-gathering and tracking system developed to address emissions of criteria pollutants is being integrated into the effort to slow global warming due to carbon dioxide ($CO_2$) emissions from fossil fuel burning. While elevated $CO_2$ does not directly affect human health, and is actually beneficial to plants, the overall projected effects of global warming are sufficiently dire that efforts to reduce $CO_2$ emissions deserve a high priority in environmental planning.

*Martha A. Sherwood, Ph.D.*

## FURTHER READING

Brimblecombe, Peter. *The Big Smoke: A History of Air Pollution in London Since Medieval Times.* 1987. New York: Routledge, 2011. A readable and fact-filled account of the history of air pollution.

Collin, Robert W. *The Environmental Protection Agency: Cleaning Up America's Act.* Westport, Conn.: Greenwood Press, 2006. Detailed history of environmental regulation, aimed at the general college-educated reader.

Committee on Air Quality Management in the United States, Board on Environmental Studies and Toxicology, Board on Atmospheric Sciences and Climate, Division on Earth and Life Studies. *Air Quality Management in the United States.* Washington, D.C.: National Academies Press, 2004. A reference for public agencies and policymakers, with much specific information.

Magoc, Chris J. *Environmental Issues in American History: A Reference Guide with Primary Documents.* Westport, Conn.: Greenwood Press, 2006. The chapters on tetraethyl lead and the Donora disaster deal with air pollution.

World Health Organization. Text edited by David Breuer. *Monitoring Ambient Air Quality for Health Impact Assessment.* Copenhagen: Author, 1999. Good coverage of efforts outside of the United States.

## WEB SITES

*Air and Waste Management Association*
http://www.awma.org/Public

*Ambient Monitoring Technology Information Center*
http://www.epa.gov/ttn/amtic

*American Academy of Environmental Engineers*
http://www.aaee.net

*National Association of Environmental Professionals*
http://www.naep.org

*United States Environmental Protection Agency*

**See also:** Environmental Engineering; Meteorology; Thermal Pollution Control.

# AIR TRAFFIC CONTROL

## FIELDS OF STUDY

Meteorology; aviation laws and regulations; instrument flying; theory of flight; aviation; crew resource management; navigation; radar fundamentals; communication; telecommunication.

## SUMMARY

Air traffic control is responsible for keeping aircraft safely separated when moving, both on the ground and in the air. It relies heavily on technology and a highly trained professional workforce. The equipment utilized in air traffic control is continually improving as technology evolves. The latest iteration of air traffic control technology involves the replacement of ground-based radar systems with satellite-based systems.

## KEY TERMS AND CONCEPTS

- **Airspace:** Air within which aircraft operate; it may be controlled or uncontrolled.
- **Automatic Terminal Information System (ATIS):** A recording giving the pilot the latest weather and airport information.
- **Clearance:** Approval issued to an individual aircraft by air traffic control for a specific activity or route.
- **Controlled Airspace:** Airspace subject to positive control by air traffic controllers.
- **Encroachment:** Violation of controlled airspace or runway by an aircraft.
- **Instrument Flight Rules:** Flight rules that require aircraft to be under positive air traffic control.
- **NextGen:** Acronym for Next Generation Air Transportation System. The latest technology in development for air traffic control, it involves satellite tracking and global positioning systems.
- **Radar:** Acronym for radio detecting and ranging; measures range, bearing, speed, and other information concerning individual aircraft.
- **Transponder:** Onboard electronic equipment that returns a radar signal to air traffic control depicting aircraft identification, altitude, speed, and course.

## DEFINITION AND BASIC PRINCIPLES

Air traffic control is the means by which separation of aircraft in flight and on the ground is maintained. This service is provided by ground-based personnel utilizing electronic systems and two-way communication. Present-day air traffic control relies primarily on radar. Radar allows air traffic controllers to identify aircraft and to determine altitude, speed, and course. This, in turn, provides the controllers the information required to maintain separation and guide aircraft to their destinations. Air traffic control is divided into three distinct entities: air traffic control towers (ATCTs); terminal radar approach control (TRACON); and air route traffic control centers (ARTCCs). Each has a distinct function, but all activities are coordinated among the sections. Flight service stations, an advisory service, are also a part of the air traffic control network.

Historically, air traffic control has been a function of government. In the United States, that responsibility falls to the Federal Aviation Administration (FAA). A number of countries, including Canada, have been experimenting with privatizing air traffic control by contracting with independent companies. In some countries, such as Brazil, air traffic control is completely under military control. In any case, air traffic controllers must coordinate the operation of thousands of aircraft every day. In the United States alone, there are as many as 5,000 aircraft operating at any moment. This can result in as many as 50,000 operations each day. Air traffic controllers have the responsibility to ensure the safety and efficiency of each operation.

## BACKGROUND AND HISTORY

In the early days of aviation, the airplane was considered a novelty that served no useful purpose. As the number of aircraft and the performance increased, accidents became more frequent. By 1925 the U.S Postal Service was attempting to establish commercial passenger service by contracting with private companies to fly mail. As this service expanded, the need for some type of air traffic control became apparent.

Until the early 1930's, aircraft operated under the "see and be seen" method of collision avoidance. As the number, size, and performance of aircraft increased, this approach proved inadequate. The earliest form of air traffic control consisted of an individual positioned at the airport with red and green flags clearing aircraft to take off or land. This system was impractical for use at night or in bad weather. The Cleveland airport was the first to establish a modern type of air traffic control system that included two-way radio communication.

In 1934, the Bureau of Air Commerce was assigned the responsibility of controlling air traffic on the newly established airways. In 1937, the Department of Commerce took over the air traffic control function. Following a number of midair collisions, the Civil Aeronautics Authority (CAA) was created. Twenty-three airway traffic control centers were established. Following World War II, the use of radar was implemented. Improved radar systems remain the primary method of controlling traffic to date, although satellite systems are being introduced.

### How It Works

The entire air traffic control system relies on radar, two-way radio communication, electronic navigation aids, and highly trained professional personnel.

**Air Route Traffic Control Centers (ARTCCs).** The air traffic control system is divided into twenty-two ARTCCs that manage traffic within specific geographical areas. The ARTCCs are responsible for all traffic other than that controlled by the terminal radar approach control (TRACON) and the control tower facilities. Primarily utilizing constant radar surveillance, the ARTCCs provide separation for aircraft operating in controlled airspace under instrument flight rules.

ARTCCs control traffic traveling between airports. When an aircraft departs the geographical area of responsibility of one ARTCC it is handed off to the next ARTCC along its flight route. The next ARTCC will control the aircraft until it is handed off to the next ARTCC or a TRACON facility.

**Terminal Radar Approach Control.** The TRACON controller accepts the aircraft from the ARTCC as it approaches within 30 to 50 miles of its destination. Again, depending primarily on radar, the TRACON controller issues instructions to the pilot, sequences aircraft for approach or departure, and provides separation between aircraft within the controlled airspace. The TRACON controller will hand off the traffic either to an ARTCC if the aircraft is departing

or to a local controller at an air traffic control tower (ATCT) if it is landing.

**Air Traffic Control Towers.** The ATCT will clear the aircraft to land or take off. The tower controller, using primarily visual contact with the aircraft, will also provide all the necessary information to the pilot, such as the runway in use, the current altimeter setting, winds, ceiling and visibility, and the aircraft's position number in the landing or takeoff sequence.

**Ground Control.** The ATCT provides additional services while the aircraft is on the ground. The tower typically hands off a landed aircraft to a ground controller. The ground controller accepts responsibility for the aircraft as soon as it clears the active runway. Even though the aircraft is on the ground, two-way communication is still required. This is important since an aircraft on the ground could taxi onto or across an active runway and cause an accident. At larger airports ground-surveillance radar is used to direct the aircraft in taxiing.

**Clearance Delivery.** The last function of the ATCT is that of clearance delivery. When an aircraft is preparing to begin a flight under an instrument flight plan, the pilot must receive clearance prior to moving the aircraft. Clearance delivery will inform the pilot of the approved route, altitude, radio frequencies, and departure procedures prior to handing the pilot over to ground control for taxi instructions.

*Air traffic control radar at an airport.* (Victor de Schwanberg/Photo Researchers, Inc.)

**Flight Service Stations.** The final component of the air traffic control structure is flight service. Flight service is an advisory-only function whereby the pilot can receive a weather briefing and file a flight plan. Flight service is also available to aircraft in flight for updated weather forecasts or to file a flight plan. Flight service can also lend assistance to aircraft in distress or pilots who become lost.

**Coordination of Services.** A typical scenario for a flight involves contacting flight service for a weather briefing and filing a flight plan. This would be followed by monitoring the airport terminal information frequency for automated weather and airport conditions. The pilot then contacts clearance delivery and receives the approved flight plan. After confirming the details of the flight plan, the pilot contacts ground control for taxi clearance. Ground control turns the aircraft over to the control tower for takeoff clearance. After takeoff the aircraft is handed off to departure control. As the aircraft approaches the limits of the area covered by departure control it is turned over to the ARTCC, which controls the aircraft until it approaches its destination. The aircraft is then handed over to approach control, followed by the control tower, and finally ground control. Throughout this entire procedure the aircraft would be under positive air traffic control and typically would be under radar surveillance.

## APPLICATIONS AND PRODUCTS

Modern air transportation could not exist without a highly developed and efficient system of air traffic control. The constantly developing technology allows controllers to guide the ever-increasing number of aircraft flying at any given time. There are differences, but virtually every country has an air traffic control system that is compatible with others. While some developing countries do not have the level of sophistication and may lack full radar coverage, the systems are standardized under international agreements.

**Radar.** With the advent of jet transport aircraft in the 1950's, an improved system of controlling these larger, faster aircraft was needed. Radar had been developed during World War II to identify approaching enemy aircraft so that fighters could be launched and directed to intercept the approaching aircraft. This system was effective from a military perspective, but the requirements for air traffic control and aircraft

separation were quite different from the military application.

**Air Route Surveillance Radar (ARSR).** In 1956, ARSR, the first radar system specifically designed for tracking and separating civilian aircraft, was developed. Twenty-three of these units were ordered for the ARTCCs. Simultaneously, the first air traffic control computer system was installed at the Indianapolis International Airport. During the following twenty years, computer-generated display systems were incorporated into the ARSR systems. This display depicted the aircraft identification, altitude, and airspeed. It also computed the aircraft flight path and indicated possible conflicts. The aircraft signal flashed as it approached the limits of the controller's airspace so it could be safely handed off to the next controller. Because of the differing requirements of the ARTCC and TRACON, two entirely different systems were designed. The system used by ARTCCs is called radar data processing, while the system used in approach control and the tower is called automated radar terminal system (ARTS).

**En Route Automation Modernization (ERAM).** A total modernization of the air traffic control system is presently under way. The FAA is committed to increasing the capacity and efficiency of the entire air traffic control system. The target date for complete implementation of this system is 2018. As a bridge between the present system and the future next generation air transportation system (NextGen), the FAA has introduced interim measures. These include new, upgraded displays and new computer systems that allow a better interface between ARTCCs and TRACONs. To bridge the changeover to NextGen, a system known as en route automation modernization (ERAM) has been developed and is operational. ERAM incorporates a modern computer language and doubles the capacity of the air traffic control system. It is also designed to integrate satellite-based navigation and communication. Along with the implementation of ERAM, both navigation and communications capabilities are being upgraded to utilize satellite-based data.

**Next Generation Air Transportation System (NextGen).** NextGen is an integrated new system and procedure that will replace the existing ground-based radar systems with a satellite-based system and use digitally transmitted information in place of air traffic control. Weather information will be embedded in transmitted information to assist the pilot with decision making. NextGen not only will increase safety and efficiency but also will eliminate most of the problems that exist with ground-based systems, because much of the present technology has limited capacity and is antiquated and difficult to maintain.

NextGen is projected to reduce aircraft delays by 21 percent, reduce fuel consumption by 1.4 billion gallons, and reduce $CO_2$ emissions by 14 million tons. The FAA estimates that the total economic benefit will exceed $22 billion.

One integral component of the new system is the aircraft-based equipment automatic dependent surveillance-broadcast (ADS-B). This system is far more accurate than ground-based radar, and it provides the pilot with terrain maps, traffic information and location, weather, and critical flight information.

The satellite-based components of NextGen are expected to provide aircraft with the capability to fly shorter, more direct, and thus more efficient routes. It will also increase the capacity of existing runways to handle traffic and reduce delays as well as aircraft noise. At Dallas/Fort Worth International Airport alone the FAA projects a 45 percent reduction in departure delays and an increase of ten additional departures per hour per runway when the system is in place. This is extremely important given the finite capacity of runways to handle traffic, particularly in poor weather, as well as the virtual impossibility of constructing new commercial airports.

Yet another component of NextGen is system-wide information management (SWIM). SWIM is a streamlined network over which NextGen information will be exchanged. SWIM will provide secure information-management architecture for sharing national airspace data utilizing off-the-shelf hardware and software.

## IMPACT ON INDUSTRY

While the majority of the research and development is initiated and funded by national governments, the actual design and production of air traffic control equipment is in the hands of private manufacturers. Government agencies around the world are working together to address the issues confronting what has become a global air traffic control network.

A number of international corporations have entered the competition for air traffic control hardware

---

### Fascinating Facts About Air Traffic Control

- The first air traffic controller was Archie W. League, who sat at the end of the runway at Lambert Airfield in St. Louis. He controlled traffic with two flags, a red one for "hold" and a checkered one for "go," while sitting under an umbrella.
- In August, 1981, PATCO, the Professional Air Traffic Controllers Organization, called for a strike after unsuccessful contract negotiations. More than 12,000 strikers were terminated by President Ronald Reagan for participating in the strike, which violated the 1955 law banning government workers from striking. Reagan gave the striking air traffic controllers 48 hours to get back to work, citing the strike a "threat to national safety." The terminated controllers were then banned by Reagan from ever being hired by the Federal Aviation Administration again.
- In August, 1993, President Bill Clinton overturned Reagan's hiring ban, and about 850 controllers who went on strike in 1981 were eventually rehired.
- Stress is a big component of an air traffic controller's job, which may be one of the reasons controllers are required to retire earlier than most other federal employees. Controllers age fifty and over are eligible for retirement, provided they have completed twenty years of active service; and all controllers who have completed twenty-five years of active service can retire regardless of age.

---

and software. Companies such as the Harris Corporation, an international communications equipment manufacturer, produce a complete air traffic management system. Japan Digital Communications provides in-flight communications access including high-speed data communications. MacDonald Dettwiler and Associates provides air traffic control software. Raytheon Canada Limited manufactures solid-state air traffic control radars as well as integrated air traffic control systems. All of these companies contribute to a compatible, worldwide air traffic control network.

## CAREERS AND COURSE WORK

While the demand for qualified air traffic controllers is increasing, there are a limited number of potential employers. In the United States, except for a few private companies staffing nonfederal control towers, the FAA is the sole employer. In some countries, the military handles all air traffic control, while some countries have contracted with private companies to run the air traffic control system. In all cases, air traffic controllers must be highly competent in areas such as communication, decision making, planning, and weather analysis.

The major portion of an air traffic controller's training typically takes place at the FAA Academy in Oklahoma City. In the United States, a number of colleges and universities are approved to provide initial training for potential controllers. This program is called the air traffic-collegiate training initiative (AT-CTI) and utilizes an approved curriculum to prepare candidates for the FAA Academy. The curricula typically include courses in theory of flight, aviation laws and regulations, aviation weather, navigation, instrument flight fundamentals, and basic air traffic control. Many also include topics in crew resource management and air traffic control computer simulations. AT-CTI candidates are required to complete at least a bachelor's degree to be eligible for employment. Historically, the FAA has hired former military controllers and even some candidates with no background or training in aviation. This process is changing, and more emphasis is being placed on the AT-CTI programs to meet the needs of training new controllers.

## SOCIAL CONTEXT AND FUTURE PROSPECTS

Given the globalization of industry and transportation, an efficient worldwide air traffic control network is critical. Every year, the number of air travelers and air freight increases. More than 2 billion passengers fly annually. Many countries, such as China, are on the cutting edge of developing high-technology air traffic control networks similar to NextGen. While the annual growth rate of air traffic in the United States is projected to be 4.5 percent, China's commercial aviation segment is projected to grow at a rate of 7.9 percent, and India's is projected at 10 percent. By 2020, 25,000 new commercial aircraft will enter service with 31 percent destined for the fast-growing Asia Pacific market. As more aircraft fill the skies the technology required to manage these aircraft successfully becomes more complex. This situation will result in increased emphasis on air traffic control and the associated technological development.

*Ronald J. Ferrara*

## FURTHER READING

Gleim, Irvin N., and Garrett W. Gleim. *FAR/AIM: Federal Aviation Regulations and the Aeronautical Information Manual*. Gainesville, Fla.: Gleim Publications, 2011. Contains the FAA regulations concerning instrument flight operations and has an extensive section detailing air traffic control and air traffic control procedures.

Illman, Paul E. *The Pilot's Air Traffic Control Handbook*. 3d ed. New York: McGraw-Hill, 1999. Although somewhat dated, this work remains an excellent overview of air traffic control from a pilot's perspective.

Montgomery, Jeff, ed. *Aerospace: The Journey of Flight*. Maxwell Air Force Base, Ala.: Civil Air Patrol National Headquarters, 2008. This text is a basic overview of the aerospace industry and includes valuable sections on careers and a well-written section on air traffic control.

Nolan, Michael S. *Fundamentals of Air Traffic Control*. 5th ed. Clifton Park, N.Y.: Delmar Cengage Learning, 2010. The most comprehensive work in print about the air traffic control system, employment requirements, and equipment in use.

Preston, Edmund, ed. *FAA Historical Chronology: Civil Aviation and the Federal Government, 1926-1996*. Washington, D.C.: Department of Transportation, 1998. A very interesting chronological snapshot of the FAA, from the inception of federal regulations to 1996.

Robbins, Billy D. *Air Cops: A Personal History of Air Traffic Control*. Bloomington, Ind.: iUniverse, 2006. This is the memoir of an air traffic controller who experienced the transition to computerized systems.

## WEB SITES

*Federal Aviation Administration*
http://www.faa.gov

*Federal Aviation Administration Academy*
http://www.faa.gov/about/office_org/headquarters_offices/arc/programs/academy

*U.S. Centennial of Flight Commission*
http://www.centennialofflight.gov/index.cfm

**See also:** Atmospheric Sciences; Avionics and Aircraft Instrumentation; Communications; Communications Satellite Technology; Maps and Mapping; Meteorology; Probability and Statistics; Risk Analysis and Management; Vertical Takeoff and Landing Aircraft.

# ALGEBRA

## FIELDS OF STUDY

College algebra; precalculus; calculus; linear algebra; discrete mathematics; finite mathematics; computer science; science; engineering; finance.

## SUMMARY

Algebra is a branch of applied mathematics that goes beyond the practical and theoretical applications of the numbers of arithmetic. Algebra has a definitive structure with specified elements, defined operations, and basic postulates. Such abstractions identify algebra as a system, so there are algebras of different types, such as the algebra of sets, the algebra of propositions, and Boolean algebra. Algebra has connections not only to other areas of mathematics but also to the sciences, engineering, technology, and other applied sciences. For example, Boolean algebra is used in electronic circuit design, programming, database relational structures, and complexity theory.

## KEY TERMS AND CONCEPTS

- **Complement Of A Set:** Group of all elements that are not in a particular group but are in a larger group; the complement of a set $A$ is the set of all elements that are not in $A$ but are in the universal set and is written $A\prime$.
- **Conjunction:** Sentence formed by connecting two statements with "and"; the conjunction of statements $p$ and $q$ is written $p \grave{U} q$.
- **Disjunction:** Sentence formed by connecting two statements with "or"; the disjunction of statements $p$ or $q$ is written $p \acute{U} q$.
- **Equal Sets:** Sets whose members are identical; if every element in set $A$ is in set $B$, and if every element in set $B$ is in set $A$, $A$ and $B$ *are equal*.
- **Equivalent Sets:** Sets that have the same cardinality.
- **Negation:** Statement that changes the truth value of a given statement to its opposite truth value; negation of statement $p$, or not $p$, is written as $\sim p$.
- **Set:** Well-defined collection of objects or elements.
- **Set Intersection:** Elements that two sets have in common; the intersection of set $A$ and set $B$ is the set of all elements in $A$ and in $B$ and is written $A \, Ç \, B$.
- **Set Union:** Set of elements that are in either or both of two sets; the union of set $A$ and set $B$ is the set of elements that are in either set $A$ or set $B$ or in both sets and is written $A \, \grave{E} \, B$.
- **Subset:** Group of elements all of which are contained within a larger group of elements; if every element of set $A$ is in set $B$, $A$ is a subset of $B$.
- **Universal Set:** Universe of discourse or fixed set from which subsets are formed; written $U$.

## DEFINITION AND BASIC PRINCIPLES

Algebra is a branch of mathematics. The word "algebra" is derived from an Arabic word that links the content of classical algebra to the theory of equations. Modern algebra includes a focus on laws of operations on symbolic forms and also provides a systematic way to examine relationships between such forms. The concept of a basic algebraic structure arises from understanding an important idea. That is, with the traditional definition of addition and multiplication, the identity, associative, commutative, and distributive properties characterize these operations with not only real numbers and complex numbers but also polynomials, certain functions, and other sets of elements. Even with modifications in the definitions of operations on other sets of elements, these properties continue to apply. Thus, the concept of algebra is extended beyond a mere symbolization of arithmetic. It becomes a definitive structure with specified elements, defined operations, and basic postulates. Such abstractions identify algebra as a system, and therefore, there are algebras of many different types, such as the algebra of sets, the algebra of propositions, and Boolean algebra.

The algebra of sets, or set theory, includes such fundamental mathematical concepts as set cardinality and subsets, which are a part of the study of various levels of mathematics from arithmetic to calculus and beyond. The algebra of propositions (logic or propositional calculus) was developed to facilitate the reasoning process by providing a way to symbolically represent statements and to perform calculations based on defined operations, properties, and truth tables. Logic is studied in philosophy, as well

## Laws and Properties

| Law or Property | Algebra of Sets (set theory) | Algebra of Propositions (logic or proposition calculus) | Boolean Algebra |
|---|---|---|---|
| | For nonempty sets $A$, $B$, and $C$ that are subsets of a universal set $U$ ($\varnothing$ designates the empty set) | For propositions $p$, $q$, and $r$ (T is a true proposition. F is a false proposition.) | For any elements $x$, $y$, and $z$ in set $B$ (The operation symbol $\times$ may be omitted.) |
| Identity property | $A \cup \varnothing = A$<br>$A \cap U = A$ | $p \vee \mathrm{F} \equiv p$<br>$p \wedge \mathrm{T} \equiv p$ | $x + 0 = x$<br>$x \times 1 = x$ |
| Complement law | $A \cup A' = U$<br>$A \cap A' = \varnothing$ | $p \vee \sim p \equiv \mathrm{T}$<br>$p \wedge \sim p \equiv \mathrm{F}$ | $x + x' = 1$<br>$x \times x' = 0$ |
| Involution law | $(A')' = A$ | $\sim(\sim p) \equiv p$ | $(x')' = x$ |
| Commutative property | $A \cup B = B \cup A$<br>$A \cap B = B \cap A$ | $p \vee q \equiv q \vee p$<br>$p \wedge q \equiv q \wedge p$ | $x + y = y + x$<br>$x \times y = y \times x$ |
| Associative property | $(A \cup B) \cup C = A \cup (B \cup C)$<br>$(A \cap B) \cap C = A \cap (B \cap C)$ | $(p \vee q) \vee r \equiv p \vee (q \vee r)$<br>$(p \wedge q) \wedge r \equiv p \wedge (q \wedge r)$ | $(x + y) + z = x + (y + z)$<br>$(x \times y) \times z = x \times (y \times z)$ |
| Distributive property | $A \cup (B \cap C) = (A \cup B) \cap (A \cup C)$<br>$A \cap (B \cup C) = (A \cap B) \cup (A \cap C)$ | $(p \vee (q \vee r) \equiv (p \vee q) \wedge (p \vee r)$<br>$(p \wedge (q \wedge r) \equiv (p \wedge q) \vee (p \wedge r)$ | $x + (y \times z) = (x + y) \times (x + z)$<br>$x \times (y + z) = (x \times y) + (x \times z)$ |
| De Morgan's laws | $(A \cup B)' = A' \cap B'$<br>$(A \cap B)' = A' \cup B'$ | $\sim(p \vee q) \equiv \sim p \wedge \sim q$<br>$\sim(p \wedge q) \equiv \sim p \vee \sim q$ | $(x + y)' = x' \times y'$<br>$(x \times y)' = x' + y'$ |
| Idempotent law | $(A \cup A) = A$<br>$(A \cap A) = A$ | $p \vee p \equiv p$<br>$p \wedge p \equiv p$ | $x + x = x$<br>$x \times x = x$ |
| Absorption law | | | $x + (x \times y) = x$<br>$x \times (x + y) = x$ |
| Domination law | $(A \cup U) = U$<br>$(A \cap \varnothing) = \varnothing$ | $p \vee \mathrm{T} \equiv \mathrm{T}$<br>$p \wedge \mathrm{F} \equiv \mathrm{F}$ | $x + 1 = 1$<br>$x \times 0 = 0$ |

as various areas of mathematics such as finite mathematics. Boolean algebra is the system of symbolic logic used primarily in computer science applications; it is studied in areas of applied mathematics such as discrete mathematics.

Boolean algebra can be considered a generalization of the algebra of sets and the algebra of propositions. Boolean algebra can be defined as a nonempty set $B$ together with two binary operations, sum (symbol +) and product (symbol ×). There is also a unary operation, complement (symbol ¢). In set $B$, there are two distinct elements, a zero element (symbol 0) and a unit element (symbol 1), and certain laws or properties hold. The laws and properties table shows how laws and properties used in the algebra of sets and the algebra of propositions relate to those of Boolean algebra.

### BACKGROUND AND HISTORY

**The Algebra of Sets.** In 1638, Italian scientist Galileo published *Discorsi e dimostrazioni matematiche: Intorno à due nuove scienze attenenti alla mecanica e i movimenti locali* (*Dialogues Concerning Two New Sciences*, 1900). In this work, Galileo recognized the basic

*British mathematician George Boole developed Boolean algebra.* (SPL /Photo Researchers, Inc.)

concept of equivalent sets and distinguishing characteristics of infinite sets. During the nineteenth century, Bohemian mathematician Bernhard Bolzano studied infinite sets and their unique properties; English mathematician George Boole took an algebraic approach to the study of set theory. However, it was German mathematician Georg Cantor who developed a structure for set theory that later led to the modernization of the study of mathematical analysis.

Cantor had a strong interest in the arguments of medieval theologians concerning continuity and the infinite. With respect to mathematics, Cantor realized that not all infinite sets were the same. In 1874, his controversial work on infinite sets was published. After additional research, he established set theory as a mathematical discipline known as *Mengenlehre* (theory of assemblages) or *Mannigfaltigkeitslehre* (theory of manifolds).

**The Algebra of Propositions and Boolean Algebra.** During the nineteenth century, Boole, English mathematician Charles Babbage, German mathematician Gottlob Frege, and Italian mathematician Giuseppe Peano tried to formalize mathematical reasoning by an "algebraization" of logic.

Boole, who had clerical aspirations, regarded the human mind as God's greatest accomplishment. He wanted to mathematically represent how the brain processes information. In 1847, his first book, *The Mathematical Analysis of Logic: Being an Essay Towards a Calculus of Deductive Reasoning*, was published with limited circulation. He rewrote and expanded his ideas in an 1854 publication, *An Investigation of the Laws of Thought: On Which Are Founded the Mathematical Theories of Logic and Probabilities*. Boole introduced the algebra of logic and is considered the father of symbolic logic.

Boole's algebra was further developed between 1864 and 1895 through the contributions of British mathematician Augustus De Morgan, British economist William S. Jevons, American logician Charles Sanders Peirce, and German mathematician Ernst Schröder. In 1904, American mathematician Edward V. Huntington's *Sets of Independent Postulates for the Algebra of Logic* developed Boolean algebra into an abstract algebraic discipline with different interpretations. With the additional work of American mathematician Marshall Stone and Polish American logician Alfred Tarski in the 1930's, Boolean algebra became a modern mathematical discipline, with connections to several other branches of mathematics, including topology, probability, and statistics.

In his 1940 Massachusetts Institute of Technology master's thesis, Claude Elwood Shannon used symbolic Boolean algebra as a way to analyze relay and switching circuits. Boole's work thus became the foundation for the development of modern electronics and digital computer technology.

Outside the realm of mathematics and philosophy, Boolean algebra has found applications in such diverse areas as anthropology, biology, chemistry, ecology, economics, sociology, and especially computer science. For example, in computer science, Boolean algebra is used in electronic circuit design, programming, database relational structures, and complexity theory.

### HOW IT WORKS

Boolean algebra achieved a central role in computer science and information theory that began with its connection to set theory and logic. Set theory, propositional logic, and Boolean algebra all share a common mathematical structure that becomes apparent in the properties or laws that hold.

**Set Theory.** The language of set theory is used in the definitions of nearly all mathematical elements, and set theory concepts are integrated throughout the mathematics curriculum from the elementary to the college level. In primary school, basic set concepts may be introduced in sorting, combining, or classifying objects even before the counting process is introduced. Operations such as set complement, union, and intersection can be easily understood in this context.

For example, let the universal set $U$ consist of six blocks, each of which is a different color. A block may be red, orange, yellow, violet, blue, or green. Using set notation, $U$ = {red, orange, yellow, violet, blue, green}. Let four of the six blocks be sorted into two subsets, $A$ and $B$, such that $A$ = {red, yellow} and $B$ = {blue, green}. The complement of set $A$ is the set of blocks that are neither red nor yellow, $A\cent$ = {orange, violet, blue, green}. The union of sets $A$ and $B$ is the set that contains all of the blocks in set $A$ or set $B$ or both, if there were any colors in common: $A$ È $B$ = {red, yellow, blue, green}. The intersection of sets $A$ and $B$ is the set of blocks that are in set $A$ and in set $B$, any color that both sets have in common. Because the two sets of blocks have no color in common, $A$ Ç $B$ = Æ.

Above the primary level, the concepts of logic are introduced. Daily life often requires that one construct valid arguments, apply persuasion, and make meaningful decisions. Thus, the development of the ability to organize thoughts and explain ideas in clear, precise terms makes the study of reasoning and the analysis of statements most appropriate.

**Logic.** In propositional algebra, statements are either true or false. A statement may be negated by using "not." Statements can be combined in a variety of ways by using connectives such as "and" and "or." The resulting compound statements are either true or false, based on given truth tables.

A compound statement such as "The First International Conference on Numerical Algebra and Scientific Computing was held in 2006 and took place at the Institute of Computational Mathematics of the Chinese Academy of Sciences in New York" can thus be easily analyzed, especially when written symbolically. The "and" connective indicates that the compound statement is a conjunction. Let $p$ be "The First International Conference on Numerical Algebra and Scientific Computing was held in 2006," a true statement; let $q$ be "(it) took place at the Institute of Computational Mathematics of the Chinese Academy of Sciences in New York," a false statement because the institute is in Beijing. The truth table for the conjunction indicates that the given compound statement is false: T Ù F º F.

## Truth Tables for Algebra of Propositions & Boolean Algebra

| Algebra of Propositions | | | Boolean Algebra | | |
|---|---|---|---|---|---|
| For propositions $p$ and $q$ (T is a true proposition; F is a false proposition) | | | For any elements $x$ and $y$ in set $B$ | | |
| Negation | | | Negation | | |
| $p$ | $\sim p$ | | $x$ | $x$ | |
| T | F | | 1 | 0 | |
| F | T | | 0 | 1 | |
| Disjunction | | | Sum | | |
| $p$ | $q$ | $p \vee q$ | $x$ | $y$ | $x + y$ |
| T | T | T | 1 | 1 | 1 |
| T | F | T | 1 | 0 | 1 |
| F | T | T | 0 | 1 | 1 |
| F | F | F | 0 | 0 | 0 |
| Conjunction | | | Product | | |
| $p$ | $q$ | $p \wedge q$ | $x$ | $y$ | $x \times y$ |
| T | T | T | 1 | 1 | 1 |
| T | F | F | 1 | 0 | 0 |
| F | T | F | 0 | 1 | 0 |
| F | F | F | 0 | 0 | 0 |

## Truth Table Verifying
$$\sim(p \vee q) \equiv \sim p \wedge \sim q$$

| $p$ | $q$ | $\sim p$ | $\sim q$ | $p \vee q$ | $\sim(p \vee q)$ | $\sim p \wedge \sim q$ |
|-----|-----|----------|----------|------------|------------------|------------------------|
| T   | T   | F        | F        | T          | F                | F                      |
| T   | F   | F        | T        | T          | F                | F                      |
| F   | T   | T        | F        | T          | F                | F                      |
| F   | F   | T        | T        | F          | T                | T                      |

Compound symbolic statements may require multistep analyses, but established properties and truth tables are still used in the process. For example, it is possible to analyze the two symbolic compound statements $\sim(p \cup q)$ and $\sim p \cup \sim q$ and also to verify that they are logically equivalent. The truth tables for each compound statement can be combined in one large table to facilitate the process. The first two columns of the table show all possibilities for the truth values of two statements, $p$ and $q$. The next three columns show the analysis of each of the parts of the two given compound statements, using the truth tables for negation, disjunction, and conjunction. The last two columns of the table have exactly the same corresponding T and F entries, showing that the truth value will be the same in all cases. This verifies that the two compound statements are logically equivalent. Note that the equivalence of these two propositions is one of De Morgan's laws: $\sim(p \cup q) \,^\circ \sim p \cup \sim q$.

**Computer Circuits.** Shannon showed how logic could be used to design and simplify electric circuits. For example, consider a circuit with switches $p$ and $q$ that can be open or closed, corresponding to the Boolean binary elements, 0 and 1. A series circuit corresponds to a conjunction because both switches must be closed for electric current to flow. A circuit where electricity flows whenever at least one of the switches is closed is a parallel circuit; this corresponds to a disjunction. Because the complement for a given switch is a switch in the opposite position, this corresponds to a negation table. When a circuit is represented in symbolic notation, its simplification may use the laws of logic, such as De Morgan's laws. The simplification may also use tables in the same way as the analysis of the equivalence of propositions,

with 1 replacing T and 0 replacing F. Other methods may use Karnaugh maps, the Quine-McCluskey method, or appropriate software.

Computer logic circuits are used to make decisions based on the presence of multiple input signals. The signals may be generated by mechanical switches or by solid-state transducers. The various families of digital logic devices, usually integrated circuits, perform a variety of logic functions through logic gates. Logic gates are the basic building blocks for constructing digital systems. The gates implement the hardware logic function based on Boolean algebra. Two or more logic gates may be combined to provide the same function as a different type of logic gate. This process reduces the total number of integrated circuit packages used in a product.

Boolean expressions can direct computer hardware and also be used in software development by programmers managing loops, procedures, and blocks of statements.

**Boolean Searches.** Boolean algebra is used in information theory. Online queries are input in the form of logical expressions. The operator "and" is used to narrow a query and "or" is used to broaden it. The operator "not" is used to exclude specific words from a query.

For example, a search for information about "algebra freeware" may be input as "algebra or freeware," "algebra and freeware," or perhaps "algebra and freeware not games." The amount of information received from each query will be different. The first query will retrieve many documents because it will select those that contain "algebra," those that contain "freeware," and those that contain both terms. The second query will retrieve fewer documents because it will select only those documents that contain both terms. The last query will retrieve documents that contain both "algebra" and "freeware" but will exclude items containing the term "games."

### APPLICATIONS AND PRODUCTS
**Logic Machines, Calculating Machines, and Computers.** The "algebraization" of logic, primarily the work of De Morgan and Boole, was important to the transformation of Aristotelian logic into modern

logic, and to the automation of logical reasoning. Several machines were built to solve logic problems, including the Stanhope demonstrator, Jevons's logic machine, and the Marquand machine. In the mid-nineteenth century, Jevons's logic machine, or logic piano, was among the most popular; it used Boolean algebra concepts. Harvard undergraduates William Burkhardt and Theodore Kalin built an electric version of the logic piano in 1947.

In the 1930's, Boolean algebra was used in wartime calculating machines. It was also used in the design of the first digital computer by John Atanasoff and his graduate student Clifford Berry. During 1944-1945, John von Neumann suggested using the binary mathematical system to store programs in computer memory. In the 1930's and 1940's, British mathematician Alan Turing and American mathematician Shannon recognized that binary logic was well suited to the development of digital computers. Just as Shannon's work served as the basis for the theory of switching and relay circuits, Turing's work became the basis for the field of automata theory, the theoretical study of information processing and computer design.

By the end of World War II, it was apparent that computers would soon replace logic machines. Later computer software and hardware developments confirmed that the logic process could be mechanized. Although research work continues to provide theoretical guidelines, automated reasoning programs such as those used in robotics development, are in demand by researchers to resolve questions in mathematics, science, engineering, and technology.

**Integrated Circuit Design.** Boolean algebra became indispensable in the design of computer microchips and integrated circuits. It is among the fundamental concepts of digital electronics that are essential to understanding the design and function of different types of equipment.

Many integrated circuit manufacturers produce complex logic systems that can be programmed to perform a variety of logical functions within a single integrated circuit. These integrated circuits include gate array logic (GAL), programmable array logic (PAL), the programmable logic device (PLD), and the complex programmable logic device (CPLD).

Engineering approaches to the design and analysis of digital logic circuits involves applications of advanced Boolean algebra concepts, including algorithmic state and machine design of sequential circuits, as well as digital logic simulation. The actual design and implementation of sizeable digital design problems involves the use of computer-aided design (CAD).

**Computer Algebra Systems.** During the 1960's and 1970's, the first computer algebra systems (CASs) emerged and evolved from the needs of researchers. Computer algebra systems are software that enable users to do tedious and sometimes difficult algebraic tasks, such as simplifying rational functions, factoring polynomials, finding solutions to a system of equations, and representing information graphically in two or three dimensions. The systems offer a programming language for user-defined procedures. Computer algebra systems have not only changed how algebra is taught but also provided a convenient tool for mathematicians, scientists, engineers, and technicians worldwide.

Among the first popular computer algebra systems were Reduce, Scratchpad, Macsyma (later Maxima), and Mu-Math. Later popular systems include MATLAB, Mathematica, Maple, and MathCAD.

In 1987, Hewlett-Packard introduced HP-28, the first handheld calculator series with the power of a computer algebra system. In 1995, Texas Instruments released the TI-92 calculator with advanced CAS capabilities based on Derive software. Manufacturers continue to offer devices such as these with increasingly powerful functions; such devices tend to decrease in size and cost with advancements in technology.

### IMPACT ON INDUSTRY

**Government and University Research.** Boolean algebra has roots and applications in many areas, including topology, measure theory, functional analysis, and ring theory. Research and study of Boolean algebras therefore includes structure theory and model theory, as well as connections to other logics. Some of the techniques for analyzing Boolean functions have been used in such areas as computational learning theory, combinatorics, and game theory.

Computer algebra, originally known as algebraic computing, is concerned with the development, implementation, and application of algorithms that manipulate and analyze mathematical expressions. Practical and theoretical research includes the development of effective and efficient algorithms for use in computer algebra systems. Research includes engineering, scientific, and educational applications.

Linear algebra begins with the study of linear equations, matrices, determinants, function spaces, eigenvalues, and orthogonality. Research and development in applied linear algebra includes theoretical studies, algorithmic designs, and implementation of advanced computer architectures. Such research involves scientific, engineering, and industrial applications.

**Engineering and Technology.** The study of associative digital network theory comprises computer science, electrical engineering digital circuit design, and number theory. Such theory is of interest to researchers at industrial laboratories and instructors and students at technical institutions. The focus is on new research and developments in modeling and designing digital networks with respect to both mathematics and engineering disciplines. The unifying associative algebra of function composition (semigoup theory) is used in the study of the three main computer functions: sequential logic (state machines), arithmetic, and combinational (Boolean) logic.

**Applied Science.** There has been a dramatic rise in the power of computation and information technology. With it have come vast amounts of data in fields such as business, engineering, and science. The challenge of understanding the data has led to new tools and approaches, such as data mining. Data mining involves the use of algorithms to identify and verify structure from data analysis. Developments in the field of data mining have brought about increased focus on higher level mathematics. Such areas as topology, combinatorics, and algebraic structures (lattices and Boolean algebras) are often included in research.

## CAREERS AND COURSE WORK

The applications of algebra are numerous, which means that those interested in algebra can pursue jobs and careers in a wide range of fields, including business, engineering, and science, particularly computer science.

**Data Analyst or Data Miner.** Data mining is a broad mathematical area that involves the discovery of patterns and hidden information in large databases, using algorithms. In applications of data mining, career opportunities emerge in e-commerce, security, forensics, medicine, bioinformatics and genomics, astrophysics, and chemical and electric power

---

## Fascinating Facts About Algebra

- Algebra has been studied since 2000 B.C.E., making it the oldest branch of written mathematics. Babylonian, Chinese, and Egyptian mathematicians proposed and solved problems in words, that is, using "rhetorical algebra."

- In 1869, British logician William S. Jevons, a student of the mathematician Augustus De Morgan, created a logic machine that used Boolean algebra. The popular machine was known as the logic piano because it had ivory keys and resembled a piano.

- English logician John Venn was heavily influenced by English mathematician George Boole, and his Venn diagrams, developed around 1880, facilitated conceptual and procedural understanding of Boolean algebra.

- In his 1936 paper, "On Computable Numbers, with an Application to the Entscheidungs problem," British mathematician Alan Turing characterized which numbers and functions in mathematics are effectively computable. His paper was an early contribution to recursive function theory, which was a topic of interest in several areas, including logic.

- With the publication of *A Symbolic Analysis of Relay and Switching Circuits* (1940) and "A Mathematical Theory of Communication" (1948), American mathematician Claude Elwood Shannon introduced a new area for the application of Boolean algebra. He showed that the basic properties of series and parallel combinations of electric devices such as relays could be adequately represented by this symbolic algebra. Since then, Boolean algebra has played a significant role in computer science and technology.

- In applied algebra, properties of groups can be used to analyze transformations and symmetry. The transformations include translating, rotating, reflecting, and dilating a pattern such as one in an M. C. Escher painting, or parts of an object such as a Rubik's cube.

- Applied algebra is used in cryptography to study codes and ciphers in problems involving data security and data integrity.

- Applied algebra is used in chemistry to study symmetry in molecular structure.

engineering. Course work should include a focus on higher level mathematics in such areas as combinatorics, topology, and algebraic structures.

**Materials Engineer.** Materials science is the study of the properties, processing, and production of such items as metallic alloys, liquid crystals, and biological materials. There are many career opportunities in research, manufacturing, and development in aerospace, electronics, biology, and nanotechnology. The design and analysis of materials depends on mathematical models and computational tools. Course work should include a focus on applied mathematics, including differential equations, linear algebra, numerical analysis, operations research, discrete mathematics, optimization, and probability.

**Computer Animator or Digital Artist.** Computer animation encompasses many areas, including mathematics, computer science, physics, biomechanics, and anatomy. Career opportunities arise in medical diagnostics, multimedia, entertainment, and fine arts. The algorithms for computer animation come from scientific relationships, statistics, signal processing, linear algebra, control theory, and computational geometry. Recommended mathematics course work includes statistics, discrete mathematics, linear algebra, geometry, and topology.

**Financial Analyst.** As quantitative methods transform the financial industry, banking, insurance, investment, and government regulatory institutions are among those relying on mathematical tools and computational models. Such tools and models are used to support investment decisions, to develop and price new securities, to manage risk, and to guide portfolio selection, management, and optimization. Course work should include a focus on the mathematics of finance, linear algebra, linear programming, probability, and descriptive statistics.

## SOCIAL CONTEXT AND FUTURE PROSPECTS

Algebra is part of two broad, rapidly growing fields, applied mathematics and computational science. Applied mathematics is the branch of mathematics that develops and provides mathematical methods to meet scientific, engineering, and technological needs. Applied mathematics includes not only discrete mathematics and linear algebra but also numerical analysis, operations research, and probability. Computational science integrates applied mathematics, science, engineering, and technology to create a multidisciplinary field developing and using innovative problem-solving strategies and methodologies.

Applied mathematics and computational science are used in almost every area of science, engineering, and technology. Business also relies on applied mathematics and computational science for research, design, and manufacture of products that include aircraft, automobiles, computers, communication systems, and pharmaceuticals. Research in applied mathematics therefore often leads to the development of new mathematical models, theories, and applications that contribute to diverse fields.

*June Gastón, B.A., M.S.Ed., M.Ed., Ed.D.*

## FURTHER READING

Barnett, Raymond A., Michael R. Ziegler, and Karl E. Byleen. *Finite Mathematics for Business, Economics, Life Sciences, and Social Sciences.* 12th ed. Boston: Prentice Hall, 2011. Covers the mathematics of finance, linear algebra, linear programming, probability, and descriptive statistics, with an emphasis on cross-discipline principles and practices. Helps develop a functional understanding of mathematical concepts in preparation for application in other areas.

Cohen, Joel S. *Computer Algebra and Symbolic Computation: Elementary Algorithms.* Natick, Mass.: A. K. Peters, 2002. Examines mathematical fundamentals, practical challenges, formulaic solutions, suggested implementations, and examples in a few programming languages appropriate for building a computer algebra system. Further reading recommendations provided.

Cooke, Roger. *Classical Algebra: Its Nature, Origins, and Uses.* Hoboken, N.J.: Wiley-Interscience, 2008. Broad coverage of classical algebra that includes its history, pedagogy, and popularization. Each chapter contains thought-provoking problems and stimulating questions; answers are provided in the appendix.

Dunham, William. *Journey Through Genius: The Great Theorems of Mathematics.* New York: Penguin, 1991. Provides historical and technical information. Each chapter is devoted to a mathematical idea and the people behind it. Includes proofs.

Givant, Steven, and Paul Halmos. *Introduction to Boolean Algebras.* New York: Springer, 2009. An informal presentation of lectures given by the authors on Boolean algebras, intended for advanced undergraduates and beginning graduate students.

Van der Waerden, B. L. *Algebra.* New York: Springer,

2003. Reprint of the first volume of the 1970 translation of van der Waerden's *Moderne Algebra* (1930), designated one of the most influential mathematics textbooks of the twentieth century. Based in part on lectures by Emmy Noether and Emil Artin.

**WEB SITES**
*American Mathematical Society*
http://www.ams.org

*Mathematical Association of America*
http://www.maa.org

*Society for Industrial and Applied Mathematics*
http://www.siam.org

**See also:** Applied Mathematics; Calculus; Computer Languages, Compilers, and Tools; Electronics and Electronic Engineering; Engineering Mathematics; Geometry; Numerical Analysis; Pattern Recognition; Probability and Statistics; Topology.

# ANESTHESIOLOGY

## FIELDS OF STUDY

Pain management; pharmacology; cardiac and pulmonary resuscitation; veterinary anesthesiology; sedation; pathophysiology; airway management; advanced life support; pediatric anesthesiology; geriatric anesthesiology; intensive care; end-of-life care; surgical anesthesiology; local anesthesiology; general anesthesiology.

## SUMMARY

Anesthesiology is a specialty of medical science concerned with the management and control of acute or chronic pain as well as the care of patients before, during, and after surgical procedures. Anesthesiologists are physicians who have specialized training in the administration of drugs or other treatments that can induce the various forms of anesthesia. The word "anesthesiology" is of Greek origin, *an* meaning "without" and *aisth'sis* meaning "sensation." Careful assessment, monitoring, and pharmaceutical treatments allow patients to undergo medical procedures without pain or distress. Anesthesia can induce temporary pain relief, amnesia, loss of responsiveness, and loss of muscle reflexes in localized areas or in the entire body.

## KEY TERMS AND CONCEPTS

- **Anesthesia:** Partial or complete loss of sensation, usually brought about by injection or inhalation.
- **Anesthesiologist:** Physician specializing in anesthesiology.
- **Anesthetist:** Person who administers anesthetic. May be a physician or specially trained nurse.
- **Emergence:** Transition from the sleep (anesthetized) state to full consciousness.
- **General Anesthesia:** Temporary anesthesia that works on the brain, affecting the entire body by creating a full loss of consciousness and responsiveness.
- **Induction:** Period from the initial introduction of an anesthetic drug, by injection or inhalation, until the optimum level of anesthesia is reached.
- **Infiltration Anesthesia:** Local anesthesia injected directly into the tissues, such as into the gums during dental procedures.
- **Local Anesthesia:** Temporary pharmacological inhibition of nerve impulses to a specific body part, typically used to treat small lacerations or perform minor surgery.
- **Nerve Block:** Regional anesthetic injected directly into a nerve (intraneural) or adjacent to the nerve (paraneural).
- **Post Anesthesia Care Unit (PACU):** Area where patients go to recover from the immediate effects of anesthesia and surgery.
- **Twilight:** State of light anesthesia.

## DEFINITION AND BASIC PRINCIPLES

Anesthesiology is a specialized division of medicine that uses drugs or other agents to cause insensibility to pain, reduced sensation, amnesia, or loss of reflexes. This branch of science involves a critical balance of biology, chemistry, physiology, and pharmacology to produce safe and effective pain management for patients. Anesthesiology has evolved beyond the operating room, affecting many other specialized areas of patient care and expanding its role in health care delivery. Anesthesiologists provide consultation and medical support in intensive care units, emergency departments, and pain management clinics, as well as for diagnostic and cardiac procedures. The goal of anesthesiology is to decrease the amount of pain and emotional distress to a patient while effectively monitoring his or her vital signs and safety. Research in this field is focused on developing more efficient administration methods, providing continuous monitoring of vital sign information, decreasing induction and emergence time, and reducing harmful side effects. Advances in anesthesiology have provided tremendous opportunities to surgeons, improving their ability to safely and effectively treat their patients.

## BACKGROUND AND HISTORY

Although the term "anesthesia" was not used until the mid-1800's, many herbal and alcoholic remedies had been used for thousands of years to dull sensation and relieve pain. Records from 1500 B.C.E. describe the use of inhaled opium preparations, and later Indian and Chinese texts encourage the use of inhaled cannabis paired with wine before medical procedures. The German physician Valerius Coruds described the synthesis of ether in 1540, and British-born American scientist Joseph Priestly discovered nitrous oxide (laughing gas) in 1772.

The impact of these discoveries and their application to medicine would not be realized until the 1800's, as ether and nitrous oxide were most commonly used for entertainment in shows known as the "ether frolics." Exactly who should be credited with the discovery of anesthesiology remains controversial. In 1844, American dentist Horace Wells attended a laughing gas show and noted that one of the performers had hurt himself during the show but did not feel any pain until the effects of the gas had worn off. Wells tested the effects of nitrous oxide by inhaling it and having one of his teeth removed. In 1846, William Thomas Green Morton, an apprentice of Wells, publicly demonstrated a painless tumor removal while his patient inhaled ether. Crawford Williamson Long argued that he had performed the same procedure using ether in 1842, but his report was not published until 1849. Physicians once used combinations of salt, ice, and ether to numb small areas, but in 1884, Austrian Karl Koller used the first cocaine-derived local anesthetic in ophthalmic surgery.

Since the scientific basics of anesthesiology were discovered in the mid-1800's, improvements in the methods of administering anesthesia and the discovery of better drugs have led to increased patient safety, improved pain control, and applications beyond the operating room.

## How It Works

Millions of patients every year, both human and animal, undergo surgical or other procedures that involve anesthesiology. The type of anesthesia and its method of administration are carefully selected based on the procedure and the patient's general state of health. The most common types of anesthesia are general, regional, and local.

General anesthesia is often described a being asleep, but it is very different and much more complicated than sleep. It is a carefully controlled state of unconsciousness, amnesia, analgesia, and paralysis that requires constant monitoring and adjustment. Surgeons require that their patients do not move

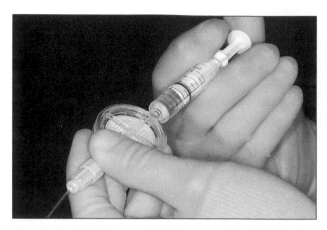

*Anesthetist injecting anesthetic into a patient's back via a catheter.* (Dr. P. Marazzi/Photo Researchers, Inc.)

during an operation, so control over voluntary and involuntary reflexes is crucial. Anesthesiologists may use a combination of three to fifteen different drugs during general anesthesia, depending on the case. These drugs, including isoflurane and desflurlane, are extremely potent, allowing the patient to feel no pain during an operation, remember nothing about the procedure, and recover safely afterward. Typically, in the first phase of general anesthesia (induction), the patient is given an initial intravenous injection that causes a drowsy or unconscious state. Oxygen is continuously provided, first through a mask fitted over the nose and mouth. When the patient reaches an unconscious state, a breathing tube is inserted through the mouth into the windpipe, and a ventilator is attached. To keep the patient in a painless, unconscious state, a carefully selected combination of narcotics, muscle relaxants, and anesthetic gases are administered. This phase is called maintenance. During the state of general anesthesia, the patient's bodily functions are also carefully monitored and controlled. When the procedure is complete, the anesthetic gases are stopped, and a combination of different drugs is given to reverse the effects of the induction drugs. This phase is called emergence. As the patient regains consciousness and is able to breathe on his or her own, the breathing tube is also removed.

Regional anesthesia, also called a nerve block, provides anesthesia to only the part of the body involved in a procedure rather than the entire body. The anesthesiologist injects local anesthetics through a needle, close to the nerves of the involved part of the body. Usually, to reduce discomfort to the patient, the skin and tissues that the needle goes through are first numbed with local anesthetic. These drugs temporarily stop the nerves from working so that no pain, sensation, or movement occurs. The most common type of regional anesthesia is spinal anesthesia, such as an epidural, which can anesthetize the abdomen and legs and is often used during childbirth.

Local anesthesia refers to injecting anesthetic into the skin to temporarily numb a small area so that a minor procedures can be done painlessly. This type of anesthetic is normally used for stitching small lacerations or for dental procedures. Drugs commonly used as local anesthetics include lidocaine and prilocaine.

### APPLICATIONS AND PRODUCTS

Anesthesiology encompasses the entire range of patients, from premature infants to the elderly, as well as individuals with complicated medical challenges. Advances in applications, drugs, and equipment have expanded the role of practitioners beyond the operating room; however, the ultimate goal of safely preventing the patient from feeling pain and distress remains the same. Anesthesiology has evolved into a number of subspecialties to address the needs of patients and health care systems. Subspecialties include critical care, pain management, and pediatric, cardiovascular, ambulatory, bariatric, geriatric, neurologic, and obstetric anesthesiology. Anesthesiology research is creating growth and development in delivery systems, vital sign and blood gas monitoring systems, respiratory accessories, pulmonary-function testing products, ventilators, and pharmaceutical agents.

Integrated anesthesia systems are at the forefront of anesthesiology research and market development. These systems combine anesthesia delivery and patient monitoring. Monitoring technologies are critical to patient safety and include medical devices with a wide array of sensors that enable earlier detection and treatment of potentially life-threatening conditions.

Critical care anesthesiology involves the care and monitoring of patients who have been admitted to the intensive care unit because of critical illness as a result of serious injury or before or after complicated major surgery. In the past, many seriously ill patients developed lung problems and died from acute respiratory failure. However, modern critical care units use mechanical ventilation to assist patients to breathe while they recover from serious injuries or major surgery. Anesthesiologists are experts in breathing assistance and resuscitation methods, and they are the attending physician in more than 30 percent of intensive care units.

Obstetric or maternity anesthesiology is a common subspecialty. Anesthesiologists are frequently consulted to provide pain relief to women in labor, anesthesia for obstetric procedures such as a cesarean section, and resuscitation to newborn infants. Safe and reliable pain relief during childbirth can be provided using epidural analgesia, a type of regional anesthesia. Small amounts of local anesthetic are injected through a small plastic tube or catheter into the woman's back, near the nerves that supply the parts of the body involved in childbirth. Almost 20 percent of births are by cesarean section, which requires anesthesia while maintaining the safety and comfort of mother and baby.

Acute pain management involves the treatment of pain following surgery or trauma, or for patients with chronic pain from terminal cancer. Anesthesiologists are able to treat most pain and provide relief to these patients. Regional anesthesia, as in epidural analgesia, was originally developed for pain during childbirth but has since been extended to a vast range of surgical procedures. A small amount of anesthesia given to patients through an epidural catheter after leg, abdominal, or chest surgery allows them to remain comfortable without the need for large amounts of narcotics, such as morphine. Advances in anesthesiology also provide other treatments for acute pain management using combinations of nonnarcotic drugs and nerve blocks to minimize the amount of narcotic medication needed and keep patients comfortable. Also, special pumps have been developed to allow patients to control the amount of narcotic medication they receive, depending on their level of pain.

### IMPACT ON INDUSTRY

Research and medical science continue to produce new and innovative surgical procedures, instrumentation, and drug treatments. To perform these procedures, patients often require some sort of sedation, pain management, or localized sensation block; therefore, anesthesiology research has had to progress to

## Fascinating Facts About Anesthesiology

- The stages of delivering general anesthesia have been compared to piloting an airplane: Taking off (induction), keeping the airplane in the air (maintenance), and landing smoothly (emergence).
- Researchers from the University of Louisville, Kentucky, found that people born with red hair require about 20 percernt more anesthesia for sedation.
- In 1853, Queen Victoria took chloroform to provide some pain relief during the birth of her seventh child.
- A monument to ether, probably the world's only monument to a drug, stands in a prominent place in Boston's Public Garden. This 40-foot-tall tribute commemorates the first use of ether as an anesthetic under the Etherdome at Massachusetts General Hospital on October 16, 1846.
- In China, acupuncture is often added to or used in place of Western anesthesia.
- It is estimated that 40 million anesthetics are administered each year in the United States.
- Bed bugs inject their saliva, containing anesthetics and anticoagulants, into their victims, making their bites initially painless.
- Patients used to decline the use of anesthesia during medical procedures for religious reasons. They thought the pain was God's will.
- Anesthesiology was the first medical specialty to specifically focus on patient safety.
- Originally ether and nitrous oxide (laughing gas) were not given to people for anesthesia or pain relief but were taken by performers to entertain an audience. These shows were called "ether frolics."

accommodate these surgical and diagnostic advances. Increased regulations from the American Society of Anesthesiologists and health care facilities necessitate the continuous evaluation and documentation of patient's oxygenation, ventilation, circulation, and temperature, which has increased the demand for efficient and effective monitoring equipment. Biotechnology and pharmaceutical companies are competing to develop drugs and devices for anesthesiology applications that will improve patient care and safety.

The future of the anesthesiology industry includes automated and feedback-controlled anesthesia workstations that are in the development and approval stages. These systems promise to allow anesthesiology professionals to focus more on the status of the patient than on the functioning of the equipment and the numerous monitoring screens. These stations are considerably more complex, and therefore, thorough training is essential. Devices to improve the economics of anesthesiology by reducing medical gas leakage and drug flow are also a focus of industry development.

### CAREERS AND COURSE WORK

Professionals in the field of anesthesiology are in high demand as more 40 million procedures involving anesthesia are performed in the United States each year. Careers in the field require extensive training but are highly transportable as the methodology is applicable in medical settings all over the world. Anesthesiology research is improving the technology and available methods, expanding the roles of anesthesiology personnel into new areas of specialty care.

The most recognized profession in anesthesiology is an anesthesiologist, a specialized medical doctor. To become an anesthesiologist, a student must complete the necessary undergraduate degree requirements to gain admission to medical school. Upon completion of medical school, the student completes an internship and a three-year anesthesiology residency. Some students complete an additional year of fellowship to specialize further. During the residency training, anesthesiologists will work toward their certification from the American Society of Anesthesiologists or the American Board of Anesthesiology by passing the required board examination. Dental and veterinary anesthesiologists are doctors of their specific professions who have specialized in the delivery of anesthetic to their patients.

Anesthetists are individuals who can administer anesthetics. They may or may not be physicians. Many anesthetists in the United States are certified registered nurse anesthetists (CRNAs) who work with other medical professionals. Educational requirements for a career as a certified registered nurse anesthetist begin with a bachelor's degree and at least one year of acute care nursing experience. This is followed by completion of a nurse anesthesia program (twenty-four to thirty-six months) leading to a master's degree, after which the student must pass the mandatory certification exam.

Anesthesiologist assistants provide anesthesia care under the direction of an anesthesiologist. These professionals have specialized master's degree training and require licensing, certification, or physician delegation, depending on where they work.

Anesthesia technicians are biomedical personnel who assist anesthesiologists, nurse anesthetists, and anesthesiologist assistants in the operating room with monitoring equipment, supplies, and patient care procedures.

## SOCIAL CONTEXT AND FUTURE PROSPECTS

Many of the advances in medical and veterinary technology, diagnosis, and treatment could not occur without parallel advancements in anesthesiology. Careers in anesthesiology have evolved beyond the operating room as professionals in this field manage a wider range of patients and provide consultation in many departments. Most anesthesiologists specialize their practice to areas such as critical care, pain management, or pediatric, geriatric, ambulatory, or cardiovascular anesthesia. Anesthesiologists are also consulted to address many societal and often controversial issues in bioethics such as addiction to pain medication and patient's right to die.

Research in anesthesiology focuses on both scientific and practical areas. Some key areas include anesthetic safety, medical quality assurance, ambulatory anesthesia, automated delivery systems, and monitoring equipment. Veterinary anesthesiology is a field of particular interest to researchers as it is extremely challenging to apply effective anesthesia techniques to the wide range of species and sizes of animals encountered by veterinarians. The great majority of anesthesiology researchers are also practicing anesthesiologists, which means that finding time and funding to conduct research in addition to their medical duties can be quite a challenge.

*April D. Ingram, B.Sc.*

## FURTHER READING

Maltby, Roger. *Notable Names in Anaesthesia.* New York: Oxford University Press, 2002. A collection of biographies of scientists who have affected and influenced anesthesiology.

Snow, Stephanie. *Blessed Days of Anaesthesia: How Anaesthetics Changed the World.* New York: Oxford University Press, 2008. Provides an interesting account of anesthesia history and how it influenced society and beliefs about enduring pain.

Stoelting, Robert K., and Ronald D. Miller. *Basics of Anesthesia.* Philadelphia: Churchill Livingstone/Elsevier, 2007. An introductory text that can help familiarize readers with the language and concepts of anesthesiology by building on basic science knowledge.

Sweeny, Frank. *The Anesthesia Fact Book: Everything You Need to Know Before Surgery.* Cambridge, Mass.: Perseus, 2003. A very experienced anesthesiologist from California writes in a readily understandable way about anesthesia. He discusses what happens before and after surgery, explains general surgery, and describes several other types of anesthesia.

## WEB SITES

*American Society of Anesthesiologists*
http://www.asahq.org

*International Anesthesia Research Society*
http://www.iars.org

*MedlinePlus*
Anesthesia
http://www.nlm.nih.gov/medlineplus/anesthesia.html

**See also:** Pediatric Medicine and Surgery; Pharmacology; Surgery.

# ANIMAL BREEDING AND HUSBANDRY

## FIELDS OF STUDY

Animal science; genetics; statistics; genomics; biotechnology; animal nutrition.

## SUMMARY

Animal husbandry is the production and care of animals. Animal husbandry is usually called animal science in universities, since academic studies involve research and the application of scientific principles. Animal breeding is often considered part of husbandry and is the application of genetic principles in the development of breeds and lines of animals for human purposes. Animal breeding principles are also used in captive breeding programs to propagate endangered wildlife species. The development of a leaner line of pigs and a strain of chickens that produces more eggs are examples of animal breeding.

## KEY TERMS AND CONCEPTS

- **Biotechnology:** Application of biological techniques to practical uses.
- **Breed:** Population of animals within a species that have similar identifying characteristics.
- **Complimentarity:** Improvement in performance of offspring from parents with different but complimentary breeding values.
- **Correlated Characters:** Traits that change together, either in the same or opposite directions.
- **Genotype:** Genetic makeup of an animal.
- **Heritability:** Measure of the relationship between phenotype and breeding value.
- **Hybrid:** Offspring of breeding different species, lines, or breeds.
- **Inbreeding:** Mating of closely related animals.
- **Line:** Group of related individuals within a breed.
- **Mating System:** Set of rules for mating selected males with selected females.
- **Outbreeding:** Mating of unrelated animals, such as animals in different breeds or lines.
- **Phenotype:** Observed physical appearance or performance of a trait.
- **Polygenic Trait:** Trait affected by many genes, with no gene having a dominating influence.
- **Population:** Group of intermating individuals within a herd, breed, or species.
- **Trait:** Any observable or measurable characteristic of an animal.

## DEFINITION AND BASIC PRINCIPLES

Animal husbandry is concerned with all aspects of the management, care, and breeding of farm animals. The goal of animal husbandry is to provide the best conditions (given economic constraints) to maximize productivity in terms of body weight, wool, milk, or eggs. The animals must remain in good health to attain this productivity and to reproduce. Animal husbandry involves the choice of proper feeds, housing, and suitable animals.

Animal breeding begins with a measurement of desirable traits (phenotype) that relate to improved animal production. The breeding value of an animal, however, is the degree to which its underlying genotype can be transmitted to its offspring. Modern methods of breeder selection combine traditional measurements of quantitative traits with the new technology of genome analysis, which aids in determining the breeder's genotype. The rate of genetic change (in animal populations) is directly related to accuracy of selection, selection intensity, and genetic variation in the population and is inversely related to generation interval. There are two primary types of breeding programs: development of breeds or lines that can be used as breeders (seedstock) and development of crossbreeds for production. Crossbreeds demonstrate improved productivity because of hybrid vigor and complimentary traits exhibited by their parents.

## BACKGROUND AND HISTORY

Animal husbandry began with the domestication of animals for human purposes from around 10,000 to 5000 b.c.e. Sheep were the first to be domesticated, followed by cattle, horses, pigs, goats, and finally chickens and turkeys. A relatively small number of species have been domesticated because they must possess several suitable characteristics that allow them to adapt to interaction with humans. Their diet must be simple to provide (the early domesticated animals depended on grazing and foraging for their food). They must be able to breed in captivity and must grow and reproduce over a relatively short time interval. They must have a calm, predictable behavior and a cooperative type of social structure.

Animal breeding started in the Roman Empire or perhaps earlier. Early breeders recognized desirable traits in animals that they wanted to propagate, so they selected those animals for mating. The characteristics of domesticated animals began to vary greatly from those of their wild cousins and became totally dependent on their human captors. Systematic selective breeding methods began with the English sheep farmer Robert Bakewell in the late 1700's. Bakewell sought to increase the growth rate of sheep so that they could be slaughtered at an earlier age, to increase the proportion of muscle, and to improve feed efficiency. The application of genetics in animal breeding began in the twentieth century. Jay Lush, a professor at Iowa State University, is considered to have been a pioneer in the application of genetic techniques in animal breeding. His *Animal Breeding Plans* (1937) advocated breeding based on quantitative measures and genetics in addition rather than just on the animal's appearance.

## How It Works

**Husbandry.** Farm animal production is an economic venture, undertaken to produce food (meat, milk, and eggs) or other animal products, such as wool, hides, hair, and pelts. Through the process of animal husbandry, growers seek to create conditions that maximize production of animal products at the lowest cost. With advancing technology and improvements in breeds, animal production has evolved from extensive systems to increasingly intensive systems.

Intensive systems put more demands on good husbandry practices, because the animals are often under more stress and depend more on humans for their well-being. Extensive systems involve keeping animals on pastures or in small pens with minimal housing. Intensive systems are most advanced in the case of poultry. Broilers (meat animals) are kept in total confinement indoors, while laying hens are kept completely in cages. Swine are also commonly kept in confinement, usually on slat or grid floors made of metal. Confinement operations require closer attention to the requirements of ventilation, sanitation, and animal interactions. Beef cattle are still grown on range or pasture, but it is more common to finish them in large feedlots concentrating thousands of animals. Dairy cattle usually have pastures for grazing but are practically always milked by machine in parlors. Sheep are still largely grazed on range or pasture for most of their growing cycle.

**Traits and Breeding Value.** The selection of animals in the early days of animal breeding depended on physical or quantitative traits exhibited by the animal without any understanding of the underlying genetic principles. These observed or measured traits are known as the phenotype of the animal. The animal's phenotype is a result of the interaction of its genotype (genetic makeup) with the environment. The goal of animal breeding is to produce animals in a herd, flock, line, strain, or breed that possess superior phenotypes that can be passed on to future generations. The degree to which observed phenotypes can be transmitted to offspring is known as heritability and is a measure of the breeding value of the animal. The selection of desired traits depends on the species of animal and the intended purposes for raising them. The selection also depends on the management practices adopted by the farmer and the relationships between farm inputs and the value of the animals. Examples of traits include calving interval for beef cattle, milk yield for dairy cattle, litter size for swine, first-year egg numbers for hens, and breast weight for meat chickens. The performance of traits can depend on the environment. A high-producing Holstein cow may not produce as well in the tropics because it is not heat tolerant.

*Two research assistants observe a sow and her piglets using a monitor linked to a camera in their sty.* (Science Source)

**Rate of Genetic Change.** Progress in a breeding program is related to the rate of genetic change in a population. There are several factors that affect this rate of change: accuracy of selection, selection intensity, genetic variation, and generation interval. The accuracy of selection relates true breeding values to their prediction for a trait under selection. Selection intensity refers to the proportion of individuals in a population that are selected. Populations selected more intensely will be genetically better than the average, leading to a faster rate of genetic change. Populations exhibiting greater genetic variation among individuals have the potential for more rapid genetic change. Finally, species having a short generation interval will have a faster rate of genetic change.

**Multiple Trait Selection.** Breeders seldom select just one trait for improvement, since a combination of traits is important for the economic success of the enterprise. In fact, selection for one trait usually affects the response to traits not selected for because of the phenomenon of correlated response. The major cause of correlated response is pleiotrophy, the situation in which one gene influences more than one trait.

Breeders practice multiple trait selection by three primary means. Tandem selection involves selecting for one trait, then another. Independent culling levels set minimum standards for traits undergoing selection, and animals are rejected that do not meet all the standards. Finally, the method of economic selection indexes assigns weighted values to the various traits.

## APPLICATIONS AND PRODUCTS

**Seedstock.** A term commonly applied to breeding stock is "seedstock." The purpose of breeding stock is to provide genes to the next generation rather than to be producers of meat, milk, wool, or eggs. Traditionally, seedstock have been purebreds, but the number of nonpurebred stock is increasing. Seedstock animals are obtained by programs of inbreeding. These programs result in an increase in homozygous or similar genotypes. As a result, seedstock have a greater tendency to pass on performance characteristics to their offspring, an ability known as prepotency. One risk of inbreeding is the expression of deleterious genes resulting in reduced performance, known as inbreeding depression. Outcrossing or linebreeding is a milder form of inbreeding and involves mating animals from different lines or strains within the same breed. This process still maintains a degree of relationship to highly regarded ancestors but is less intense than breeding first-degree relatives. Outcrossing allows for the return of vigor that can be lost by inbreeding, while still maintaining the genetic gains obtained by inbreeding.

**Crossbred Animals.** Mating animals from different species is known as crossbreeding. The resultant offspring are known as hybrids. In modern animal husbandry, even hybrids are commonly used in crossbreeding systems. Crossbred animals are used for production and are designed to take advantage of hybrid vigor and breed complimentarity. Hybrid vigor is the increased performance of hybrid offspring over either purebred parent, especially in traits such as fertility and survivability. A classic example of complimentarity is the crossing of specialized male and female lines of broiler chickens. Individuals from male lines are heavily muscled and fast growing but not great egg producers, while individuals from the female line are outstanding egg producers.

**Artificial Insemination.** Artificial insemination is a reproductive technology that has been used for a long time. Semen is collected from males and is used to breed females. Because semen can be frozen, it can be used to eventually sire thousands of offspring. This expanded use of superior males can markedly increase the rate of genetic change. Estrus synchronization facilitates artificial insemination by ensuring that a group of females come into estrus at the same time.

**Embryo Transfer.** Embryo transfer involves collecting embryos from donor females and transferring them to recipient females. Although the motive for embryo transfer is to propagate valuable genes from females, the number of progeny is much fewer, and the procedure is more difficult and costly than artificial insemination.

A variation of embryo transfer is the emerging technology of in vitro fertilization. This technology involves collecting eggs from donor females, which are then matured, fertilized, and cultured in the laboratory. The embryos can then be transferred to recipient females or frozen for later use. The procedure is very expensive and time-consuming. However, it has the potential to aid genetic selection and crossbreeding programs. The genotype of the embryo could be determined before pregnancy. Knowing the genotype could be particularly important for dairy cattle, which frequently have fertility problems.

**Cloning.** Cloning is the production of genetically identical animals. Cloning allows the breeder to predict the characteristics of offspring, to increase uniformity of breeding stock, and to preserve and extend superior genetics. The preferred method of cloning, somatic cell nuclear transfer, involves removing the nuclei from multiple unfertilized eggs, followed by the transfer of somatic cells from the animal to be cloned. If the process is successful, the resulting embryo is placed in the uterus of a surrogate mother for development.

**Genetic Marker Technology.** Genetic marker technology was made possible with the development of reasonably inexpensive and efficient genomic analysis of farm animals. The term commonly used in the genetic marker field is quantitative trait loci (QTI).

---

### Fascinating Facts About Animal Breeding and Husbandry

- Genomic estimated breeding values are expected to revolutionize dairy breeding programs. The calculation combines traditional parent average evaluation programs with the new genetic marker discoveries.
- Using estimated breeding values, a farmer can select for cows that calve easily or have calves with lower birth weight.
- The Animal Improvement Programs Laboratory of the U.S. Department of Agriculture conducts research into the genetic evaluation of dairy cattle and goats, directed at improving yield traits and nonyield traits that affect health and profitability.
- Since 1950, milk production per cow has risen from 5,313 to 16,400 pounds; age to market weight of broiler chickens has decreased from 12 weeks to 7.3 weeks; eggs per hen per year has risen from 174 to 254 eggs.
- Holstein dairy cows originated in Europe and were imported into the United States in the 1800's. Early breeding was for cows that would make the best use of grass.
- Captive breeding has saved many wild animal species from extinction. These species include wolves, the bison, the Peregrine falcon, the California condor, and the whooping crane.
- Most poultry and swine grown for meat are crossbred, while purebred stock are used as breeders.

---

Animal breeders select for traits of economic importance that are largely quantitative traits. These traits are usually controlled by a large number of genes, even thousands of genes. Each gene can contribute a small portion to the total genetic variation of the trait. Since the location (locus) and identity of these genes on the DNA molecule is frequently unknown, the use of genetic markers has become important. Genetic markers are associated with quantitative genes and can be identified in the laboratory.

**Single Nucleotide Polymorphisms.** Another term associated with genetic markers is single nucleotide polymorphisms (SNPs). Nucleotides are the building blocks of DNA, and polymorphism means "many forms." Nucleotides are made of one of four different bases. Genes are made of many nucleotides. The exchange of one base for another in a nucleotide is an SNP, which can change the expression of a gene. Instead of analyzing the entire genome of an animal, the dense SNP array test measures around 50,000 SNPs, which is then related to the genetic merit of the animal. With traditional breeding programs, each offspring is assumed to have inherited an average sample of genes from his or her sire (father) and dam (mother). Full siblings have equal parent average (PA) but are expected to share only half of their genes as copies of the same genes in their parents. Considerable improvements in breeding value have been demonstrated by the use of a genomic predicted transmitting ability (gPTA) calculation, which combines genomic data with the traditional parent average data.

### IMPACT ON INDUSTRY

In 1966, the World Congress on Genetics Applied to Livestock Production was established to provide an avenue for researchers to present their research findings. The areas of research reflect the concerns or problems of the respective countries. Most of the papers presented at the congress every four years have been on breeding for meat or milk production, estimating genetic parameters, and designing sustainable breeding programs. In spite of widespread dissemination of research findings worldwide, there is a large gap in biotechnology applications between developing and developed countries. Research partnerships between developed and developing countries are much fewer than those between developed countries. The only exception is artificial insemination, which is not really a new technology. The more complex technologies such as embryo transfer and genetic markers are adopted less frequently.

**Government and University Research.** Animal breeding research at the federal level takes place within the Agricultural Research Service (ARS), a branch of the U.S. Department of Agriculture (USDA). Research takes place at one hundred stations located throughout the country as well as in some foreign countries. The service welcomes collaboration with businesses, state and local governments, and universities. Many of its accomplishments are a result of these joint efforts. ARS researchers adopted quantitative measures for evaluating breeding stock early on, leading to calculations that predict the average performance of offspring, such as "expected progeny difference" for beef cattle and "predicted transmitting ability" for dairy cattle. They have been deeply involved in genome sequencing of farm animals. By the use of the Illumina Bovine SNP 50 Bead Chip, the genotypes of more than 40,000 animals have been determined. In 2007, the USDA Animal Genomics Strategic Planning Task Force, consisting of members of the ARS and the Cooperative Research, Education, and Extension Service, as well as university collaborators, developed the "Blueprint for USDA Efforts in Agricultural Animal Genomics." The blueprint identifies three major areas of focus: outreach, discovery, and infrastructure.

The land-grant college system inaugurated a close relationship between the federal government and the states. The Morrill Act (1862) and Hatch Act (1887) provided federal funds for the establishment of state colleges of agriculture and for associated agricultural research stations. The Cooperative Extension Service was also established to disseminate research information to producers and consumers of agricultural products. Practically all animal science departments have faculty that have had assignments in foreign countries, with the largest number being in Latin America.

**Industry Research and Applications.** The new technology of SNP markers is revolutionizing the selection of animal breeders. This technology was developed in a partnership among the companies Illumina and Merial, the Agricultural Research Service, the National Association of Animal Breeders, and researchers at several universities and institutes. The researchers found that the breeding value of an animal could be determined by association of these genetic markers with production traits. The genetic evaluation of breeding stock has two advantages—speed and lower costs—over the traditional parent average method. Instead of waiting for proof of breeding value through progeny performance, the animals can be selected very early in their lifetimes. Selection by genetic markers is particularly useful for identifying males with low heritability traits such as fertility and longevity.

The nature of animal breeding companies is changing, and many do not maintain their own breeding stock. They also provide services such as semen collection or embryo transfer, and maintain and sell the semen and embryos. The private sector is playing an increasingly important role in livestock genetic improvement. Specialized breeding firms supply virtually all commercial poultry breeding stock as well as increasing amounts of genetic material for swine, beef, and dairy cattle. Private investment in livestock breeding is affected by demand from producers, market structure, intellectual property protection, new technologies, and market globalization.

## CAREERS AND COURSE WORK

The field of animal husbandry is called animal science in colleges and universities to reflect scientific study and applications in the field. Students specifically interested in animal husbandry should concentrate on course work and experience related to production and management. Animal science also encompasses areas such as agribusiness, government, and research and teaching. Job titles in animal husbandry can include livestock or dairy herdsperson, stable manager, veterinary technician, feed mill supervisor, or farm manager. Previously, on-farm experience was enough to work in animal husbandry, but it has become a more complex field. A two-year associate's degree should be considered minimal for the field, while a four-year bachelor's degree would be beneficial for managerial positions. Course work can include animal production, biology, chemistry, animal growth and development, physiology, animal nutrition, biotechnology, farm management, and economics.

A career in animal breeding and genetics requires a doctoral degree, whether employment is in academia or industry. The careers can include such specialties as quantitative or molecular genetics, bioinformatics, immunogenetics, and functional genomics. The prerequisites for graduate studies typically include undergraduate course work in animal science. Graduate courses can include animal breeding, statistics, endocrinology, genome analysis, population and quantitative genetics, animal breeding strategies, statistical methods, and physiology and metabolism. Specific course work varies depending on the school.

## SOCIAL CONTEXT AND FUTURE PROSPECTS

Some controversy has arisen because state agricultural experiment stations (SAES) are increasingly entering into collaboration with and receiving funding from private firms. Because the agricultural experiment stations are public institutions, some people feel that they may be compromising their independence and objectivity. However, these stations are primarily involved in basic research, and the private firms conduct the necessary practical research leading to commercialization of new products.

The modern factory farming methods, with poultry, swine, and other farm animals kept in confinement and in crowded conditions, has been condemned by animal rights activists. They claim that because the animals are often not able to perform their natural and instinctive behaviors, they are suffering. Animal science departments have been aware of these criticisms and have developed a new field of farm animal welfare. Faculty positions in the emerging field have been filled, and students are being introduced to the concepts. Animal welfare is being studied academically in a manner that is validated and measured objectively and, therefore, is reliable. The discipline considers the relationship of farm animal welfare to the animals' environment in three areas: how the animals feel, their fitness and health, and their natural behaviors.

One issue addressed by the Food and Agriculture Organization of the United Nations is animal diversity. As globalization of agriculture encourages breeding for high-input, high-output animals, some breeds of livestock are becoming extinct. This leads to the existence of fewer breeds, which means less flexibility when confronted with an emerging disease or changed environmental conditions. Another problem is that genetic breeding takes place mostly in advanced countries, under the conditions of intensive agriculture and the local environment. These conditions and farming methods are not the same as in some lesser developed countries, and animals that produce well in one country may not do as well in another because of environmental factors.

The potential benefits and profits from improvements in animal breeding and husbandry are great, and the field is likely to remain active. Genetic engineering is likely to be part of animal breeding, despite social concerns, although these areas of concern may affect how genetic engineering is used.

*David Olle, B.S., M.S.*

## FURTHER READING

Bourdon, Richard. *Understanding Animal Breeding.* 2d ed. Upper Saddle River, N.J.: Prentice Hall, 2000. An excellent basic text on animal breeding that presents concepts with a minimum of mathematics.

Herren, Ray V. *The Science of Animal Agriculture.* 3d ed. Clifton, N.Y.: Thomson Delmar Learning, 2007. Covers all aspects of the sciences involved in animal agriculture, including breeding.

Taylor, Robert, and Thomas Field. *Scientific Farm Animal Production.* 9th ed. Upper Saddle River, N.J.: Pearson Prentice Hall, 2009. Provides an excellent introduction to all aspects of animal husbandry and production of all major farm animals.

Turner, Jacky. *Animal Breeding, Welfare, and Society.* Washington, D.C.: Earthscan, 2010. Examines how the trend toward human intervention in animal breeding is affecting animal behavior, health, and well-being.

## WEB SITES

*American Livestock Breeds Conservancy*
http://www.albc-usa.org

*Food and Agriculture Association of the United Nations*
http://www.fao.org

*Sustainable Animal Production*
http://www.agriculture.de/acms1/conf6/index.htm

*U.S. Department of Agriculture*
Agricultural Research Service
http://www.ars.usda.gov/main/main.htm

*World Congress on Genetics Applied to Livestock Production*
http://www.wcgalp2010.org

**See also:** Agricultural Science; Egg Production; Genetically Modified Organisms; Genetic Engineering; Genomics; Veterinary Science.

# ANTHROPOLOGY

## FIELDS OF STUDY

Biological anthropology; cultural anthropology; linguistic anthropology; archaeological anthropology; biology; archaeology; linguistics; psychology; sociology; social psychology; forensics; epistemology; evolutionary biology; political science; religion; ethnomusicology; history; geography; geology; genetics; anatomy; economics; public health.

## SUMMARY

Anthropology is the scientific study of all aspects of the human species, across the whole geographic and temporal span of human existence. It covers languages, customs, family structures, social behavior, politics, morality, health, and biology. As such, possible studies range from comparative analyses of human and chimpanzee family units to historical examinations of how languages change and to neurobiological studies of the roots of altruism in the brain. At its core, anthropology is concerned with what makes human beings who they are and why.

## KEY TERMS AND CONCEPTS

- **Assimilation:** Assumption of the social behaviors and beliefs of a different group to the point where one's own cultural traits are eliminated.
- **Consanguinity:** Being related by blood or descended from a common ancestor.
- **Cosmology:** Set of beliefs about the origin of the universe.
- **Cultural Transmission:** Process by which patterns of thought and behavior are passed down from generation to generation; also known as enculturation and socialization.
- **Descent Group:** People who share an ancestor; relationships may be defined through members' mothers (matrilineal descent), their fathers (patrilineal descent), or some other means.
- **Ethnocentrism:** Using the practices of one's own culture to appraise the practices of another culture; the opposite of cultural relativism.
- **Ethnography:** Study of a particular culture, made through observing its members.

- **Ethnology:** Study of cultures in order to note their differences and similarities.
- **Etic Approach:** Method in which a culture is studied from outside in order to create objectivity; the opposite of an emic approach.
- **Kinesics:** Study of body movements and gestures used to communicate.
- **Modal Personality:** Cohesive set of personality traits, common to the majority of the people in a culture.
- **Morphology:** Study of morphemes, or meaningful units of sound within a given language.
- **Syncretism:** Tendency for different cultures, religions, belief systems, or varieties of a word within a language to converge and become one.

### DEFINITION AND BASIC PRINCIPLES

Anthropology is concerned with the origin, evolution, behavior, beliefs, culture, and physical features of humankind. Rather than focusing on a single facet of human existence, anthropology is characterized by an inclusive approach that treats each feature of culture and society as interrelated parts of a whole. An anthropological study of psychology, for instance, might place it in the context of language and culture; a study of medicine might include an analysis of politics and human adaptation. This approach is known as holism. In addition, no matter how large or small the question an anthropologist asks—a research topic might be as specific as the types of wedding ornamentation used in a particular culture or as broad as how language evolved in different societies over millennia—its underlying aim will be to gain a deeper understanding of the species as a whole. Therefore, cross-cultural comparisons, or the practice of analyzing the similarities and differences between distinct human communities, is an essential component of anthropological research.

A few other basic principles distinguish anthropology as a science. One is an emphasis on immersion fieldwork, in which the scientist enters into the community being studied in order to observe it from the inside rather than the outside. Similarly, anthropology—in this way greatly influenced by feminism and postmodernism—breaks down the notion of scientists as objective observers, acknowledging that their mental models of the world are bounded by their own cultures. In other words, scientists' particular social backgrounds, experiences, and worldviews are inseparable from the perspectives they bring to ethnography. Anthropologists strive to overcome this limitation by consciously applying the principle of cultural relativism, judging each society by its own internal system of ethical and social guidelines. For example, an anthropological study of the ancient Chinese practice of foot binding (tightly tying cloth around female babies' feet in order to keep them small as the girl grows) would refrain from making a moral judgment about the tradition and instead would focus on describing its origins and the rationale for it within Chinese society.

## BACKGROUND AND HISTORY

The word "anthropology," meaning "study of man," predates the development of anthropology as a modern scientific field. It was first used in the sixteenth century to describe a philosophical or theological examination of the soul. The term was later used by nineteenth-century German scientist Johann Friedrich Blumenbach in a sense closer to its modern meaning, to denote the study of both the physical and the psychological aspects of humankind. In the late nineteenth-century, the science of anthropology was dominated by an ethnocentric approach that peered at global cultures through the lens of strict Victorian mores and that largely labeled these cultures as primitive curiosities. Early anthropologists believed that cultural differences could be traced to genetic variances in the human species that resulted in moral and mental disparities.

German American anthropologist Franz Boas is given credit for formulating the principle of cultural relativism at the beginning of the twentieth century. Boas insisted that culture was not passed on from generation to generation through the genes but instead consisted of learned behaviors acquired over time through immersion in a group and eventual habituation to the behavior of others. He rejected the idea that culture developed toward a destination and that any single culture was inherently superior to or more advanced than another. Although anthropology has changed in the years since Boas first laid down his ideas, these principles continue to inform the field as a whole.

Over the course of the field's development, the focus has shifted away from broad descriptions of

an entire culture or way of life and toward narrower examinations of specific features of community life. Thus, a cultural anthropologist may specialize in studies of specific topics such as music, marriage practices, or taboos.

## HOW IT WORKS

**As in other fields of science, empirical research— research based on observed and verifiable data— forms a cornerstone of anthropology.** The classic type of anthropological study is fieldwork, in which the scientist travels to and lives among a group of people with the intent of documenting their cultural practices. Other tools of archaeological investigation, common to the four main subfields of anthropology, include surveys, interviews, archival research, recordings, and statistical analysis.

**Cultural Anthropology.** Cultural anthropology is the most widely practiced and well-known subfield of anthropology. It is the study of human thought, knowledge, and practices—any behavioral trait, in other words, that is passed on not through the genes but through language, art, and ritual. The most important tool of the cultural anthropologist is the ethnographic study. Researchers directly observe members of a group as they go about their daily lives. They interview subjects, record oral histories, and make detailed reports of all that they see and hear. The goal of these observations is to make sense, in a deep way, of the reasoning behind the cultural practices of a given society. For example, it can be difficult for Western minds to apprehend the facial tattooing rituals that exist in many native cultures, including that of the Maori of New Zealand. Rather than seeing these practices as masochistic or stigmatizing, cultural anthropologists seek to uncover the principles that motivate them. In the case of the Maori, tattooing serves as a status symbol—the more complex one's facial tattoos, the higher one's rank—and as a sign of affiliation between group members.

**Archaeological Anthropology.** Archaeological anthropology is the study of human cultures through the material artifacts they produce, including such objects as mechanical devices, toys, writings, paintings, pottery, religious icons, buildings, and funerary items. Archaeological anthropologists collect, categorize, and describe these artifacts, then use them to piece together plausible theories about the belief systems and traditional practices of the societies from which they came. Anthropologists specializing in societies that have long passed out of existence, such as ancient Greek, Roman, Indian, Egyptian, and Mayan cultures, set up stations known as digs to unearth buried fragments and objects. Those anthropologists interested in the behavior of prehistoric hunter-gatherer societies or even the protohuman ancestors of the human species focus their attention on clues from fossilized skeletons and primitive tools. However, later human cultures provide equally valid subjects for archaeological-anthropological study. Archival maps, photographs, and other historical documents, for instance, can help scientists build up a clearer impression of the habits and ways of life practiced in communities such as those of nineteenth-century American coal miners or rural fishermen in colonial Malaya (now Malaysia).

**Linguistic Anthropology.** Linguistic anthropologists are concerned with the origin, development, and structure of the roughly 6,000 human languages that exist in the world, as well as with how language shapes and is shaped by culture. They study technical aspects of language such as phonetics, phonology, grammar, syntax, vocabulary, and semiotics. More broadly, researchers investigate how language is used to define social groups and transmit meaning from generation to generation, how the grammatical forms of a community's language affect common patterns of reasoning and thought, and how natural and artificial phenomena are represented or symbolized in language. For example, the vocabulary of some Chinese dialects includes an elaborate set of words describing specific family relations—terms that distinguish older siblings from younger ones, paternal from maternal relatives, and relatives by blood or marriage. The existence of such words in the language reflects an intense focus in Chinese culture on questions of kinship links and rankings.

Other fruitful areas of study include the relationships among languages within and across linguistic families and the question of how the pronunciation and usage of words evolved over time. Computational linguists, for example, may use software to conduct complex statistical analyses of thousands of instances of a word to reconstruct how different speakers meant and interpreted it. Linguistic anthropologists approach such issues through analyses of written texts and audio and video recordings and seek out language that occurs naturally in social discourse

rather than language used in the context of formal interviews.

**Biological Anthropology.** Biological anthropology seeks to understand humankind by studying its physical anatomy. It examines the origin and evolution of the human brain, body, and nervous system; the physical diversity of individuals and groups within the human species; the place of humankind in relation to other animal species, both living and extinct, within the natural world; and the physiological bases of psychological processes and behaviors. There are various subfields within biological anthropology. Researchers who focus on human evolution conduct comparative analyses of human and nonhuman primates, making use of fossil evidence, DNA studies, and field research of animals in the wild. Medical anthropologists study issues of epidemiology (how diseases and health-related behaviors spread across human populations) and ethnomedicine (how different human societies think about, explain, and treat disease and health).

## APPLICATIONS AND PRODUCTS

**Medicine and Public Health.** The work of medical anthropologists leads to many useful applications in the world of public health. For example, an anthropologist who spent nearly two years immersed in the daily lives of families with children with the genetic disease cystic fibrosis produced a detailed set of practical clinical guidelines for physicians working with such families. The guidelines help doctors communicate effectively with children and their parents about cystic fibrosis symptoms and treatment. Another common medical application of anthropological research is the development of public health campaigns targeted to the specific worldviews and cultural traditions of a population. For example, anthropologists have worked with community health workers in Malaysia to educate locals about steps they can take to prevent the spread of the deadly virus responsible for dengue fever. Through door-to-door surveys and interviews, the anthropologists determined the Malaysian public's common misconceptions about dengue fever and helped create a more accessible list of recommendations for local inhabitants.

**Crime Investigation.** Forensic anthropologists, who apply the tools of anthropological study to skeletal remains such as bones and teeth, are indispensable members of any crime investigation team.

Since the 1930's, for instance, physical anthropologists employed at the Smithsonian Institution in Washington, D.C., have served as consultants to the Federal Bureau of Investigation (FBI) on a large number of criminal cases. The anthropologists help with tasks such as recovering physical evidence from the scene of the crime, determining victims' age and gender, and estimating the time and mechanisms of their deaths. One of the most crucial anthropological tools involved in crime investigation is facial reproduction. In facial reproduction, a forensic anthropologist uses computer simulation to reconstruct a picture of a deceased person's facial features that have been degraded by trauma or decomposition. This is done using scientific knowledge about the relationships between the shape of the hard tissue of the skull and the soft tissues that make up the face.

**Military Applications.** Anthropologists began assisting in military operation in the second half of the twentieth century. In the United States, for instance, cultural anthropologists working in the Human Terrain Team—a program under the umbrella of the U.S. Army Training and Doctrine Command—travel with soldiers to the field of warfare, interviewing local inhabitants and collecting data about the culture, sentiments, and needs of the people who live in battle zones in Afghanistan and Iraq. Their expertise helps the military improve security, target its aid and reconstruction efforts more effectively, work better with local governments and organizations, and conduct better counterterrorism programs. For example, army officials hope to use knowledge gained from the cultural advice of field anthropologists to persuade local tribal leaders to work with the Afghan police. Such assistance, however, comes at a cost. In 2008, two anthropologists lost their lives while working with the military, and the American Anthropological Association has expressed serious ethical concerns about the use of anthropology as a tool to aid the army in determining specific targets for military operations.

**Social Problems.** Anthropological studies are used by governments and various organizations across the globe to analyze and address a variety of social problems, such as malnutrition, poverty, teenage parenthood, unemployment, and drug abuse. For example, anthropological studies of population trends in countries such as Swaziland have shown that the number

of children a woman has is inversely related to the amount of education she receives. Formal education raises women's awareness of birth control, opens doors to new jobs, and reduces their willingness to participate in practices such as arranged marriages and polygyny (two or more wives at the same time). As a result, one simple initiative applied in many developing nations with overpopulation problems is to increase the educational opportunities available to girls and young women.

Similarly, anthropological studies have been conducted of microcredit banks such as India's Grameen Bank. Microcredit agencies lend very small amounts of money to individuals in deep poverty, without requiring collateral, to help fund income-generating activities and housing. Anthropological research reveals that, despite its good intentions, microcredit can lead to the serious problem of debt cycling (paying off previous loans with new ones) among people who are already in financial difficulty. Anthropological research can thus be essential in evaluating the effectiveness of social programs that have already been put into place and in making recommendations for the future.

## IMPACT ON INDUSTRY

**Government and University Research.** Governments around the world use anthropological studies to help them evaluate the needs of their populations and plan public policy. They also rely on anthropologists to help them preserve information about national history. In the United States, the National Council for the Humanities and the National Science Foundation are the two major government agencies that provide funding to support anthropological research. In 2009, for example, the National Science Foundation approved funding for studies on the use of racial imagery depicting Asian Americans in advertising, how culture and national policy affect the safety of birthing, and how community television programs affect citizen participation in local, state, and national government.

**Product Development.** Anthropologists frequently work for companies that produce appliances, electronics, cosmetics, packaged foods, and a host of other consumer products. Their job is to use tools such as questionnaires and in-home observations to survey customers' needs and frustrations. They identify areas of potential for product development,

## Fascinating Facts About Anthropology

- Ann Dunham, the mother of U.S. president Barack Obama, was a trained anthropologist whose doctoral dissertation was a study of the agricultural blacksmiths of Indonesia.

- Anthropological studies have documented a few cases of cannibalism (the eating of human flesh) in tribal societies such as the ancient Aztecs, usually as a ritual associated with war or funerals.

- Many early cultures subscribed to the belief, known as animism, that all natural objects or forces—such as rivers, mountains, and winds—are animated by a divine spirit.

- The language of the Plains Indian tribe known as the Assiniboine, with only fifty native speakers left, would be lost forever if anthropologists were not working to preserve it through recordings of stories, songs, and oral histories.

- Retail anthropologists are tracking the growth of self-service kiosks selling everything from food to newspapers to portable MP3 players. They say more and more consumers are drawn to these devices because they prefer not to have to interact with someone at a checkout.

- In 2008, the National Science Foundation awarded $100,000 to a group of anthropologists at the University of California, Irvine, to conduct an ethnographic study of the online role-playing game World of Warcraft.

- According to linguistic anthropologists, the claim that Eskimos have a hundred (or more) words for "snow" is a myth. The most snow words any scholar can come up with for an Inuit or Aleut language is two dozen, no more than the number of snow-related words in English—which include such terms as "frost," "hail," "ice," "slush," and "sleet."

suggest how products can be made more usable, and evaluate the impact and effectiveness of items that are already on the market. Anthropologists are a huge asset to companies wishing to ensure that the telephone, video game, or photocopier they have spent millions of dollars researching and developing is not one of the three-quarters of all new products that fail. Among the companies that have used the services of anthropologists are Xerox, Intel, Nokia, IBM, and Motorola. To fill this need, a new breed of consulting company has emerged that focuses on conducting anthropological studies of the retail market for clients. The Brazilian-based Anthropos Consulting group, which specializes in corporate anthropology, has developed products for pet food companies, airlines, hotels, and banks.

**Marketing and Retail.** Retail anthropologists, who study people's shopping habits, help clients such as grocery stores, shopping malls, mail-order companies, and Internet businesses increase the number and quality of the sales they make. Anthropologists study, for instance, the walking patterns and behavior of customers in stores, making observations such as the fact that mirrors or other shiny surfaces cause people to slow down, or that narrow aisles may deter a customer from spending time looking at products. They watch how people read their mail, noting which kinds of envelopes they tend to open and which they throw away. They also conduct statistical analyses of large amounts of purchasing data, looking for trends—such as whether women buy more shoes at a certain time of year or whether shoppers tend to purchase certain products one after the other—that can be used by marketers.

### CAREERS AND COURSE WORK

A typical anthropology degree at the undergraduate level will involve a comprehensive series of courses in anthropological theories and methodologies, classes on broad topics such as human evolution and migration, and specific course work pertaining to a particular subfield. For example, a student interested in linguistic anthropology might enroll in classes on the structure of language, the neurobiology of language, and the sociological impact of language. No matter what subspecialty he or she chooses, an anthropology student should also gain a strong grounding in research methodologies, statistics, computational analysis, interview skills, and formal writing. Work experience or internships are helpful not only for gaining practical experience but also for making contacts in the field who can serve as professional mentors and advisers later in the student's career. Although such opportunities certainly exist at archaeological digs for those interested in archaeological anthropology, many other types of placements are also appropriate. For instance, a student might conduct an ethnographic study of breast-feeding among minority mothers at a community health clinic or study the history of the

English language in the dictionary department of an academic publishing house.

Once a student has graduated, his or her training in anthropology serves as an excellent preparation for virtually any career in which the understanding of the principles behind human behavior would be an asset. Possible career paths include archaeology, museum curation, linguistics, forensic science, marketing, human resources, social work, and consulting of all kinds. Anthropologists' ability to place problems within the context of human communities makes them valuable interpreters between organizations (corporations or governments) and the individuals they serve. For example, professionals with anthropology backgrounds have been hired to design more user-friendly software, create public awareness campaigns around pressing health issues, and interview families to discover the ways in which they use greeting cards to communicate with one another. Also, government agencies frequently employ anthropologists as policy researchers, program evaluators, planners, and analysts of all kinds.

## SOCIAL CONTEXT AND FUTURE PROSPECTS

In an ever-changing world, anthropology is a means of preserving the social, artistic, linguistic, and cultural heritage of different human communities for curious future generations. The significance of anthropological research, however, is far from merely intellectual. Increasingly, it is as common to find anthropologists engaged in solving practical problems for corporations, hospitals, or government agencies as it is to find them writing academic research papers or studying remote tribes. For example, the American Anthropological Association has a membership of about 10,000 professional anthropologists; the number of members who subscribe to its mailing list for applied anthropologists—scientists employed outside of the academic setting—totals more than 6,000. Also, contemporary anthropologists are more likely than ever before to turn their attention toward the beliefs and practices of their own cultures rather than conducting studies far from home. One example is British anthropologist Kate Fox's 2004 study of the thought patterns, rules, behaviors, and cultural rituals of modern British society.

*M. Lee, B.A., M.A.*

## FURTHER READING

Brenneis, Donald. "A Partial View of Contemporary Anthropology." *American Anthropologist* 106, no. 3 (2004): 580-588. Addresses issues and practices within the field and makes proposals for sustaining the value of anthropological studies in the future. Refers to specific anthropological sites in the United States and abroad.

Brown, Peter, and Ron Barrett, eds. *Understanding and Applying Medical Anthropology.* 2d ed. Boston: McGraw-Hill, 2010. Each chapter includes an introduction to its author, thought questions, and citations.

Haviland, William A., et al. *Anthropology: The Human Challenge.* Belmont, Calif.: Thomson Wadsworth, 2008. A comprehensive textbook of cultural anthropology. Each chapter has a glossary, reflection questions, and sidebars such as "Anthropologists of Note" and "Biocultural Connections."

Metcalf, Peter. *Anthropology: The Basics.* New York: Routledge, 2005. A brief introduction to the field of cultural anthropology. Includes photographs and text boxes highlighting important concepts and case studies.

Park, Michael Alan. *Biological Anthropology.* 6th ed. Boston: McGraw-Hill, 2010. This discussion of biological anthropology contains chapter summaries, charts, drawings, a glossary, a pronunciation guide, and suggested readings.

Stephens, W. Richard, and Elliot M. Fratkin. *Careers in Anthropology.* Boston: Allyn and Bacon, 2003. A practical guide to careers in anthropology and related fields, organized as a series of real-life case studies describing the paths that sixteen different professionals followed after graduation.

## WEB SITES

*American Anthropological Association*
What Is Anthropology?
http://www.aaanet.org/about/WhatisAnthropology.cfm

*National Science Foundation*
Cultural Anthropology Program Overview
http://www.nsf.gov/sbe/bcs/anthro/cult_overview.jsp

**See also:** Archaeology; DNA Analysis; Forensic Science; Game Theory; Psychiatry; Urban Planning and Engineering.

# ANTIBALLISTIC MISSILE DEFENSE SYSTEMS

## FIELDS OF STUDY

Astronautical engineering; physics; electronics engineering; electrical engineering; mechanical engineering; optical engineering; software engineering; systems engineering.

## SUMMARY

To protect people and their possessions from harm by incoming missiles launched by a potential enemy, antiballistic missile defense systems have been designed to detect, track, and destroy incoming missiles. These systems are designed to fire guided defensive missiles to hit the incoming missiles before they strike their targets.

## KEY TERMS AND CONCEPTS

- **Ballistic Missile:** Missile that is powered only during the first part of its flight; includes missiles powered and steered during the end of their flight.
- **Boost Phase:** Initial part of a missile's flight, when it is under power.
- **Electromagnetic Pulse (EMP):** Burst of electromagnetic radiation given off by a nuclear explosion that sends a burst of energy through electronic devices and may severely damage them.
- **Exoatmospheric:** Outside or above the atmosphere.
- **Hit To Kill:** Destroying a warhead by hitting it precisely, often compared to hitting a bullet with a bullet.
- **Kill Vehicle:** Speeding mass that smashes into a warhead using its energy of motion (kinetic energy) to destroy the warhead.
- **Midcourse Phase:** Middle portion of a missile's flight, as it coasts through space.
- **Terminal Phase:** End of a missile's flight, as it reenters the atmosphere and nears its target.
- **Warhead:** Payload of the missile that does the desired damage; it can consist of high explosives, a nuclear bomb, or a kill vehicle.

## DEFINITION AND BASIC PRINCIPLES

An intercontinental ballistic missile (ICBM) can deliver enough nuclear explosives to devastate a city. Because ICBMs plunge from the sky at very high speeds, unseen by the eye until they hit, they are difficult to defend against. That is the purpose of an antiballistic missile (ABM) system. Such systems have several essential tasks: to detect a missile launch or ascent, to track the missile during its midcourse and its terminal flight, and to calculate an intercept point for the defensive missile, fire a defensive missile, track the defensive missile, guide it to its target, and verify that the target has been destroyed.

Using multiple independently targeted reentry vehicle (MIRV) technology, ten to twelve independently targeted warheads and decoys can be delivered by a single ICBM (which is against the signed but unratified 1993 Strategic Arms Limitations Talks II treaty). Each deployed warhead requires a defensive missile to destroy it. Countries with MIRV technology could attack with enough warheads and decoys to saturate any ABM system, allowing at least some nuclear warheads to reach their targets. If the targeted country retaliates, the situation is likely to escalate into full-scale nuclear war.

Although ABM systems cannot prevent nuclear war, they do have some uses. For example, a limited system could defend against a few accidental launches of ICBMs or against a few missiles from rogue nations.

## BACKGROUND AND HISTORY

The evolution of ABM systems has been driven by politics and perceived need. In the 1960's, the strategy of "mutual assured destruction," or MAD, was articulated by U.S. secretary of defense Robert McNamara. If either side launched a first strike against the other, the nonaggressor would still have enough warheads left to devastate the aggressor. An effective ABM system would have upset this balance.

The ABM treaty of 1972 was signed by U.S. president Richard Nixon and the Soviet Union's general secretary, Leonid Brezhnev, and was amended in 1974 to allow each side to have only one ABM site. The Soviet Union elected to place its single ABM system around Moscow, a system still maintained by Russia.

The United States built the Safeguard ABM system at the Grand Forks Air Force Base in North Dakota, where it guarded Minuteman III missiles. Safeguard operated for only a few months before it was closed down. The U.S. Missile Defense Agency was formed in 1983 (under a different name) and given the charge to develop, test, and prepare missile defense systems. Also in 1983, President Ronald Reagan announced a change: The offensive MAD would become the defensive building of an impenetrable shield under the Strategic Defense Initiative, popularly dubbed "Star Wars." Although Edward Teller, chief architect of the hydrogen bomb, assured President Reagan that the Strategic Defense Initiative could be implemented, it proved to be unfeasible.

After the Cold War ended with the dissolution of the Soviet Union in 1991, both the United States and Russia eventually concluded that the most likely use of their ABM systems would be against a limited strike by a country such as North Korea or Iran. To guard against Iranian and North Korean missiles, the United States has twenty-six ground-based interceptor (GBI) missiles at Fort Greeley, Alaska, and four at Vandenberg Air Force Base in California. (Two sites would not have been allowed under the ABM treaty, but the United States withdrew from the ABM treaty in June, 2002.) Warships equipped with the Aegis Combat System (a system that uses radar and computers to guide weapons aimed at enemy targets), with Standard Missile-3's (SM-3's), are stationed in the Mediterranean and the Black Sea to protect Europe.

## HOW IT WORKS

An ABM system must successfully perform several different functions to work properly. Initial detection may be done by a remote ground radar, by an airborne radar plane, or even by infrared sensors in space. Infrared sensors are particularly effective at spotting the hot rocket plume of a launching ICBM. Normal radar is line of sight and cannot detect targets beyond the curve of the Earth. Therefore, airborne radar is used; being higher, it can see farther.

To guard the United States from attack by ICBMs or submarine-launched missiles, three PAVE PAWS (Precision Acquisition Vehicle Entry Phased-Array Warning System) radars look outward from American borders. PAVE is a U.S. Air Force program, and the three radar systems are located on Air Force bases in Massachusetts, Alaska, and California. The systems track satellites and can spot a car-sized object in

---

### Fascinating Facts About Antiballistic Missile Defense Systems

- The first ABM system deployed by the United States, the Safeguard system, had a fatal flaw that was known before it was built. Exploding warheads on the first salvo of Sprint missiles would blind Sprint's tracking radar.
- In 2008, the United States used a Standard Missile-3 to shoot down an orbiting satellite.
- American ABM systems can defend against an accidental launch or a limited attack by a rogue nation but not a full-scale attack by Russia or China.
- The United States can field a layered missile defense, where each successive layer deals with missiles that survived the previous layer. It would start with the long-range Ground-Based Interceptor, then the Aegis standard missiles, the THAAD missile, and finally the short-range Aegis missiles.
- The "sizzler" is a Russian-built cruise missile with a 300-kilometer maximum range. It flies at Mach 0.8 until it nears its target, when it accelerates to nearly Mach 3 and takes a zigzag path to its target—making it difficult to shoot it down. China, India, Algeria, and Vietnam have all purchased it.
- It may not be possible to defend against a full-scale attack by a determined aggressor who could place ten warheads and decoys in a single missile (banned by treaty). Each warhead and possibly each decoy would need to be targeted by a separate missile, unless several kill vehicles could fly on a single ABM (also banned by treaty).

---

space 5,500 kilometers away. The initial detection of a long-range missile would probably come from a PAVE PAWS radar system. The radar and associated computers must also classify the objects they detect and determine if the objects are threatening.

If PAVE PAWS spots a suspicious object, a defensive missile site will be notified and will begin tracking the object with its on-site phased-array radar. If the object is still deemed a threat, permission to fire on it must be given either by direct command or by standing orders. Aided by a fire-control computer, the operator selects the target and fires one or two missiles at it. The missiles are guided using ground radar until they approach the target, when the missile's own radar or infrared sensors assume the tracking duties.

Modern missiles are usually hit-to-kill missiles, although some carry conventional explosives to ensure the kill. Next, the radar and computer must see if the target has been destroyed, and if not, defenses closer to the ICBM's target must be activated. These actions are all coordinated through the "command, control, and communications" resources of the on-site unit.

Initially, antiballistic missiles were designed to approach their targets and detonate a nuclear warhead, so great accuracy was not required. Close was good enough. The Safeguard ABM system had a long-range Spartan missile with a 5-megaton yield and the short-range Sprint missile, which carried a neutron bomb (an "enhanced radiation bomb"). The Spartan was to engage targets in space, and any surviving targets would be destroyed in the atmosphere by the Sprint. Unfortunately, the nuclear explosions would produce electromagnetic pulses, which would blind the Spartan/Sprint guiding radar. This problem encouraged designers to work toward a hit-to-kill technology. Russia has an ABM system around Moscow and has removed the nuclear warheads from these missiles and replaced them with conventional explosives.

## APPLICATIONS AND PRODUCTS

The hardware of an ABM system includes several key parts. Radars and infrared sensors are the eyes of an ABM system. Defensive missiles destroy the invading missiles, and computers calculate trajectories and direct the defensive missiles.

**Radar.** A phased-array radar antenna is a key component of any modern ABM system. It consists of an array of hundreds or thousands of small antennas mounted in a regular array of rows and columns on a wall. The radar can project a beam in a certain direction or receive a return echo from a particular direction by activating the small antennas in certain patterns. Because this is all done electronically without the radar antenna moving, many scan patterns can be run simultaneously. The Aegis SN/SPY-1 radar can simultaneously track more than one-hundred targets at a distance of up to 190 kilometers. Some multimission Navy destroyers have the AN/SPY-3 radar. It combines the functions of several radars into one, requires fewer operators, and is less visible to other radars. An advanced radar can also interrogate an IFF (identify friend or foe) device. Incoming missiles can also be identified by their flight paths and radar signatures.

**Missiles.** The Patriot system began as an antiaircraft system but was upgraded to defend against tactical missiles. It seemed to do well against scud missiles during the 1990-1991 Gulf War, but later analysis showed that most of the claimed kills were actually ripped apart by air resistance. The Patriot system is highly mobile because all its modules are truck or trailer mounted. One hour after the unit arrives on site, it can be up and running. The system uses the Patriot Advanced Capability-2 (PAC-2) missile with a range of 160 kilometers, a ceiling of 24 kilometers, a speed of Mach 5, and an explosive warhead. Its phased-array radar is difficult to jam.

The PAC-3 missile is smaller; four of them will fit in the space taken by one PAC-2 missile. This gives a missile battery more firepower. The missile travels at Mach 5, has a ceiling of 10 to 15 kilometers, and a maximum range of 45 kilometers. The warhead is hit-to-kill, backed up with an exploding fragmentation bomb on a proximity fuse. Two missiles are fired at a target, with the second missile firing a few seconds after the first. The second missile targets whatever might be a warhead left in the debris from the first warhead's impact. Under test conditions, the PAC-3 missile has scored twenty-one intercepts out of thirty-nine attempts.

The Theater High Altitude Area Defense missile system (THAAD) is designed to shoot down short, medium, and intermediate range missiles during their terminal phases. It uses the AN/TPY-2 radar. THAAD missiles also have some limited capability against ICBMs. Their effective range is about 200 kilometers with a peak altitude of 150 kilometers. Their warhead is a kinetic kill vehicle (KKV). When the missile nears its target, the KKV is explosively separated from its spent rocket. Guided by an advanced infrared sensor, steering rockets adjust the KKV's course so that it will hit dead on. In the six tests since 2006, the THAAD missile hit its target all six times.

The Ground-Based Midcourse Defense (GMD) system is designed to defend against a limited attack by intermediate- and long-ranged missiles. Its missile is the Ground-Based Interceptor (GBI), a three-stage missile with an exoatmospheric kill vehicle (EKV). The GBI is not mobile but is fired from an underground silo. Although there is on-site tracking radar, the GMD can receive early warnings from radars hundreds of kilometers away. The GBI travels at about 10 kilometers per second and has a ceiling of about

2,000 kilometers. Out of fourteen tests, GBIs have hit their targets eight times.

**Aegis.** Ticonderoga-class cruisers and Arleigh Burke-class destroyers all have Aegis Combat Systems (ACSs). Aegis was built to counter short- and medium-range ballistic missiles, aircraft, and other ships. Aegis combines several key parts: the AN/SPY-1 phased-array radar, the MK 99 Fire Control System, the Weapons Control System, the Command and Decision Suite, and the Standard Missile-2 (SM-2).

The SM-2's speed is Mach 3.5, and its range is up to 170 kilometers. The missile has radar and an infrared seeker for terminal guidance, and it has a blast fragmentation warhead. The SM-2 is being replaced by the SM-6, which has twice the range and better radar and is more agile so that it can better deal with the Russian "sizzler" cruise missile.

The third Aegis missile is the Standard Missile-3 (SM-3). It has four stages, a range of more than 500 kilometers, a ceiling of more than 160 kilometers, and a kinetic kill vehicle (KKV). It is guided by ground radar and by onboard infrared sensors. On February 21, 2008, an SM-3 missile was used to shoot down a failed U.S. satellite. The satellite was 240 kilometers above the ground, and the missile approached it at 36,667 kilometers per hour. The satellite had never reached its proper orbit and was coming down because of air resistance. The reason given for shooting it down was the large amount of toxic hydrazine fuel still aboard. Many viewed it as an excuse to test the antisatellite capability of the SM-3 because it was likely that the hydrazine would have been dispersed and destroyed when the satellite reentered the atmosphere. When equipped with the SM-3, Aegis can serve as an ABM system for assets within range. As has been noted, Aegis-equipped warships with the Standard Missile-3 (SM-3) are stationed in the Mediterranean and the Black Sea to protect Europe from Iranian missiles.

**Lasers.** The Airborne Laser Test Bed (ALTB) is mounted in a modified Boeing 747 designated the Boeing YAL-1. It has a megawatt-class chemical laser that gets its energy from the chemical reaction between oxygen and iodine compounds. It has successfully destroyed target missiles in flight, but they were not far away. It is unlikely that the laser's range will ever exceed 300 kilometers, and if the aircraft must loiter that close to the launch site, it is in danger of being shot down by the enemy nation's air defense. In 2009, Secretary of Defense Robert Gates recommended that the ALTB project be cut back to limited research.

Another laser project that showed promise was the Tactical High Energy Laser (THEL). The THEL is a deuterium fluoride chemical laser with a theoretical power of 100 kilowatts. It was a joint project with the United States and Israel and was able to shoot down Katyusha rockets but nothing larger. Although lasers show promise, it seems unlikely that they will be used in an ABM system anytime soon.

## IMPACT ON GOVERNMENTS AND INDUSTRY

ABM systems have encouraged governments to cooperate. The Patriot system is to be replaced by the Medium Extended Air Defense System (MEADS), a joint project of the United States, Germany, and Italy. It is designed for quick setup so that it will be ready almost as quickly as it is unloaded. Although it is more capable than the Patriot system, MEADS has been streamlined so that it requires only one-fifth of the cargo flights to deliver it to its operation site. Its purpose is to protect against tactical ballistic missiles, unmanned aerial drones, cruise missiles, and aircraft. The ceiling of the PAC-3 MSE (missile segment enhancement) is 50 percent higher and its range is twice the range of the PAC-3. The United States together with Israel developed the Arrow missile to protect Israel from Iranian missiles. The United States, Russia, Israel, Japan, China, the Republic of China (Taiwan), India, North Korea, and Iran all have vigorous ABM programs and active defense industries. Other countries have smaller programs and use hardware manufactured elsewhere. For example, Patriot missile systems manufactured by Raytheon are in thirteen countries.

In the following list of the top defense companies involved in ballistic missile defense, no distinction has been made between companies that are the prime contractor or a subcontractor, and employee numbers and revenues (as opposed to profits) are for 2009. The numbers in parentheses are the ranks, by revenue, of these U.S. defense contractors.

1.  Boeing Company is involved with the Airborne Warning and Control System (AWACS radar), Aegis SM-3, Arrow Interceptor, Ground-based Midcourse Defense (GMD) system, Patriot Advanced Capability-3 (PAC-3), and Strategic Missile and Defense Systems. Boeing has

about 160,000 employees and a revenue of $68.3 billion.

2. Lockheed Martin has 140,000 employees and revenue of $45 billion. It produces the Terminal High Altitude Area Defense (THHAD) weapon system, the Medium Extended Air Defense System (MEADS), Airborne Laser Test Bed, Space-Based Infrared System (SBIRS), and various other missiles and satellites.

3. Northrop Grumman has developed a Kinetic Energy Interceptor (KEI), is developing a satellite that will spot and track ICBMs, and is modernizing the Minuteman III missiles. The company has about 120,000 employees and revenue of about $32 billion.

4. General Dynamics is involved with Aegis and several satellite systems with infrared sensors that can track ballistic missiles. Its revenue is about $32 billion, and it has 91,200 employees.

5. Raytheon builds exoatmospheric kill vehicles, Standard Missiles-2, -3, and -6, airborne radar, the Patriot missile system, and other defense-related equipment. The company had more than 72,000 employees and earned $27 billion.

6. L-3 Communications is involved with the Ground-Based Midcourse Defense system, the Standard Missile-3, and the Aegis system. It has a revenue of $14 billion and a workforce of 64,000.

7. Orbital Sciences manufactures the GMD three-stage boost vehicle. Orbital has 3,100 employees and a revenue of $1.2 billion per year.

Altogether, these companies have around 650,000 employees and about $220 billion in revenue. Research shows that for every job in the defense industry, one to four more jobs are created in the community to meet the needs of the defense workers and their families. Reasonable guesses are that 20 percent of the employees work on antiballistic missile projects, and that each job draws one other job to the area. The effect on the economy of antiballistic missile programs is then 260,000 jobs and $88 billion each year.

**CAREERS AND COURSE WORK**

Many defense industry jobs in the United States require a security clearance; this means the applicant must be a U.S. citizen. A strong background in the physical sciences is necessary for the aerospace industry. High school students should take all the courses in physics, chemistry, computer science, and mathematics that they can. At least a bachelor's degree in science or engineering is required. Employees will need to write reports and make presentations, so students should take some classes in writing and speech. They may eventually become a team or unit leader; if so, they may wish they had taken a simple business management course. Those who are involved with research and development need a feel for how things work. It helps if they like to build or repair things. They should be creative and be able to think of new ways to do things.

Bachelor's degrees are sufficient for a number of aerospace positions: astronautical, computer, electrical, mechanical, optical engineer or physicist. Employees generally start as junior members of a team, but as time passes, they can become senior members and then perhaps team leaders with more responsibility. Astronautical engineers design, test, and supervise the construction of rockets, missiles, and satellites. Computer engineers interface hardware with computers, writing and debugging programs that instruct the hardware to do what is wished. Electrical engineers design, develop, and produce radio frequency data links for missile applications. Mechanical engineers design, analyze, and integrate cryogenic components and assemblies. Optical engineers develop solutions to routine technical problems and work with signal processing analysis and design as well as sensor modeling and simulation. Systems engineers design systems for missile guidance and control, computational fluid mechanics analysis, and wind tunnel testing. Physicists who work with ABM systems test electrical and mechanical components, measure radiation effects, and mitigate them if necessary.

**SOCIAL CONTEXT AND FUTURE PROSPECTS**

People have always wondered whether money spent on an ABM system could be better spent elsewhere. Some maintain that even an excellent system would be unlikely to protect against some city destroyers. ICBMs with a dozen warheads and decoys could overwhelm any ABM system. Furthermore, if nation A thought that nation B was installing an effective ABM system, nation A might launch a preemptive strike before the ABM system became operational. At the very least, nation A would probably build more ICBMs and escalate the international arms race. Because of such considerations, it is generally conceded that a limited ABM system to deal with accidental launches or a few missiles from rogue nations makes sense.

The United States proposed defending Europe against Iranian missiles by placing defensive missiles and radar in Poland and the Czech Republic. Russia saw this move as a means to blunt a Russian attack on the United States. It threatened to respond with nuclear weapons if Poland or the Czech Republic allowed the installations. In 2009, President Barack Obama scrapped that plan and announced that Aegis-equipped warships would be stationed in the Mediterranean and Black Seas, where they could defend Europe. Russia welcomed this change, which makes it plain that simply building ABM installations may have serious political consequences.

Another consideration is that sooner or later another asteroid will hit Earth, and humans might want to do something about it. For example, asteroid (29075) 1950 DA has a 0.0033 percent (one-third of 1 percent) chance of hitting Earth on March 16, 2880. Experience and technology developed from the various ABM programs will most likely be of some use in dealing with errant asteroids, as their technology of guiding missiles toward a target may be adapted to deflect an incoming asteroid by just enough to prevent it from destroying the planet.

*Charles W. Rogers, B.A., M.S., Ph.D.*

## FURTHER READING

Burns, Richard Dean. *The Missile Defense Systems of George W. Bush: A Critical Assessment.* Santa Barbara, Calif.: Praeger, 2010. A critical look at the fiscal and political costs to deploy a ground-based ABM system and the effects of trying to extend it to Europe.

Denoon, David. *Ballistic Missile Defense in the Post-Cold War Era.* Boulder, Colo.: Westview Press, 1995. An overview of various proposed ABM systems along with ways to judge if they would be worth the expense of constructing them.

Hey, Nigel. *The Star Wars Enigma: Behind the Scenes of the Cold War Race for Missile Defense.* Dulles, Va.: Potomac Books, 2006. An interesting story of who pushed for the Strategic Defense Initiative (SDI) and other ABM systems.

O'Rourke, Ronald. *Navy Aegis Ballistic Missile Defense (BMD) Program: Background and Issues for Congress.* Washington, D.C.: Congressional Research Service, 2010. O'Rourke, a specialist in naval affairs, discusses the past, present, and future of Aegis and especially the politics behind decisions affecting it.

Payne, Keith B. *Strategic Defense: "Star Wars" in Perspective.* Mansfield, Tex.: Hamilton Press, 1986. A wide-ranging examination of the various issues of SDI.

Sloan, Elinor C. "Space and Ballistic Missile Defense." In *Security and Defense in the Terrorist Era: Canada and the United States Homeland.* Montreal: McGill-Queen's University Press, 2010. How Canada's response to the threat of ballistic missiles is, and should be, different from that of the United States.

## WEB SITES

*Boeing*
Defense, Space, and Security
http://www.boeing.com/bds

*Federation of American Scientists*
Military Analysis Network
http://www.fas.org/programs/ssp/man/index.html

*Lockheed Martin*
Missiles and Fire Control
http://www.lockheedmartin.com/mfc/products.html

*Northrop Grumman*
Missile Defense
http://www.northropgrumman.com/missiledefense/index.html

*Raytheon*
Missile Systems
http://www.raytheon.com/businesses/rms

*Union of Concerned Scientists*
Nuclear Weapons and Global Security
http://www.ucsusa.org/nuclear_weapons_and_global_security

*U.S. Department of State*
Arms Control and International Security
http://www.state.gov

**See also:** Aeronautics and Aviation; Long-Range Artillery; Military Sciences and Combat Engineering.

# APIOLOGY

## FIELDS OF STUDY

Apiculture; melittology; entomology; agriculture; agricultural geography; biology; zoology; biochemistry; chemistry; physical science; statistics; botany; insect and pest management; food production; animal science; ecology.

## SUMMARY

Apiology is the scientific study of the honeybee. It is a subdiscipline of melittology, which is the study of all bees and is a branch of entomology. Apiologists study the evolution of the honeybee and answer questions surrounding its biology, particularly its social behavior and the ecological role the insect plays in its habitat. Other areas of study include the reproduction cycle of the honeybee, its proficiency to gather nectar, and its production of honey. A critical topic for the apiologist is diseases of the honeybee. Given the migratory nature of commercial beekeeping, the spread of bee diseases and their treatment have become a major challenge for beekeepers.

## KEY TERMS AND CONCEPTS

- **Apis Mellifera:** Genus and species name for the honeybee.
- **Bee Behavior:** Actions a bee takes in its instinctual inclination to survive.
- **Beehive:** Container that can house a colony of bees.
- **Brood:** Bees not yet developed out of the egg, larval, or pupal state.
- **Colony Collapse Disorder:** Collapse of a colony after a seemingly healthy colony of worker bees falls into a sudden, fatal decline.
- **Comb:** Wax material produced by bees and formed into hexagonal cells to store brood, nectar, honey, and pollen.
- **Drone:** Male bee.
- **Honey:** Product of the hive produced by the bees through a process of the fermentation of nectar.
- **Parthenogenesis:** State of being born of virgin female from an unfertilized egg.
- **Pollination:** Process of plant fertilization in which bees carry pollen from plant to plant.
- **Queen Bee:** Fertile female bee.
- **Swarm:** Process whereby part of a colony of honeybees leaves the hive in the company of a queen to propagate a new colony.
- **Worker Bee:** Sterile female bee.

### DEFINITION AND BASIC PRINCIPLES

The terms "apiology" and "apiculture" (another word for apiology) are derived from the Latin word for bee, *apis*. The science to which these terms refer focuses specifically on the honeybee. Although beekeepers are required to manage and maintain their colonies using up-to-date methods and, in effect, are practicing apiculture, in contrast, it is the outcome of the research of apiologists and those scientifically investigating the well-being of the honeybee that sets the standard for good practice by beekeepers. Managers of commercial bee operations that might contain thousands of colonies are highly dependent on good science to help them stay productive as well as competitive.

### BACKGROUND AND HISTORY

Apiary research and development emerged from a basic understanding of the behavior of bees. The outcome of this research has had considerable practical impact.

The standard removable-frame beehive—a design that was patented in 1852—was developed by observing bee behavior. Patent holder Lorenzo Langstroth, a beekeeper and Congregationalist preacher living in Ohio, noticed that bees maintained a prescribed space before they sealed over the frames of the hive. His response to this bee space was to develop a structure to function as the hive body that could be completely taken apart for harvesting honey and inspecting the bees. The outcome of this observation radically changed the way bees were housed and maintained.

The science of apiology has greatly contributed to understanding the genetic character and reproductive behaviors of bees. By the 1800's, it was generally accepted that the drone mated with the queen. The queen had the potential of laying eggs that became worker bees or drones. The Austrian monk Gregor Mendel, known for his genetic experiments with pea plants, also explored heredity by keeping and breeding bees. However, his bees were housed in the traditional stationary bee house, and he had no control over which drones bred with the queen.

In the early 1900's, apiology researchers had perfected the technique of instrumental insemination of the queen bee. This method made it possible to control the breeding process and has allowed for dramatic advances in the crossing of bees. The practical outcome of this science for the beekeeper is a well-tempered and productive bee.

Apiology also has provided insight into the diseases and enemies of the honeybee. Extensive research is conducted on mites and their control within the colony. Out of the study of bacterial problems such as foulbrood (a disease that attacks honeybee larvae), new practices in apiary management have emerged that can control these devastating honeybee enemies. Likewise, apiology provides an understanding of the symbiotic relationships that exist between bees and other insects as well as other animals.

## HOW IT WORKS

**Beekeeping.** The process of keeping bees requires an understanding of the biology of bees and the instinctual behaviors they exhibit within their colony. The colony's ability for honey production, fecundity (potential for reproduction), and the tendency to swarm, are also factors that must be managed by the beekeeper to maintain the viability of the colony. The beekeeper must become aware of the environmental

*Beekeeper holding tray of sealed honeycomb, deliberately scratched to show the honey.* (Ken Cavanagh/Photo Researchers, Inc.)

conditions that are conducive to pollination and nectar gathering. This requires an understanding of the blooming cycles of various flora within a region as well as weather conditions best suited for flying times by the bees. The outcome of the proper application of this knowledge will be expressed in the form of stronger and more productive colonies.

**Bee Diseases.** Diseases within the hive take a variety of forms. Some of these are found exclusively in the brood, thereby weakening the colony by lowering its overall population. In contrast, some of these diseases affect the adult bee, destroying the working population. The impact of these maladies can be substantial. Some diseases are the result of bacteria, such as American and European foulbrood; a virus, such as sacbrood; or a parasite such as *Nosema apis*. Mites such as *Varroa jacobsoni* are considered to be the number one killer of honeybees globally. These huge ectoparasites live on the outside of the bodies of the bees and the brood. They survive by attaching themselves to the insect and drawing out its hemolymph, a fluid similar to blood. During the early 1900's, the acarine mite (*Acarapis woodii*), also known as the tracheal mite, was responsible for killing all the black honeybees in England. These mites live in the trachea of the bee, cutting off its airway and choking it. Understanding the biology of this mite is the result of extensive research in apiology. Another pest that can wreak havoc on a hive is the wax moth (*Galleria mellonella*), which destroys the comb and the brood. Because of the nature of commercial migratory beekeeping, these diseases and pests can spread rapidly as colonies from one apiary come into contact with colonies from other apiaries.

Research into the diseases impacting honeybees is crucial to the survival and management of the honeybee. Apiologists identify these diseases by noting symptoms, tracking their spread geographically, and finding methods for their control. Integrated pest management systems using both cultural mechanical controls (such as hive modifications) and genetic controls (such as breeding bees that are resistant to a particular kind of disease) have all emerged from apiologists' research.

**Bee Genetics.** The domestication of the honeybee has been an ongoing process in societies as far back as those in ancient Egypt. In the twenty-first century, the crossbreeding of honeybees is controlled through scientific methods such as the instrumental insemination of the queen. This requires surgical-grade laboratory instruments, which hold and prepare the queen and the selection of drone bees that contribute semen for the insemination. This technique of breeding ensures that the traits desired in these bees will be secured. Techniques such as these have allowed for the quick development of new hybrid strains of the honeybee. Benedictine monk Karl Kehrle, known as Brother Adam, of Buckfast Abbey in England, is best known for his contribution to honeybee breeding. By selecting queens with particular traits and behaviors from around the world, he created what is now known as the Buckfast bee. The Buckfast bee was bred for good temper, honey production, and resistance to the acarine mite. Apiology also has genetically produced queens that can combat varroa mites. For example, varroa-sensitive hygiene (VSH) queens have been developed with the ability to detect the presence of varroa within a colony. Upon detection, they begin to clean the brood chambers of larvae that contain the mite.

## Applications and Products

**Honey.** Honey production is the primary commercial output of the honeybee colony. About 200 million pounds of honey are produced yearly within the United States. Consumption in the United States is twice that amount. Thus, honey is imported from sources around the world. The floral sources for nectar, the substance from which bees produce honey, is variable and determines the type and color of honey produced. Perhaps as many as three hundred varieties of honey can be produced in the United States alone, all having their own unique taste and color. Lighter-grade honeys and honey still in the comb are consumed directly and are usually sold at relatively high prices. Darker-grade honey may be used to sweeten baked goods. Popular floral sources for honey include clover, apple blossom, orange blossom, and huckleberry. Honey is sold in dehydrated form, with fruit flavors added to it, in comb form with capped and uncapped cells filled with honey, and as either organic or nonorganic regular or creamed honey. Aside from its use for humans as the world's oldest sweetener, honey is used in the production of paints, adhesives, cosmetics, and medicinal products.

**Pollination.** A major source of value for the beekeeper and the grower, and hence the public at large,

is the immense contribution of pollination that bees make to agriculture. Honeybees are used to pollinate the blossoms of many vegetables and fruits. The distribution of bees to locations timed to coincide with the seasonal arrival of blossoms is known as migratory beekeeping and constitutes a large and important industry. Some research suggests that nearly one-third of a human's diet is composed of products pollinated by insects, and about 80 percent of this pollination is accomplished by honeybees.

**Medicine from the Hive.** Substances coming from the beehive as well as the sting of the bee itself have been considered medicinal products. "Apitherapy" is a collective term referring to the application of bee venom and other products from the hive for medical purposes. Although modern apitherapy takes a scientific approach, it has a rich history in folk medicine. It began as a healing art practiced by the early Greeks, Egyptians, and Chinese. In the 1800's, the idea of bee-venom therapy was introduced to the United States. Bee venom, taken either directly from being stung or by injection with a syringe, is used in the treatment of arthritis, rheumatism, tendinitis, and multiple sclerosis; it has also been used to treat the pain from gout, shingles, and other maladies. It is believed that bee-venom therapy can increase circulation of the blood. Apitherapy is also used to decrease inflammation of tissues and to stimulate immune responses.

Royal jelly is the substance made by worker bees to feed potential queen bees within the hive. This substance can also be used to treat open wounds and as an energy tonic. The extraction of royal jelly is a labor-intensive process. Propolis is a bee-produced substance reddish to brown in color with the consistency of sticky clay. The bees utilize propolis to bridge over areas within the hive that exceed bee space so as to seal in the hive. This product has antifungal and antibacterial properties, so it can be used medicinally.

Bees collect plant pollen as a nutritional source for the colony. For some people, pollen is an allergen. But some studies suggest that human consumption of raw pollen from particular locales, as well as the raw honey that contains these pollen spores, can build immunity to pollen allergies from the same locales.

**Beeswax.** The worker bees of the hive secrete wax through glands on the sides of their bodies. In the colony, this wax is used to make comb, which is the base for building up hexagonally shaped cells. After

## Fascinating Facts About Apiology

- Honeybees provide about 80 percent of crop pollination. Some of the crops they pollinate include apples, pears, cherries, plums, almonds, avocados, cantaloupes, blueberries, cranberries, cucumbers, watermelons, sunflowers, and pumpkins.
- The first successful instrumental insemination of a queen bee was performed in 1926.
- Some studies suggest that the economic contribution through pollination by honeybees is greater than $14 billion annually.
- Apiologists studying the foraging habits of honeybees have found that a bee makes as many as ten flower stops per minute collecting pollen and nectar from as many as six hundred flowers before returning to its colony.
- Africanized bees made their way north through South and Central America following an accidental release in Brazil in 1957. The bees entered North America in the early 1990's and have come to reside in Texas, California, Louisiana, Arkansas, New Mexico, Mississippi, Alabama, and Florida. The massive number of stings these bees inflict can be fatal to both animals and humans, which is why they are also called killer bees.
- Varroa mites are the number one killer of honeybees worldwide and are one of the largest ectoparasites (parasites that live on the outside of the host). If this mite's size was considered on a human scale, it would as big as a basketball.
- The acarine mite, also known as the tracheal mite, resides within the trachea of the bee. These parasites lay their eggs in the airway of the bee, where they mature. Only mated mites leave the trachea of the host bee to find another bee in which to reside. Eventually, all the mites will leave a host bee after the infested bee dies. These mites must reside within a host and will die if they cannot find another.

the advent of the removable-comb by Langstroth, frames within the hive contained a foundation made from beeswax with a hexagonal imprint. From this paper-thin wax sheet, the bees produce the honeycomb that is used in rearing brood and storing food for the colony.

Beeswax finds its way into many common substances and products. Beeswax is well known in its use in candle production. In art, melted beeswax is used as a base for batik designs on fabric and in producing encaustic paints. Additionally, beeswax can be carved into figures that are used as models in the production of metal casting.

Many cosmetics, as well as some foods, use beeswax. Beeswax is also used as a seal in tree grafting.

### IMPACT ON INDUSTRY

**Migratory Beekeeping Industry.** The process of commercial pollination by bees translates into millions of bees doing labor that would otherwise have to be done by hand or by spraying pollen with an airplane. Both of these processes are costly. Bees—by virtue of both their efficiency, in visiting each flower in an orchard or field, and the fact that they do not need to be paid for their time—translate into millions of dollars saved in the production of food for human and animal consumption. Pollination by honeybees is limited only when growers misjudge blossom times, do not set sufficient numbers of colonies in the field, or confront weather that does not cooperate with the bees' flying time.

A symbiotic and economic relationship exists with honeybee pollinators and the plants they pollinate. Bees pollinating plants derive food and nectar for their colonies to survive for the winter. This can translate into an economic savings to the beekeeper. If plants or trees to be pollinated are profuse with pollen and nectar, the bees can store this food and feed themselves over the winter, requiring limited supplemental feeding from the beekeeper. Additionally, many migratory beekeepers are in business for pollination only and not honey production. This savings is critical to the migratory beekeeper's economic survival.

Perhaps the most attractive aspect of this industry is that it is among the few agricultural practices that have, for centuries, maintained the ecological principles of sustainability and have caused limited environmental degradation by avoiding mechanical means of pollination.

**Research.** Both public and private organizations spearhead the research efforts in apiology. At the national level, the U.S. Department of Agriculture maintains research stations in Tucson, Arizona; Baton Rouge, Louisiana; Logan, Utah; Beltsville, Maryland; and Weslaco, Texas. A number of universities have conducted ongoing research into areas such as colony collapse syndrome, Africanized bees or killer bees (an aggressive strain that began infiltrating the southern United States during the early 1990's), bee diseases, bee pests, and bee genetics, as well as apiary economics. At the state and local levels, multiple bee-keeping associations in all fifty states, as well as the District of Columbia and Puerto Rico, assist in disseminating information and in some cases collecting and providing data to scientists conducting research.

The International Bee Research Association in the United Kingdom was established in 1949 and publishes a series of research journals, books, and pamphlets for the world beekeeping community. It also sponsors international conferences dealing with beekeeping issues. In the United States, publishing outlets for scientific research include the *American Bee Journal*, established in 1861 and published by Dadant and Sons, and Bee Culture, now online but originally established in 1873 and published by A. I. Root and Sons.

**Honeybee Genome Project.** Funded by the National Human Genome Research Institute and the U.S. Department of Agriculture, the effort to sequence the genes of *Apis mellifera* began in 2003 and was completed three years later. Interest in tracing the genes of the honeybee was based on its importance in agricultural production and its social structure. However, the project may also provide insight into human health conditions, such as immunity to disease and allergies.

**Colony Collapse Disorder.** Sometimes referred to as CCD, colony collapse disorder has made international news and describes the demise of millions of honeybee colonies. Given the economic benefit that bees provide in agricultural production, research into this problem is crucial. Although researchers think that a virus might be the cause of this bee malady, scientists are quick to point out that since the mid-twentieth century the varroa mite has been the most damaging pest to the bee industry worldwide. Many researchers suggest that varroa or a combination of this mite and some other factor might be the cause of CCD. Fighting the varroa-mite outbreak has required both integrated pest management systems and treatment with pesticides.

## CAREERS AND COURSE WORK

Pursuing a career as an apiologist or a bee researcher requires a formal education that includes at least a bachelor of science (B.S.) degree. A master of science (M.S.), and in some cases a Ph.D., is required in most research and teaching posts at the university level. In all cases, a curriculum that develops the student's ability to establish systematic scientific studies, utilize a variety of computer programs that model environmental and biological conditions, and apply statistics is foundational for a career in apiology. In the United States, only a handful of colleges offer degrees in the field of apiology or apiculture. A degree in biology or entomology might be the key to employment. Ideally, such academic training would include apiary management or some related field of agriculture. Practical experience is also important and in some cases might pave the way to a related college program or even a career.

Researchers are found in both the private and public sectors. Chemical companies dealing with apiary pesticides and other chemicals seek trained individuals for laboratory and field research. At the federal level, the U.S. Department of Agriculture employs research scientists in their apiary units around the country. Additionally, some states have active apiary programs within their own departments of agriculture. At the state level, the trained apiologist may find duties both in a laboratory and in the field, inspecting hives.

Depending on the size of the apiary, migratory beekeepers individually maintain the seasonal requirements of their apiary operation as well as own and operate semitruck transportation for the movement of their bees. In some cases, transportation is leased out to companies that are specialists in handling loads of bees. Forklifts and tractors are used to move palletized beehives from flatbed trucks to positions within the field of pollination and then back again to flatbed semitrailers for hauling to the next pollination or nectar-gathering site. A number of skills are employed to maintain the apiary. These might include building and repairing bee boxes and frames, "pulling of supers" (boxes loaded with honey), operating honey-harvesting equipment, and working in the bee yard re-queening (introducing a new queen) and checking hives for disease.

## SOCIAL CONTEXT AND FUTURE PROSPECTS

Although mason bees, such as the orchard bee (*Osmia lignaria*), have garnered some attention as an alternative pollinator, as a honey producer and pollinator the honeybee still reigns supreme. The impact of honeybee pests such as varroa has already exposed the public to the importance of the honeybee in the production of food and to the potential of rising food costs if the demise of the honeybee continues. Further research into the interaction of bees with pesticides and herbicides as well as other environmental agents will continue to be critical research topics. Bee management systems will continue to be updated to accommodate environmental changes. Such changes are already being observed in some areas where reestablishing new colonies on a yearly basis is common practice. This requires the raising of more bees as well as new queens. Techniques such as instrumental insemination, which speeds the process of genetic modifications outside simple breeding programs, will also continue to develop as demand for bees with resistance to pests and diseases increases. Out of economic necessity, research will continue to develop around the honeybee. Its integration into the economic fabric of agriculture is far too great to ignore.

*M. Mustoe, Ph.D.*

## FURTHER READING

Bradbear, Nicola. *Beekeeping and Sustainable Livelihoods: FAO Diversification Booklet No. 1.* Rome, Italy: Food and Agriculture Organization of the United Nations, 2004. This short treatise discusses apiary management as a sustainable agricultural enterprise.

Connor, Lawrence John, ed. *Asian Apiculture: Proceedings of the First International Conference on the Asian Honey Bees and Bee Mites.* Cheshire, Conn.: Wicwas Press, 1993. Overview of the impact of Asian mites such as varroa.

_____. *Queen Rearing Essentials.* Kalamazoo, Mich.: Wicwas Press, 2009. A discussion of the process of grafting and selecting queen bees to improve production.

Cook, Albert John. *Manual of the Apiary.* 3d ed. Chicago: T. G. Newman & Son, 1878. This text is an interesting view on apiary science from the past, much of which still has applications.

Graham, Joe M., ed. *The Hive and the Honey Bee: A New Book on Beekeeping Which Continues the Tradition of "Langstroth on the Hive and the Honeybee."* Rev. ed. Hamilton, Ill.: Dadant, 1992. Offers history and background on modern foundations of apiculture.

Root, A. I., and E. R. Root. *The ABC and XYZ of Bee Culture: A Cyclopedia of Everything Pertaining to the Care of the Honey-Bee: Bees, Hives, Honey, Implements, Honey Plants, Etc.* 41st ed. Medina, Ohio: A. I. Root, 2007. As the rest of the subtitle of this venerable source states, this volume contains "facts gleaned from the experience of thousands of beekeepers and afterward verified in our apiary." A key resource for apiologists.

Seeley, Thomas D. *Honeybee Democracy.* Princeton, N.J.: Princeton University Press, 2010. An entomologist surveys the collective behavior of honeybees.

Spivak, Marla, and Gary S. Reuter. *Honey Bee Diseases and Pests.* St. Paul: University of Minnesota, 2010. This book is an overview of bee diseases and treatment methods.

## WEB SITES

*Bee Culture*
http://www.beeculture.com

*International Bee Research Association*
http://www.ibra.org.uk

*National Human Genome Research Institute*
Honey Bee Genome Sequencing
http://www.genome.gov/11008252

*U.S. Department of Agriculture*
http://www.usda.gov

**See also:** Agricultural Science; Computer Science; Plant Breeding and Propagation; Wildlife Conservation.

# APPLIED MATHEMATICS

Mechanical design; mechanical engineering; fluid dynamics; hydraulics; pneumatics; electronic engineering; physics; process modeling; physical chemistry; chemical kinetics; geologic engineering; geographic information systems; computer science; statistics; actuarial science; particle physics; epidemiology; investments; game theory; game design.

## SUMMARY

Applied mathematics is the application of mathematical principles and theory in the real world. The practice of applied mathematics has two principal objectives: One is to find solutions to challenging problems by identifying the mathematical rules that describe the observed behavior or characteristic involved, and the other is to reduce real-world behaviors to a level of precise and accurate predictability. Mathematical rules and operations are devised to describe a behavior or property that may not yet have been observed, with the goal of being able to predict with certainty what the outcome of the behavior would be.

## KEY TERMS AND CONCEPTS

- **Algorithm:** Effective method for solving problems that uses a finite series of specific instructions.
- **Boundary Conditions:** Values and properties that are required at the boundary limits of a mathematically defined behavior.
- **Markov Chain:** Probability model used in the study of processes that are considered to move through a finite sequence of steps, with repeats allowed.
- **Modeling:** Representation of real-world events and behaviors by mathematical means.
- **Particle-In-A-Box Model:** Classic mathematical analogy of the motion of an electron in an atomic orbital, subject to certain conditions, such as the requirement to have specific values at the limiting boundaries of the box.
- **Sparse System:** Any system whose parameter coefficients can be represented in a matrix that contains mostly zeros. Such systems are described as loosely coupled and typically represent members that are linked together linearly rather than as a network.
- **String Theory:** Mathematical representation of all subatomic particles as having the structures of strings rather than discrete spheres.
- **Wavelet Minimization:** Mathematical process whereby small random variations in frequency-dependent measurements (background noise) are removed to allow a clearer representation of the measurements.

### DEFINITION AND BASIC PRINCIPLES

Applied mathematics focuses on the development and study of mathematical and computational tools. These tools are used to solve challenging problems primarily in science and engineering applications and in other fields that are amenable to mathematical procedures. The principal mathematical tool is calculus, often referred to as the mathematics of change. Calculus provides a means of quantitatively understanding how variables that cannot be controlled directly behave in response to changes in variables that can be controlled directly. Thus, applied mathematics makes it possible to make predictions about the behavior of an environment and thus gain some mastery over that environment.

For example, suppose a specific characteristic of the behavior of individuals within a society is determined by the combination of a large number of influencing forces, many of which are unknown and perhaps unknowable, and therefore not directly controllable. Study of the occurrence of that characteristic in a population, however, allows it to be described in mathematical terms. This, in turn, provides a valid means of predicting the future occurrence and behavior of that characteristic in other situations. Applied mathematics, therefore, uses mathematical techniques and the results of those techniques in the investigation or solving of problems that originate outside of the realm of mathematics.

The applications of mathematics to real-world phenomena rely on four essential structures: data structures, algorithms, theories and models, and computers and software. Data structures are ways of organizing information or data. Algorithms are specific methods of dealing with the data. Theories and models are used in the analysis of both data and ideas and represent the rules that describe either the way the data were formed or the behavior of the data. Computers and software are the physical devices that are used to manipulate the data for analysis and application. Algorithms are central to the development of software, which is computer specific, for the manipulation and analysis of data.

## BACKGROUND AND HISTORY

Applied mathematics, as a field of study, is newer than the practices of engineering and building. The mathematical principles that are the central focus of applied mathematics were developed and devised from observation of physical constructs and behaviors and are therefore subsequent to the development of those activities. The foundations of applied mathematics can be found in the works of early Egyptian and Greek philosophers and engineers. Plane geometry is thought to have developed during the reign of the pharaoh Sesostris, as a result of agricultural land measurements necessitated by the annual inundation of the Nile River. The Greek engineer Thales of Miletus is credited with some of the earliest and most profound applications of mathematical and physical principles in the construction of some of his devices, although there is no evidence that he left a written record of those principles. The primary historical figures in the development of applied mathematics are Euclid and Archimedes. It is perhaps unfortunate that the Greek method of philosophy lacked physical experimentation and the testing of hypotheses but was instead a pure thought process. For this reason, there is a distinction between the fields of pure mathematics and applied mathematics, although the latter depends strictly on the former.

During the Middle Ages, significant mathematical development took place in Islamic nations, where *al geber*, which has come to be known as algebra, was developed, but the field of mathematics showed little progress in Europe. Even during the Renaissance period, mathematics was the realm almost exclusively of astronomers and astrologers. It is not certain that even Leonardo da Vinci, foremost of the Renaissance engineers and artists, was adept at mathematics despite the mathematical brilliance of his designs. The major historical development of applied mathematics began with the development of calculus by Sir Isaac Newton and Gottfried Wilhelm Leibniz in the seventeenth century. The applicability of mathematical principles in the development of scientific pursuits during the Industrial Revolution brought applied mathematics to the point where it has become essential for understanding the physical universe.

## HOW IT WORKS

Applied mathematics is the creation and study of mathematical and computational tools that can be broadly applied in science and engineering. Those tools are used to solve challenging problems in these and related fields of practice. In its simplest form, applied mathematics refers to the use of measurement and simple calculations to describe a physical condition or behavior.

A simple example might be the layout or design of a field or other area of land. Consider a need to lay out a rectangular field having an area of 2,000 square meters ($m^2$) with the shortest possible perimeter. The area ($A$) of any rectangular area is determined as the product of the length ($l$) and the width ($w$) of the area in question. The perimeter ($P$) of any rectangular area is determined as the sum of the lengths of all four sides, and in a rectangular area, the opposite sides are of equal length. Thus, $P = 2l + 2w$, and $A = l \times w = 2000$ $m^2$. By trial and error, pairs of lengths and widths whose product is 2,000 may be tried out, and their corresponding perimeters determined. A field that is 2,000 meters long and 1 meter wide has an area of 2,000 square meters and a perimeter of 4,002 meters. Similarly, a field that is 200 meters long and 10 meters wide has the required area, and a perimeter of only 420 meters. It becomes apparent that the perimeter is minimized when the area is represented as a square, having four equal sides. Thus, the length of each side must be equal to the square root of 2,000 in magnitude. Having determined this, the same principles may be applied to the design of any rectangular area of any size.

The same essential procedures as those demonstrated in this simple example apply with equal validity to other physical situations and are, in fact, the very essence of scientific experimentation and research. The progression of the development of mathematical models and procedures in many different areas of application is remarkably similar. Development of a mathematical model begins with a simple expression to which refinements are made, and the

## Fascinating Facts About Applied Mathematics

- Applied mathematics research toward the development of the quantum computer, which would operate on the subatomic scale, has led some researchers to postulate the simultaneous existence of multiple universes.
- Applied mathematics may have begun as geometry in ancient Egypt during the reign of the pharaoh Sesostris, who instituted the measurement of changes in land area for the fair assessment of taxes in response to the annual flooding of the Nile River.
- Ancient engineers apparently used certain esoteric principles of mathematics in the construction of the pyramids in Egypt and elsewhere, as the dimensions of these structures reflect the golden ratio ($\phi$).
- The flow of traffic along a multilane highway can be described by the same mathematics that describes the flow of water in a system of pipes.
- The behavior of an electron bound to a single proton is described mathematically as a particle in a box, because it must conform to specific conditions determined by the confines of the box.
- The seemingly uniform movement of individual birds in flocks of flying birds demonstrates the chaos theory, in which discrete large-scale patterns arise from unique small-scale actions. This is also known as the butterfly effect.
- The growth rates of bananas in the tropics and other natural behaviors can be described exactly by differential equations.
- Three-dimensional MRI can be used to practice and perfect surgical techniques without ever touching an actual patient.

results of the calculation are compared with the actual behavior of the system under investigation. The changes in the difference between the real and calculated behaviors are the key to further refinements that, ideally, work to bring the two into ever closer agreement. When mathematical expressions have been developed that adequately describe the behavior of a system, those expressions can be used to describe the behaviors of other systems.

A key component to the successful application of mathematical descriptions or models is an understanding of the different variables that affect the behavior of the system being studied. In fluid dynamics, for example, obvious variables that affect a fluid are the temperature and density of the fluid. Less obvious perhaps are such variables as the viscosity of the fluid, the dipolar interactions of the fluid atoms or molecules, the adhesion between the fluid and the surface of the container through which it is flowing, whether the fluid is flowing smoothly (laminar flow) or turbulently (nonlaminar flow), and a number of other more obscure variables. A precise mathematical description of the behavior of such a system would include corrective terms for each and every variable affecting the system. However, a number of these corrective terms may be considered together in an approximation term and still produce an accurate mathematical description of the behavior.

An example of such an approximation may be found in the applied mathematical field of quantum mechanics, by which the behavior of electrons in molecules is modeled. The classic quantum mechanical model of the behavior of an electron bound to an atomic nucleus is the so-called particle-in-a-box model. In this model, the particle (the electron) can exist only within the confines of the box (the atomic orbital), and because the electron has the properties of an electromagnetic wave as well as those of a physical particle, there are certain restrictions placed on the behavior of the particle. For example, the value of the wave function describing the motion of the electron must be zero at the boundaries of the box. This requires that the motion of the particle can be described only by certain wave functions that, in turn, depend on the dimensions of the box. The problem can be solved mathematically with precision only for the case involving a single electron and a single nuclear proton that defines the box in which the electron is found. The calculated results agree extremely well with observed measurements of electron energy.

For systems involving more particles (more electrons and more nuclear protons and neutrons), the number of variables and other factors immediately exceeds any ability to be calculated precisely. A solution is found, however, in a method that uses an approximation of the orbital description, known as a Slater-type orbital approximation, rather than a precise mathematical description. A third-level Gaussian treatment of the Slater-type orbitals, or STO-3G analysis, yields calculated results for complex molecular structures that are in excellent agreement with the observed values measured in physical experiments. Although the level of mathematical technique is vastly more complex than in the simple area example, the basic method of finding an applicable method is almost exactly identical.

## APPLICATIONS AND PRODUCTS

Applied mathematics is essentially the application of mathematical principles and theories toward the resolution of physical problems and the description of behaviors. The range of disciplines in which applied mathematics is relevant is therefore very broad. The intangible nature of mathematics and mathematical theory tends to restrict active research and development to the academic environment and applied research departments of industry. In these environs, applied mathematics research tends to be focused rather than general in nature. Applied mathematics is generally divided into the major areas of computational mathematics: combinatorics and optimization, computer science, pure mathematics, and statistical and actuarial science. The breadth of the research field has grown dramatically, and the diversity of subject area applications is indicated by the applied mathematics research being conducted in atmospheric and biological systems applications, climate and weather, complexity theory, computational finance, control systems, cryptography, pattern recognition and data mining, multivariate data analysis and visualization, differential equation modeling, fluid dynamics, linear programming, medical imaging, and a host of other areas.

**Computational Mathematics.** Simply stated, computational mathematics is the process of modeling systems quantitatively on computers. This is often referred to in the literature as *in silico*, indicating that the operation or procedure being examined is carried out as a series of calculations within the silicon-based electronic circuitry of a computer chip and not in any tangible, physical manner. Research in computational mathematics is carried out in a wide range of subject areas.

The essence of computational mathematics is the development of algorithms and computer programs that produce accurate and reliable models or depictions of specific behaviors. In atmospheric systems, for example, one goal would be to produce mathematical programs that precisely depict the behavior of the ozone layer surrounding the planet. The objective of such a program would be to predict how the ozone layer would change as a result of alterations in atmospheric composition. It is not feasible to observe the effects directly and would ultimately be counterproductive if manifesting an atmospheric change resulted in the destruction of the ozone layer. Modeling the system *in silico* allows researchers to institute virtual changes and determine what the effect of each change would be. The reliability of the calculated effect depends directly on how accurately the model describes the existing behavior of the system.

**Medical Imaging.** An area in which applied mathematics has become fundamentally important is the field of medical imaging, especially as it applies to magnetic resonance imaging (MRI). The MRI technique developed from nuclear magnetic resonance (NMR) analysis commonly used in analytical chemistry to determine molecular structures. In NMR, measurements are obtained of the absorption of specific radio frequencies by molecules held within a magnetic field. The strength of each absorption and specific patterns of absorptions are characteristic of the structure of the particular molecule and so can be used to determine unequivocally the exact molecular structure of a material.

One aspect of NMR that has been greatly improved by applied mathematics is the elimination of background noise. A typical NMR spectrum consists of an essentially infinite series of small random signals that often hide the detailed patterns of actual absorption peaks and sometimes even the peaks themselves. The Fourier analysis methodology, in which such random signals can be treated as a combination of sine and cosine waves, also known as wavelet theory, eliminates a significant number of such signals from electromagnetic spectra. The result is a much more clear and precise record of the actual absorptions. Such basic NMR spectra are only one dimensional, however. The second generation modification of NMR systems was developed to produce a two-dimensional representation of the NMR absorption spectrum, and from this was developed the three-dimensional NMR imaging system that is known as MRI. Improvements that make the Fourier analysis technique ever more effective in accord with advances in the computational abilities of computer hardware are the focus of one area of ongoing applied mathematics research.

**Population Dynamics and Epidemiology.** Population dynamics and epidemiology are closely related fields of study. The former studies the growth and movements of populations, and the latter studies the growth and movements of diseases and medical conditions within populations. Both rely heavily for their mathematical descriptions on many areas of applied mathematics, including statistics, fluid dynamics, complexity theory, pattern recognition, data visualization, differential equation modeling, chaos theory, risk management, numerical algorithms and techniques, and statistical learning.

In a practical model, the movements of groups of individuals within a population are described by many of the same mathematical models of fluid dynamics that apply to moving streams of particles. The flow of traffic on a multilane highway or the movement of people along a busy sidewalk, for example, can be seen to exhibit the same gross behavior as that of a fluid flowing through a system of pipes. In fact, any population that can be described in terms of a flow of discrete particles, whether molecules of water or vehicles, can be described by the same mathematical principles, at least to the extent that the variables affecting their motion are known. Thus, the forces of friction and adhesion that affect the flow of a fluid within a tube are closely mimicked by the natural tendencies of drivers to drive at varying speeds in different lanes of a multilane highway. Window-shoppers and other slow-moving individuals tend to stay to the part of the sidewalk closest to the buildings, while those who walk faster or more purposefully tend to use the part of the sidewalk farthest away from the buildings, and this also follows the behavior of fluid flow.

The spread or movement of diseases through a population can also be described by many of the same mathematical principles that describe the movements of individuals within a population. This is especially true for diseases that are transmitted directly from person to person. For other disease vectors, such as animals, birds, and insects, a mathematical description must describe the movements of those particular populations, while at the same time reflecting the relationship between those populations and the human population of interest.

**Statistical Analysis and Actuarial Science.** Perhaps the simplest or most obvious use of applied mathematics can be found in statistical analysis. In this application, the common properties of a collection of data points, themselves measurements of some physical property, are enumerated and compared for consistency. The effectiveness of statistical methods depends on the appropriately random collection of representative data points and on the appropriate definition of a property to be analyzed.

Statistical analysis is used to assess the consistency of a common property and to identify patterns of occurrence of characteristics. This forms the basis of the practice of statistical process control (SPC) that has become the primary method of quality control in industry and other fields of practice. In statistical process control, random samples of an output stream are selected and compared to their design standard. Variations from the desired value are determined, and the data accumulated over time are analyzed to determine patterns of variation. In an injection-molding process, for example, a variation that occurs consistently in one location of the object being molded may indicate that a modification to the overall process must be made, such as adjusting the temperature of the liquid material being injected or an alteration to the die itself to improve the plastic flow pattern.

In another context, one that is tied to epidemiology, the insurance and investment industries make very detailed use of statistical analysis in the assessment of risk. Massive amounts of data describing various aspects of human existence in modern society are meticulously analyzed to identify patterns of effects that may indicate a causal relationship. An obvious example is the statistical relationship pattern between healthy lifestyle and mortality rates, in which obese people of all age groups have a higher mortality rate than their counterparts who maintain a leaner body mass. Similarly, automobile insurance rates are set much higher for male drivers between the ages of sixteen and twenty-five than for female drivers in that age group and older drivers because statistical analysis of data from accidents demonstrates that this particular group has the highest risk of involvement in a traffic accident. This type of data mining is a continual process as relationships are sought to describe every factor that plays a role in human society.

## IMPACT ON INDUSTRY

**Statistical Process Control.** The practice of statistical process control has had an unprecedented effect on modern society, especially in the manufacturing industry. Before the development of mass-production methods, parts and products were produced by skilled and semiskilled craftspeople. The higher the level of precision required for a production piece, the more skilled and experienced an artisan or craftsperson had to be. Manpower and cost of production quickly became the determining factors in the availability of goods.

The advent of World War II, however, ushered in an unprecedented demand for materials and products. The production of war materials required the development of methods to produce large numbers of products in a short period of time. Various accidents and mistakes resulting from the rapid pace of production also indicated the need to develop methods of quality control to ensure that goods were produced to the expected level of quality and dependability. The methods of quality assurance that were developed at that time were adopted by industries in postwar Japan and developed into a rigorous protocol of quality management based on the statistical analysis of key features of the goods being produced. The versatility of the system, when applied to mass-production methods, almost instantly eliminated the dependency on skilled tradespeople for the production of precision goods.

In the intervening half century, statistical process control became a required component of production techniques and an industry unto itself, spawning such rigorous programs as the International Organization for Standardization's ISO 9000, ISO 9001, and ISO 14000; the Motorola Corporation's Six Sigma; and Lean Manufacturing (derived from the Toyota Production System), which have become global standards.

**Economics.** The analysis of data by mathematical means, especially data mining and pattern recognition in population dynamics, has provided much of the basis of the economic theory that directs the conduct of business on a global scale. This has been especially applicable in regard to the operation of the stock market. In large part, the identification of patterns and trends in the historical data of trade and business provides the basic information for speculation on projected or future consumer trends. Although this is important for the operation of publicly traded businesses on the stock market, it is more important in the context of government. The same

trends and patterns play a determining role in the design and establishment of government policies and regulations for both domestic and international affairs.

**Medicine and Pharmaceuticals.** Applied mathematics, particularly the use of Fourier analysis and wavelet theory, has sparked explosive growth in the fields of medicine and pharmaceutical development. The enhanced analytical methods available to chemists and bioresearchers through NMR and Fourier transform infrared (FTIR) spectroscopy facilitate the identification and investigation of new compounds that have potential pharmaceutical applications. In addition, advanced statistical methods, computer modeling, and epidemiological studies provide the foundation for unprecedented levels of research.

The science of medical diagnostic imaging has grown dramatically because of the enhancements made available through applied mathematics. Computer science, based on the capabilities of modern digital computer technology, has replaced the physical photographic method of X-ray diagnostics with a digital version that is faster and more efficient and has the additional capability of zooming in on any particular area of interest in an X-ray image. This allows for more accurate diagnostics and is especially important for situations in which a short response time is essential.

The methodology has enabled entirely new procedures, including MRI. Instead of relatively plain, two-dimensional images, real-time three-dimensional images can be produced. These images help surgeons plan and rehearse delicate surgical maneuvers to perfect a surgical technique before ever touching a patient with a scalpel. MRI data and modifications of MRI programming can be used to control the sculpting or fabrication of custom prosthetic devices. All depend extensively on the mathematical manipulation of data to achieve the desired outcome.

## CAREERS AND COURSE WORK

The study of applied mathematics builds on a solid and in-depth comprehension of pure mathematics. The student begins by taking courses in mathematics throughout secondary school to acquire a solid foundational knowledge of basic mathematical principles before entering college or university studies. A specialization in mathematics at the college and university level is essential to practically all areas of study. As so many fields have been affected by applied mathematics, the methodologies being taught in undergraduate courses are a reflection of the accepted concepts of applied mathematics on which they are constructed. As the depth of the field of applied mathematics indicates, career options are, for all intents and purposes, unlimited. Every field of endeavor, from anthropology to zoology, has a component of applied mathematics, and the undergraduate must learn the mathematical methods corresponding to the chosen field.

Students who specialize in the study of applied mathematics as a career choice and proceed to postgraduate studies take courses in advanced mathematics and carry out research aimed at developing mathematics theory and advancing the application of those developments in other fields.

Particular programs of study in applied mathematics are included as part of the curriculum of other disciplines. The focus of such courses is on specific mathematical operations and techniques that are relevant to that particular field of study. All include a significant component of differential calculus, as appropriate to the dynamic nature of the subject matter.

## SOCIAL CONTEXT AND FUTURE PROSPECTS

Human behavior is well suited and highly amenable to description and analysis through mathematical principles. Modeling of population dynamics has become increasingly important in the contexts of service and regulation. As new diseases appear, accurate models to predict the manner in which they will spread play an ever more important role in determining the response to possible outbreaks. Similarly, accurate modeling of geological activities is increasingly valuable in determining the best ways to respond to such natural disasters as a tsunami or earthquake. The Earth itself is a dynamic system that is only marginally predictable. Applied mathematics is essential to developing models and theories that will lead to accurate prediction of the occurrence and ramifications of seismic events. It will also be absolutely necessary for understanding the effects that human activity may be having on the atmosphere and oceans, particularly in regard to the issues of global warming and the emission of greenhouse gases. Climate models that can accurately and precisely predict the effects of such activities will continue to

be the object of a great deal of research in applied mathematics.

The development of new materials and engineering new applications with those materials is an ongoing human endeavor. The design of extremely small nanostructures that employ those materials is based on mathematical principles that are unique to the realm of the very small. Of particular interest is research toward the development of the quantum computer, a device that operates on the subatomic scale rather than on the existing scale of computer construction.

*Richard M. J. Renneboog, M.Sc.*

## FURTHER READING

Anton, Howard. *Calculus: A New Horizon.* 6th ed. New York: John Wiley & Sons, 1999. Offers a clear and progressive introduction to many of the basic principles of applied mathematics.

De Haan, Lex, and Toon Koppelaars. *Applied Mathematics for Database Professionals.* New York: Springer, 2007. Focuses on the use of set theory and logic in the design and use of databases in business operations and communications.

Howison, Sam. *Practical Applied Mathematics: Modelling, Analysis, Approximation.* New York: Cambridge University Press, 2005. Provides a basic introduction to several practical aspects of applied mathematics, supported by in-depth case studies that demonstrate the applications of mathematics to actual physical phenomena and operations.

Kurzweil, Ray. *The Age of Spiritual Machines: When Computers Exceed Human Intelligence.* New York: Penguin Books, 2000. Presents a thought-provoking view of the nature of artificial intelligence that may arise as a direct result of applied mathematics.

Moore, David S. *The Basic Practice of Statistics.* 5th ed. New York: Freeman, 2010. Introduces statistical analysis at its most basic level and progresses through more complex concepts to an understanding of statistical process control.

Naps, Thomas L., and Douglas W. Nance. *Introduction to Computer Science: Programming, Problem Solving, and Data Structures.* 3d ed. St. Paul, Minn.: West Publishing, 1995. Presents a basic introduction to the principles of computer science and the development of algorithms and mathematical processes.

Rubin, Jean E. *Mathematical Logic: Applications and Theory.* Philadelphia: Saunders College Publishing,

1990. An introduction to the principles of mathematical logic that requires the reader to work closely through the material to obtain a good understanding of mathematical logic.

Washington, Allyn J. *Basic Technical Mathematics with Calculus.* 9th ed. Upper Saddle River, N.J.: Pearson Prentice Hall, 2009. Provides thorough and well-explained instruction in applied mathematics for technical pursuits in a manner that allows the reader to grasp many different mathematical concepts.

**WEB SITES**

*Mathematical Association of America*
http://www.maa.org

*Society for Industrial and Applied Mathematics*
http://www.siam.org

**See also:** Algebra; Calculus; Computed Tomography; Computer Science; Demography and Demographics; Engineering Mathematics; Game Theory; Geometry; Numerical Analysis; Pattern Recognition; Probability and Statistics.

# APPLIED PHYSICS

## FIELDS OF STUDY

Physics; mathematics; applied mathematics; mechanical engineering; civil engineering; metrics; forensics; biological physics; photonics; microscopy and microanalysis; biomechanics; materials science; surface science; aeronautics; hydraulics; nanotechnology; robotics; medical imaging; radiology.

## SUMMARY

Applied physics is the study and application of the behavior of condensed matter, such as solids, liquids, and gases, in bulk quantities. Applied physics is the basis of all engineering and design that requires the interaction of individual components and materials. The study of applied physics now extends into numerous other fields, including physics, chemistry, biology, engineering, medicine, geology, meteorology, and oceanography.

## KEY TERMS AND CONCEPTS

- **Biomechanics:** The science of physical movement in living systems.
- **Condensed Matter:** Matter is a physical quantity that is normally recognized as having the form of a solid, a liquid, or a gas, also known as a bulk quantity.
- **Forensics:** The science of analysis to determine cause and effect after an event has occurred, highly dependent upon the application of laws of physics in the case of physical events.
- **Magnetic Levitation:** A technology that utilizes the mutual repulsion of like magnetic fields to maintain a physical separation between an object and a track (such as a high-speed train and its track bed), while keeping the two in virtual contact.
- **Magnetic Resonance Imaging (MRI):** A noninvasive diagnostic imaging technology based on nuclear magnetic resonance that provides detailed images of internal organs and the structures of living systems.
- **Nanoprobe:** A device designed to perform a specific function, constructed of nanometer (10-9 meter) scale components.
- **Newtonian Mechanics:** The behavior of condensed matter as described by Newton's laws of motion and inertia.
- **Nuclear Magnetic Resonance (NMR):** A property of certain atomic nuclei having unpaired electrons to absorb energy of a specific electromagnetic wavelength within a constant magnetic field.
- **Prosthetics:** The branch of applied physics that deals with the design of artificial components that replace or augment natural biological components; the artificial components that are designed and constructed to replace normal biological components.
- **Quantum Mechanics:** The physical laws describing the behavior of matter in quantities below the nanometer (10-9 meter) scale, especially of individual atoms and molecules.
- **Spectrometry:** An analytical technology that utilizes the specific electromagnetic energy of specific wavelengths to identify or quantify specific components of a material or solution.

### DEFINITION AND BASIC PRINCIPLES

Applied physics is the study of the behavior and interaction of condensed matter. Condensed matter is commonly recognized as matter in the phase forms of solids, liquids, and gases. Each phase has its own unique physical characteristics, and each material has its own unique physical suite of properties that derive from its chemical identity. The combination of these traits determines how something consisting of condensed matter interacts with something else that consists of condensed matter.

The interactions of constructs of condensed matter are governed by the interaction of forces and other vector properties that are applied through each construct. A lever, for example, functions as the medium to transmit a force applied in one direction so that it operates in the opposite direction at another location. Material properties and strengths are also an intimate part of the interaction of condensed matter. A lever that cannot withstand the force applied to it laterally will bend or break rather than function as a lever. Similarly, two equally hard objects, such as a train car's wheels and the steel tracks upon which they roll, rebound elastically with little or no generation of friction. If, however, one of the materials is less hard than the other, as in the case of a steel rotor and a composition brake pad, a great deal of friction characterizes their interaction.

## BACKGROUND AND HISTORY

Applied physics is of necessity the oldest of all practical sciences, dating back to the first artificial use of an object by an early hominid. The basic practices have been in use by builders and designers for many thousands of years. With the development of mathematics and measurement, the practice of applied physics has grown apace, relying as it still does upon the application of basic concepts of vector properties (force, momentum, velocity, weight, moment of inertia) and the principles of simple machines (lever, ramp, pulley).

The modern concept of applied physics can be traced back to the Greek philosopher Aristotle (c. 350 B.C.E.), who identified several separate fields for systematic study. The basic mathematics of physics was formulated by Pythagoras about two hundred years before, but was certainly known to the Babylonians as early as 1600 B.C.E. Perhaps the greatest single impetus for the advance of applied physics was the development of calculus by Sir Isaac Newton and the fundamental principles of Newtonian mechanics in the seventeenth century. Those principles describe well the behaviors of objects composed of condensed matter, only failing as the scale involved becomes very small (as below the scale of nanotechnology). Since the Industrial Revolution and into the twenty-first century, the advancing capabilities of technology and materials combine to enable even more advances and applications, yet all continue to follow the same basic rules of physics as the crudest of constructs.

## HOW IT WORKS

As the name implies, applied physics means nothing more or less than the application of the principles of physics to material objects. The most basic principles of applied physics are those that describe and quantify matter in motion (speed, velocity, acceleration, momentum, inertia, mass, force). These principles lead to the design and implementation of devices and structures, all with the purpose of performing a physical function or action, either individually or in concert.

A second, and equally important, aspect of applied physics is the knowledge of the physical properties and characteristics of the materials being used. This includes such physical properties as melting and boiling points, malleability, thermal conductivity, electrical resistivity, magnetic susceptibility, density, hardness, sheer strength, tensile strength, compressibility, granularity, absorptivity, and a great many other physical factors that determine the suitability of materials for given tasks. Understanding these factors also enables the identification of new applications for those materials.

**Design.** Applied physics is the essential basis for the design of each and every artificial object and construct, from the tiniest of nanoprobes and the simplest of levers to the largest and most complex of machines and devices. Given the idea of a task to be performed and a device to perform that task, the design process begins with an assessment of the physical environment in which the device must function, the forces that will be exerted against it, and the forces that it will be required to exert in order to accomplish the set task. The appropriate materials may then be selected, based on their physical properties and characteristics. Dimensional analyses determine the necessary size of the device, and also play a significant role in determining the materials that will be utilized.

All of these factors affect the cost of any device, an important aspect of design. Cost is an especially important consideration, as is the feasibility of replacing a component of the designed structure. For example, a device operating in a local environment in which replacement of a component, such as an actuating lever, is easily carried out may be constructed using a simple steel rod. If, however, the device is something like the Mars Rover, for which replacement of worn parts is not an option, it is far more reasonable and effective to use a less-corrosive, stronger, but more costly material, such as titanium, to make movable parts. The old design tenet that form follows function is an appropriate rule of thumb in the field of applied physics.

## APPLICATIONS AND PRODUCTS

In many ways, applied physics leads to the idea of creating devices that build other devices. At some point in time, an individual first used a handy rock to break up another rock or to pound a stick into the ground. This is a simple example of applied physics, when that individual then began to use a specific kind of rock for a specific purpose; this action advanced again when someone realized that attaching the rock to a stick provided a more effective tool. Applied physics has advanced far beyond the crude rock-on-a-stick hammer, yet the exact same physical principles apply now with the most elegant of impact devices. It would not be possible to itemize even a small percentage of the applications and products that have resulted from applied physics, and new physical devices are developed each day.

**Civil Society.** Applied physics underpins all physical aspects of human society. It is the basis for the design and construction of the human environment and its supporting infrastructure, the most obvious of which are roads and buildings, designed and engineered with consideration of the forces that they must withstand, both human-made and natural. In this, the highest consideration is given to the science and engineering of materials in regard to the desired end result, rather than to the machines and devices that are employed in the actual construction process. The principles of physics involved in the design and construction of supporting structures, such as bridges and high-rise towers, are of primary importance. No less important, however, are the physical systems that support the function of the structure, including such things as electrical systems, internal movement systems, environmental control systems, monitoring systems, and emergency response systems, to name a few. All are relevant to a specific aspect of the overall physical construct, and all must function in coordination with the other systems. Nevertheless, they all also are based on specific applications of the same applied physics principles.

**Transportation.** Transportation is perhaps the single greatest expenditure of human effort, and it has produced the modern automotive, air, and railway transportation industries. Applied physics in these areas focuses on the development of more effective and efficient means of controlling the movement of the machine while enhancing the safety of human and nonhuman occupants. While the physical principles by which the various forms of the internal combustion engine function are the same now as when the devices were first invented, the physical processes used in their operation have undergone a great deal of mechanical refinement.

A particular area of refinement is the manner in which fuel is delivered for combustion. In earlier gasoline-fueled piston engines, fuel was delivered via a mechanical fuel pump and carburetor system that functions on the Venturi principle. This has long since been replaced by constant-pressure fuel pumps and injector systems, and this continues to be enhanced as developers make more refinements based on the physical aspects of delivering a liquid fuel to a specific point at a specific time. Similarly, commercial jet engines, first developed for aircraft during World War II, now utilize the same basic physical principles as did their earliest counterparts. Alterations and developments that have been achieved in the interim have focused on improvement of the operational efficiency of the engines and on enhancement of the physical and combustion properties of the fuels themselves.

In rail transport, the basic structure of a railway train has not changed; it is still a heavy, massive tractor engine that tows a number of containers on very low friction wheel-systems. The engines have changed from steam-powered behemoths made up of as many as one-quarter-million parts to the modern diesel-electric traction engines of much simpler design.

Driven by the ever-increasing demand for limited resources, this process of refinement for physical efficiency progresses in every aspect of transportation on land, sea, and air. Paradoxically, it is in the area of railway transport that applied physics offers the greatest possibility of advancement with the development of nearly frictionless, magnetic levitation systems upon which modern high-speed bullet trains travel. Physicists continue to work toward the development of materials and systems that will be superconducting at ambient temperatures. It is believed that such materials will completely revolutionize not only the transportation sector but also the entire field of applied physics.

**Medical.** No other area of human endeavor demonstrates the effects of applied physics better than the medical field. The medical applications of applied physics are numerous, touching every aspect of medical diagnostics and treatment and crossing over into many other fields of physical science. The most obvious of these is medical imaging, in which X-ray diagnostics, magnetic resonance imaging (MRI), and other forms of spectroscopic analysis have become primary tools of medical diagnostics. Essentially, all of the devices that have become invaluable as medical tools began as devices designed to probe the physical nature of materials.

Spectrographs and spectrometers that are now routinely used to analyze the specific content of various human sera were developed initially to examine the specific wavelengths of light being absorbed or emitted by various materials. MRI developed from the standard technique of physical analytical chemistry called nuclear magnetic resonance spectrometry, or NMR. In this methodology, magnetic signals are recorded as they pass through a material sample, their patterns revealing intimate details about the three-dimensional molecular structure of a given compound. The process of MRI that has developed from this simpler application now permits diagnosticians to view the internal structures and functioning of living systems in real time. The diagnostic X ray that has become the single most common method of examining internal biological structures was first used as a way to physically probe the structure of crystals, even though the first practical demonstration of the existence of X rays by German physicist Wilhelm Röntgen in 1895 was as an image of the bones in his wife's hand.

**Military.** It is not possible to separate the field of applied physics from military applications and martial practice, and no other area of human endeavor so clearly illustrates the double-edged sword that is the nature of applied physics. The field of applied physics at once facilitates the most humanitarian of endeavors and the most violent of human aggressions. Ballistics, which gets its name from the Roman Latin word for "war," is the area of physics that applies to the motion of bodies moving through a gravitational field. The mathematical equations that describe the motion of a baseball being thrown from center field equally describe the motion of an arrow or a bullet in flight, of the motion of the International Space Station in its orbit, and of the trajectory of a warhead-equipped rocket. The physical principles that have permitted the use of nuclear fission as a source of reliable energy are the same principles that define the functioning of a nuclear warhead.

In the modern age it is desired that warfare be carried out as quickly and efficiently as possible as the means to restore order and peace to an embattled area, with as little loss of life and destruction of property as possible. To that end, military research relies heavily on the application of physics in the development of weapons, communications, and surveillance systems.

**Digital Electronics.** Applied physics has also led to the development of the transistor, commonly attributed to William Shockley, in 1951. Since that time, the development of ever-smaller and more efficient systems based on the semiconductor junction transistor has been rapid and continuous. This has inexorably produced the modern technology of digital electronics and the digital computer. These technologies combine remarkably well to permit the real-time capture and storage of data from the many and various devices employed in medical research and diagnostics, improve the fine control of transportation systems and infrastructure, and advance a veritable host of products that are more efficient in their use of energy than are their nondigital counterparts.

## IMPACT ON INDUSTRY

There is a common misconception that persons with degrees and expertise within the field of physics are not presented with sufficient opportunities for employment after graduating. This misconception reflects a difference in academic terminology versus business terminology only, and not reality. It is true that few physics graduates actually work as traditional physicists. Given the training and ability in mathematics and physical problem-solving that physicists (anyone with a degree in physics) have developed, they tend to acquire positions as industrial physicists instead, holding positions with industrial titles as managers, engineers, computer scientists, and technical staff.

Data from the American Institute of Physics indicate that physicists represent a significantly higher-than-average income group. About one-fifth of all physicists enter military careers, while about one-half of those with a doctorate remain in academia and teaching and about one-half of those with master's degrees make their careers in industry.

Given that applied physics is the fundamental basis of all human technology upon which modern society has been built, it is essentially impossible to assign a monetary value to the overall field. The direct economic value that derives from any individual aspect of applied physics is often measured in billions of U.S. dollars, and in some cases the effects are so far-reaching as to be immeasurable. The amount spent on space research, for example, by the U.S. government through the National Aeronautics and Space Administration (NASA), is less than 1 percent of the annual national budget of approximately $2.5 trillion. The return on this investment in terms of corporate and personal taxes, job creation, and economic growth, however, is conservatively estimated to be no less than seven-to-one. That is to say, every $1 million invested in space research generates a minimum of $7 million in tax revenues and economic growth.

Not the least significant of effects is the spin-off that enters the general economy as a direct result of applied physics research and development. Space research has provided a long list of consumer products and technical enhancements that touch everything from golf balls and enriched baby foods to industrial process modeling, virtual reality training, and education aids. The list of spin-off technologies and products from this source alone is very long, and it continues to grow as research advances. It is interesting to note that many of the products of space research that were developed with the intent of bettering a human presence on other planets have proved eminently applicable to bettering the human presence on Earth.

The wireless communications industry is one of the most obvious spin-offs from space research, and it now represents a significant segment of the overall economy of nations around the world. The value of this industry in Canada, for example, provides a good indication of the value of wireless communications to other nations. The wireless communications industry in Canada serves a country that has four persons per square kilometer, as compared with thirty persons per square kilometer in the United States. In 2010, the economic value of the wireless communications industry in Canada was approximately US$39 billion, of which $16.9 billion was a direct contribution to the gross domestic product (GDP) through the sale of goods and services. This contribution to the GDP compared favorably with those of the automotive manufacturing industry, the food manufacturing industry, and agricultural crop production.

Wireless communications produced up to 410,000 Canadian jobs, or about 2.5 percent of all jobs in Canada; this marks a significantly higher average salary than the national average. The value-added per employee in the wireless telecommunications industry was also deemed to be more than 2.5 times higher than the national average. While a direct extrapolation to the effect of this industry on the economy of the United States, and other nations, is not appropriate because of the differences in population distribution and geography, it is nevertheless easy to see that the wireless communications industry now represents a significant economic segment and has far-reaching effects.

Furthermore, wireless communication can greatly reduce unproductive travel time and the associated economic costs, while significantly improving the logistics of business operations. Wireless communications empowers small businesses, especially in rural

## Fascinating Facts About Applied Physics

- When the semiconductor junction transistor was invented in 1951, it initiated the digital electronics revolution and formed the functional basis of the personal computer. In 1954, scientists at RAND envisioned a home computer in the year 2000, complete with a teletype Fortran-language programming station, a large CRT television monitor, rows of blinking lights and dials, and twin steering-wheels.

- Nuclear magnetic resonance (NMR) spectrometry is an analytical method that is sensitive enough to differentiate and measure the magnetic environments of individual atoms in molecules.

- Creating a Stone Age hammer by fastening a rock to a stick is as much a product of applied physics as is the creation of the International Space Station.

- A rock thrown through the air obeys exactly the same laws of ballistics as does a bullet fired from a high-powered rifle or a satellite orbiting a planet.

- The B2 Spirit stealth bomber, a U.S. military aircraft representing sophisticated applied physics, has a unit cost of more than US$1.15 billion (as of 2003).

- Spectrography, initially invented to identify specific wavelengths of emitted or absorbed light, is the basis for all analytical methods that rely on electromagnetic radiation.

- The formal identification of the science known as applied physics can be traced back to Aristotle, to circa 350 B.C.E.

- All human technology is the product of applied physics. Spin-off from such advanced activities as space research translates into a great number of everyday products, including cell phones, home computers, and digital egg timers, with an overall economic value of trillions of U.S. dollars annually.

areas. Medical devices represent another large economic segment that has grown out of applied physics research and development. Recent surveys (2008) suggest that it is one segment that is relatively unaffected by economic upswings and downswings. Given that medical methodology continues to advance on the strength of diagnostic devices, while the development of medical technology continues to advance in step with advancing medical methodology, it is logical that the effects of economic changes are mitigated in this field.

As of 1999, medical research in the United States represented an investment of more than $35 billion. The economic value of medical research was estimated in 1999 to be approximately $48 trillion for the elimination of death from heart disease and $47 trillion for a cure for cancer. Simply reducing cancer mortality by a mere 1 percent would generate an economic return of $500 billion. These estimates are all based on the assumed economic "value of life" as benefits to persons deriving from an extended ability to participate in society. The research and development of medical knowledge that would generate these economic returns cannot be called applied physics. Rather, the various technologies by which the overall medical research program is supported are the direct result of applied physics.

In the realm of military expenditure, it is estimated that approximately 10 percent of the U.S. GDP is directly related to military purposes. This includes research and development in applied physics technologies.

## CAREERS AND COURSEWORK

Programs of study for careers in applied physics or in fields that are based on applied physics include advanced mathematics, physics, and materials science. Depending on the area of career specialization, coursework may also include electronics engineering, mechanical engineering, chemistry and chemical engineering, biology and biomechanics, instrumentation, and more advanced courses in mathematics and physics. The range of careers is broad, and the coursework required for each will vary accordingly.

Credentials and qualifications required for such careers range from a basic associate's degree from a community college to one or more postgraduate degrees (master's or doctorate) from a top-tier specialist university. An advanced degree is not a requirement

for a career in an applied physics discipline, and the level of education needed for a successful career will depend on the nature of the discipline chosen.

Each discipline of applied physics has a core knowledge base of mathematics and physics that is common to all disciplines. Students should therefore expect to take several courses in these areas of study, including advanced calculus and a comprehensive program of physics. Other courses of study will depend on the particular discipline that a person chooses to pursue. For disciplines that focus on medical applications, students will also take courses in chemistry and biology, because understanding the principles upon which biological systems operate is fundamental to the effective application of physics and physical devices within those systems. Engineering disciplines, the most directly related fields of applied physics, require courses in materials science, mechanics, and electronics. As may be expected, there is a great deal of overlap of the study requirements of the different disciplines in the field of applied physics.

It is also worth noting that all programs require familiarity with the use of modern computers and general software applications, and that each discipline will make extensive use of specialized applications, which may be developed as proprietary software rather than being commercially available.

## SOCIAL CONTEXT AND FUTURE PROSPECTS

Applied physics has become such a fundamental underpinning of modern society that most people are unaware of its relevance. A case in point is the spin-off of technologies from the NASA space program into the realm of everyday life. Such common household devices as microwave ovens and cellular telephones, medical devices such as pacemakers and monitors, and the digital technology that permeates modern society were developed through the efforts of applied physics in exploring space. For most laypersons, the term "applied physics" conjures images of high-energy particle accelerators or interplanetary telescopes and probes, without the realization that the physical aspect of society is based entirely on principles of applied physics. As these cutting-edge fields of physical research continue, they will also continue to spawn new applications that are adapted continually into the overall fabric of modern society.

The areas of society in which applied physics has historically played a role, particularly those involving modern electronics, will continue to develop. Personal computers and embedded microcontrollers, for example, will continue to grow in power and application while decreasing in size, as research reveals new ways of constructing digital logic circuits through new materials and methods. The incorporation of the technology into consumer products, transportation, and other areas of the infrastructure of society is becoming more commonplace. As a result, the control of many human skills can be relegated to the autonomous functioning of the corresponding device. Consider, for example, the now-built-in ability of some automobiles to parallel park without human intervention.

The technologies that have been developed through applied physics permit the automation of many human actions, which, in turn, drives social change. Similarly, the ability for nearly instantaneous communication between persons in any part of the world, which has been provided through applied physics, has the potential to facilitate the peaceful and productive cooperation of large numbers of people toward resolving significant problems.

*Richard M. Renneboog, M.Sc.*

**FURTHER READING**

Askeland, Donald R. *The Science and Engineering of Materials.* 3d ed. London: Chapman & Hall, 1998. Includes chapters on the atomic nature of matter, the microstructure and mechanical properties of materials, the engineering of materials, the physical properties of materials, and the failure modes of materials.

*The Britannica Guide to the One Hundred Most Influential Scientists.* London: Constable & Robinson, 2008. Provides brief biographical and historical information about renowned people of science, from Thales of Miletus to Tim Berners-Lee. Includes chapters on Bacon, da Vinci, Kepler, Descartes, Watt, Galvani, Volta, Faraday, Kelvin, Maxwell, Tesla, Planck, Rutherford, Schrödinger, Fermi, Turing, Feynman, Hawking, and many other physical scientists.

Clegg, Brian. *Light Years: An Exploration of Mankind's Enduring Fascination with Light.* London: Judy Piatkus, 2001. Examines the history and relevance of the various concepts of philosophy and science in attempting to understand the physical nature of light.

Giancoli, Douglas C. *Physics for Scientists and Engineers with Modern Physics* Toronto, Ont.: Pearson Prentice Hall, 2008. Provides fundamental instruction in the basic physical laws, including Newtonian mechanics, thermodynamic properties, magnetism, wave mechanics, fluidics, electrical properties, optics, and quantum mechanics.

Kaku, Michio. *Visions: How Science Will Revolutionize the Twenty-first Century.* New York: Anchor Books, Doubleday, 1997. Examines the author's view of three great influences of the twenty-first century: computer science, biomolecular science, and quantum science.

Kean, Sam. *The Disappearing Spoon.* New York: Little, Brown, 2010. Provides a thoughtful and entertaining account of the persons and histories associated with each element on the periodic table, and how they have variously affected the course of society.

Sen, P. C. *Principles of Electric Machines and Power Electronics.* 2d ed. New York: John Wiley & Sons, 1997. Includes chapters on magnetic circuits, transformers, DC machines, asynchronous and synchronous machines, and special machines such as servomotors and stepper motors.

Valencia, Raymond P., ed. *Applied Physics in the Twenty-first Century.* New York: Nova Science, 2010. Includes chapters on nano metal oxide thin films, spectroscopic analysis, computational studies, magnetically modified biological materials, and plasma technology.

**WEB SITES**

*American Institute of Physics*
www.aip.org

**See also:** Aeronautics and Aviation; Applied Mathematics; Bionics and Biomedical Engineering; Biophysics; Civil Engineering; Communication; Computer Science; Digital Logic; Microscopy; Military Sciences and Combat Engineering; Radiology and Medical Imaging; Space Science; Transportation Engineering.

# ARCHAEOLOGY

## FIELDS OF STUDY

History; geography; geology; ecology; anthropology; sociology; economics; chemistry; mineralogy; sedimentology; geophysics; geographic information systems; evolutionary science; geomorphology; paleontology; ethnography; environmental studies; zooarchaeology (or archaeozoology); archaeobotany (or paleoethnobotany); paleobotany; palynology; paleoclimatology; archaeoastronomy; geoarchaeology; Egyptology; Phoeniciology; classical archaeology; Assyriology; historical archaeology; prehistoric archaeology; industrial archaeology; experimental archaeology; ethnoarchaeology; archaeometry; mathematics; statistics.

## SUMMARY

Archaeology is the field of applied science concerned with the techniques and practice of collecting, preserving, and analyzing the artifacts and physical remains left behind by human civilizations. The central purpose of archaeology is to study the lives and cultural practices of past societies, both ancient and modern. Archaeology has a deep intangible impact on human experience, as a means of understanding and maintaining the multifaceted cultural heritage produced by the human species. It also has many practical applications; for example, archaeological studies of rural agriculture can help farmers increase the yield of their crops, and urban archaeologists can advise municipal administrators on issues such as transit patterns and garbage disposal.

## KEY TERMS AND CONCEPTS

- **Absolute Dating:** Dating an artifact using a time scale (years B.C.E. and C.E.); also known as chronometric dating.
- **Arbitrary Level:** Predetermined depth to which all digging at any given level of an archaeological excavation will descend.
- **Artifact:** Physical object that was made or altered by human activity; one that is not naturally occurring.
- **Ecofact:** Artifact consisting of organic or environmental remains; often something used by humans (for example, food remains).

- **Feature:** Artifacts and ecofacts that together form an identifiable entity (such as a burial ground).
- **Judgmental Sampling:** Process of excavating and analyzing only artifacts from certain areas of an archaeological site that have been selected based on the judgment of an archaeologist.
- **Lithics:** Study of stone artifacts, such as tools or weapons.
- **Matrix:** Physical surroundings or materials in which an artifact is found, such as soil.
- **Probabilistic Sampling:** Process of excavating and analyzing only artifacts from certain areas of an archaeological site that have been selected, at least partially, by chance.
- **Radiocarbon Dating:** Method of absolute dating based on the radioactive decay of carbon in organic materials.
- **Relative Dating:** Process of determining how old an artifact or feature is in relation to some other object or event.
- **Stratigraphy:** Study of stratifications, or layers of deposits that have built up sequentially over time.
- **Typology:** Practice of classifying artifacts into groups with similar characteristics; the study of such groups through time.
- **Use-Wear Analysis:** Technique of determining how an artifact was used by examining it microscopically, looking for marks indicating wear or damage.

## DEFINITION AND BASIC PRINCIPLES

Archaeology is the study of past human cultures, both historic and prehistoric, through the systematic excavation, inspection, and interpretation of material remains such as tools, toys, clothing, bones, buildings, and other artifacts. In the United States, archaeology is considered a subfield of anthropology, which is the science concerned with the origin, evolution, behavior, beliefs, culture, and physical features of humankind.

The traditional image of archaeology held by many people (and reinforced by television programs and motion pictures featuring adventurous archaeologists such as Indiana Jones) is that it is a field largely concerned with examining ancient artifacts drawn from the extremely distant past. This is certainly true of many subfields of archaeology, such as Egyptology,

Assyriology (study of ancient Assyria), classical archaeology, and prehistoric archaeology. However, archaeological tools and techniques can be and frequently are applied to the analysis of human cultures across all segments of time, from the very beginning of the species to much later epochs. For example, industrial archaeologists study the development and use of industrial methods, most of which did not truly come into being on a large scale until the eighteenth century, while urban archaeologists study the patterns of life revealed by the material past of metropolises—such as New York or Paris—that still exist and that have a long history of human habitation.

Archaeology as practiced in the modern world is characterized by several key principles. First, the field is not simply descriptive but rather explanatory. Archaeologists are concerned not just with the question of what happened in the past but also how and why it happened. To that end, they attempt to interpret the artifacts and features they uncover for clues about the belief systems, behaviors, traditions, and social, political, and economic lives of the cultures at hand. For example, archaeological excavations of the artifacts buried in the tombs of ancient Egyptian pharaohs have revealed a great deal of specific knowledge about what Egyptians believed to be true about the afterlife. Second, it is multiscalar, meaning that the examination of a particular culture takes place simultaneously on many scales, each of which is intertwined with the others. For instance, an archaeologist may combine the small-scale analysis of individual pots, vases, and other clay artifacts—which give insight into the specific production processes used by the artisans who created them—with large-scale evidence about how those objects were handled on trade routes and marketplace procedures—which give insight into the overall historical trajectory of an entire civilization. By the same token, no matter which geographic region an archaeologist is studying, he or she is likely to examine collected data in the light of wider global trends.

### BACKGROUND AND HISTORY

People have always been fascinated by the cultures that came before them. Even in ancient Babylon, for example, relics such as ruined temples were objects of interest. During the European Renaissance, an early form of archaeology arose that was mainly focused on classical antiquities and the investigation

---

### Fascinating Facts About Archaeology

- Some archaeologists spend their days working in scuba-diving suits. The field of underwater archaeology focuses on physical remains of human culture found in the sea or other bodies of water, such as those created by shipwrecks or airplane crashes.

- Archaeologists make use of any scientific tool that can help them uncover information about artifacts—including sophisticated medical imaging. In 2009, computed tomography (CT) scans of 3,500-year-old mummies showed that many of them suffered from heart disease, something that was believed to be a modern affliction.

- Some archaeologists use incredibly tiny artifacts to construct huge ideas. Palynologists study preserved pollen and spores—which can last for thousands of years—to come up with ideas about what the environment was like in ancient times.

- Many elaborate archaeological hoaxes have been perpetrated over the course of the field's history. For more than two decades, for instance, the amateur Japanese archaeologist Shinichi Fujimura buried and then unearthed, to great fanfare, hundreds of objects that he claimed were genuine artifacts from Japanese prehistory.

- Food archaeologists may try to recreate the diet of a culture long past, using the physical evidence that remains from plant and animal meals, as well as written recipes that survive. In 2009, archaeologists managed to analyze and reproduce an ancient Chinese wine using chemical analyses of the wine residue found in pottery jars.

- It may not be the most flattering nickname, but the private-sector archaeologists who work for CRM firms in the United States bear it with pride. What do they call themselves? Shovelbums.

- Archaeologists do not always have to dig deep to unearth artifacts. Aerial archaeology involves conducting bird's-eye view reconnaissance missions using cameras mounted on kites, remote-controlled parachutes, hot-air balloons, model airplanes, and helicopters.

---

of large prehistoric sites and monuments, such as Stonehenge. The eighteenth century witnessed the first formal archaeological excavations, one of which was conducted by future president Thomas Jefferson on an Indian burial mound located on his Virginia estate. In the nineteenth century, the site of the ancient city Pompeii, destroyed by a sudden volcanic eruption, became the focus of intensive archaeological investigations. Innovative techniques, such as the use of plaster of Paris to create molds of bones and other artifacts, were developed by early practitioners such as Italian scholar Giuseppe Fiorelli.

During the nineteenth century, archaeology began to develop the formality of a scientific discipline. The simultaneous maturation of the field of geology, with its theory of the stratification of rocks, helped establish similar ideas in archaeology. In addition, archaeologists began to work with the principle of uniformitarianism for the first time; this was the idea that past and presents societies had more commonalities with each other than differences. Finally, the nineteenth century was the age in which naturalist Charles Darwin's theory of evolution, combined with the accumulating weight of physical evidence, led archaeologists to search for material evidence of human activity in the very distant past, far beyond the age of the Earth as defined by biblical scholars.

Archaeology underwent a transformation during the twentieth century, when many useful field techniques were developed that helped transform the field into a true science. Detailed records were kept at every excavation, including the location and description of each artifact. Sites were also divided into grids and searched systematically, and drawings and models were made of every dig. Advances were also made in absolute dating; particularly significant was the invention of radiocarbon dating by chemist Willard Libby in the United States. Partly as a result of these techniques, archaeology became far more quantitative, or driven by data and statistical analysis, than it had previously been. Hypotheses and theories were carefully tested before being accepted.

## How It Works

**Archaeological Surveys.** The first step in conducting any archaeological investigation is to identify and take stock of the location in which the study will be held. This is known as a survey. The choice of location can be determined in several ways. Documentary sources, such as old maps or other written materials, can sometimes be used to accurately discover the location of an archaeological site. For example, clues in the writings of the ancient Greek poet

Homer helped scientists find the ruined city of Troy (in northwest Turkey). In other cases, archaeologists attempt to survey sites on which new developments such as roads or buildings are planned to unearth important remains before they are destroyed. This is known as salvage archaeology. On occasion, developers have integrated remains found by salvage archaeologists into their construction plans, as in the case of the Aztec temple dug up in Mexico City and used as part of the subway station that was built on that site. Other archaeological sites of interest are discovered through ground reconnaissance, in which either a judgmental or probabilistic sampling method is used to search for physical evidence, or aerial reconnaissance, which can sometimes turn up traces of human activity that cannot be seen from the ground.

Once a site has been identified as being of archaeological interest and an excavation begins, a deeper survey is conducted of the selected area. The same kinds of sampling methods used in ground reconnaissance are available to archaeologists at this point. Judgmental sampling is based solely on an archaeologist's judgment about which locations within the site should be searched for evidence. For example, since humans tend to settle near sources of water, judgmental sampling might concentrate the excavation along a river bed that runs through the site. Probabilistic sampling, in contrast, is subject to chance and variation, and may help archaeologists turn up unexpected discoveries or make more accurate predictions about areas not sampled. Probabilistic sampling can take the form of simple random, stratified random, or systematic sampling. Simple random sampling is where any given location within a site has an equal chance of being sampled. Stratified random sampling is where the site is divided into naturally occurring regions, such as forest versus cultivated land, and the number of searches conducted in each region is based on its size in relation to the entire site. Systematic sampling is where a site is divided into equal parts and then sampled at consistent intervals (for example, a search is made every 100 meters).

**Types of Archaeological Evidence.** The evidence used by archaeologists to form a picture of an ancient or past society can be classified into four major categories: material, environmental, documentary, and oral evidence. Of the four, material evidence is perhaps the most fundamental. It can consist of either organic or inorganic remains that have been used by or constructed by people. Material evidence includes buildings, tools, pottery, toys, textiles, baskets, and food remains. Environmental evidence can take many forms, including soil samples, minerals, pollen, spores, animal bones, shells, and fossils.

Because the Earth has undergone many significant geological and climate-related changes over the course of its history, environmental evidence can help reveal what the world looked like during the time of a particular group's existence in an area. Environmental evidence can also offer clues about people's relationships with the landscape around them. For example, if the materials in a lower layer of a dig consisted mostly of wood, charcoal, and pollen, while those in a higher layer of a dig consisted mostly of cattle bones, one hypothesis might be that a region that had once been wooded had at some point been cleared by the inhabitants of the area and used for the rearing of animals.

Documentary evidence, or written records, can be extremely useful to archaeologists whenever it is available. Depending on the epoch, or span of history, from which these documents arise, they may consist of text inscribed on stone slabs, clay tablets, papyrus, or other types of materials. They may record laws, serve as proof of legal contracts (such as marriages or business agreements), list births and deaths in a city, or inventory commodities such as grain or shells. Finally, oral and ethnographic evidence—interviews and oral histories with modern-day inhabitants of a particular area—can help supplement the physical remains collected in an excavation. Certain practices such as the techniques of rural architecture or traditions of marriage, for instance, may have been preserved without much variation over time. In such cases, living members of a culture that has been around for many generations may be able to provide key information about the habits of their ancestors.

**Dating Archaeological Evidence.** Archaeologists have a plethora of techniques for placing objects in time. Relative dating methods identify artifacts as being older or younger than others, thus placing them into a rough inferred sequence. One of the most straightforward techniques of relative dating is to note the depth at which a particular object was found at a site; that is, in which strata, or deposit layer, it was contained. This method, known as stratigraphy, is an effective way to gauge the relative age of objects

because layers build up on top of other layers over time. Other means of relative dating include pollen analysis, ice-core sampling, and seriation. Seriation uses the association of artifacts with known dates of use, along with knowledge about how the frequency of use of those artifacts changed over time, as markers for other items found in close proximity.

Absolute dating methods are far more accurate than relative dating techniques (which can be confused by the effects of natural disasters or animal activity on the organization of strata). They also produce more specific results. The absolute dating technique that is used more than any other by archaeologists is radiocarbon dating, which is capable of assigning an accurate age to artifacts of biological origin (including bone, charcoal, leather, shell, hair, textiles, paper, and glues) that are 50,000 years old or younger. Radiocarbon dating relies on the fact that the element carbon has a particular isotope—carbon 14—which is radioactive. Radiocarbon dating is also known as carbon-14 dating. This radioactive carbon combines with oxygen in the atmosphere to form carbon dioxide, which is absorbed by plants as they photosynthesize. Because plants form the base of every food chain, all living things on Earth contain some amount of radiocarbon within their bodies. Because scientists know the half-life of radiocarbon (the time it takes for half the amount in a given sample to decay), they are able to measure the radiocarbon in any organic material and calculate how long it has been around. Besides radiocarbon dating, other absolute dating techniques include dendochronology, which relies on the annual growth rings present in the trunks of long-lived trees, and thermoluminescent dating, in which trace amounts of radioactive atoms in rock, soil, and clay can be heated to produce light, the intensity of which varies depending on the age of the object. This method is primarily used to date pottery and other human-created artifacts.

**Interpreting Evidence.** Three basic concepts dominate the approach an archaeologist takes when he or she approaches an artifact, a feature, or a site: context, classification, and chronology. Context, or where an object is found and what other items surround it, can reveal more of a complete story than any single artifact. For example, a decorated container found within a tomb could serve one purpose; the same container found in the kitchen of a dwelling could have been intended for an entirely different one. Classification (also known as typology) also helps place unearthed artifacts within a particular frame of reference. For example, determining that all the items found at a particular location belong to the same type—such as cooking utensils, hunting supplies, or bathing vessels—could indicate how a location was used or who used it. Finally, chronology is essential to understanding the relationship between the elements of any archaeological investigation. For example, unless they have been disturbed, objects located at lower levels, or strata, of a particular excavation site will be older than those located at higher levels. This can help archaeologists trace the development of a particular tool over time or to see and identify important changes in the cultural practices of a given society.

## APPLICATIONS AND PRODUCTS

**Cultural Resource Management.** The major practical application of archaeological tools and techniques can be found in the area of cultural resource management (CRM). In a sense, CRM is a form of institutionally legislated salvage archaeology. It involves the use of archaeological skills to identify, preserve, and maintain important features of historic and prehistoric culture to benefit the public interest. In most countries, including the United States and Canada, any planned development of any significance, including the building of new gas or oil pipelines, residences, highways, golf courses, or any other construction, requires the developer—whether private or public—to conduct a CRM survey to comply with legal requirements. The goal of CRM is to ensure that important pieces of a nation's past are conserved. In the United States, the major pieces of legislation mandating CRM include the National Historic Preservation Act, the National Environmental Policy Act, the Archaeological Resources Protection Act, and the Native American Graves Protection and Repatriation Act. CRM studies are generally performed by trained archaeologists serving as consultants and involve reconnaissance and sampling intended to scan the area for the presence of significant archaeological sites, features, or artifacts. If any are found, the developers are responsible for safely excavating and preserving them. If this is impossible or the site is identified as especially historically significant, the project may be terminated.

**Waste Management.** The field of garbage archaeology, or garbology, was pioneered by American urban

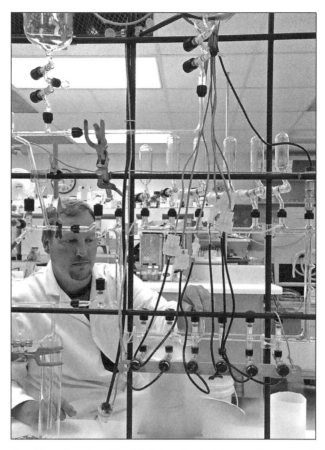

*Samples of carbon dioxide being loaded into a linear accelerator (linac) for radiocarbon dating.* (Enrico Sacchetti/ Photo Researchers, Inc.)

archaeologist William Rathje in the 1970's. By applying the tools of archaeology to the waste dumped in landfills, Rathje and other scientists have been able to provide many practical insights that are useful to waste management specialists and city, state, and federal administrators concerned with reducing waste and conserving energy. For instance, they showed that even supposedly biodegradable artifacts, such as paper, wood, and even food, are preserved for far longer when packed tightly together in a landfill than they otherwise would be. This is because the dense conditions reduce the amount of oxygen available for decomposition to take place. One of the implications of this discovery is that simply switching from nonbiodegradable to biodegradable materials in the production process—banning plastic disposable cups in favor of paper ones, for example—will not be enough to reduce the amount of effectively permanent waste that is generated by consumers and filling landfills to the point of overflowing.

**Agriculture.** In some parts of the world, environmental archaeologists have discovered evidence that areas of land that are marked by barren soil in fact used to be rich and fertile. For example, the Negev Desert in Israel, with its high temperatures and meager rainfall, used to be the site of an ancient urban society that cultivated crops such as wheat and grapes. Archaeologists also established that the climate in the area had not changed over the 2,000-year period in between. Using aerial reconnaissance techniques, scientists were able to show that the ancient farmers had employed a water-delivery system composed of terraces and cisterns to collect and redirect rainwater from the infrequent flash floods that occur in the area. This insight has proved to be of immense importance to modern-day agricultural scientists and farmers in Israel and has played a role in the country's successful efforts to "green" the Negev. Similarly, local farmers in rural Peru have begun using prehistoric field technologies unearthed by environmental archaeologists—primarily elevated fields that improve drainage and help protect plants from chilly nights—to dramatically increase their crop yields.

**Crime Investigation.** Forensic archaeologists apply the methodologies of archaeology to the investigation of crimes, working closely with coroners and police officers to collect and analyze physical evidence from the scene. They investigate fragments of bone, teeth, soil, fabric, jewelry, or other artifacts, while more subtle clues such as disturbed soil or markings from tools or weapons may also provide important leads. Just as in other forms of archaeology, the forensic archaeologist first establishes the boundaries of the site to be surveyed, divides it into grids, and performs a thorough excavation of the material and environmental evidence found in the area. Stratigraphy is an important element of forensic archaeology, particularly so when a body has been buried before being found; the deeper into the soil digging proceeds, the older the remains, personal artifacts, or other evidence that is found will be.

## IMPACT ON INDUSTRY

**Public Archaeological Research.** Governments across the world invest in archaeological studies designed to identify, excavate, and preserve the cultural and geological history of their nations. This kind of archaeological research is especially vital in countries with an extremely long history of human habitation, such as Israel, Egypt, China, and Greece. In the United States, federal entities such as the Forest Service, the National Park Service, the Bureau of Land Management, and the U.S. Army Corps of Engineers all engage in regular archaeological field studies. For example, in 2003, the National Park Service completed a survey of Yellowstone National Park that included fieldwork on how American Indians used land before the arrival of Europeans. In addition, each state has an office dedicated to historical preservation that employs archaeologists to perform similar research within state lines. In the United States, the National Council on the Humanities and the National Science Foundation are the two major government agencies that provide funding to support archaeological research. Important public research also takes place under the auspices of museums. Among the most active are the Smithsonian Institution, the Penn Museum, and the Field Museum of Chicago.

**Industry and Business.** Through the mechanism of cultural resource management (CRM), archaeological research has a major impact on both private and public sector development. Any municipality or state wishing to build new infrastructure, as well as any company wishing to engage in construction, whether building new homes or offices or installing telecommunications cables, must hire a consulting archaeologist or CRM firm to ensure that full compliance with legislation regarding the preservation of cultural resources. Unlike academic research, this type of archaeological research is not only robust but also growing. It is estimated to have an industry worth of about $1 billion per year and to employ a total of about 14,000 people—the majority of them trained archaeologists—within the United States alone. CRM research is especially active in the western and southwestern parts of the country, largely as a result of interest in developing these areas to exploit their natural oil and gas reserves.

**Major Corporations.** Hundreds of small to mid-sized CRM firms operate within the United States alone. Because knowledge of local geography, geology, and history are essential, in general the market tends to be fragmented, with small firms each controlling only a tiny portion of the total available work.

Some CRM companies, however, have managed to grow into nationwide firms serving multiple regions, including Statistical Research Inc. (SRI), and engineering and construction firms such as the Louis Berger Group and Parsons offer CRM services.

## CAREERS AND COURSE WORK

The typical archaeology career path begins with an undergraduate degree in anthropology, with an emphasis on course work covering the fields of biological, linguistic, and cultural anthropology. In addition, the aspiring archaeologist should be sure to take a set of courses in related sciences—such as geology, chemistry, ecology, environmental studies, evolutionary science, geophysics, anatomy, and paleontology. Mathematics and statistics are equally important subjects, since much archaeological analysis requires a keen understanding of these areas. In the humanities, history courses—preferably focused on a particular geographic area or epoch in time—are essential, as is an advanced knowledge of at least one foreign language. In addition, business management and technical writing skills will assist any archaeologist who chooses to go into the field of CRM. Finally, if at all possible, students should attempt to pursue archaeological internships, fieldwork placements, or other forms of practical work experience during their undergraduate careers. Such experiences not only are a means of gaining hands-on knowledge of the field techniques used in archaeology but also provide students with professional contacts.

A bachelor's degree will serve as sufficient minimum qualification for many professional archaeological positions in the private sector, particularly in consulting firms hired by developers to perform CRM surveys—in such settings, practical work experience is more important. However, additional graduate study at the master's or doctoral level is necessary to obtain a role as a crew supervisor and enter higher pay brackets. For archaeologists who wish to become faculty in academic institutions, a doctoral degree in archaeology and evidence of original research is required. Other organizations for which archaeologists work include museums, city and state governments, and the federal government. Typical roles for archaeologists include conducting field investigations, performing analyses of found artifacts, curating museum archaeology collections, teaching courses, and publishing research papers.

## SOCIAL CONTEXT AND FUTURE PROSPECTS

Archaeology is one of the most important and influential lenses through which scientists, historians, and other scholars view pieces of the past that otherwise would be lost. Its findings often reveal stunning insights into the nature of human civilization and the growth of human culture. For example, archaeologists piecing together skeletons found in Africa have been able to uncover the point at which the proto-human species Neanderthal died out and *Homo sapiens* (anatomically identical to modern humans) replaced it, while the ancient cave paintings discovered in Lascaux, an area in the southwest of France, demonstrate that human beings were creating art as long ago as 15,000 B.C.E. In other words, archaeology is a science dedicated to telling the story of humanity itself.

Although archaeology is at its core a science concerned with the investigation of material objects, the future of archaeology may be surprisingly metaphysical: A small but growing movement exists that is determined to treat the World Wide Web as a treasure trove of archaeological artifacts. Internet archaeology, as this nascent subfield of archaeology is called, seeks to archive and interpret Web sites, Web pages, and graphics that were created during the early years of the Internet and are no longer being updated by their owners.

*M. Lee, B.A., M.A*

## FURTHER READING

Fagan, Brian. *Before California: An Archaeologist Looks at Our Earliest Inhabitants.* Lanham, Md.: Rowman & Littlefield, 2004. Examines the indigenous cultures that existed in California before the arrival of Europeans. Organized chronologically, contains complete notes and references.

Gamble, Clive. *Archaeology: The Basics.* 2d ed. New York: Routledge, 2008. A succinct, accessible guide for the beginning archaeology student that includes figures, boxes, and a comprehensive index.

Kelly, Robert L., and David Hurst Thomas. *Archaeology.* 5th ed. Belmont, Calif.: Wadsworth, 2010. A comprehensive introductory textbook for the undergraduate level. Each chapter includes full-color photographs, a running glossary, a summary, additional reading, and themed sidebars.

Murray, Tim. *Milestones in Archaeology: A Chronological Encyclopedia.* Santa Barbara, Calif.: ABC-CLIO, 2007.

Covers five centuries of archaeological discovery through two hundred entries. Includes maps, photographs, drawings, and bibliographies.

Renfrew, Colin, and Paul G. Bahn, eds. *Archaeology: The Key Concepts.* New York: Routledge, 2005. Presents the central ideas in archaeology, including concepts of time, experimental archaeology, and the archaeology of gender.

**WEB SITES**

*Archaeological Institute of America*
http://www.archaeological.org

*National Parks Service*
Federal Archaeology Program
http://www.nps.gov/archeology/sites/fedarch.htm

*Smithsonian Tropical Research Institute*
Archaeology and Anthropology
http://www.stri.org/english/research/programs/archeology_anthropology/index.php

**See also:** Anthropology; Applied Mathematics; Civil Engineering; Forensic Science; Maps and Mapping; Mineralogy; Paleontology; Plane Surveying; Probability and Statistics.

# ARCHITECTURE AND ARCHITECTURAL ENGINEERING

## FIELDS OF STUDY

Physics; geometry; calculus; chemistry; electric circuitry; lighting system design; urban planning and development; computer-assisted design; business and environmental law; accounting.

## SUMMARY

Architecture and architectural engineering are fields involved in the design and construction of buildings. Working in close association with architects, architectural engineers translate people's needs and desires into physical space by applying a wide range of engineering and other technologies to provide building systems that are functional, safe, economical, environmentally healthy, and in harmony with the architect's aesthetic intent.

## KEY TERMS AND CONCEPTS

- **Air Handler:** Device that blows air across a split-unit central air-conditioning evaporator coil into the air ducts of a building.
- **Building Envelope:** Technical term that refers to the components of a building, including the foundation, walls, roof, windows, and doors, and separating indoor and outdoor environments.
- **Building Regulations:** Building codes and permits developed in keeping with the best practices of the construction industry.
- **Computer-Aided Geometric Design (CAGD):** Use of computers to construct and represent free-form curves, surfaces, and volumes in the design process.
- **Concurrent Engineering:** Project design conducted by teams who work together rather than making serial contributions.
- **Design Charrette:** Intense multidisciplinary workshop lasting several days and held early in the planning process to engage the ideas and concerns of those responsible for the key components of a building project, including the architects, designers, engineers, contractors, managers, and consultants.
- **Fluid Mechanics:** Study of fluids and their properties.

- **Green Building Movement:** Organized effort toward sustainability in building design that emphasizes the life cycle of a building and its relationship to the natural landscape, energy efficiencies, indoor environmental quality, the reduction of toxins, and the use of recyclable, earth-friendly materials.
- **Heating, Ventilation, And Air-Conditioning Systems (HVAC):** Technologies providing heat, ventilation, and cooled air, essential components of built environments.
- **Life-Cycle Assessment:** Assessment of a product's life from raw materials to useful productivity to disposal.
- **Schematic Design:** Materials detailing the design structures and components of a project. It includes drawings, specifications, and cost projections.
- **Whole Building Design:** Design and building concept engaging architecture and engineering professionals in a team process to plan and implement the design and construction of highly integrated building and construction projects.

### DEFINITION AND BASIC PRINCIPLES

Architecture and architectural engineering are complex and highly skilled fields. Architects develop the graphic design of buildings or dwellings and are often directly involved in their construction. Architectural engineers are certified professionals specializing in the application of engineering principles and systems technology to the design and function of a building. The term "architecture" sometimes includes the engineering technologies of building design. Architecture applies aesthetics, measurement, and design to the cooperative organization of human life. An essential dynamic of this process is the creation of technology. The engineering profession designs a magnitude of objects, structures, and environments to meet human needs and purposes. Building engineers specialize in the technologies that benefit human health and well-being within a built environment. These technologies are the direct result of exponential advances in the manipulation of chemical, hydraulic, thermal, electrical, acoustical, computational, and mechanical systems. Solar, wind, and nuclear power play an important role in the creation of sustainable buildings and environments, and the innovative use of new and recycled construction materials is an essential part of efficient planning and design.

## BACKGROUND AND HISTORY

As early as about 10,000 B.C.E., small human settlements began to form along the fertile banks of the Nile River in Egypt, the Tigris and the Euphrates rivers in the Middle East, the Indus River in India, and the Yellow and Yangtze rivers in China. These settlements grew, becoming prosperous city-states that supported the erection of large buildings. Temples, pyramids, and public forums served as religious and political symbols of increasing private and public wealth and power. Elaborate structures of a grand scale required specialized instruments such as levers, rollers, inclined planes, saws, chisels, drills, cranes, and right-angle tools for surveying land surfaces and for masonry construction.

In ancient Egypt, individuals noted for their ability to provide shelter, public works, water supplies, and transportation infrastructure for their communities were known as "master builders." The ancient Greeks called these individuals *architektons*. In Roman armies, military engineers termed "architects" designed siege engines and artillery and built roads, bridges, baths, aqueducts, and military fortifications.

**Developing Technologies.** During the Middle Ages, craftsmen and artisans contributed considerable expertise and know-how to building construction and helped refine the tools of the trade. With the advent of the printing press, classical Greek, Roman, and Arabic treatises covering topics in architecture, mathematics, and the sciences were disseminated, and their content found its way into practical uses. Advances in graphic representation (such as the use of perspective) and the geometric manipulation of the Cartesian coordinate system enhanced the conceptual art of two- and three-dimensional spatial translations. Technologies flourished, creating the impetus that came to a head during the Industrial Revolution.

**The Modern Era.** Population pressures on urban centers during the eighteenth and nineteenth centuries accelerated the forward momentum of numerous technologies. The transition to fossil fuels, the development of steam power, and the distribution of electric power, as well as improvements in the production of iron and steel, transformed societies.

These technologies had a profound impact on the principles of architectural design and building systems.

During the nineteenth century, the availability of energy and the technologies of mass production accelerated the standardization and distribution of construction materials worldwide. Heating flues, air ducts, elevator shafts, internal plumbing, and water and sewage conduits were developed in Great Britain and installed in hospitals, mills, factories, and public buildings. These innovative technologies were quickly adopted for private and commercial use in the United States, the British Commonwealth, and the continental nations.

New technologies for lighting, internal heating, and ventilation were first greeted with misgiving, but over time, building environments took on a life of their own in the minds of city health experts, architects, and engineers. A robust educational system developed to enable architects to master the increasing complexity and diversity of building design and construction technologies. Professional trade associations were formed to administer programs of certification and to determine public standards of health and safety that were codified and set within applicable statutory frameworks.

In the second half of the twentieth century, computers have enabled architects and architectural engineers to function with a greater degree of precision. Computer-aided drafting and design (CAD) software is an essential tool for architects. Computers also can be used to create simulations and animations, as well as detailed presentation graphics, tables, charts, and models, in a relatively short period of time.

## HOW IT WORKS

Architects and architectural engineers play prominent roles in the planning, design, construction, maintenance, and renovation of buildings. All processes of a project are initiated and directed to reflect the needs and desires of the project owner. These directives are modified by detailed sets of construction documents and best practices. Building codes are enforced to safeguard the health and safety of the owner, users, and residents; environmental standards are followed to protect property and to ensure the efficient use of natural resources; and attractive design features are chosen to enhance the community. The choice of construction materials, the visual and spatial relationship of the building to the environment and community, traffic safety, fire prevention, landscaping, and mechanical, electrical, plumbing, ventilation, lighting, acoustical, communication, and security systems are all essential features that determine the quality, functionality, and character of a building.

**Articulating the Owner's Intent.** The contract process begins with the owner's interest in a residential, public, or commercial building. The selection of an architect for the project will be based on the type of construction (light or heavy) and the particular purpose the building is intended to serve. The plans for simple one- and two-family dwellings frequently follow well-established homebuilder association guidelines and designs that can be modified to meet the future owner's budget and lifestyle. Larger, more complex designs (such as a church or a school) will be managed by a team of architects and engineers carefully chosen for their expertise in key aspects of the project. The selections of the architectural engineering team and the project manager are considered the most important steps in the design process.

The careful and deliberate coordination of effort and timely verbal and written communication are essential at all stages of the project. At the outset, the owner and those directly involved in the corporate administration of the project must work closely together to establish the values and priorities that will guide the design. Detailed feedback loops of self-assessment are conducted throughout the different phases of the project to ensure multiple levels of engagement and evaluation. In the introductory phase, questions are raised to clarify the owner's intentions and to begin addressing key issues such as the scope of the project, the projected use and life cycle of the building, the level of commitment to environmental quality issues, and health and energy requirements. Once these issues have been formulated, a site evaluation is conducted. This process further refines the relationship of the project to the natural surroundings and includes a thorough review of the environmental integrity of the land; local codes governing transportation, water supply, sewage, and electrical infrastructure; health and safety issues; and the relationship of the surroundings to the projected indoor environment.

**Project Design.** The architect, working with other professionals, uses the information obtained in the introductory phase to create a design that fits the owner's needs and desires, complies with all building codes, and suits the site.

In situations where project turnaround is rapid, a design charrette may be scheduled. This is an intense, rapid-fire meeting of key professionals to quickly develop a schematic design that meets the primary objectives of the project. Participants include administrative representatives from all the internal offices involved in the project. In addition, utilities managers, key community and industrial partners, technology experts, financial institutions, and other stakeholders in the project are invited to set goals, to discuss problems, and to resolve differences. The process is a demanding one, challenging participants to think beyond the boundaries of their professional biases to understand and bring to focus a building design that reflects the highest standards of their respective trades.

**Project Documentation.** Project documentation is a highly technical orchestration of graphics, symbols, written correspondence, schematic sequencing and scheduling, and audiovisual representation. Construction drawings are a precise, nonverbal map that translates physical spatial relationships into intelligible two- and three-dimensional projections and drawings. The sequence of presentation and the lexicographical notations used to represent and to communicate the dimensions, systems, materials, and objects within a constructed space are highly refined, requiring years of careful study and practice. Project documentation also includes all the legal and civic certifications, permits, and reviews required before a certificate of occupancy is finally issued.

The documents of the project must be recorded in a narrative that can be understood by the owner and the different teams of workers involved. Documentation is particularly important for those project teams working toward LEED (Leadership in Energy and Environmental Design) certification. The LEED Green Building Rating System is an internationally recognized, voluntary certification program that assists commercial and private property owners to build and maintain buildings according to the best practices of green building design. It is administered by the U.S. Green Building Council, a nonprofit organization whose members adhere to industry standards set to maximize cost savings, energy efficiencies, and the health of the indoor environment.

**Systems Analyses and Energy Efficiencies.** Whole-building performance is an essential concept for achieving energy efficiencies and sustainable mechanical systems. Critical decisions regarding the choice of materials, the orientation of the building and its effect on lighting, the dimensions of the space in question, and the requirements for heating, cooling, mechanical, electrical, and plumbing systems demand a high level of technical analysis and integration among a number of building professionals to create a viable base design amenable for human use and habitation. Other factors incorporated into quality building designs include accessibility for the disabled, the mitigation of environmental impacts, and legal compliance with building codes and requirements.

As a result of the tremendous advantages of computational technologies, systems engineers have an expanding pool of resources available to assist in the optimization of energy efficiencies. Simulation software packages such as EnergyPlus, DOE-2, and ENERGY-10 make it possible to evaluate the potential for using renewable energy sources, the effects of daylight, thermal comfort, ventilation rates, water conservation, potential emissions and contaminant containment strategies, and scenarios for recycling building materials. Landscape factors that affect a building's energy use include irrigation systems, the use of chemical insecticides and fertilizers, erosion control factors, the conservation of native vegetation, and adequate shade and wind protection. Engineers use simulations to evaluate a series of alternatives to provide mechanical, solar, electrical, hydraulic, and heating, ventilation, and air-conditioning systems that optimize critical energy features of the design.

**Statement of Work and Contract Bid.** The final construction documents include all the drawings, specifications, bidding information, contract forms, and cost features of the project. Certifications of compliance with all regulatory standards and applicable codes are included in the documentation and are verified and signed by the project architects and engineers. Once a bid is secured and a final contract is signed, the design team maintains a close relationship with the contractors and subcontractors to monitor the details of construction and systems installation. The final commissioning process serves to verify the realization of the owner's intent and to thoroughly test the systems installed in the building. The integrity of the building envelope, the function of all mechanical systems, and the stability of energy efficiency targets are validated by a third party.

## APPLICATIONS AND PRODUCTS

The primary applications of architecture and architectural engineering are residential, commercial, and public structures. The Bauhaus school, founded in 1919 in Weimar, Germany, expanded the traditional view of architecture by envisioning the construction of a building as a synthesis of art, technology, and craft. Walter Gropius, the founder of the school, wanted to create a new architectural style that reflected the fast-paced, technologically advanced modern world and was more functional and less ornate. Another architect associated with the school, Ludwig Mies van der Rohe developed an architectural style that is noted for its clarity and simplicity. The S. R. Crown Hall, which houses the College of Architecture at the Illinois Institute of Technology in Chicago, is considered to be the finest example of his work.

Modern architecture, characterized by simplicity of form and ornamentation that arises from the building's structure and theme, became the dominant style in the mid-1900's and has continued into the twenty-first century, despite the rise of postmodern architecture in the 1970's. Postmodernism incorporated elements such as columns strictly for aesthetic reasons. American architect Robert Venturi argued that ornamentation provided interest and variation. Venturi's architectural style is represented by the Guild House, housing for the elderly in Philadelphia, that originally featured a television antenna as a decorative element.

Architect Christopher Alexander has combined technology with architecture, incorporating inventions in concrete and shell design into aesthetically pleasing works, including the San Jose Shelter for the Homeless in California and the Athens Opera House in Greece. He argues for an organic approach to architecture, with people participating in the designs of the buildings that they will use and of the environments in which they will live.

**New Paths.** Alexander's work concerns itself with the environment in which buildings are constructed, which is a major part of architecture in the twenty-first century. Innovations in architecture and architectural engineering involve creating sustainable buildings or communities or creating buildings in unusual environments.

The Eden Project houses a series of eight large geodesic domes constructed in 2001 at St. Austell in Cornwall. These domes, or biomes, provide self-sustaining environments housing thousands of plants gathered from around the world. The tubular steel and thermoplastic structures use active and passive sources of heat, and innovative ventilation and water systems collect rainwater and groundwater for recirculation in the building envelope. The project's architects, Nicholas Grimshaw and Partners, have developed a system that helps other building professionals understand natural impacts on human environments.

Outside the city limits of Shanghai, Chinese architects, engineers, and planners are working with the London-based firm Arup to design and build the world's first eco-city on Chongming Island, the third largest island in China. Located on the edges of the Yangtze River, the new city, Dongtan, is unusual for the wetlands and the bird sanctuary on its periphery. The first phase of the project will provide a living community for 5,000 inhabitants on a plot that measures 1 square kilometer. By 2050, 500,000 people are to work and live in a city 30 square kilometers in size.

In the Netherlands, innovative designers are experimenting with amphibious housing, or homes that rise and fall with water levels. Along the banks of the Meuse River in Maasbommet, the internationally recognized construction company Dura Vermeer opened a small community of amphibious and floating homes in 2006. Designed by architect Ger Kengen, the amphibious models rest on a hollow foundation filled with foam and anchored on two mooring poles located at the front and back of the structure. The poles allow the house to float on rising water to a height of 18 feet. Floating houses remain on water year round.

Similar structures were designed in New Orleans, Louisiana, following Hurricane Katrina. In 2010, the Special No. 9 House, designed by John C. Williams Architects for the Brad Pitt Make It Right Foundation, won the American Institute of Architects Committee on the Environment (AIA/COTE) Top Ten Green Projects Award. The FLOAT house, also in New Orleans, is another Make It Right Foundation project developed by Morphosis Architects under the direction of architect and University of California, Los Angeles, professor Thom Mayne.

**Habitats in Space.** Perhaps the most spectacular applications of the methods of architectural engineering involve the manned space flight program. Many of the mechanical systems that make homes and public places so livable are essential for creating similar environments in space. The International Space Station of the National Aeronautics and Space Administration (NASA) is a monumental project involving the efforts of fifteen nations to develop sustainable human habitats in space. Orbiting nearly 250 miles above the Earth at a speed of 22,000 miles per hour, the space station uses eight massive solar and radiator panels to control heat and provide energy for the ship's modular systems. Heating and cooling systems are essential for the maintenance of the ship, the internal environment, and the space suits worn during maneuvers outside the spaceship. Construction materials are in continuous development to find products that can withstand severe cold, radiation, and heat during reentry into Earth's atmosphere. In 2007, NASA developed a lightweight ceramic ablater material able to withstand temperatures of 5,000 degrees Fahrenheit Other insulation materials are used for electric wires, paints, and protective cladding for the ship. Many of the thermal fabrics designed for space have been adapted by athletes and health providers worldwide.

Inside the ship, mechanical systems generate heat and other emissions. Air quality is regularly monitored for carbon dioxide and oxygen levels. Waste materials are carefully recycled or packaged for shuttle return to Earth for disposal. Transportation routes, cycles of delivery, points of entry and egress, fire safety, lighting and electrical systems, oxygen generators, efficient waste disposal systems, water distiller and filtration systems for the storage, treatment, and recycling of water—all the technologies that add so much to the quality of human life have been engineered to meet the needs of space travel. In turn, these novel space systems have enormous applicability in the design of products for use in homes and places of work on Earth.

### IMPACT ON INDUSTRY

Architecture and architectural engineering are inextricably linked to the trades and industries that support the manufacture and distribution of building materials. The choices of building materials and the systems employed to provide water, energy, lighting, and other elements vary according to architectural trends, regulations, and owners' preferences. The energy crisis of the 1970's deepened public concern for the conservation of the Earth's limited resources. Rising standards of health and consumer safety, stringent environmental regulations, booming urban and suburban populations, and an increased awareness of the potentially disastrous effects of climate change on built environments and communities had a cumulative effect on already struggling manufacturing sectors during the second half of the twentieth century. This intensified interest in sustainable products, systems, and architectural design.

Since the first global modeling studies of climate conditions were created for the 1972 United Nations Conference on the Human Environment, urban land managers, building architects, and engineers worldwide have responded by studying ecologically responsible building designs and renewable building technologies. The interest in sustainable architecture has resulted in the emergence of architectural firms specializing in green construction or in the preservation of historic structures, finding opportunities to adapt these buildings to modern uses. Building and construction supply companies also have developed ways to mitigate environmental concerns. For example, Swisstrax Corporation created a process that transforms rubber particles from old tires into modular flooring tiles.

### CAREERS AND COURSE WORK

A career in architecture requires a five-year bachelor's degree in architecture or a four-year bachelor's degree plus a two-year master's degree in architecture. Those with undergraduate degrees in other areas may attend a three-year master's program in architecture. Usually students gain experience through an internship in an architectural firm. Architects with a degree from an accredited school and some practical experience must pass the Architect Registration Examination to obtain a license to practice.

The successful completion of a general program of study in architectural engineering requires demonstrated competencies in the mechanics of heating, ventilation, air-conditioning, plumbing, fire protection, electrical, lighting, transportation, and structural systems. Students may pursue a five-year program that terminates with a bachelor's degree in architectural engineering and generally leads to certification or a four-year program, followed by a master's degree. Students may also obtain a degree in architectural engineering technology, which specializes in the technology of building design.

Architectural engineering graduates must pass a series of exams administered by the National Council of Examiners for Engineering and Surveying to obtain a license to practice.

### SOCIAL CONTEXT AND FUTURE PROSPECTS

From the earliest records of human history, architects and engineers have advanced human civilization. The practical and aesthetic dimensions of architecture and engineering technologies have had profound social, political, economic, and religious

---

## Fascinating Facts About Architecture

- In 1972, shortly after the sixty-floor John Hancock Tower was built in Boston, its 4×11-foot windows began popping out and crashing to the sidewalk below. The 10,334 windows were replaced by 0.5-inch tempered glass, but the problem persisted.

- When 108-story, 1,451-foot-tall Sears Tower (renamed the Willis Tower in 2009) was built in 1974, it was the tallest building in the world. It held this position until 1998, when the 1,482.6-foot Petronas Twin Towers were built in Kuala Lumpur, Malaysia.

- Canadian-American architect Frank Gehry designed the Guggenheim Museum in Bilbao, Spain, completed in 1997. The museum features a curvaceous, free-form sculptural style and consists of a steel frame with titanium sheathing.

- In 2009, Custom-Bilt Metals developed a roofing product that combines traditional standing seam metal roofing with thin film solar technology. The lightweight system can be installed by a roofing contractor and an electrical contractor rather than a solar expert.

- Ancient Romans used underground heaters (hypocausts) to distribute warmth from fires through the double-layered floors and walls of public baths. This technology is considered one of the earliest heating, ventilation, and air-conditioning systems.

- The Pharos lighthouse of Alexandria, one of the seven wonders of the ancient world, was built in the third century B.C.E. by Sostratos of Cnidus. It stood about 400 feet tall, the height of a modern 40-story building, and its light was said to be visible from 100 miles away. It was destroyed in an earthquake in 1303.

impacts. The practices of architecture and engineering are just as much a product of cooperative human evolution as they are catalysts of change. Indeed, no thorough study of these fields is complete without a thoughtful and rigorous understanding of the environmental and cultural forces that stimulated their advances. Topographies, geographies, climates, political economies, known technologies, the commodities of exchange, and the natural resources available for human use stimulated human imagination and invention in novel and diverse ways in particular times and places. These innovations are reflected in the range of human responses to the need for food, shelter, safety, hygiene, and social interaction.

In modern society, these needs are modified by larger global concerns including the efficient use of energy and natural resources, pollution containment, and the need for sustainable life systems. Green architecture is likely to become an increasingly prominent part of the field. Architects will explore sustainable construction that looks at not only the building but also the environment and community, and they will pursue construction in unusual environments, such as housing on the water or in space. They will also look at ways to reuse existing buildings and structures, particularly those worthy of preservation. In choosing materials and systems, they will be investigating ways to recycle building materials and to incorporate systems that reduce pollution and minimize energy usage.

*Victoria M. Breting-García, M.A.*

## FURTHER READING

Allison, Eric, and Lauren Peters. *Historic Preservation and the Livable City*. Hoboken, N.J.: John Wiley & Sons, 2011. Examines ways to preserve historical buildings and create livable city centers.

Farin, Gerald, Joseph Josef Hoschek, and Myung-Soo Kim. *Handbook of Computer-Aided Geometric Design*. Boston: Elsevier. 2002. A thorough introduction to computer-aided design and architectural drawing.

Fisanick, Christina, ed. *Eco-architecture*. Farmington Hills, Mich.: Greenhaven Press, 2008. A collection of essays written by advocates and opponents of green architecture.

Garrison, Ervan G. *A History of Engineering and Technology: Artful Methods*. 2d ed. Boca Raton, Fla.: CRC Press, 1999. An engaging historical survey documenting scientific, technological, architectural, and mechanical innovations that contributed to the fields of engineering and architectural design.

Goldberger, Paul. *Why Architecture Matters*. New Haven, Conn.: Yale University Press, 2009. Examines the psychological effects that buildings have on people.

Huth, Mark W. *Understanding Construction Drawings*. 5th ed. Clifton Park, N.Y.: Delmar Learning, 2010. Examines the construction process. Includes exercises for the design and interpretation of construction drawings for residential and commercial buildings.

Pohl, Jens G. *Building Science: Concepts and Applications*. Oxford, England: Wiley-Blackwell, 2011. Discusses environmental engineering and sustainable architecture.

## WEB SITES

*American Institute of Architects*
http://www.aia.org

*American Institute of Building Design*
http://www.aibd.org

*National Institute of Building Sciences*
http://www.nibs.org/index.php/nibshome.htm

*Royal Institute of British Architects*
http://www.architects.com

*Society of American Registered Architects*
http://www.sara-national.org

*U.S. Department of Energy*
Energy Efficiency and Renewable Energy, Buildings
http://www.eere.energy.gov/topics/buildings.html

**See also:** Civil Engineering; Environmental Engineering; Landscape Architecture and Engineering.

# AREOLOGY

## FIELDS OF STUDY

Atmospheric chemistry; astronomy; biology; climatology; engineering; cosmochemistry; global physiography; astrobiology; cartography; biogeophysics; chemistry; robotics; computer science; geochemistry; petrology; geodesy; mineralogy; geomorphology; geophysics; glaciology; hydrology; meteorology; planetary engineering; space exploration; space physics; soil science; volcanology.

## SUMMARY

Areology, from the words *areo* (Ares, the Greek god of war) and *logy* (theory), is the interdisciplinary study of Mars. Most of the earth science disciplines can be applied to areology. As an interdisciplinary endeavor, areology also includes the study of the technologies for Mars exploration, both by robotic and manned craft, and the history of human speculation concerning the prospects for life on Mars, including the scientific principles, expectations, and designs for human colonization, and the engineering of the Martian planetary surface to support human life.

## KEY TERMS AND CONCEPTS

- **Absolute Age:** Age of a geological unit measured in years.
- **Bombardment:** Repeated collision of a planet with asteroids, usually over geologic time scales.
- **Chaotic Terrain:** Low region within heavily cratered uplands that appears to consist of irregular, blocky, fractured landscape.
- **Chasma:** Canyonlike feature on Mars, from the Latin for "large canyon or gorge"; the plural is *chasmata.*
- **Crustal Dichotomy:** Pronounced hemispheric contrast in physical characteristics of a planet's crust.
- **Ejecta:** Material blasted loose during the formation of an impact crater and deposited around that crater.
- **Flyby:** Mission procedure in which a spacecraft on its way to another destination examines a planet as it flies past that planet.
- **Gravity Map:** A map that shows variations in gravitational attraction across a planetary surface that results from variations found in the internal density of the planet.
- **Mascon:** Acronym for "mass concentration," which describes a zone of anomalously high density within Mars.
- **Mons:** Term used in names of mountainous features on Mars, from the Latin for "mountain"; the plural is *montes.*
- **Planitia:** Term used to indicate Martian regions composed of plains, from the Latin for "plains"; the plural is *planitiae.*
- **Province:** Region of similar terrain or a grouping of geological units with similar or related origins.
- **Relative Age:** Age of a feature or geological unit in relation to other features or geological units.
- **Rover:** Self-propelled, robotically operated vehicle used for exploring the surface of a body distant in space.
- **Terraforming:** Transformation of an alien landscape into one more suitable for human beings.
- **Vallis:** Used in naming valleylike features on the surface of Mars, from the Latin for "valley"; the plural is *valles.*

### DEFINITION AND BASIC PRINCIPLES

Areology is sometimes narrowly defined as the study of the geology of Mars, but it more properly involves not only most of the other earth sciences (from meteorology to hydrology to mineralogy) but also space physics, cosmochemistry, and astrobiology. Given that it deals with largely speculative prospects for life on Mars—both indigenous and imported, in the past, present, or future—areology must also take into account both the history of science and the literature of science fiction.

Although the term "areology" was in fact popularized by science fiction author Kim Stanley Robinson in his Mars trilogy (*Red Mars, Green Mars, Blue Mars*), the debate in the scientific community has largely swung between the poles of "wet Mars" (Mars once had water) and "white Mars" (Mars never had water). American astronomer Percival Lowell, who claimed to see through his telescope visions of supposedly water-filled canals on Mars, established one pole of the debate: Mars was a dynamic planet warm and wet enough to support life at the present time.

From its zenith in Lowell's work of the 1890's, this vision of Mars declined to its nadir after the Mariner 4 flyby in 1965. Mariner 4 showed a cratered, dusty ball clad in only the most diaphanous of atmospheres—one whose white polar regions were declared to be most likely covered in dry ice (carbon dioxide rather than water). As the data from the Viking landers of the 1970's proved inconclusive and controversial, the vision of dry, white Mars dominated discussion of the planet for decades.

After the failures of several probes, the successes of a growing armada of orbiters, landers, and rovers began to suggest in the 1990's and 2000's that, cold as it might be, Mars was not as dry and white as many in the planetology community had long contended. The notion that water ice was an important component on the Martian surface made a comeback, along with physical and chemical evidence of a potentially watery past.

The successes and failures of these unmanned spacecraft, along with the discoveries made possible by their successes, have set the parameters for the continuing discussion of the efficacy, expense, and likelihood of manned missions to Mars and eventual human colonization of the planet.

## BACKGROUND AND HISTORY

Scientific interest in Mars goes back to the seventeenth century and the work of Galileo Galilei, Johannes Kepler, and Giovanni Domenico Cassini—the last of whom, in 1666, observed the Martian polar caps and calculated the length of the Martian day. The apparent Earth-like nature of Mars led French author Bernard le Bouvier de Fontenelle in 1688 and British astronomer William Herschel in 1784 to speculate on the nature of life on Mars.

Despite this, in the scientific community before the end of the nineteenth century Mars was generally not seen as the best candidate for a second life-supporting world in the solar system. Venus—significantly closer to Earth in terms of size, mass, gravity, distance from the Sun, and actual travel time—at first seemed the more likely choice, and was still argued to be such until the advent of radar and radio telescopy, which pierced the thick Venusian atmosphere. Probes then confirmed the planet's merciless heat.

In literary history, too, the case was similar. Lucian of Samosata wrote of a trip to the Moon in his *True History* as early as the second century, and in the

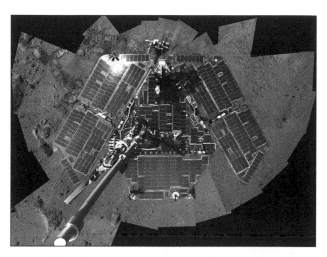

*Spirit Mars Exploration Rover on Mars.* (NASA/JPL-Caltech/Cornell University/Photo Researchers, Inc.)

eighteenth century both Jonathan Swift (in *Gulliver's Travels*) and Voltaire (in *Micromégas*) hypothesized the existence of two as-yet-undiscovered Martian moons. It was not until 1877, after Italian astronomer Giovanni Schiaparelli claimed to see on Mars a network of straight lines he called *canali* (canals), that writers began to examine Mars as the solar system's other main abode of life. This shift began with Percy Greg's *Across the Zodiac* in 1880 and continued most prominently with H. G. Wells's *War of the Worlds* in 1898.

Since Wells, the scientific understanding of Mars has been reflected in—and shaped by—the writings of Aleksandr Bogdanov, Edgar Rice Burroughs, J. H. Rosny, Stanley G. Weinbaum, Ray Bradbury, Leigh Brackett, Robert Heinlein, Isaac Asimov, Philip K. Dick, Frederik Pohl, Kim Stanley Robinson, and many more. Nowhere is the relationship between scientific speculation and speculative fiction clearer than the future Mars projects and programs put forward by space scientists from Wernher von Braun to Robert Zubrin.

## HOW IT WORKS

**Telescopy.** Although areology is a relatively new term, the roots of a general discipline of Mars studies stretch back nearly four centuries. For the first three and a half centuries, however, these studies were exclusively telescopic. Mars was an object viewed through an eyepiece from Earth. The power of telescopes and the levels of resolution they offered grew steadily over time, and telescopic studies remain very important in areological research, but recognition of the inherent limitations of such studies led the push to move scientific instrumentation closer to Mars via flyby, then linger in orbit to gather more detailed data. Eventually this led to landing scientific payloads on the planet's surface, then to making those payloads capable of self-propulsion across that surface.

**Flyby.** Mars 1, also known as Sputnik 23, was launched on November 1, 1962, and was intended to fly past Mars at a distance of about 11,000 kilometers or 7,000 miles. It was the first Soviet Mars probe and carried a package of scientific instrumentation that included television photographic equipment, a magnetometer probe, a spectral reflectometer, a spectrograph, a micrometeoroid impact instrument, and radiation sensors. Data from this instrumentation package were to be broadcast back to Earth via radio and television transmitters. Although Mars 1 lost contact with Earth before accomplishing its flyby, the configuration of its scientific instrumentation package (for collecting data) and transmission capabilities (for returning that collected data to Earth) became the standard for all Mars flyby missions.

The American craft Mariner 4, launched on November 28, 1964, completed the first successful flyby of Mars. Mariner 4's television pictures of the Martian surface were the first images of another planet sent back from deep space and changed the way the scientific community viewed the possibility of life on Mars. Mariner 6 and Mariner 7, in 1969, were similarly successful flyby missions, making closer approaches and providing more photographic and other data to that already compiled by Mariner 4.

**Orbiter.** In 1971, Mariner 9 was launched and, once inserted into orbit around Mars, became the first spacecraft to orbit another planet. It was followed soon after by two successful Russian orbiters, Mars 2 and Mars 3.

In orbiting the planet, Mariner 9 photographed 100 percent of the Martian surface and was able to wait out a prolonged dust storm that obscured much of the planet's surface—something a flyby mission could not have done. Mariner 9's successful data collection laid the groundwork not only for the later Viking orbiter/lander missions but also for successful later-generation orbiters with more advanced instrument packages, including the Mars Global Surveyor in 1996, and Mars Odyssey, Mars Express, and Mars

Reconnaissance Orbiter during the first decade of the twenty-first century.

**Lander.** The Soviet Mars 3, whose orbiter was successful, also had a partially successful lander component in that its descent module, which contained both a lander and a rover, was able to utilize aerodynamic braking, parachutes, and retro-rockets to make a soft landing. Unfortunately, twenty seconds after touching down the lander stopped transmitting, and it was unable to deploy its rover component.

Considerably more successful were the American Viking 1 and Viking 2 craft, whose orbiters achieved orbit and whose landers, again through a combination of aerodynamic braking, parachutes, and retro-rockets, landed softly and stayed in operation for years, completing scientific objectives that included not only photographic imaging at the planet's surface but also soil analysis and biological-assay experiments for evidence of organic compounds and, potentially, the presence of life.

Later successful U.S. landers included the Mars Pathfinder lander/rover mission (which utilized air bags rather than retro-rockets during the last phase of its landing) and the Phoenix, which studied the geologic history of water on Mars, its involvement in Martian climate change, and the planet's past or future habitability.

**Rover.** Although Russian Mars 2 and Mars 3 descent modules brought rovers with them as early as 1971, no rover was successfully deployed on Mars until the U.S. Pathfinder mission of 1997 deployed its Sojourner rover. Able to travel about a half kilometer, or one-third of a mile from the lander, the Sojourner rover returned 550 photographs to Earth and the data from chemical analyses of sixteen locations on the Martian surface.

Mars Exploration Rovers (MER) Spirit and Opportunity landed on opposite sides of Mars in 2004. Both vehicles were intended to engage in geologic, hydrologic, and biologic assessment activities: to examine rocks and soils for evidence of past water activity, as well as assess whether the environments that prevailed when water was present were conducive to life.

The Spirit and Opportunity rovers have been tremendously successful, their missions lasting more than twenty times the planned duration. Opportunity is still operational and holds the record for longest Mars surface mission. The two rovers have covered

## Fascinating Facts About Areology

- The launch patch for Mars Exploration Rover Spirit features Marvin the Martian from the Looney Tunes cartoons.
- The launch patch for Mars Exploration Rover Opportunity features Duck Dodgers, an avatar of Daffy Duck.
- A dog killed in Nakhla, Egypt, in 1911 is reported to be the only known casualty of a Martian meteorite.
- In 1984, a meteorite of Martian origin (ALH84001) was discovered in Antarctica and contained what looked like fossil bacteria. The evidence remains inconclusive and controversial.
- A Viking orbiter's photograph of a low knoll on the Cydonia plateau caused scientists on the imaging team to joke that the image looked like a human face, but the Cydonia landform, in fact, depicts what looks like a human face.
- Writers Jonathan Swift and Voltaire both independently "predicted" that Mars had two moons, considerably more than a century before they were discovered.
- The seemingly high percentage of Mars mission failures has been attributed to a Mars curse (alternatively called the "Galactic Ghoul" and the "Mars Triangle") but is most likely caused by more prosaic circumstances, such as the use of complex advanced technologies in unprecedentedly severe environments.

far more terrain and have provided far more data than any previous mission.

## APPLICATIONS AND PRODUCTS

The National Aeronautics and Space Administration (NASA) lists more than 2,000 applications and products on its spin-off database. These spin-offs from space research contribute to national security, the economy, productivity, and lifestyle not only in the United States but also throughout the world. These spin-offs are so numerous and ubiquitous that people are scarcely aware of them and too often take them for granted. Below is a sampling of those specifically related to Mars research, many of which were developed in response to areological studies of Martian surface conditions.

**Sensors.** NASA research into detecting biological traces on Mars has resulted in biosensor technology monitoring water quality. Sensors incorporating carbon nanotubes tipped with single strands of nucleic acid from waterborne pathogens can detect minute amounts of disease-causing bacteria, viruses, and parasites and be used to alert organizations to potential biological hazards in water used for agriculture, food and beverages, showers, and at beaches and lakes.

NASA's Jet Propulsion Laboratory (JPL) developed a bacterial spore-detection system for Mars-bound spacecraft that can also recognize anthrax and other harmful, spore-forming bacteria on Earth and alert people of the impending danger.

JPL also developed a laser diode-based gas analyzer as part of the 1999 Mars Polar Lander mission to explore the possibility of life-giving elements on Mars. It has since been used on aircraft and on balloons to study weather and climate, global warming, emissions from aircraft, and numerous other areas where chemical-gas analysis is needed.

**Computing and Imaging.** NASA Advanced Supercomputing (NAS) division, which includes the Columbia supercomputer, is responsible for a wide range of products, from the development of computational fluid dynamics (CFD) computer codes to novel immersive visualization technologies used to pilot the Spirit and Opportunity rovers. Wide-screen panoramic photography technologies developed for the Mars rovers' Pancam robotic platform is in production as a GigaPan platform for automating the creation of highly detailed digital panoramas in consumer cameras.

**Materials.** Multilayer textiles developed for the air bags used in the Mars Pathfinder and Exploration Rovers are being used in Warwick Mills' puncture- and impact-resistant TurtleSkin product line of metal flex armor (MFA) vests, which are comparable to rigid steel plates but far more comfortable.

The thin, shiny insulation material used extensively in the Mars rover missions—a strong lightweight plastic, vacuum-metallized film that minimizes weight impact on vehicle payload while also protecting spacecraft, equipment, and personnel from the extreme temperature fluctuations of space—is found in applications ranging from reflective thermal blankets to party balloons.

## IMPACT ON INDUSTRY

The annual budget for the entire American space program is about $19 billion, or 0.8 percent of the $2.4 trillion budget. Collectively, however, secondary applications (spin-offs) represent a substantial return on the national investment in aerospace research: $7 from come back from spin-offs for every $1 spent on research. This surplus is generated by taxes from increased jobs in aerospace as well as all the other fields that produce spin-off goods.

Although it is difficult to sort out the actual worldwide spending on research and development relating specifically to Mars exploration and to separate out space-related research from other aerospace and defense spending, the NASA budget for Mars exploration in the first part of the twenty-first century has generally averaged around $500 million per year. Given multiplier effects and the share of Mars-related research in many aerospace-industrialized nations worldwide, including Japan, France, United Kingdom, Canada, Belgium, Russia, and China, the total value of global Mars-related research and development is estimated at roughly $13 billion.

**Government and University Research.** In Mars exploration-related research, NASA has a robust international relationship with agencies like the European Space Agency and the Japanese Aerospace Exploration Agency and with governmental and university scientific researchers from the United Kingdom, France, Italy, Australia, Belgium, Canada, Japan, Sweden, and Switzerland.

In Mars exploration-related research within the United States government, NASA maintains close ties with many Defense Department units, including the Naval Research Lab but particularly Defense Advanced Research Projects Agency (DARPA) and the U.S. Army, whose work involving tracked vehicles and robotics have paralleled JPL's work with rovers.

In Mars exploration-related research within NASA itself, JPL is the most important of NASA's dozen nationwide centers. JPL, which was established by the California Institute of Technology, has formed strategic relationships with ten schools that have major commitments to space exploration: Arizona State University; Carnegie Mellon University; Dartmouth College; Massachusetts Institute of Technology; Princeton University; Stanford University; University of Arizona; University of California, Los Angeles; University of Michigan; and University of Southern California. Through such relationships, JPL and its university collaborators facilitate joint access to particular capabilities in science, technology, and engineering and encourage better understanding of the state of research in the broader scientific community. Such collaborations also support students in space exploration topics, including graduate research on topics of interest to JPL/NASA, student participation in JPL summer programs, and input regarding courses of strong interest to NASA. Such collaborations also cultivate JPL's future workforce and ensure a pipeline to meet future technical challenges.

**Industry and Business.** Mars exploration-related research is most closely connected to aerospace and robotics. A single NASA/JPL program in development as of this writing, Distributed Spacecraft Technology Program for Precision Format Flying, involves companies as varied as Guidance Dynamics Corporation, DI-TEC International, Ball Aerospace, Applied Physics Technologies, Tera Semicon Corporation, and Pacific Wave Industries. The true impact, however, is less direct and found largely through the role secondary applications or spin-offs play in the wide variety of industries that make use of sensing, computing, imaging, and advanced materials.

## CAREERS AND COURSE WORK

Courses in astronomy, biology, chemistry, computer science, engineering, geology, and mathematics are foundational for students wishing to pursue careers in areology.

Master's and doctoral degrees are often the necessary minimum qualification for more advanced academic, governmental, or industrial careers in Mars exploration-related science. More specialized courses may include astrobiology, biochemistry, geophysics, climatology, hydrology, geodesy, and robotics, as well as a number of specializations within engineering, particularly mechanical, electrical, human factors, or systems.

Although areology is geological at its root, it is also the general study of a world other than that known to humans and at this point in its development is strongly interdisciplinary, so background in a diversity of fields, including the history of science and the study of literature concerning Mars, can also prove very helpful.

## SOCIAL CONTEXT AND FUTURE PROSPECTS

For areology, the twentieth century was shaped by two important movements. One was the transition from an understanding of Mars based on telescopy to one characterized by spacecraft with scientific instrument payloads flying by, orbiting, landing, and discharging mobile quasi-autonomous vehicles onto the surface to "follow the water" and look for evidence of life. The other was the movement from an understanding of Mars based primarily in fictional speculation to one increasingly based in science.

The question of past or present life on Mars, however, remains in the realm of speculation. The great debates in this century for areology will begin with whether remotely controlled or increasingly autonomous robotic vehicles can conclusively decide the question of past or present life or whether that question can be conclusively decided only through expensive, potentially dangerous (and perhaps infeasible) manned missions to Mars. That in itself, however, presents a problem: If there is no life on Mars, should the planet be preserved in its pristine state? Conversely, if there is life on Mars, should people risk causing the extinction of that life through contamination from Earth—or humanity's own extinction, through contamination from something on Mars?

These sound more and more like the speculations of science fiction, and matters become only more speculative as people contemplate the efficacy and feasibility of expensive, dangerous, and longer-term effects of colonization and terraforming of Mars by humans.

In trying to find Mars analogues on Earth, scientists are learning more about the limits to life on the world. By setting up microbial observatories in environments that may be in at least one way or another like certain environments on Mars, people have broadened their understanding of the diversity of life on Earth, ultimately serving to make Mars and Earth look more like each other at their extremes than previously assumed.

*Howard V. Hendrix, B.A., M.A., Ph.D.*

## FURTHER READING

Brandenburg, John E., and Monica Rix Paxson. *Dead Mars, Dying Earth.* Freedom, Calif.: Crossing Press, 1999. Controversial text arguing that Mars was much warmer and wetter until roughly a half billion years ago, when catastrophic climate change ended its ability to support life, and how understanding the death of Mars may save Earth.

Chapman, Mary G., and Laszlo P. Keszthelyi. *Preservation of Random Megascale Events on Mars and Earth: Influence on Geologic History.* Boulder, Colo.: Geological Society of America, 2009. Illustrated, multicontributor volume of essays by professional geologists regarding the preservation of large-scale geologic events on Earth and Mars.

Harland, David M. *Water and the Search for Life on Mars.* Chichester, England: Praxis, 2005. Richly illustrated examination of the "follow the water" approach to Mars exploration and the implications of the possibility of life existing (or having existed) on the planet.

Kargel, Jeffrey S. *Mars: A Warmer Wetter Planet.* Chichester, England: Praxis, 2004. A well-researched, thoroughly illustrated, and extensive examination of the wet-Mars hypothesis and what it means to human expectations and realities concerning the Red Planet.

Morton, Oliver. *Mapping Mars: Science, Imagination, and the Birth of a World.* New York: Picador, 2002. Cartography meets philosophy in this illustrated text that explores the natural history and topography of Mars.

Tokano, Tetsuya, ed. *Water on Mars and Life.* New York: Springer-Verlag, 2005. This collection of essays by professional scientists details the role of water in the planetary evolution of early Mars, water reservoirs on Mars, and the possible astrobiological importance of terrestrial analogues of putative aqueous environments on Mars.

Turner, Martin J. L. *Expedition Mars.* Chichester, England: Praxis, 2004. A thoroughly illustrated history of space exploration (both manned and robotic) tending toward a future Mars landing and the challenges inherent in undertaking an expedition to the planet.

Zubrin, Robert. *Mars on Earth: The Adventures of Space Pioneers in the High Arctic.* New York: Jeremy P. Tarcher/Penguin, 2003. An account of the Flashline Mars Arctic Research Station, a Mars analogue habitat on Devon Island in the Canadian Arctic, which was inhabited by volunteers during a simulation of human habitation on Mars that began in 2001.

**WEB SITES**

*Association of Mars Explorers*
http://marsexplorers.org

*Jet Propulsion Laboratory*
http://www.jpl.nasa.gov

*The Mars Society*
http://www.marssociety.org

*National Aeronautics and Space Administration*
Spinoffs
http://www.sti.nasa.gov/tto

**See also:** Atmospheric Sciences; Climatology; Computer Science; Hydrology and Hydrogeology; Mechanical Engineering; Mineralogy; Planetology; Robotics; Space Science; Space Stations; Telescopy.

# ARTIFICIAL INTELLIGENCE

## FIELDS OF STUDY

Expert systems; knowledge engineering; intelligent systems; computer vision; robotics; computer-aided design and manufacturing; computer programming; computer science; cybernetics; parallel computing; electronic health record; information systems; mobile computing; networking; business; physics; mathematics; neural networks; software engineering.

## SUMMARY

Artificial intelligence is the design, implementation, and use of programs, machines, and systems that exhibit human intelligence, with its most important activities being knowledge representation, reasoning, and learning. Artificial intelligence encompasses a number of important subareas, including voice recognition, image identification, natural language processing, expert systems, neural networks, planning, robotics, and intelligent agents. Several important programming techniques have been enhanced by artificial intelligence researchers, including classical search, probabilistic search, and logic programming.

## KEY TERMS AND CONCEPTS

- **Automatic Theorem Proving:** Proving a theorem from axioms, using a mechanistic procedure, represented as well-formed formulas.
- **Computer Vision:** Technology that allows machines to recognize objects by characteristics, such as color, texture, and edges.
- **First-Order Predicate Calculus:** System of formal logic, including Boolean expressions and quantification, that is rich enough to be a language for mathematics and science.
- **Game Theory:** Technology that supports the development of computer programs or devices that simulate one or more players of a game.
- **Intelligent Agent:** System, often a computer program or Web application, that collects and processes information, using reasoning much like a human.
- **Logic Programming:** Programming methodology that uses logical expressions for data, axioms,

and theorems and an inference engine to derive results.
- **Natural Language Processing:** How humans use language to represent ideas and to reason.
- **Neural Network:** Artificial intelligence system modeled after the human neural system.
- **Planning:** Set of processes, generally implemented as a program in artificial intelligence, that allows an organization to accomplish its objectives.
- **Robotics:** Science and technology used to design, manufacture, and maintain intelligent machines.

## DEFINITION AND BASIC PRINCIPLES

Artificial intelligence is a broad field of study, and definitions of the field vary by discipline. For computer scientists, artificial intelligence refers to the development of programs that exhibit intelligent behavior. The programs can engage in intelligent planning (timing traffic lights), translate natural languages (converting a Chinese Web site into English), act like an expert (selecting the best wine for dinner), or perform many other tasks. For engineers, artificial intelligence refers to building machines that perform actions often done by humans. The machines can be simple, like a computer vision system embedded in an ATM (automated teller machine); more complex, like a robotic rover sent to Mars; or very complex, like an automated factory that builds an exercise machine with little human intervention. For cognitive scientists, artificial intelligence refers to building models of human intelligence to better understand human behavior. In the early days of artificial intelligence, most models of human intelligence were symbolic and closely related to cognitive psychology and philosophy, the basic idea being that regions of the brain perform complex reasoning by processing symbols. Later, many models of human cognition were developed to mirror the operation of the brain as an electrochemical computer, starting with the simple Perceptron, an artificial neural network described by Marvin Minsky in 1969, graduating to the backpropagation algorithm described by David E. Rumelhart and James L. McClelland in 1986, and culminating in a large number of supervised and nonsupervised learning algorithms.

When defining artificial intelligence, it is important to remember that the programs, machines, and models developed by computer scientists, engineers, and cognitive scientists do not actually have human intelligence; they only exhibit intelligent behavior. This can be difficult to remember because artificially intelligent systems often contain large numbers of facts, such as weather information for New York City; complex reasoning patterns, such as the reasoning needed to prove a geometric theorem from axioms; complex knowledge, such as an understanding of all the rules required to build an automobile; and the ability to learn, such as a neural network learning to recognize cancer cells. Scientists continue to look for better models of the brain and human intelligence.

## BACKGROUND AND HISTORY

Although the concept of artificial intelligence probably has existed since antiquity, the term was first used by American scientist John McCarthy at a conference held at Dartmouth College in 1956. In 1955-1956, the first artificial intelligence program, Logic Theorist, had been written in IPL, a programming language, and in 1958, McCarthy invented Lisp, a programming language that improved on IPL. *Syntactic Structures* (1957), a book about the structure of natural language by American linguist Noam Chomsky, made natural language processing into an area of study within artificial intelligence. In the next few years, numerous researchers began to study artificial intelligence, laying the foundation for many later applications, such as general problem solvers, intelligent machines, and expert systems.

In the 1960's, Edward Feigenbaum and other scientists at Stanford University built two early expert systems: DENDRAL, which classified chemicals, and MYCIN, which identified diseases. These early expert systems were cumbersome to modify because they had hard-coded rules. By 1970, the OPS expert system shell, with variable rule sets, had been released by Digital Equipment Corporation as the first commercial expert system shell. In addition to expert systems, neural networks became an important area of artificial intelligence in the 1970's and 1980's. Frank Rosenblatt introduced the Perceptron in 1957, but it was *Perceptrons: An Introduction to Computational Geometry* (1969), by Minsky and Seymour Papert, and the two-volume *Parallel Distributed Processing: Explorations in the Microstructure of Cognition* (1986),

by Rumelhart, McClelland, and the PDP Research Group, that really defined the field of neural networks. Development of artificial intelligence has continued, with game theory, speech recognition, robotics, and autonomous agents being some of the best-known examples.

## HOW IT WORKS

The first activity of artificial intelligence is to understand how multiple facts interconnect to form knowledge and to represent that knowledge in a machine-understandable form. The next task is to understand and document a reasoning process for arriving at a conclusion. The final component of artificial intelligence is to add, whenever possible, a learning process that enhances the knowledge of a system.

**Knowledge Representation.** Facts are simple pieces of information that can be seen as either true or false, although in fuzzy logic, there are levels of truth. When facts are organized, they become information, and when information is well understood, over time, it becomes knowledge. To use knowledge in artificial intelligence, especially when writing programs, it has to be represented in some concrete fashion. Initially, most of those developing artificial intelligence programs saw knowledge as represented symbolically, and their early knowledge representations were symbolic. Semantic nets, directed graphs of facts with added semantic content, were highly successful representations used in many of the early artificial intelligence programs. Later, the nodes of the semantic nets were expanded to contain more information, and the resulting knowledge representation was referred to as frames. Frame representation of knowledge was very similar to object-oriented data representation, including a theory of inheritance.

Another popular way to represent knowledge in artificial intelligence is as logical expressions. English mathematician George Boole represented knowledge as a Boolean expression in the 1800's. English mathematicians Bertrand Russell and Alfred Whitehead expanded this to quantified expressions in 1910, and French computer scientist Alain Colmerauer incorporated it into logic programming, with the programming language Prolog, in the 1970's. The knowledge of a rule-based expert system is embedded in the if-then rules of the system, and because each if-then rule has a Boolean representation, it can be seen as a form of relational knowledge representation.

Neural networks model the human neural system and use this model to represent knowledge. The brain is an electrochemical system that stores its knowledge in synapses. As electrochemical signals pass through a synapse, they modify it, resulting in the acquisition of knowledge. In the neural network model, synapses are represented by the weights of a weight matrix, and knowledge is added to the system by modifying the weights.

**Reasoning.** Reasoning is the process of determining new information from known information. Artificial intelligence systems add reasoning soon after they have developed a method of knowledge representation. If knowledge is represented in semantic nets, then most reasoning involves some type of tree search. One popular reasoning technique is to traverse a decision tree, in which the reasoning is represented by a path taken through the tree. Tree searches of general semantic nets can be very time-consuming and have led to many advancements in tree-search algorithms, such as placing bounds on the depth of search and backtracking.

Reasoning in logic programming usually follows an inference technique embodied in first-order predicate calculus. Some inference engines, such as that of Prolog, use a back-chaining technique to reason from a result, such as a geometry theorem, to its antecedents, the axioms, and also show how the reasoning process led to the conclusion. Other inference engines, such as that of the expert system shell CLIPS, use a forward-chaining inference engine to see what facts can be derived from a set of known facts.

Neural networks, such as backpropagation, have an especially simple reasoning algorithm. The knowledge of the neural network is represented as a matrix of synaptic connections, possibly quite sparse. The information to be evaluated by the neural network is represented as an input vector of the appropriate size, and the reasoning process is to multiply the connection matrix by the input vector to obtain the conclusion as an output vector.

**Learning.** Learning in an artificial intelligence system involves modifying or adding to its knowledge. For both semantic net and logic programming systems, learning is accomplished by adding or modifying the semantic nets or logic rules, respectively.

Although much effort has gone into developing learning algorithms for these systems, all of them, to date, have used ad hoc methods and experienced limited success. Neural networks, on the other hand, have been very successful at developing learning algorithms. Backpropagation has a robust supervised learning algorithm in which the system learns from a set of training pairs, using gradient-descent optimization, and numerous unsupervised learning algorithms learn by studying the clustering of the input vectors.

### APPLICATIONS AND PRODUCTS

There are many important applications of artificial intelligence, ranging from computer games to programs designed to prove theorems in mathematics. This section contains a sample of both theoretical and practical applications.

**Expert Systems.** One of the most successful areas of artificial intelligence is expert systems. Literally thousands of expert systems are being used to help both experts and novices make decisions. For example, in the 1970's, Dell developed a simple expert system that allowed shoppers to configure a computer as they wished. In the 2010's, a visit to the Dell Web site offers a customer much more than a simple configuration program. Based on the customer's answers to some rather general questions, dozens of small expert systems suggest what computer to buy. The Dell site is not unique in its use of expert systems to guide customer's choices. Insurance companies, automobile companies, and many others use expert systems to assist customers in making decisions.

There are several categories of expert systems, but by far the most popular are the rule-based expert systems. Most rule-based expert systems are created with an expert system shell. The first successful rule-based expert system shell was the OPS 5 of Digital Equipment Corporation (DEC), and the most popular modern systems are CLIPS, developed by the National Aeronautics and Space Administration (NASA) in 1985, and its Java clone, Jess, developed at Sandia National Laboratories in 1995. All rule-based expert systems have a similar architecture, and the shells make it fairly easy to create an expert system as soon as a knowledge engineer gathers the knowledge from a domain expert. The most important component of a rule-based expert system is its knowledge base of rules. Each rule consists of an if-then statement

*This robot, called Adam, can think up scientific theories and test them with no human help. Adam has taken artificial intelligence to a new level.* (AP Photo)

and a simple rule could be if (red-wine) then (it-tastes-good). The expert system also has an inference engine that can apply multiple rules in an orderly fashion so that the expert system can draw conclusions by applying its rules to a set of facts introduced by a user. Although it is not absolutely required, most rule-based expert systems have a user-friendly interface and an explanation facility to justify its reasoning.

**Theorem Provers.** Most theorems in mathematics can be expressed in first-order predicate calculus. For any particular area, such as synthetic geometry or group theory, all provable theorems can be derived from a set of axioms. Mathematicians have written programs to automatically prove theorems since the 1950's. These theorem provers either start with the axioms and apply an inference technique, or start with the theorem and work backward to see how it can be derived from axioms. Resolution, developed in Prolog, is a well-known automated technique that can be used to prove theorems, but there are many others. For Resolution, the user starts with the theorem, converts it to a normal form, and then mechanically builds reverse decision trees to prove the theorem. If a reverse decision tree whose leaf nodes are all axioms is found, then a proof of the theorem has been discovered.

Gödel's incompleteness theorem (proved by Austrian-born American mathematician Kurt Gödel) shows that it may not be possible to automatically prove an arbitrary theorem in systems as complex as the natural numbers. For simpler systems, such as group theory, automated theorem proving works if the user's computer can generate all reverse trees or a suitable subset of trees that can yield a

with multiple antecedents, multiple consequences, and possibly a rule certainty factor. The antecedents of a rule are statements that can be true or false and that depend on facts that are either introduced into the system by a user or derived as the result of a rule being fired. For example, a fact could be red-wine proof in a reasonable amount of time. Efforts have been made to develop theorem provers for higher order logics than first-order predicate calculus, but these have not been very successful.

Computer scientists have spent considerable time trying to develop an automated technique for proving the correctness of programs, that is showing that any valid input to a program produces a valid output. This is generally done by producing a consistent model and mapping the program to the model. The first example of this was given by English mathematician Alan Turing in 1931, by using a simple model now called a Turing machine. A formal system that is rich enough to serve as a model for a typical programming language, such as C++, must support higher order logic to capture the arguments and parameters of subprograms. Lambda calculus, denotational semantics, von Neuman geometries, finite state machines, and other systems have been proposed to provide a model onto which all programs of a language can be mapped. Some of these do capture many programs, but devising a practical automated method of verifying the correctness of programs has proven difficult.

**Intelligent Tutor Systems.** Almost every field of study has many intelligent tutor systems available to assist students in learning. Sometimes the tutor system is integrated into a package. For example, in Microsoft Office, an embedded intelligent helper provides popup help boxes to a user when it detects the need for assistance and full-length tutorials if it detects more help is needed. In addition to the intelligent tutors embedded in programs as part of a context-sensitive help system, there are a vast number of stand-alone tutoring systems in use.

The first stand-alone intelligent tutor was SCHOLAR, developed by J. R. Carbonell in 1970. It used semantic nets to represent knowledge about South American geography, provided a user interface to support asking questions, and was successful enough to demonstrate that it was possible for a computer program to tutor students. At about the same time, the University of Illinois developed its PLATO computer-aided instruction system, which provided a general language for developing intelligent tutors with touch-sensitive screens, one of the most famous of which was a biology tutorial on evolution. Of the thousands of modern intelligent tutors, SHERLOCK, a training environment for electronic troubleshooting, and PUMP, a system designed to help learn algebra, are typical.

**Electronic Games.** Electronic games have been played since the invention of the cathode-ray tube for television. In the 1980's, games such as Solitaire, Pac-Man, and Pong for personal computers became almost as popular as the stand-alone game platforms. In the 2010's, multiuser Internet games are enjoyed by young and old alike, and game playing on mobile devices is poised to become an important application. In all of these electronic games, the user competes with one or more intelligent agents embedded in the game, and the creation of these intelligent agents uses considerable artificial intelligence. When creating an intelligent agent that will compete with a user or, as in Solitaire, just react to the user, a programmer has to embed the game knowledge into the program. For example, in chess, the programmer would need to capture all possible configurations of a chess board. The programmer also would need to add reasoning procedures to the game; for example, there would have to be procedures to move each individual chess piece on the board. Finally, and most important for game programming, the programmer would need to add one or more strategic decision modules to the program to provide the intelligent agent with a strategy for winning. In many cases, the strategy for winning a game would be driven by probability; for example, the next move might be a pawn, one space forward, because that yields the best probability of winning, but a heuristic strategy is also possible; for example, the next move is a rook because it may trick the opponent into a bad series of moves.

### IMPACT ON INDUSTRY

United States government support has been essential in artificial intelligence research, including funding for the 1956 conference at which McCarthy introduced the term "artificial intelligence." The Defense Advanced Research Projects Agency (DARPA) was a strong early supporter of artificial intelligence, then reduced support for a number of years before again providing major support for research in basic and applied artificial intelligence. Industry support for development of artificial intelligence has generally emphasized short-range projects, and university research has developed both theory and applications. Although estimates of the total value of the goods and services produced by artificial intelligence technology in a year are impossible to determine, it is clear that it is in the range of billions of dollars a year.

**Government, Industry, and University Research.** Many government agencies have provided support for basic and applied research in artificial intelligence. In 1985, NASA released the CLIPS expert system shell, and it remains the most popular shell. The National Science Foundation (NSF) supports a wide range of basic research in artificial intelligence, and the National Institutes of Health (NIH) concentrates on applying artificial intelligence to health systems. Important examples of government support for artificial intelligence are the many NSF and NIH grants for developing intelligent agents that can be

---

### Fascinating Facts About Artificial Intelligence

- In 1847, George Boole developed his algebra for reasoning that was the foundation for first-order predicate calculus, a logic rich enough to be a language for mathematics.
- In 1950, Alan Turing gave an operational definition of artificial intelligence. He said a machine exhibited artificial intelligence if its operational output was indistinguishable from that of a human.
- In 1956, John McCarthy and Marvin Minsky organized a two-month summer conference on intelligent machines at Dartmouth College. To advertise the conference, McCarthy coined the term "artificial intelligence."
- Digital Equipment Corporation's XCON, short for eXpert CONfigurer, was used in house in 1980 to configure VAX computers and later became the first commercial expert system.
- In 1989, international chess master David Levy was defeated by a computer program, Deep Thought, developed by IBM. Only ten years earlier, Levy had predicted that no computer program would ever beat a chess master.
- In 2010, the Haystack group at the Computer Science and Artificial Intelligence Laboratory at the Massachusetts Institute of Technology developed Soylent, a word-processing interface that lets users edit, proof, and shorten their documents using Mechanical Turk workers.

---

embedded in health software to help doctors identify at-risk patients, suggest best practices for these patients, manage their health care, and identify cost savings. DARPA also has a very active program in artificial intelligence, including the Deep Learning project, which supports basic research into "hierarchical machine perception and analysis, and applications in visual, acoustic and somatic sensor processing for detection and classification of objects and activities." The goal is to develop a better biological model of human intelligence than neural networks and to use this to provide machine support for high-level decisions made from sensory and learned information.

**Industry and Business Sectors.** Many, if not all, software companies use artificial intelligence in their software. For example, the Microsoft Office help system is based on artificial intelligence. Microsoft's Visual Studio programming environment, uses IntelliSense, an intelligent code completion system, and Microsoft's Xbox, like all game systems, uses many artificial intelligence techniques.

Medical technology uses many applications from artificial intelligence. For example, MEDai (an Elsevier company) offers business intelligence solutions that improve health care delivery for clinics and hospitals. Other businesses that make significant use of artificial intelligence are optical character recognition companies, such as Kurzweil Technologies; speech recognition companies, such as Dragon Speaking Naturally; and companies involved in robotics, such as iRobot.

### CAREERS AND COURSE WORK

A major in computer science is the most common way to prepare for a career in artificial intelligence. One needs substantial course work in mathematics, philosophy, and psychology as a background for this degree. For many of the more interesting jobs in artificial intelligence, one needs a master's or doctoral degree. Most universities teach courses in artificial intelligence, neural networks, or expert systems, and many have courses in all three. Although artificial intelligence is usually taught in computer science, it is also taught in mathematics, philosophy, psychology, and electrical engineering. Taking a strong minor in any field is advisable for someone seeking a career in artificial intelligence because the discipline is often applied to another field.

Those seeking careers in artificial intelligence generally take a position as a systems analyst or programmer. They work for a wide range of companies, including those developing business, mathematics, medical, and voice recognition applications. Those obtaining an advanced degree often take jobs in industrial, government, or university laboratories developing new areas of artificial intelligence.

## SOCIAL CONTEXT AND FUTURE PROSPECTS

After artificial intelligence was defined by McCarthy in 1956, it has had a number of ups and downs as a discipline, but the future of artificial intelligence looks good. Almost every commercial program has a help system, and increasingly these help systems have a major artificial intelligence component. Health care is another area that is poised to make major use of artificial intelligence to improve the quality and reliability of the care provided, as well as to reduce its cost by providing expert advice on best practices in health care.

Ethical questions have been raised about trying to build a machine that exhibits human intelligence. Many of the early researchers in artificial intelligence were interested in cognitive psychology and built symbolic models of intelligence that were considered unethical by some. Later, many artificial intelligence researchers developed neural models of intelligence that were not always deemed ethical. The social and ethical issues of artificial intelligence are nicely represented by HAL, the Heuristically programmed ALgorithmic computer, in Stanley Kubrick's 1968 film *2001: A Space Odyssey*, which first works well with humans, then acts violently toward them, and is in the end deactivated.

Another important ethical question posed by artificial intelligence is the appropriateness of developing programs to collect information about users of a program. Intelligent agents are often embedded in Web sites to collect information about those using the site, generally without the permission of those using the Web site, and many question whether this should be done.

*George M. Whitson III, B.S., M.S., Ph.D.*

## FURTHER READING

Giarratano, Joseph, and Peter Riley. *Expert Systems: Principles and Programming*. 4th ed. Boston: Thomson Course Technology, 2005. Provides an excellent overview of expert systems, including the CLIPS expert system shell.
Minsky, Marvin, and Seymour Papert. *Perceptrons: An Introduction to Computational Geometry*. Rev. ed. Boston: MIT Press, 1990. Originally printed in 1969, this work introduced many to neural networks and artificial intelligence.
Rumelhart, David E., James L. McClelland, and the PDP Research Group. *Parallel Distributed Processing: Explorations in the Microstructure of Cognition*. 1986. Reprint. 2 vols. Boston: MIT Press, 1989. Volume 1 gives an excellent introduction to neural networks, especially backpropagation. Volume 2 shows many biological and psychological relationships to neural nets.
Russell, Stuart, and Peter Norvig. *Artificial Intelligence: A Modern Approach*. 3d ed. Upper Saddle River, N.J.: Prentice Hall, 2010. The standard textbook on artificial intelligence, it provides a complete overview of the subject, integrating material from expert systems and neural networks.
Shapiro, Stewart, ed. *Encyclopedia of Artificial Intelligence*. 2d ed. New York: John Wiley & Sons, 1992. Contains articles covering the entire field of artificial intelligence.

## WEB SITES

*Association for the Advancement of Artificial Intelligence*
http://www.aaai.org/home.html

*Computer Society*
http://www.computer.org

*Defense Advanced Research Projects Agency*
Deep Learning
http://www.darpa.mil/i2o/programs/deep/deep.asp

*Institute of Electrical and Electronics Engineers*
http://www.ieee.org

*Massachusetts Institute of Technology*
Computer Science and Artificial Intelligence Laboratory
http://www.csail.mit.edu

*Society for the Study of Artificial Intelligence and Simulation of Behaviour*
http://www.aisb.org.uk

**See also:** Computer-Aided Design and Manufacturing; Computer Engineering; Computer Languages, Compilers, and Tools; Computer Science; Human-Computer Interaction; Parallel Computing; Pattern Recognition; Robotics; Video Game Design and Programming.

# ARTIFICIAL ORGANS

## FIELDS OF STUDY

Biology; anatomy; biophysics; chemistry; physics; mathematics; physiology; genetics; immunology; molecular biology; organ transplantation; biomedical engineering.

## SUMMARY

Artificial organs are complex systems of natural or manufactured materials used to supplement failing organs while they recover, sustain failing organs until transplantation, or replace failing organs that cannot recover. Some whole organs have artificial counterparts: heart, kidneys, liver, lungs, and pancreas. Smaller body parts also have artificial counterparts: blood, bones, heart valves, joints, skin, and teeth. In addition, there are mechanical support systems for circulation, hearing, and breathing. Artificial organs are composed of biomaterials, biological or synthetic materials that are adapted for use in medical applications.

## KEY TERMS AND CONCEPTS

- **Biocompatibility:** Absence of immune reaction or rejection against biological or synthetic materials.
- **Biohybrid:** Interfacing of a biological material with a synthetic material.
- **Biomaterial:** Biological or synthetic material that is adapted for medical use.
- **Extracorporeal:** Outside the body; often used to describe large mechanical support systems.
- **Hemodialysis:** Removal of metabolic waste products and extra water from the blood in cases of kidney failure.
- **Hemoperfusion:** Removal of toxins from the blood in cases of liver failure.
- **Immunomodulation:** Exerting an affect on the immune system, either stimulation or suppression.
- **Organ Failure:** State in which an organ does not perform its natural functions.

## DEFINITION AND BASIC PRINCIPLES

Artificial organs are complex systems that assist or replace failing organs. The human body is composed of ten major organ systems: nervous, circulatory, respiratory, digestive, excretory, reproductive, endocrine, integumentary (skin), muscular, and skeletal. The nervous system transmits signals between the brain and the body via the spinal cord and nerves. The circulatory system transports blood to deliver oxygen and nutrients to the body and to remove waste products. Its organs are the heart, blood, and blood vessels. It works closely with the respiratory system, in which the lungs and trachea perform oxygen exchange between the body and the environment. The digestive system breaks down food and absorbs its nutrients. Its organs include the esophagus, stomach, intestinal tract, and liver. The excretory system rids the body of metabolic waste in the forms of urine and feces. The reproductive system provides sex cells and in females the organs to develop and carry an embryo to term. The endocrine system consists of the pituitary, parathyroid, and thyroid glands, which secrete regulatory hormones. The integumentary system is the body's external protection system. Its organs include skin, hair, and nails. The muscular system recruits muscles, ligaments, and tendons to move the parts of the skeletal system, which consists of bones and cartilage.

## BACKGROUND AND HISTORY

While he was still a medical student in 1932, renowned cardiac surgeon Michael E. DeBakey introduced a dual-roller pump for blood transfusion. It has since become the most widely used type of clinical pump for cardiopulmonary bypass and hemodialysis. Physician John H. Gibbon, Jr., of Philadelphia, developed the first clinically successful heart-lung pump. He initially demonstrated it in 1953, when he closed a hole between the atria of an eighteen-year-old girl.

In 1954, American physician Joseph Murray performed the first successful human kidney transplant from one identical twin to the other in Boston. In 1962, he performed the first kidney transplant in unrelated persons. In 1967, surgeon Christiaan Barnard performed the first successful human heart transplant in Cape Town, South Africa. The patient, a fifty-four-year-old man, lived another eighteen days.

Physician Willem J. Kolff is considered to be "the father of the artificial organ." In 1967, he emigrated

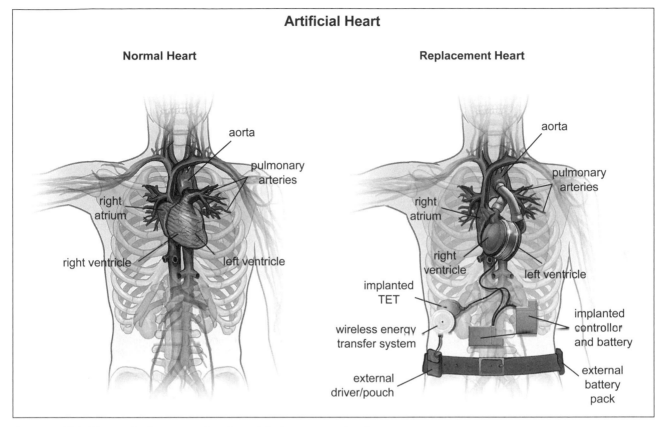

**Artificial Heart**

**Normal Heart**

aorta

pulmonary arteries

right atrium

right ventricle

left ventricle

**Replacement Heart**

aorta

pulmonary arteries

right atrium

right ventricle

left ventricle

implanted TET

wireless energy transfer system

implanted controller and battery

external driver/pouch

external battery pack

*In an artificial heart, the lower two chambers of the heart are replaced.*

from the Netherlands and spent a good deal of his career at the University of Utah, where he became a distinguished professor emeritus of internal medicine, surgery, and bioengineering. He led the designing of numerous inventions, including the modern kidney dialysis machine, the intra-aortic balloon pump, an artificial eye, an artificial ear, and an implantable mechanical heart.

American physician Robert K. Jarvik refined Kolff's design into the Jarvik-7 artificial heart, intended for permanent use. In 1982, at the University of Utah, American surgeon William C. DeVries implanted it into retired dentist Barney Clark, who survived 112 days.

#### HOW IT WORKS

The existence and performance of artificial organs depend on the collaboration of scientists, engineers, physicians, manufacturers, and regulatory agencies. Each of these groups provides a different

perspective of pumps, filters, size, packaging, and regulation.

**Hemodynamics.** The human heart acts as a muscular pump that beats an average of 72 times a minute. Each of the two ventricles pumps 70 milliliters of blood per beat or 5 liters per minute. Blood pressure is measured and reported as two numbers: the systolic pressure exerted by the heart during contraction and the diastolic pressure, when the heart is between contractions. Hemodynamics is the study of forces related to the circulation of the blood. The hemodynamic performance of artificial organs must match that of the natural body to operate efficiently without resulting in damage. Calculations may be made using computational fluid dynamics (CFD); relevant parameters include solute concentration, density, temperature, and water concentration. In addition to artificial hearts, which are intended to perform all cardiac functions, there is a mechanical circulatory implement called a ventricular assist device (VAD) that supports the function of the natural heart while it is recovering from a heart attack or surgery. Its pumping action may be pulsatile, in rhythmic waves matching those of the beating heart, or continuous.

**Mass Transfer Efficiency.** The human kidney acts as a filter to remove metabolic waste products from the blood. A person's kidneys process about 200 quarts of blood daily to remove two quarts of waste and extra water, which are converted into urine and excreted. Without filtration, the waste would build to a toxic level and cause death. Patients with kidney failure may undergo dialysis, in which blood is withdrawn, cleaned, and returned to the body in a periodic, continuous, and time-consuming process that requires the patient to remain relatively stationary. Portable artificial kidneys, which the patient wears, filter the blood while the patient enjoys the freedom of mobility. Filtration systems may involve membranes with a strict pore size to separate molecules based on size or columns of particle-based adsorbents to separate molecules by chemical characteristics. Mass transfer efficiency refers to the quality and quantity of molecular transport.

**Scale.** The development of artificial organs requires that biological processes that can be duplicated in the laboratory be scaled up to work within the human body without also magnifying the weaknesses. Biological functions occur at the organ, tissue, cellular, and molecular levels, which are on micro- and nanoscales. In addition, machines that work in the engineering laboratory must be scaled down to work within the human body without crowding the other organs. Novel power sources and electronic components have facilitated miniaturization. Size must also be balanced with efficiency and cost. Computer-aided design software is being used to create virtual three-dimensional models before fabrication.

**Biomaterials.** Artificial organs are made of natural and/or manufactured materials that have been adapted for medical use. The properties of these materials must be controlled down to the nanometer scale. The biological components may serve in gene therapy, tissue engineering, and the modification of physiological responses. The synthetic materials must be biocompatible, which means that they do not trigger an adverse physiological reaction such as blood clotting, inflammatory response, scar-tissue formation, or antibody production. The

biomechanics of the artificial organ, such as friction and wear, must be known and parts must be sterile before use. Biomaterials have been developed for subspecialties such as orthopedics and ophthalmics.

**Regulation.** The body has natural feedback systems that allow the exchange of information with the brain for optimal regulation. Artificial organs that communicate directly with the brain are still in development. The present models require sensors and data systems that may be monitored by physicians. Implanted devices must be able to be inspected without direct observation. Another aspect of regulation is the uniform manufacturing of artificial organs in compliance with performance and patient safety specifications.

## APPLICATIONS AND PRODUCTS

The collective knowledge of scientists, engineers, physicians, manufacturers, and regulatory agencies has produced the applications and products in the interdisciplinary realm of artificial organs.

**Hemodynamics.** Knowledge of hemodynamics, the study of blood-flow physics, has led to the development of artificial circulatory assistance. The ventricular assist device (VAD) supplements the contraction of the two lower chambers of the heart so the heart muscle does not have to work as hard while it is healing. The cardiopulmonary bypass pump, also known as a heart-lung machine, provides blood oxygenation and circulating pressure during open-heart surgery when the heart is stopped. A similar application called extracorporeal membrane oxygenation (ECMO) is used to assist neonates and infants in the intensive care unit and to maintain the viability of organs pending transplantation. The natural pressure generated by a healthy heart is used to send blood through versions of artificial lungs and kidneys without batteries.

**Mass Transfer Efficiency.** Information about molecular transport and delivery, known as mass transfer efficiency, has been applied to separation and secretion functions of artificial organs. In hemodialysis, toxins are removed from circulating blood that passes through a filter called a dialyzer. This process also removes excess salts and water to maintain a healthy blood pressure. The dialyzer is composed of a semipermeable membrane or cylinder of hollow synthetic fibers that separates out the metabolic-waste solutes in the incoming blood by diffusion into dialysate solution, leaving cleaner outgoing blood. Hemofiltration is a similar process; however, the filtration occurs without dialysate solution because instead of diffusion, the solutes are removed more quickly by hydrostatic pressure. Another separation technique in medical applications is apheresis, in which the constituents of blood are isolated. This may be achieved by gradient density centrifugation or absorption onto specifically coated beads. The

---

## Fascinating Facts About Artificial Organs

- Aviator Charles Lindbergh worked with physician Alexis Carrel, a noted pioneer of vascular surgery, to find a means of oxygenating blood other than the lungs. They created a basic oxygen-exchange device that led to the development of the heart-lung bypass machine for artificial circulation.
- The first successful human organ transplant was performed in 1954. A kidney was transplanted from an identical twin donor.
- In the United States, the need for organ replacement therapies increases by 10 percent each year. In 2008, there were nearly 100,000 people on the national waiting list for organ transplants. Three of every four individuals on the list were waiting for a kidney, with an average wait exceeding five years. One patient waiting for an organ transplant dies every seventy-three minutes.
- About 350,000 Americans are reliant on hemodialysis because of kidney damage associated with diabetes and hypertension. Medicare spends $25 billion annually on treatments for kidney failure, which translates to spending 6 percent of its budget on 1 percent of its recipients.
- The National Organ Transplant Act of 1984 bans "valuable consideration" in exchange for providing an organ for transplantation. Thus, human organs cannot be bought or sold in the United States.
- In Spain, the law allows for the presumption of organ donation after death with consent unless the individual stated otherwise before death. As a result, the procurement rate of organ harvesting from cadavers is 35 percent higher than that in the United States. If the United States had the same presumed-consent law, nearly 14,000 more organs could be procured annually.

---

therapeutic application is the absorptive removal of a specific blood component that is causing an adverse reaction in a patient, with the remaining components returned to the patient's circulatory system. The pathogenic blood component might be malignant white blood cells, excess platelets, low-density lipoprotein, autoantibodies, or plasma. The second application of apheresis is the separation of components following blood donation. Concentrated red blood cells are administered in the treatment of sickle-cell crisis or malaria. Plasmapheresis is used to collect fresh frozen plasma as well as rare antibodies and immunoglobulins.

**Scale.** Miniaturization of artificial organs has been facilitated by the application of smaller, more efficient batteries, transistors, and computer chips. For example, hearing aids once had to be worn with cumbersome amplifiers and batteries disguised in a purse or camera case with a carrying strap. Existing models fit completely in the ear canal and a computer chip facilitates digital rather than analogue processing for crisper sound. The artificial kidney has evolved into a wearable model that weighs 10 pounds and is seventeen times smaller than a conventional dialysis machine. Its hollow-fiber filter must be replaced once a week and its dialysate solution must be replenished daily. However, this maintenance is a trade-off that many patients are willing to make for freedom of movement. On the horizon is an artificial retina that depends on a miniature camera to transmit images. Conversely, research is under way to produce large-scale cultures of tissues on biohybrid matrices and scaffolding for transplantation.

**Biomaterials.** Synthetic materials are used in artificial organs. Dacron (polyethylene terephthalate) is a polyester fiber with high tensile strength and resistance to stretching whether wet or dry, chemical degradation, and abrasion. Patches of it are sewn to arteries to repair aneurysms. When tubing of it is used as an aortic valve bypass, the patient will not require subsequent blood-thinning medications. Gore-Tex (expanded polytetrafluoroethylene) is an especially strong microporous material that is waterproof. Vascular grafts made from it are supple and resist kinks and compression. It is also used for replacing torn anterior and posterior cruciate ligaments in the knee.

Perfluorocarbon fluids are synthetic liquids that carry dissolved oxygen and carbon dioxide with negligible toxicity, no biological activity, and a short retention time in the body. These features make them ideal for medical applications. One of these fluids, perfluorodecalin, is typically used as a blood substitute (also called a blood extender) because it mixes easily with blood without changing the hemodynamics. It increases the oxygen-carrying capacity of the blood and penetrates ischemic (oxygen-deprived) tissues especially easily because of its small particle size. This makes it particularly useful in the healing of ulcers and burns. It is also used in conjunction with ECMO in the life support of preterm infants to increase oxygenation and to keep the lungs inflated, reducing exertion. Furthermore, it is used in the preservation of harvested organs and cultured tissue for transplantation, extending their viable storage time.

**Regulation.** The application of regulatory systems has allowed artificial organs to be adjusted while they are in use. Artificial cardiac pacemakers, which supplement the natural electrical pacemaking capabilities of the heart to normalize a slow or irregular heartbeat, are externally programmable so that cardiologists are able to establish the optimal pacing parameters for each patient. Adjustments are made with radio frequency programming, so no further surgery is required. Contemporary hearing aids have volume controls that the wearer can adjust to suit changing surroundings. The inability to detect high- or low-pitch sounds is not a function of volume, yet pitch range can be adjusted in a hearing aid by an audiologist. Other parameters are also adjustable and the audiologist can reprogram the hearing aid as a person's hearing loss changes.

## IMPACT ON INDUSTRY

**Professional Societies.** The International Federation for Artificial Organs (IFAO) was founded in 1977 and serves three member societies: the American Society for Artificial Internal Organs (ASAIO) founded in 1955, the European Society for Artificial Organs (ESAO) founded in 1974, and the Japanese Society for Artificial Organs (JSAO) founded in 1962. The first president of the IFAO was "the father of the artificial organ," Willem J. Kolff.

**Government Regulation.** Artificial organs fall under the federal regulation of the United States Food and Drug Administration (FDA), Center for Devices and Radiological Health, Office of Science and Technology, Division of Physical Sciences. Artificial organs are considered to be medical devices, but they are some of the most complicated that the FDA evaluates. The Office of Science and Technology is charged with determining the scientific parameters that are most relevant to product evaluation and identifying valid techniques for measuring those parameters. The parameters for artificial organs include damage to blood components and neurological consequences. To establish quality-control standards for the production of artificial organs, the Division of Physical Sciences frequently consults national and international standards groups.

**University Research.** Several universities are front-runners in the research and development of artificial organs. The University of California, San Francisco (UCSF) schools of pharmacy and medicine have a joint department, the Department of Bioengineering and Therapeutic Sciences, that has recently introduced a prototype of the first implantable artificial kidney. The University of Pittsburgh School of Medicine has partnered with the University of Pittsburgh Medical Center to establish the McGowan Institute for Regenerative Medicine, where they seek solutions to the repair or replacement of tissues and organs through the manipulation of cells, genes, other biomaterials, and bioengineering technologies.

**Major Corporations.** Corporations worldwide are involved with the production and marketing of artificial organs. Some specialize in a particular technology applicable to several organ systems and some specialize in a particular organ system and offer a variety of products. Some of these companies are: Abbott Laboratories; Abiomed Inc.; F. Hoffmann-La Roche Inc.; Alliqua, Inc.; Medtronic Inc.; SynCardia Systems, Inc.; Thoratec Corporation; Ventracor Ltd.; WorldHeart Corporation; and Xenogenics Corporation.

## CAREERS AND COURSE WORK

Universities offer various undergraduate and graduate programs related to artificial organs. The University of Pittsburgh's Swanson School of Engineering offers undergraduate degrees in bioengineering with concentration choices of cellular and medical product engineering, biomechanics, and biosignals and imaging. Brown University Department of Molecular Pharmacology, Physiology, and Biotechnology offers master's and doctoral degrees in artificial organs, biomaterials, and cellular technology.

Medical schools support researchers in the field of biotechnology development. Within the Michael

E. DeBakey Department of Surgery at Baylor College of Medicine in Houston is the division of Transplant Surgery and Assist Devices. The University of Maryland School of Medicine's Department of Surgery has an Artificial Organs Laboratory as one of the surgical research laboratories. This is a collaborative field and scientists, biomedical engineers, physicians, and businesspeople work together in commercial ventures to design and fabricate functional artificial organs. Because artificial organs fall under the regulatory domain of the FDA as medical devices, manufacturers must undergo rigorous product development, clinical trials, and patent protection prior to FDA approval. Components are then made to custom specifications. Some companies produce a multitude of medical devices, while some specialize in specific technologies such as biotransport, dialysis, perfusion, and cell culture matrices.

## SOCIAL CONTEXT AND FUTURE PROSPECTS

The number of Americans older than sixty-five years of age is expected to double within the next twenty-five years. The fastest growing age group is people older than eighty-five years of age. Increasing life span of the general population is a direct result of improved health care. The shortage of donor organs is also increasing. As of 2010, more than 16,000 people were waiting for liver transplants, but each year, only about 6,500 kidneys become available. For the 93,000 people waiting for kidney transplants in 2010, only 17,000 donated kidneys became available for transplant. The need for artificial organs as a bridge to transplantation or even as a permanent substitute for failed organs is becoming increasingly urgent.

Once only made of synthetic components, artificial organs are becoming biohybrid organs: a combination of biological and synthetic components. Examples include functionally competent cells enveloped within immuno-protective artificial membranes and tissues cultured on chemically constructed matrices. Experiments are underway to develop an antibacterial agent that can be incorporated into biomaterials to reduce the risk of infection from these organ surfaces. Emerging technologies also involve sensors and intelligent control systems, biological batteries and alternate power sources, and innovative delivery systems.

Other areas of research include the miniaturization of artificial organs for pediatric use and the development of smaller and more efficient batteries and sensors that will be capable of more accurate communication between the artificial organ and the brain. Another goal is to incorporate wireless capabilities so the artificial organ may be programmed, monitored, and recharged remotely so the patient has increased freedom of mobility.

*Bethany Thivierge, B.S., M.P.H.*

## FURTHER READING

Fox, Renée C., and Judith P. Swazey. *Spare Parts: Organ Replacement in American Society*. New York: Oxford University Press, 1992. Discusses not only the progression of organ transplantation methods but also the emotional significance attached to the human body and its parts as well as the ethical concerns regarding organ replacement.

Hench, Larry L., and Julian R. Jones, eds. *Biomaterials, Artificial Organs, and Tissue Engineering*. Boca Raton, Fla.: CRC Press, 2005. Provides multiple essays and introductory topics on artificial organs and tissue engineering.

McClellan, Marilyn. *Organ and Tissue Transplants: Medical Miracles and Challenges*. Berkeley Heights, N.J.: Enslow, 2003. Explores the history of organ transplantation as well as the ensuing medical, ethical, and financial issues.

Sharp, Lesley A. *Bodies, Commodities, and Biotechnologies: Death, Mourning, and Scientific Desire in the Realm of Human Organ Transfer*. New York: Columbia University Press, 2008. Explores how organ transplantation and artificial organs have changed cultural attitudes toward the body.

## WEB SITES

*American Society for Artificial Internal Organs*
http://asaio.com

*International Federation for Artificial Organs*
http://www.ifao.org

*Society for Biomaterials*
http://www.biomaterials.org

**See also:** Audiology and Hearing Aids; Bioengineering; Biomechanical Engineering; Cardiology; Cell and Tissue Engineering; Nephrology; Stem Cell Research and Technology; Xenotransplantation.

# ATMOSPHERIC SCIENCES

## FIELDS OF STUDY

Climatology; meteorology; climate change; climate modeling; hydroclimatology; hydrometeorology; physical geography; chemistry, physics.

## SUMMARY

Atmospheric sciences includes the fields of physics and chemistry and the study of the composition and dynamics of the layers of air that constitute the atmosphere. Related topics include climatic processes, circulation patterns, chemical and particulate deposition, greenhouse gases, oceanic temperatures, interaction between the atmosphere and the ocean, the ozone layer, precipitation patterns and amounts, climate change, air pollution, aerosol composition, atmospheric chemistry, modeling of pollutants both indoors and outdoors, and anthropogenic alteration of land surfaces that in turn affect conditions within the ever-changing atmosphere.

## KEY TERMS AND CONCEPTS

- **Atmosphere:** Gaseous layer that surrounds the Earth and is held in place by gravity.
- **Atmospheric Pressure:** Gravitational force caused by the weight of overlying layers of air.
- **Atmospheric Water:** Water in the atmosphere; it can be in a gaseous, liquid, or frozen form.
- **Coriolis Effect:** Rightward (Northern Hemisphere) and leftward (Southern Hemisphere) deflection of air caused by the Earth's rotation.
- **Energy Balance:** Return to space of the Sun's energy that was received by the Earth.
- **Front:** Boundary between airmasses that differ in temperature, moisture, and pressure.
- **Greenhouse Effect:** Process by which longwave radiation is trapped in the atmosphere and then returned to the Earth's surface by counterradiation.
- **Occluded Front:** Front produced when a cold front overtakes a warm front and forces the air upward.
- **Troposphere:** Lowest level of the Earth's atmosphere that contains water vapor that can condense into clouds.

- **Urban Heat Island:** Urbanized area that experiences higher temperatures than surrounding rural lands.

## DEFINITION AND BASIC PRINCIPLES

Atmospheric sciences is the study of various aspects of the nature of the atmosphere, including its origin, layered structure, density, and temperature variation with height; natural variations and alterations associated with anthropogenic impacts; and how it is similar to or different from other atmospheres within the solar system. The present-day atmosphere is in all likelihood quite dissimilar from the original atmosphere. The form and composition of the present-day atmosphere is believed to have developed about 400 million years ago in the late Devonian period of the Paleozoic era, when plant life developed on land. This vegetative cover allowed plants to take in carbon dioxide and release oxygen as part of the photosynthesis process.

The atmosphere consists of a mixture of gases that remain in place because of the gravitational attraction of the Earth. Although the atmosphere extends about 6,000 miles above the Earth's surface, the vast proportion of its gases (97 percent) are located in the lower 19 miles. The bulk of the atmosphere consists of nitrogen (78 percent) and oxygen (21 percent). The last 1 percent of the atmosphere contains all the remaining gases, including an inert gas (argon), which accounts for 0.93 percent of the 1 percent, and carbon dioxide ($CO_2$), which makes up a little less than 0.04 percent. Carbon dioxide has the ability to absorb longwave radiation leaving the Earth and shortwave radiation from the Sun; therefore, any increase in carbon dioxide in the atmosphere has profound implications for global warming.

## BACKGROUND AND HISTORY

Evangelista Torricelli, an Italian physicist, mathematician, and secretary to Galileo, invented the barometer, which measures barometric pressure, in 1643. The first attempt to explain the circulation of the global atmosphere was made in 1686 by Edmond Halley, an English astronomer and mathematician. In 1735, George Hadley, an English optician, described a pattern of air circulation that became known as a Hadley cell. In 1835, Gustave-Gaspard Coriolis, a French engineer and mathematician, analyzed the movement of air on a rotating Earth, a pattern that became known as the Coriolis effect. In 1856, William Ferrel, an American meteorologist, developed a model of hemispheric circulation of the atmosphere that became known as a Ferrel cell. Christophorus Buys Ballot, a Dutch meteorologist, explained the relationship between the distribution of pressure, wind speed, and direction in 1860.

Manned hot-air balloon flights beginning in the mid-nineteenth century facilitated high-level observations of the atmosphere. For example, in 1862, English meteorologist James Glaisher and English pilot Henry Coxwell reached 29,000 feet, at which point Glaisher became unconscious and Coxwell was partially paralyzed so that he had to move the control valve with his teeth. In 1902, Léon Teisserenc de Bort of France was able to determine that air temperatures begin to level out at 39,000 feet and actually increase at higher elevations. In the twentieth century, additional information about the upper atmosphere became available through radio waves, rocket flights, and satellites.

## HOW IT WORKS

A knowledge of the basic structure and dynamics of the atmosphere are a necessary foundation for understanding applications and practical uses based on atmospheric science.

**Layers of the Atmosphere.** The heterosphere and the homosphere form the two major subdivisions of the Earth's atmosphere. The uppermost subdivision (heterosphere) extends from about 50 miles above the Earth's surface to the outer limits of the atmosphere at about 6,000 miles. Nitrogen and oxygen, the heavier elements, are found in the lower layers of the heterosphere, and lighter elements such as hydrogen and helium are found at the uppermost layers of the atmosphere. The homosphere (or lowest layer) contains gases that are more uniformly mixed, although their density decreases with height. Some exceptions to this statement occur with the existence of an ozone layer at an altitude of 12 to 31 miles and with variations in concentrations of carbon dioxide, water vapor, and air pollutants closer to the Earth's surface.

The atmosphere can be divided into several zones based on decreasing or increasing temperatures as elevation increases. The lowest zone is the troposphere, where temperatures decrease from sea level up to an altitude of 10 miles in equatorial and tropical regions and up to an altitude of 4 miles at the poles. This lowermost zone holds substantial amounts of water vapor, aerosols that are very small, and light particles that originate from volcanic eruptions, desert surfaces, soot from forest and brush fires, and industrial emissions. Clouds, storms, and weather systems occur in the troposphere.

The tropopause marks the boundary between the troposphere and the next higher layer, the stratosphere, which reaches an altitude of 30 miles above the Earth's surface. Circulation in this layer occurs with strong winds that move from west to east. There is limited circulation between the troposphere and the stratosphere. However, manned balloons, certain types of aircraft (Concorde and the U-2), volcanic eruptions, and nuclear bomb tests are able to break through the tropopause and enter the stratosphere.

The gases in the stratosphere are generally uniformly mixed, with the major exception of the ozone layer, which is found at an altitude range of 12 to 31 miles above the Earth. This layer is extremely important because it shields life on Earth from the intense and harmful ultraviolet radiation from the Sun. The ozone layer has been diminishing because of the release of chlorofluorocarbons (CFCs), organic compounds containing chlorine, fluorine, and carbon, used as propellants in aerosol sprays and in synthetic chemical compounds used for refrigeration purposes. In 1978, the use of CFCs in aerosol sprays was banned in the United States, but they are still used in some refrigeration systems. Other countries continue to use CFCs, which eventually get into the ozone layer and result in ozone holes of considerable size. In 1987, members of the international community took steps to reduce CFC production through the Montreal protocol, and by 2003, the rate of ozone depletion had began to slow down. Although the manufacture and use of CFCs can be controlled, natural events that are detrimental to the ozone layer cannot be prevented. For example, the 1991 eruption of Mount Pinatubo in the Philippines reduced the ozone layer in the midlatitudes nearly 9 percent.

Temperatures decrease with elevation at the stratopause, where the mesosphere layer begins at about 30 miles and continues to an altitude of about 50 miles. The mesopause at about 50 miles marks the beginning of the thermosphere, where the density of the air is very low and holds minimal amounts of heat. However, even though the atmospheric density is minimal at altitudes above 155 miles, there is enough atmosphere to have a drag effect on spaceships.

**Atmospheric Pressure.** The gas molecules in the atmosphere exert a pressure due to gravity that amounts to about 15 pounds per square inch on all surfaces at sea level. As the distance from the Earth gets larger, in contrast to the various increases and decreases in atmospheric temperature, atmospheric pressure decreases at an exponential rate. For example, air pressure at sea level varies from about 28.35 to 31.01 inches of mercury, averaging 29.92 inches. The pressure at the top of Mount Everest at 20,029 feet can get as low as 8.86 inches. This means that each inhalation of air at this altitude is about one-third of the pressure at sea level, producing severe shortness of breath.

**Earth's Global Energy Balance.** The Earth's elliptical orbit about the Sun ranges from 91.5 million miles at perihelion (closest point to the Sun) on January 3 to 94.5 million miles at aphelion (furthest from the Sun) on July 4, averaging 93 million miles. The Earth intercepts only a tiny fraction of the total energy output of the Sun. Upon reaching the Earth, part of the incoming radiation is reflected back into space, and part is absorbed by the atmosphere, land, or oceans. Over time, the incoming shortwave solar radiation is balanced by a return to outer space of longwave radiation.

**Earth-Moon Differences.** Scientists believe that the moon's surface has a large number of craters formed by the impact of meteorites. In contrast, there are relatively few meteorite craters on the Earth, even though, based simply on its size, the Earth is likely to have been hit by as many or even more meteorites than the Moon. This notable difference is attributed to the Earth's atmosphere, which burns up incoming meteorites, particularly small ones (the Moon does not have an atmosphere). Larger meteorites can pass through the Earth's atmosphere, but their impact craters may have been filled in or washed away over millions of years. Only the more recent ones, such as Meteor Crater in northern Arizona, with a diameter of 4,000 feet and a depth of 600 feet, remain easily recognizable.

**Air Masses.** Different types of air masses within the troposphere, the lowest layer of the atmosphere, can be delineated on the basis of their similarity in temperature, moisture, and to a certain extent, air pressure. These air masses develop over continental and maritime locations that strongly determine their physical characteristics. For example, an air mass starting in the cold, dry interior portion of a continent develops thermal, moisture, and pressure differences that can be substantially different from an air mass that develops over water. Atmospheric dynamics also allow air masses to modify their characteristics as they move from land to water and vice versa.

Air mass and weather front terminology were developed in Norway during World War I. Norwegian meteorologists were unable to get weather reports from the Atlantic theater of operations; consequently, they developed a dense network of weather stations that led to impressive advances in atmospheric modeling that are still being used.

**The Radiation Budget.** The incoming solar energy that reaches the Earth is primarily in the shortwave (or visible) portion of the electromagnetic spectrum. The Earth's energy balance is attained by about one-third of this incoming energy being reflected back to space and the other two-thirds leaving the Earth as outgoing longwave radiation. This balance between incoming and outgoing energy is known as the Earth's radiation budget. The decades-long, ongoing National Aeronautics and Space Administration (NASA) program known as Clouds and the Earth's Radiant Energy System (CERES) is designed to measure how much shortwave and longwave radiation leaves the Earth from the top of the atmosphere.

Clouds play a very important role in the global radiation balance. For one thing, they constantly change over time and in type. Some clouds, such as high cirrus clouds found near the top of the troposphere at 40,000 feet, can have a substantial impact on atmospheric warming. Accordingly, the value of CERES is based on its ability to observe if human or natural changes in the atmosphere can be measured even if they are smaller than large-scale energy variations.

**Greenhouse Effect.** Selected gases in the lower parts of the atmosphere trap heat and then radiate some of that heat back to Earth. If there was no natural greenhouse effect, the Earth's overall average temperature would be close to 0 degrees Fahrenheit, rather than the existing 57 degrees Fahrenheit.

The burning of coal, oil, and gas makes carbon dioxide ($CO_2$) the major greenhouse gas, accounting

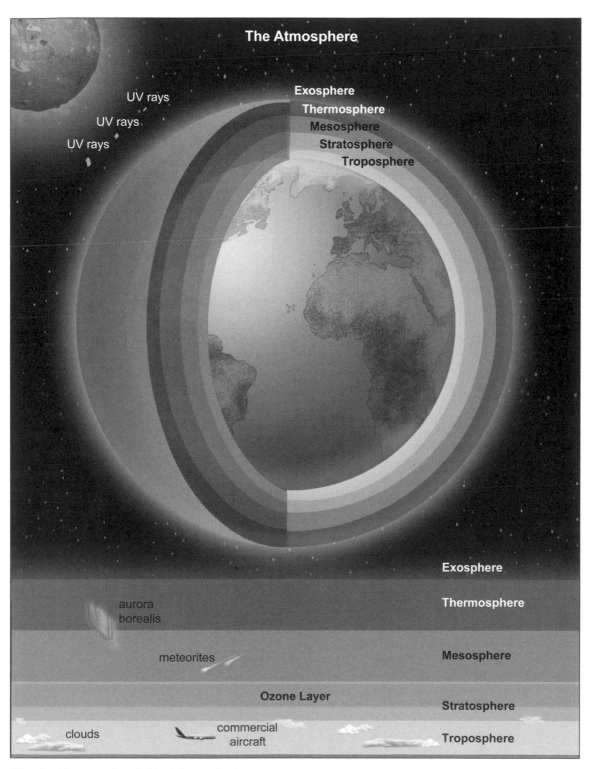

The Earth's atmosphere is divided into five layers.

for nearly half of the total amount of heat-producing gases in the atmosphere. Before the Industrial Revolution in Great Britain in the mid-eighteenth century, the estimated level of carbon dioxide in the atmosphere was about 280 parts per million by volume (ppmv). Estimates for the natural range of carbon dioxide for the past 650,000 years range from 180 to 300 ppmv. All these values are less than the 391 ppmv recorded in January, 2011. Carbon dioxide levels have been increasing since 2000 at a rate of 1.9 ppmv each year. The radiative effect of carbon dioxide accounts for about one-half of all the factors that affect global warming. Estimates of carbon dioxide levels at the end of the twenty-first century range from 490 to 1,260 ppmv.

The second most important greenhouse gas is methane ($CH_4$), which accounts for about 14 percent of all global warming factors. The origin of this gas is attributed to the natural decay of organic matter in wetlands, but anthropogenic activity—rice paddies, manure from farm animals, the decay of bacteria in sewage and landfills, and biomass burning (both natural and human induced)—results in a doubling of the amount of this gas over what would be produced solely by wetland decay.

Chlorofluorocarbons (CFCs) absorb longwave energy (warming effect), but they also have the ability to destroy stratospheric ozone (cooling effect). The warming radiative effect is three times greater than the cooling effect. CFCs account for about 10 percent of all global warming factors. Tropospheric ozone from air pollution and nitrous oxide ($N_2O$) from motor vehicle exhaust and bacterial emissions from nitrogen fertilizers account for about 10 percent and 5 percent, respectively, of all global warming factors.

Several kinds of human actions lead to a cooling of the Earth's climate. For example, the burning of fossil fuels results in the release of tropospheric aerosols, which acts to scatter incoming solar radiation back into space, thereby lowering the amount of solar energy that can reach the Earth's surface. These aerosols also lead to the development of low and bright clouds that are quite effective in reflecting solar radiation back into space.

## APPLICATIONS AND PRODUCTS

Atmospheric science is applied in many ways. It is used to help people better understand their global and interplanetary environment and to make it possible for them to live safely and comfortably within that environment. By using the principles of this field, researchers, engineers, and space scientists have developed a vast number of applications. Among the most important are those used to track and predict weather cycles and climate.

**Remote Sensing Techniques.** Oceans cover about 71 percent of the Earth's surface, which means that large portions of the world do not have weather stations or places where precipitation can be measured with standard rain gauges. To provide more information about precipitation in the equatorial and tropical parts of the world, NASA and the Japan Aerospace Exploration Agency initiated the Tropical Rainfall Monitoring Mission (TRMM) in 1997. The orbit of the TRMM satellite monitors the Earth between 35 degrees north and 35 degrees south latitude. The goal of the study is to obtain information about the extent of precipitation, along with its intensity and length of occurrence. The major instruments on the satellite are radar to detect rainfall, a passive microwave imager that can acquire data about precipitation intensity and the extent of water vapor, and a scanner that can examine objects in the visible and infrared portions of the electromagnetic spectrum. The goal of data collection is to obtain the necessary climatological information about atmospheric circulation in this portion of the Earth to develop better mathematical models for determining large-scale energy movement and precipitation.

**Geostationary Satellites.** Geostationary operational environmental satellites (GOES) enable researchers to view images of the planet from what appears to be a fixed position above the Earth. The satellites are actually circling the globe at a speed that is in step with the Earth's rotation. This means that a satellite at an altitude of 22,200 miles will make one complete revolution in the same twenty-four hours and direction that the Earth is turning above the equator. At this height, the satellite is in a position to view nearly one-half of the planet at any time. On-board instruments can be activated to look for special weather conditions such as hurricanes, flash floods, and tornadoes. On-board instruments are also used to make precipitation estimates during storm events.

**Doppler Radar.** Doppler radar was first used in England in 1953 to pick up the movement of small storms. The basic principle guiding this type of radar is that back-scattered radiation frequency detected at a certain location changes over time as the target, such as a storm, moves. A transmitter is used to send short but powerful microwave pulses. When a foreign object (or target) is intercepted, some of the outgoing energy is returned to the transmitter, where a receiver can pick up the signal. An image (or echo) from the target can then be enlarged and shown on a screen. The target's distance is revealed by the time that elapses between transmission and return. The radar screen cannot only indicate where the precipitation is taking place but also reveal the intensity of the rain by the amount of the echo's brightness. In short, Doppler radar has become a very useful device for determining the location of a storm and the intensity of its precipitation and for obtaining good estimates of the total amount of precipitation.

**Responses to Climate Change.** Since the 1970's, many scientists have pointed out the possibility that human activity is having more than a short-term impact on the atmosphere and therefore on weather and climate. Although much debate continues on the full impact of human activities and greenhouse gas emissions, the atmospheric sciences have led to conferences, United Nations conventions, and agreements among nations on ways that human beings can alter their behavior to halt or at least mitigate the possibility of global climate change. The impact of these agreements, still in their infancy, remains unknown—as does the overall effect of human activity on weather and climate (the models for which are highly complex). However, the insights contributed by the atmospheric sciences to the overall debate on whether climate change is primarily anthropogenic (human caused)—and whether global warming is actually taking place—have caused many nations and individuals to modify their attitudes toward human relationships with the global environment, resulting in national and intergovernmental changes in policies concerning carbon emissions, as well as personal decisions ranging from the consumption of "green" building materials to the purchase of vehicles fueled by noncarbon sources of energy.

## IMPACT ON INDUSTRY

**Global Perspective.** The World Meteorological Organization, headquartered in Geneva, Switzerland, was established to encourage weather station networks to acquire many types of atmospheric data. Accordingly, in 2007, its members decided to expand the Global Observing System (GOS) and other related observing systems, including the Global Ocean Observing System (GOOS), Global Terrestrial Observing System (GTOS), and the Global Climate Observing System (GCOS). Data are being collected from some 10,000 manned and automatic surface weather stations, 1,000 upper-air stations, more than 7,000 ships, 100 moored and 1,000 floating buoys that can drift with the currents, several hundred radars, and more than 3,000 commercial airplanes that can acquire key data on aspects of the atmosphere, land, and ocean surfaces on a daily basis.

**Government Research.** About 180 countries maintain meteorological departments. Although many of these departments are small, the larger countries tend to have well-established governmental organizations. The major U.S. agency involved in the atmospheric sciences is the National Oceanic and Atmospheric Administration (NOAA). The agency's National Climatic Data Center (NCDC) in Asheville, North Carolina, has meteorological records going back to 1880 for both the world and the United States. These records provide invaluable historical information. For example, sea ice in the Arctic Ocean typically reaches its maximum extent in March. The coverage at the end of March, 2010, was 5.8 million square miles, which marked the seventeenth consecutive March with below-average areal extent. The National Climatic Data Center issues monthly temperature and precipitation summaries for all of the states in addition to many specialized climate data publications. It also publishes monthly mean climatic data for temperature, precipitation, barometric pressure, sunshine, and vapor pressure (that portion of atmospheric pressure that is attributed to water vapor at the time of measurement) on a global scale for about 2,000 surface sites. In addition, monthly mean upper air temperatures, dew point depressions, and wind velocities are collected and published for about 500 locations scattered around the world.

**University Research.** Forty-eight states (all but Rhode Island and Tennessee) have either a state climatologist or someone with comparable responsibility. Most of the state climatologists are connected with state universities, in particular, the land grant institutions. The number of cooperative weather stations established since the late nineteenth century to take daily readings of temperature and precipitation in each of the forty-eight states has varied over time.

These cooperative weather stations include public and private water supply facilities, colleges and universities, airports, and interested citizens.

**Industry and Business Sectors.** The number, size, and capability of private consulting firms has increased since the latter part of the twentieth century. Perhaps among the earliest entrants into this market were frost-warning service providers for citrus and vegetable growers in Arizona, Florida, and California. These private companies expanded as better forecasting and warning techniques were developed. One example of a private enterprise using atmospheric sciences is AccuWeather.com, which has seven global and fourteen regional forecast models for the United States and North America and prepares a daily weather report for *The New York Times*.

## CAREERS AND COURSE WORK

The study of the physical characteristics of the atmosphere falls within the purview of atmospheric scientists. Those interested in a career in this technical area should recognize that there are several categories of specialization. The major group of specialists are operational meteorologists, who are responsible for weather forecasts. They have to carefully study the temperature, humidity, wind speed, and barometric pressure from a number of weather stations to make daily and long-range forecasts. They use data from weather satellites, radar, special sensors, and observation stations in other locations to make forecasts.

In contrast to meteorologists, who focus on short-term weather forecasts, the study of changes in weather over longer periods of time such as months, years, and in some cases centuries, is handled by climatologists. Other atmospheric scientists concentrate on research. For example, physical meteorologists are concerned with various aspects of the atmosphere, such as its chemical and physical properties, energy transfer, severe storm mechanics, and the spread of air pollutants over urbanized areas. The growing interest in air pollution and water shortages has led to another group of research scientists known as environmental meteorologists.

Given the importance of weather forecasting on a daily basis, operational meteorologists who work in weather stations may work on evenings, weekends, and holidays. Research scientists who are not engaged in weather forecasts may work regular hours.

In 2009, the American Meteorological Society estimated that there are about one hundred undergraduate and graduate atmospheric science programs in the United States that offer courses is such departments as physics, earth science, environmental science, geography, and geophysics. Entry-level positions usually require a bachelor's degree with at least twenty-four semester hours in courses covering atmospheric science and meteorology. The acquisition of

---

## Fascinating Facts About Atmospheric Sciences

- In April, 1934, a surface wind gust of 231 miles per hour was recorded at Mount Washington Observatory in New England.

- The coldest temperature in the world, −127 degrees Fahrenheit, was recorded at the Russian weather station at Vostok in Antarctica in 1958.

- The world's lowest barometric pressure (25.69 inches of mercury) was recorded on October 12, 1979, during Typhoon Tip in the western Pacific Ocean, and the highest (32.06 inches) was recorded in Tosontsengel, Mongolia, on December 19, 2001.

- One study found that cloud formation and rain occurred more frequently on weekdays than weekends, because of higher levels of air pollutants, at least in humid regions with cities.

- Acid rain, a mixture containing higher than normal amounts of nitric and sulfuric acid, occurs when emissions of sulfuric acid and nitrogen dioxide (released from power plants and decaying vegetation) react with oxygen, water, and other chemicals, forming acidic compounds.

- Mean sea levels for the world increased 0.07 inches per year in the twentieth century, an amount that is much larger than the average rate of increase for the last several thousand years.

- The specific chemicals in particulate matter, an air pollutant, depend on the source and the geographic location. They include inorganic compounds such as sulfate, nitrate, and ammonia; organic compounds formed by the incomplete burning of wood, gasoline, and diesel fuels; and secondary organic aerosols, new compounds formed when pollutants combine.

- Winds that reach the center of a low-pressure area rise, causing condensation, cloud formation, and storms. Dry winds are drawn down the center of a high-pressure area, resulting in fair weather.

a master's degree enhances the chances of employment and usually means a higher salary and more opportunities for advancement. A doctorate is required only for those who want a research position at a university.

In 2008, excluding research positions in colleges and universities, about 9,400 atmospheric scientists were working, and about one-third were employed by the federal government. Above-average employment growth is projected until 2018, representing a 15 percent increase from 2008. The median annual average salary for atmospheric scientists in May, 2008, was $81,290. The middle 50 percent had earnings between $55,140 and $101,340.

## SOCIAL CONTEXT AND FUTURE PROSPECTS

Climate change may be caused by both natural internal/external processes in the Earth-Sun system or by human-induced changes in land use and the atmosphere. Article 1 of the United Nations Framework Convention on Climate Change (entered into force March, 1994) states that the term "climate change" should refer to anthropogenic changes that affect the composition of the atmosphere rather than natural causes, which should be referred to "climate variability." An example of natural climate variability is the global cooling of about 0.5 degrees Fahrenheit in 1992-1993 that was caused by the 1991 eruption of Mount Pinatubo in the Philippines. The 15 million to 20 million tons of sulfuric acid aerosols ejected into the stratosphere reflected incoming radiation from the sun, thereby creating a cooling effect. Many suggest that the above-normal temperatures experienced in the first decade of the twenty-first century provide evidence of climate change caused by human activity. Based on a variety of techniques that allow scientists to estimate the temperature in previous centuries, the year 2005 was the warmest in the last thousand years. A 2009 article published by the American Geophysical Union suggests that human intervention in Earth systems has reached a point where the Holocene epoch of the past 12,000 years is becoming a new Anthropocene epoch in which human systems have become primary Earth systems rather than simply influencing natural systems.

Numerous observations strongly suggest a continuing warming trend. Snow and ice have retreated from areas such as Mount Kilimanjaro in Tanzania, which at 19,340 feet is the highest mountain in Africa.

Glaciated areas in Switzerland also provide evidence of this warming trend. The Special Report on Emission Scenarios issued in 2001 by the Intergovernmental Panel on Climate Change (IPCC) examined the broad spectrum of possible concentrations of greenhouse gases by considering the growth of population and industry along with the efficiency of energy use. The IPCC computer climate models were used to estimate future trends. For example, a global temperature increase of 35.2 to 39.2 degrees Fahrenheit by the year 2100 is a IPCC standard estimate.

*Robert M. Hordon, B.A., M.A., Ph.D.*

## FURTHER READING

Christopherson, Robert W. *Geosystems: An Introduction to Physical Geography*. 8th ed. Upper Saddle River, N.J.: Pearson Prentice Hall, 2012. Covers many topics in atmospheric sciences. Color illustrations.

Coley, David A. *Energy and Climate Change: Creating a Sustainable Future*. Hoboken, N.J.: John Wiley & Sons, 2008. A detailed review of energy topics and their relationship to climate change and energy technologies.

Ellis, Erle C., and Peter K. Haff. "Earth Science in the Anthropocene: New Epoch, New Paradigm, New Responsibilities." *EOS, Transactions, American Geophysical Union* 90, no. 49 (2009): 473. Makes the point that human systems are no longer simply influencing the natural world but are becoming part of it.

Gautier, Catherine. *Oil, Water, and Climate: An Introduction*. New York: Cambridge University Press, 2008. A good discussion of the impact of fossil fuel burning, particularly on climate change.

Lutgens, Frederick K., and Edward J. Tarbuck. *The Atmosphere: An Introduction to Meteorology*. 11th ed. Upper Saddle River, N.J.: Prentice Hall, 2010. A very good textbook written with considerable clarity.

Strahler, Alan. *Introducing Physical Geography*. 5th ed. Hoboken, N.J.: John Wiley, & Sons 2011. Covers weather and climate; contains superlative illustrations, clear maps, and lucid discussions of the material.

Wolfson, Richard. *Energy, Environment, and Climate*. New York: W. W. Norton, 2008. Provides an extensive discussion of the relationship between energy and climate change.

## WEB SITES

*American Geophysical Union*
http://www.agu.org

*American Meteorological Society*
http://www.ametsoc.org

*Intergovernmental Panel on Climate Change*
http://www.ipcc.ch

*International Association of Meteorology and Atmospheric Sciences*
http://www.iamas.org

*National Oceanic and Atmospheric Administration*
National Climatic Data Center
http://www.Ncdc.noaa.gov

*National Weather Association*
http://www.nwas.org

*U.S. Environmental Protection Agency*
Air Science
http://www.epa.gov/airscience

*U.S. Geological Survey*
Climate and Land Use Change
http://www.usgs.gov/climate_landuse

*World Meteorological Organization*
Climate
http://www.wmo.int/pages/themes/climate/index_en.php

**See also:** Barometry; Climate Engineering; Climate Modeling; Climatology; Measurement and Units; Meteorology; Remote Sensing; Temperature Measurement.

# AUDIO ENGINEERING

## FIELDS OF STUDY

Acoustics; analogue signal processing; digital signal processing; electronic sound; electronics; live sound; microphone design and use; mixing and mastering; postproduction editing; recording software; sound compression; sound synchronization.

## SUMMARY

Audio engineering is the capture, enhancement, and reproduction of sounds. It requires an aesthetic appreciation of music and sound quality, a scientific understanding of sound physics, and a technical familiarity with recording equipment and computer software. This applied science is essential to the music industry, film, television, and video game production, live television and radio broadcasting, and advertising. In addition, it contributes to educational services for the visually impaired and to forensic evidence analysis.

## KEY TERMS AND CONCEPTS

- **Acoustic Sound:** Natural sound that is not produced or enhanced electronically.
- **Analogue Signal:** Information transmitted as a continuous signal that is responsive to change.
- **Compression:** Removing portions of a digital signal to store audio data in less space.
- **Digital Signal:** Information transmitted as a string of binary code without distortion.
- **Electronic Sound:** Sound that is produced or enhanced with technological manipulation, such as computer software.
- **Equalization:** Boosting or weakening bass or treble frequencies to achieve a desired sound.
- **Mastering:** Polishing the final audio product and compressing it for commercial reproduction.
- **Mixing:** Blending multiple recording tracks to achieve a desired effect.
- **Syncing:** Synchronizing an audio track with a video sequence.

## DEFINITION AND BASIC PRINCIPLES

Audio engineering, also known as sound engineering and audio technology, is the recording, manipulation, and reproduction of sound, especially music. Audio engineers run recording sessions, work the equipment, and collaborate on the finished product. They are simultaneously technicians, scientists, and creative advisers. They work in recording studios, producing the following: instrumental and vocal music recordings; film and television soundtracks, syncing, and sound effects; music and voice-overs for radio and television commercials; and music and sound effects for video games.

A recording studio is a specialized environment designed to capture sounds accurately for enhancement and reproduction. The acoustic sounds produced by instruments and vocalists are picked up by strategically placed microphones and transmitted as analogue electrical signals to recording equipment, where they may be converted into digital data. Signals may be modified by the use of a mixing console, also called a mixing board or sound board, that changes the characteristics and balance of the input, which may be coming from multiple microphones or signals recorded in different sessions. The final product then undergoes mastering for commercial reproduction and compression for distribution in a digital format.

Audio engineers are not acoustic engineers: graduates of a formal university program in engineering who work with architects and interior designers to plan and install audio systems for large venues such as churches, school auditoriums, and concert halls. Audio engineers are engineers in the sense that they are needed to devise a creative solution to a complex sound challenge and oversee its implementation.

## BACKGROUND AND HISTORY

Sound capture and reproduction, and thus audio engineering, began with the invention of the phonograph by Thomas Alva Edison in 1877. Sound was recorded on cylinders; the first were wrapped in tin foil and later ones in wax. By 1910, cylinders were replaced with disks, which held longer recordings, were somewhat louder, and could be more economically mass-produced. The disks were spun on a turntable at standard speeds—initially 78 revolutions per minute (rpm). Larger disks were played at 33 rpm and smaller disks were played at 45 rpm. Discs were originally made of shellac and later made of vinyl. They were played with needles (styli) made of industrial diamond, which held a point.

Concurrently, RCA was creating microphones that improved recorded sound quality. In the 1940's, sound began being recorded on magnetic tape and could be reproduced in stereo and as mixed multiple tracks. Digital technology appeared in the 1980's and by the turn of the century, digital recordings were produced with computer technology. Using data compression, digital audio recordings can produce quality replication of the original music in a format that requires less data storage (computer memory); MP3 is one such format and is popular for portable consumer music systems.

### How It Works

**Sound.** Sound is the waves of pressure a vibrating object emits through air or water. The three most meaningful characteristics of sound waves are wavelength, amplitude, and frequency. The wavelength is the distance between equivalent points on consecutive waves, such as peak to peak. A short wavelength means that more waves are produced per second, resulting in a higher sound. The amplitude is the strength of the wave; the greater the amplitude, the greater the volume (loudness). The frequency is the number of wavelengths that occur in one second; the greater the frequency, the higher the pitch because the sound source is vibrating quickly.

**Hearing.** Hearing is the ability to receive, sense, and decipher sounds. To hear, the ear must direct the sound waves inside, sense the sound vibrations, and translate the sensations into neurological impulses that the brain can recognize. The outer ear funnels sound into the ear canal. It also helps the brain determine the direction from which the sound is coming.

When the sound waves reach the ear canal, they vibrate against the eardrum. These vibrations are amplified by the eardrum's movement against three tiny bones (the malleus, incus, and stapes) located behind it. The stapes rests against the cochlea, and when it transmits the sound, it creates waves in the fluid of the cochlea.

The cochlea is a coiled, fluid-filled organ that contains 30,000 hairs of different lengths that resonate at different frequencies. Vibrations of these hairs trigger complex electrical patterns that are transmitted along the auditory nerve to the brain, where they are interpreted.

The frequency of sound waves is measured in hertz (Hz). Humans have a hearing range from 20 to 20,000 Hz. Another name for the frequency of a sound wave is the musical pitch. Pitches are often referred to as musical notes, such as middle C.

The relative loudness of a sound compared with the threshold of human hearing is measured in decibels (dB). Conversation is usually conducted at 40 to 60 dB, while a car passing at 10 meters may be 80 to 90 dB, a jet engine 100 meters away may be 110 to 140 dB, and a rifle fired 1 meter away is 150 dB. Long-term (not necessarily continuous) exposure to sounds greater than 85 dB may cause hearing loss.

The human ear can discern between two musical instruments playing the same note at the same volume by the recognition of a sound characteristic called timbre. Often described by adjectives such as bright versus dark, smooth versus harsh, and regular versus random or erratic, timbre is often what distinguishes music from noise.

**Sound Capture.** Transducers are devices that change energy from one form into another. A microphone changes acoustical signals into electrical signals, while a speaker changes electrical signals into acoustical signals. The source of the incoming electrical signals may be immediate, such as a microphone or electrical musical instrument, or it may be a recording, such as a magnetic tape or compact disc.

Microphones come in many varieties, such as dynamic, ribbon,

## Sound Waves

source
receiver
"hello"

*Sound is transmitted through the air as a sound wave.*

condenser, parabolic, and lavaliere. They also vary by their polar patterns, that is, their area of sensitivity to sounds coming in from different directions relative to the receiving membrane. They may be omnidirectional (sensitive to sounds coming from all directions), unidirectional (intended for directed sound reception), or cardioid (having a heart-shaped area of sensitivity). The choices of variety, polar pattern, and placement affect the quality and quantity of sound capture.

**Signal Processing.** The auditory electrical signal from a microphone is relatively weak, so it must be amplified before the sound can be deliberately modified through signal processing. Incoming sound may be modified in its analogue form or converted to digital data before alteration. Sound mixing is the process of blending sounds from multiple sources into a desired end product. It often starts with finding a balance between vocal and instrumental music or dissimilar instruments so that one does not overshadow the other. It involves the creation of stereo or surround sound from the placement of sound in the sound field to simulate directionality (left, center, or right). Equalizing adjusts the bass and treble frequency ranges. Effects such as reverberation may be added to create dimension. Signals may undergo gating and compression to remove unwanted noise and extraneous data selectively.

**Sound Output.** Auditory electrical signals may then be sent to speakers, where they are converted into acoustical signals to be heard by a live audience. Digital signals may be broadcast in real time over the Internet as streaming audio. Otherwise, the processed signals may be stored for future reproduction and distribution. Analogue signals may be stored on magnetic tape. Digital signals may be stored on a compact disc or subjected to MP3 encoding for storage on a computer or personal music player.

## APPLICATIONS AND PRODUCTS

**Instrumental and Vocal Music.** As specialists in the capture, enhancement, and reproduction of sound, audio engineers are crucial to successful recording sessions. They collaborate with producers and performers technically to generate the shared artistic vision. They determine the choice and placement of microphones and closely scrutinize the parameters of the incoming signals to collect sufficient data with which to work. They manage the scheduling of studio sessions to keep all participants working efficiently, especially when multiple tracks are being recorded and mixed at different times. They act professionally and deliver the finished product with the highest quality possible.

Audio engineers are also responsible for the restoration of classic recordings that would otherwise be lost. They rescue the raw data that was captured in the first recording, strengthen the sound while preserving the style of the original period, and return it to audiences in a contemporary format.

Because musicians go on concert tours, audio engineers accompany them to provide optimum live sound quality in each different venue. They conduct sound checks before performances and make adjustments for conditions such as wind on outdoor stages.

**Film and Television.** In the recording studio, audio engineers oversee the production of music soundtracks for films and television shows. Unlike songs that stand alone, the music must be carefully synchronized to the action of the film. It must also swell and ebb with precision to arouse audience emotion.

Foley recording is the production of sound effects that are inserted into videos after they are filmed to add realism and dramatic tension. Foley recording can be synchronized efficiently to video footage because the sound effects are produced in real time, not modified from stock recordings. In addition, sounds that do not exist in reality and so would not be catalogued in a prerecorded audio library must be created.

**Live Broadcasting.** Audio engineers may be seen sitting at mixing consoles or computers monitoring and adjusting the audio input and output quality at church services, lectures, theatrical performances, and events held in large auditoriums. They may similarly be found as part of a broadcasting team at live sporting events held outdoors, such as football or baseball games, golf tournaments, and the Olympic Games.

**Radio and Television Commercials.** Audio engineers are instrumental in the production of radio and television commercials, not only for their recording and sound-processing skills but also for their production skills. Because advertising time is sold in specific brief allotments, engineers must encourage the actors to perform at an accelerated pace and later edit the audio to fit within the time allowed. They may also be asked to recruit or audition competent musicians and voice actors to meet the client's needs.

**Video Games.** The skills of audio engineers enhance the production of popular video games. In addition to providing sound effects such as explosions and gunfire, engineers must create appropriate imaginary sounds such as spaceships landing, ambient sound effects such as slot machine bells and crowd murmurs, and realistic situational sounds, such as footsteps going from grass to gravel. In some cases, they may be called on to provide minor character voices or record spoken instructions.

**Forensic Evidence Analysis.** Police may seek the assistance of an experienced audio engineer to remove unnecessary background noise from covert recordings of suspected criminals and to make voiceprint comparisons with known exemplars. Voiceprints, vocal qualities that can be demonstrated on a sound spectrograph, are personal because each person's oral and pharyngeal anatomy is distinctive; however, they are not unique like fingerprints because children often sound like their parents and share similar voiceprints. Research has shown that the error rates of misidentifying suspects (false positives) and improperly eliminating suspects (false negatives) are respectably low.

**Audio Books.** Recordings of books originated in 1932 under the auspices of the American Foundation for the Blind as educational tools for the visually impaired. Books were recorded on shellac discs and played on a turntable. Books on audio cassettes came along twenty years later, and later audio books could be listened to on CDs or portable digital music devices. Audio engineers are responsible for processing the audio signal to optimize the clarity of human speech and editing numerous recitations into one continuous, flawless performance.

## IMPACT ON INDUSTRY

**Employment Statistics.** According to the U.S. Bureau of Labor Statistics, nearly 200,000 people are presently employed in the audio engineering field. Further growth is expected, with a 17 percent increase predicted to occur between 2006 and 2016. Although the Consumer Electronics Association reported that from 2000 to 2009, American consumers spent 35 percent less on home stereo components (about $960 million), in that same time period, sales of portable digital music devices exploded to $5.4 billion. People are still willing to spend money on

---

### Fascinating Facts About Audio Engineering

- Foley recording is the real-time recording of sound effects that are inserted into videos in post-production to add realism and dramatic tension. The field originated in 1927 and is named for Jack Foley, who worked on films at Universal Studios, adding sounds other than dialogue to "talkies."

- Late actor Don LaFontaine, best known for his commercial and film trailer voice-overs, was originally an audio engineer. He began his long, successful voice-over career in 1964, when he had to fill in at the microphone for a missing actor in an important presentation to MGM.

- The Moog synthesizer became popular in 1967 after it was played at the Monterey International Pop Festival. Its first commercial success as an electronic instrument was on the 1968 Wendy Carlos classical music album *Switched-On Bach.*

- In 1975, the University of Miami's Frost School of Music was the first American university to offer a bachelor of music degree in music engineering technology.

- Just as sound studio equipment can be used to clean up extraneous noise from old recordings for remastering, it can also be used in forensic voice analysis, also called voiceprinting. In 2002, former audio engineer Tom Owen enhanced a foreign radio broadcast of Osama bin Laden and compared it with a 1998 interview to determine that bin Laden was still alive.

- The proprietary computer software Auto-Tune, which adjusts recorded singing to the right note or pitch, has been described as "Photoshop for the human voice." It was creatively misused in the recording of Cher's 1998 hit "Believe."

---

music, keeping audio engineers in business. As of May, 2006, the median annual income of audio engineers was $43,010; the top 10 percent were earning more than $90,770. Though audio engineers are needed in various industries (radio, television, film, and advertising), there is a great deal of competition for audio engineering jobs in large cities.

**Professional Organizations.** The more than 14,000 members of the Audio Engineering Society, an international group founded in 1948 in the United States, include recording engineers, broadcast technicians, acousticians, mixing engineers, equipment designers, and mastering engineers. They share creative and scientific information about audio standards and technologies. Sustaining members from manufacturing, research, and fields related to audio engineering include Bose, Dolby Laboratories, Klipsch Group, Motorola, and Sennheiser.

## CAREERS AND COURSE WORK

Technical and vocational schools offer diploma and certificate programs in audio engineering along with internships for hands-on experience. The Musicians Institute in Los Angeles offers a certificate in audio engineering with subspecialties such as postproduction or live-sound production. Other schools, such as the Conservatory of Recording Arts and Sciences in Gilbert, Arizona, provide certification for demonstrated proficiency in software and equipment related to audio engineering.

Some universities, such as Indiana University, offer a bachelor of arts in music degree with a concentration in the music industry and a track within that concentration in sound engineering. Texas State University School of Music offers a bachelor of science degree in sound recording technology. The Peabody Institute at Johns Hopkins University offers a bachelor of music degree in recording arts and sciences and a master of arts degree in audio sciences.

Many audio engineers are self-taught. Audio engineering is a multifaceted discipline, requiring a creative appreciation of music and sound quality, a scientific understanding of sound-wave physics, and precise technical familiarity with recording equipment and computer software.

Careers vary with expertise and experience. An assistant audio engineer is typically responsible for the setup and breakdown of a recording session, including the placement of the microphones. A staff engineer records the sound. A mixing engineer coordinates multiple recording tracks to produce the desired effect. A mastering engineer adds the finishing touches to the final product and compresses it for mass duplication. A chief engineer works with the record producer, making the technical decisions that help achieve the artistic vision for the project.

## SOCIAL CONTEXT AND FUTURE PROSPECTS

Audio engineering is a combination of technology, science, and art. Advancements in audio engineering will come in all three areas. On the technical front, classic (especially pre-1920) recordings will continue to be found, researched, and digitally restored. Improved transducer materials are being sought and new computer software applications for signal processing are being developed. Surround sound is being refined to accompany three-dimensional and high-definition television programs and films as well as video games.

Scientific research into psychoacoustics, the study of sound perception, is expanding. The eventual understanding of how music affects a person's brain will advance the field of music therapy, which seems to touch every facet of a person's being to restore and maintain health. Researchers are also exploring the connections between sound characteristics and the perceptions of timbre and spatial placement and between these perceived attributes and listening preference.

The artistic manipulation of sound is broadening the definition of music and musical instruments. Computer-mediated music has inspired the creation of mobile phone and laptop orchestras. Music enhancement by selectively masking undesired frequencies of instruments and highlighting others is introducing new sound combinations previously not experienced.

*Bethany Thivierge, B.S., M.P.H.*

## FURTHER READING

Friedman, Dan. *Sound Advice: Voiceover from an Audio Engineer's Perspective*. Bloomington, Ind.: AuthorHouse, 2010. An audio engineer who is passionate about his work presents basic yet often overlooked information, especially about recording equipment and studio etiquette.

Hampton, Dave. *The Business of Audio Engineering*. New York: Hal Leonard, 2008. Schools offer practical training, but this widely respected expert provides strategies for maneuvering within the industry of audio engineering, including professional presentation and customer relations.

_____. *So, You're an Audio Engineer: Well, Here's the Other Stuff You Need to Know*. Parker, Colo.: Outskirts Press, 2005. With years of experience working with musicians, singers, and other entertainment personalities, the author provides insights into the customer-service aspects of audio engineering.

Powell, John. *How Music Works: The Science and Psychology of Beautiful Sounds, from Beethoven to the Beatles and Beyond*. New York: Little, Brown, 2010. This book, which comes with a CD, presents the components of music and explains what makes music more than sound.

Talbot-Smith, Michael. *Sound Engineering Explained.* 2d ed. Woburn, Mass.: Focal Press, 2001. This book explains the basic principles of audio engineering.

_____, ed. *Audio Engineer's Reference Book.* 2d ed. Woburn Mass.: Focal Press, 1999. A complete resource that explains the scientific principles of sound and provides technical information about audio equipment.

**WEB SITES**
*Acoustical Society of America*
http://acousticalsociety.org

*Audio Engineering Society*
http://www.aes.org

*Society of Professional Audio Recording Services*
http://spars.com

**See also:** Acoustics; Music Technology.

# AUDIOLOGY AND HEARING AIDS

## FIELDS OF STUDY

Biology; chemistry; physics; mathematics; anatomy; physiology; genetics; pharmacology; neurology; head and neck anatomy; acoustics.

## SUMMARY

Audiology is the study of hearing, balance, and related ear disorders. Hearing disorders may be the result of congenital abnormalities, trauma, infections, exposure to loud noise, some medications, and aging. Some of these disorders may be corrected by hearing aids and cochlear implants. Hearing aids amplify sounds so that the damaged ears can discern them. Some fit over the ear with the receiver behind the ear, and some fit partially or completely within the ear canal. Cochlear implants directly stimulate the auditory nerve, and the brain interprets the stimulation as sound.

## KEY TERMS AND CONCEPTS

- **Analogue Signal Processing:** Process in which sound is amplified without additional changes.
- **Audiologist:** Licensed professional who assesses hearing loss and oversees treatment.
- **Auditory Nerve:** Nerve that carries stimuli from the cochlea to the brain.
- **Cochlea:** Coiled cavity within the ear that contains nerve endings necessary for hearing.
- **Deafness:** Full or partial inability to detect or interpret sounds.
- **Digital Signal Processing:** Process in which sound is received and mathematically altered to produce clearer, sharper sound.
- **Hearing:** Ability to detect and process sounds.
- **Ototoxicity:** Damage to the hair cells of the inner ear, resulting in hearing loss.
- **Sound:** Waves of pressure that a vibrating object emits through air or water.

## DEFINITION AND BASIC PRINCIPLES

Audiology is the study of hearing, balance, and related ear disorders. Audiologists are licensed professionals who assess hearing loss and related sensory input and neural conduction problems and oversee treatment of patients.

Hearing is the ability to receive, sense, and decipher sounds. To hear, the ear must direct the sound waves inside, sense the sound vibrations, and translate the sensations into neurological impulses that the brain can recognize.

The outer ear funnels sound into the ear canal. It also helps the brain determine the direction from which the sound is coming. When the sound waves reach the ear canal, they vibrate the eardrum. These vibrations are amplified by the eardrum's movement against three tiny bones behind the eardrum. The third bone rests against the cochlea; when it transmits the sound, it creates waves in the fluid of the cochlea.

The cochlea is a coiled, fluid-filled organ that contains 30,000 hairs of different lengths that resonate at different frequencies. Vibrations of these hairs trigger complex electrical patterns that are transmitted along the auditory nerve to the brain, where they are interpreted.

## BACKGROUND AND HISTORY

The first hearing aids, popularized in the sixteenth century, were large ear trumpets. In the nineteenth century, small trumpets or ear cups were placed in acoustic headbands that could be concealed in hats and hairstyles. Small ear trumpets were also built into parasols, fans, and walking sticks.

Electrical hearing devices emerged at the beginning of the twentieth century. These devices, which had an external power source, could provide greater amplification than mechanical devices. The batteries were large and difficult to carry; the carrying cases were often disguised as purses or camera cases. Zenith introduced smaller hearing aids with vacuum tubes and batteries in the 1940's.

In 1954, the first hearing aid with a transistor was introduced. This led to hearing aids that were made to fit behind the ear. Components became smaller and more complex, leading to the marketing of devices that could fit partially into the ear canal in the mid-1960's and ones that could fit completely into the ear canal in the 1980's.

**HOW IT WORKS**

Audiologists are concerned with three kinds of hearing loss: conductive hearing loss, in which sound waves are not properly transmitted to the inner ear; sensorineural hearing loss, in which the cochlea or the auditory nerve is damaged; and mixed hearing loss, which is a combination of these.

**Conductive Hearing Loss.** Otosclerosis, inefficient movement of the three bones in the middle ear, results in hearing loss from poor conduction. This disease is treatable with surgery to replace the malformed, misaligned bones with prosthetic pieces to restore conductance of sound waves to the cochlea.

Meniere's disease is thought to result from an abnormal accumulation of fluid in the inner ear in response to allergies, blocked drainage, trauma, or infection. Its symptoms include vertigo with nausea and vomiting and hearing loss. The first line of treatment is motion sickness and antinausea medications. If the vertigo persists, treatment with a Meniett pulse generator may result in improvement. This device safely applies pulses of low pressure to the middle ear to improve fluid exchange.

Hearing loss may result from physical trauma, such as a fracture of the temporal bone that lies just behind the ear or a puncture of the eardrum. These injuries typically heal on their own, and in most cases, the hearing loss is temporary.

A gradual buildup of earwax may block sound from entering the ear. Earwax should not be removed with a cotton swab or other object inserted in the ear canal; that may result in infection or further impaction. Earwax should be softened with a few drops of baby oil or mineral oil placed in the ear canal twice a day for a week and then removed with warm water squirted gently from a small bulb syringe. Once the wax is removed, the ear should be dried with rubbing alcohol. In stubborn cases, a physician or audiologist may have to perform the removal.

Foreign bodies in the ear canal, most commonly toys or insects, may block sound. If they can be seen clearly, they may be carefully removed with tweezers. If they cannot be seen clearly or moved, they may be floated out. For toys, the ear canal is flooded with warm water squirted gently from a small bulb syringe. For insects, the ear canal should first be filled with mineral oil to kill the bug. If the bug still cannot be seen clearly, the ear canal may then be flooded with warm water squirted gently from a small bulb syringe.

Once the object is removed, the ear canal should be dried with rubbing alcohol.

**Sensorineural Hearing Loss.** Some medications have adverse effects on the auditory system and may cause hearing loss. These medications include large doses of aspirin, certain antibiotics, and some chemotherapy agents. Doses of antioxidants such as vitamin C, vitamin E, and ginkgo biloba may ameliorate these ototoxic effects.

Exposure to harmful levels of noise, either over long periods or in a single acute event, may result in hearing loss. If the hair cells in the cochlea are destroyed, the hearing loss is permanent. Hearing aids or cochlear implants may be necessary to compensate.

An acoustic neuroma is a noncancerous tumor that grows on the auditory nerve. It is generally surgically removed, although patients who are unable to undergo surgery because of age or illness may undergo stereotactic radiation therapy instead.

Presbycusis is the progressive loss of hearing with age as a result of the gradual degeneration of the cochlea. It may be first noticed as an inability to hear high-pitched sounds. Hearing aids are an appropriate remedy.

**Mixed Hearing Loss.** Infections of the inner ear by the viruses that cause mumps, measles, and chickenpox and infections of the auditory nerve by the viruses that cause mumps and rubella may cause permanent hearing loss. Because fluid builds up and viruses do not respond to antibiotics, viral ear infections may require surgical treatment. In a surgical procedure called myringotomy, a small hole is created in the eardrum to allow the drainage of fluid. A small tube may be inserted to keep the hole open long enough for drainage to finish.

Some children are born with hearing loss as the result of congenital abnormalities. Screening to determine the nature and severity of the hearing loss is difficult. Children are not eligible for cochlear implants before the age of twelve months. Young children often do well with behind-the-ear hearing aids, which are durable and can grow with them.

**APPLICATIONS AND PRODUCTS**

Audiological products consist mainly of hearing aids and cochlear implants.

**Hearing Aids.** Although hearing aids take many forms, basically, all of them consist of a microphone that collects sound, an amplifier that magnifies the sound, and a speaker that sends it into the ear canal. They require daily placement and removal, and must be removed for showering, swimming, and battery replacement. Most people do not wear them when sleeping.

Hearing aids that sit behind the ear consist of a plastic case, a tube, and an earmold. The case contains components that collect and amplify the sound, which is then sent to the earmold through the tube. The earmold is custom-made to fit comfortably. This type of hearing aid is durable and well suited for children, although their earmolds must be replaced as they grow.

Hearing aids that sit partially within the ear canal are self-contained and custom-molded to the ear canal, so they are not recommended for growing children. Because of the short acoustic tube that channels the amplified sound, they are prone to feedback, which makes them less than ideal for people with profound hearing loss. Newer models offer feedback cancellation features.

Hearing aids that sit completely within the ear canal are self-contained and custom-molded. However, thin electrical wires replace the acoustic tube, so this type of hearing aid is free of feedback and sound distortion. They are not well suited for elderly people because their minimal size limits their volume capabilities.

The first hearing aids were analogue, a process that amplified sound without changing its properties. Although amplification was adjustable, the sound was not sorted, so background noise was also amplified. New hearing aids are digital, a process in which sound waves are converted into binary data that can be cleaned up to deliver clearer, sharper sounds.

The choice of hearing aid depends on the type and severity of hearing loss. Hearing aids will not restore natural hearing. However, they increase a person's awareness of sounds and the direction from which they are coming and heighten discernment of words.

**Cochlear Implants.** The U.S. Food and Drug Administration approved cochlear implants in the mid-1980's. A cochlear implant differs from a hearing aid in its structure and its function. It consists of a microphone that collects sound, a speech processor that sorts the incoming sound, a transmitter that relays the digital data, a receiver/stimulator that converts the sound into electrical impulses, and an electrode array that sends the impulses from the stimulator to the auditory nerve. Hearing aids compensate for damage within the ear; cochlear implants avoid the damage within the ear altogether and directly stimulate the auditory nerve.

Whereas hearing aids are completely removable, cochlear implants are not. The internal receiver/stimulator and electrode array are surgically implanted. Three to six weeks after implantation, when the surgical incision has healed, the external portions are fitted. The transmitter and receiver are held together through the skin by magnets. Thus, the external portions may be removed for sleep, showering, and swimming.

### IMPACT ON INDUSTRY

Audiology is practiced in many clinics and hospitals throughout the United States, and research in the field is conducted at governmental agencies and facilities associated with universities and medical institutions. Hearing aids, cochlear implants, and other devices designed to assist people with hearing loss are the focus of many medical equipment manufacturers.

**University Research.** A Massachusetts Institute of Technology research team, led by electrical engineering professor Dennis M. Freeman, is studying how the ear sorts sounds in an effort to create the next generation of hearing aids. This team is studying how the tectorial membrane of the cochlea carries sound waves that move from side to side, while the basilar membrane of the cochlea carries sound waves that move up and down. Together, these two membranes detect individual sounds. This team has genetically engineered mice that lack a crucial gene and protein of the tectorial membrane to study the compromised hearing.

Second-year audiology students at the University of Western Australia in Perth are conducting research projects in the development of new hearing assessment tests for children of all ages. These tests include electrophysiological studies for infants, efficient screening for children beginning school, and audiovisual games for grade school students.

**Government Research.** Researchers funded by the National Institute on Deafness and Other Communication Disorders (NIDCD) are looking at ways that hearing aids can selectively enhance sounds to improve the comprehension of speech. A related research team is studying ears on animals to fine-tune directional microphones that may make conversations more easily understood. Another related research team is learning to use computer-aided design programs to create and fabricate better fitting and performing hearing aids.

The U.S. Department of Defense oversees the Military Audiology Association, an organization that strives to provide hearing health care, education, and research to American fighting forces. Its Committee

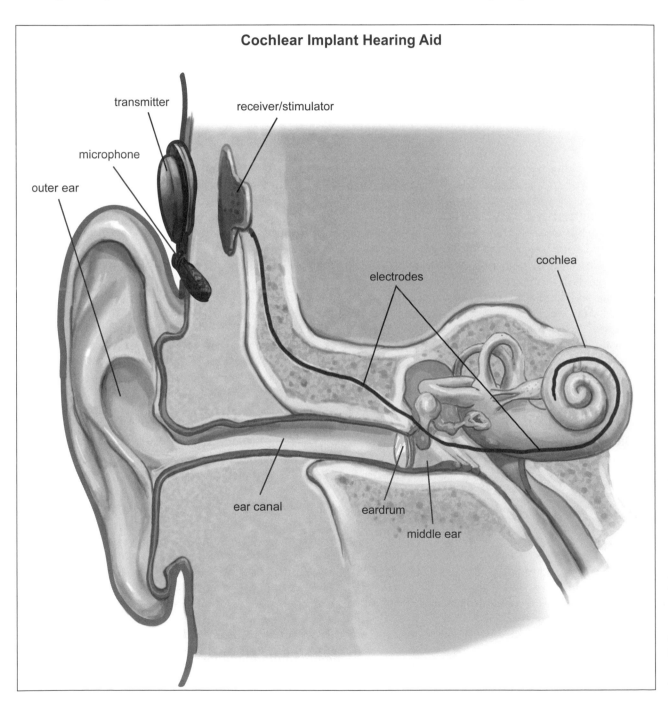

### Cochlear Implant Hearing Aid

transmitter

receiver/stimulator

microphone

outer ear

cochlea

electrodes

ear canal

eardrum

middle ear

*A cochlear implant differs from a hearing aid, delivering a utilitarian representation of sound.*

for Audiology Research is investigating alternative helmet designs to protect hearing without blocking auditory cues and also developing a bone conduction microphone for insertion into helmets to facilitate hands-free radio operation.

**Medical Clinics.** In addition to providing standard audiological care, medical clinics put into practice the knowledge obtained by research. William H. Lippy, founder of the Lippy Group for Ear, Nose, and Throat in Warren, Ohio, is one of the few surgeons in the world who has perfected the stapedectomy, an operation in which the stapes (the bone closest to the cochlea) is replaced with a prosthesis to restore hearing. He created the artificial stapes and many surgical instruments for this procedure. He is sharing his knowledge through peer-reviewed articles, book chapters, and an online video library.

Researchers at the Mayo Clinic are working to improve hearing aid use in younger persons, who are less likely to wear amplification devices than older people. They are developing remote programming of digital hearing aids over the Internet, disposable hearing aids (because children break or lose things easily), and devices that wearers can adjust themselves.

**Assistive Device Manufacturers.** Companies are continuing to design, manufacture, and market assistive devices for individuals with hearing loss. These devices can be divided into two categories: listening devices and alerting devices. Listening devices include amplified telephones and cell phones, wireless headsets for listening to television, and FM or infrared receivers for use in theaters and churches. Alerting devices signal users through flashing lights, vibration, or increased sound. They work in combination with smoke and carbon monoxide detectors, baby monitors, alarm clocks, and doorbells.

Some companies offer patch cords that connect the speech processors of cochlear implants to assistive listening devices for improved effect. Such patch cords make it easier to use cell phones and enjoy music on MP3 players. They are available with and without volume controls.

## CAREERS AND COURSE WORK

Audiologists are licensed professionals but not medical doctors. Physicians who specialize in ears, nose, and throat disorders are called otorhinolaryngologists. Audiologists must have a minimum of a master's degree in audiology; a doctoral (Au.D.) degree is becoming increasingly desirable in the workplace and is required for licensure in eighteen states. State licensure is required to practice in all fifty states. To obtain a license, an applicant must graduate from an accredited audiology program, accumulate 300 to 375 hours of supervised clinical experience, and pass a national licensing exam. To get into a graduate program in audiology, applicants must have had undergraduate courses in biology, chemistry, physics, anatomy, physiology, math, psychology, and communication. Forty-one states have continuing education requirements to renew a license. An audiologist must pass a separate exam to mold and place hearing aids. The American Board of Audiology and the American Speech-Language-Hearing Association (AHSA) offer professional certification programs for licensed audiologists.

Audiologists can go into private practice as a sole proprietor or as an associate or partner in a larger practice. They may also work in hospitals, outpatient clinics, and rehabilitation centers. Some are employed in state and local health departments and school districts. Some teach at universities and conduct academic research, and others work for medical device manufacturers. Audiologists may specialize in working with specific age groups, conducting hearing protection programs, or developing therapy programs for patients who are newly deaf or newly hearing.

## SOCIAL CONTEXT AND FUTURE PROSPECTS

In 2008, there were 12,800 audiologists working in the United States. About 64 percent were employed in health care settings and 14 percent in educational settings. The number of audiologists was expected to grow by 25 percent until 2018. The projected increased need can be traced to several factors.

As the population of older people continues to grow, so will the incidence of hearing loss from aging. In addition, the market demand for hearing aids is expected to increase as devices become less noticeable and existing wearers switch from analogue to digital models. At the same time, advances in medical treatment are increasing the survival rates of premature infants, trauma patients, and stroke patients, populations who may experience hearing loss.

Hearing aid manufacturers are on their fourth generation of products, and digital devices are becoming

## Fascinating Facts About Audiology and Hearing Aids

- More than 31.5 million Americans have some degree of hearing loss.
- Three out of every 1,000 babies in the United States are born with hearing loss.
- Only one in five people who would benefit from a hearing aid actually wears one.
- Approximately 188,000 people around the world have cochlear implants. In the United States, 41,500 adults and 25,500 children have had the device implanted.
- Noise exposure is the most common cause of hearing loss. One study has shown that people who eat substantial quantities of salt are more susceptible to hearing damage from noise.
- Addiction to Vicodin (hydrocodone with acetaminophen) can result in complete deafness.
- After its discovery in 1944, streptomycin was used to treat tuberculosis with great success; however, many patients experienced irreversible cochlear dysfunction as a result of the drug's ototoxicity.
- Medications that can cause hearing loss in adults are aspirin in large quantities, antibiotics such as streptomycin and neomycin, and chemotherapy agents such as cisplatin and carboplatin. Taking antioxidants such as vitamin C, vitamin E, zinc, and ginkgo biloba helps combat ototoxicity.
- The first nationally broadcast television program to show open captioning was *The French Chef with Julia Child*, which appeared on station WGBH from Boston on August 5, 1972.
- The National Captioning Institute broadcast the first closed-captioned television series on March 16, 1980. Real-time closed captioning was developed in 1982.
- The Television Decoder Circuitry Act of 1990 mandated that by mid-1993 all television sets 13 inches or larger must have caption-decoding technology.

and practices. Researchers are continuing to study the genes and proteins related to specialized structures of the inner ear. They hope to discover the biological mechanisms behind hearing loss in order to interrupt them or compensate for them on the molecular level.

*Bethany Thivierge, B.S., M.P.H.*

### FURTHER READING

Dalebout, Susan. *The Praeger Guide to Hearing and Hearing Loss: Assessment, Treatment, and Prevention.* Westport, Conn.: Praeger, 2009. Guides those who are experiencing hearing loss through the process of assessment and describes possible treatments and assistive devices.

DeBonis, David A., and Constance L. Donohue. *Survey of Audiology: Fundamentals for Audiologists and Health Professionals.* 2d ed. Boston: Pearson/Allyn and Bacon, 2008. Provides excellent coverage of audiology, focusing on assessment and covering pediatric audiology.

Gelfand, Stanley A. *Essentials of Audiology.* 3d ed. New York: Thieme Medical Publishers, 2009. A comprehensive introductory textbook with abundant figures and study questions at the end of each chapter.

Kramer, Steven J. *Audiology: Science to Practice.* San Diego, Calif.: Plural, 2008. Examines the basics of audiology and how they relate to practice. Contains a chapter on hearing aids by H. Gustav Mueller and Earl E. Johnson and a chapter on the history of audiology by James Jerger.

Lass, Norman J., and Charles M. Woodford. *Hearing Science Fundamentals.* Philadelphia: Mosby, 2007. Covers the anatomy, physiology, and physics of hearing. Contains figures, learning objectives, and chapter questions.

Moore, Brian C. J. *Cochlear Hearing Loss: Physiological, Psychological and Technical Issues.* 2d ed. Hoboken, N.J.: John Wiley & Sons, 2007. Comprehensive coverage of issues associated with cochlear hearing loss, including advances in pitch and speech perception.

Roeser, Ross J., Holly Hosford-Dunn, and Michael Valente, eds. *Audiology.* 2d ed. 3 vols. New York: Thieme, 2008. Volumes in this set cover diagnosis, treatment, and practice management. Focus is on clinical practice.

smaller and providing increasingly better sound processing. Neurosurgical techniques also are improving. Public health programs are promoting hearing protection. Excessive noise is the most common cause of hearing loss, and one-third of noise-related hearing loss is preventable with proper protective equipment

**WEB SITES**

*American Academy of Audiology*
http://www.audiology.org

*American Speech-Language-Hearing Association*
http://www.asha.org

*Audiological Resource Association*
http://www.audresources.org

*Educational Audiology Assocation*
http://www.edaud.org

*Military Audiology Association*
http://www.militaryaudiology.org

*National Institute on Deafness and Other Communication Disorders*
http://www.nidcd.nih.gov

**See also:** Geriatrics and Gerontology; Occupational Health; Otorhinolaryngology; Pediatric Medicine and Surgery; Prosthetics; Rehabilitation Engineering; Speech Therapy and Phoniatrics.

# AUTOMATED PROCESSES AND SERVOMECHANISMS

Electronics; hydraulics and pneumatics; mechanical engineering; computer programming; machining and manufacturing; millwright; quality assurance and quality control; avionics; aeronautics.

## SUMMARY

An automated process is a series of sequential steps to be carried out automatically. Servomechanisms are systems, devices, and subassemblies that control the mechanical actions of robots by the use of feedback information from the overall system in operation.

## KEY TERMS AND CONCEPTS

- **CNC (Computer Numeric Control):** An operating method in which a series of programmed logic steps in a computer controls the repetitive mechanical function of a machine.
- **Control Loop:** The sequence of steps and devices in a process that regulates a particular aspect of the overall process.
- **Feedback:** Information from the output of a process that is fed back as an input for the purpose of automatically regulating the operation.
- **Proximity Sensor:** An electromagnetic device that senses the presence or absence of a component through its effect on a magnetic field in the sensing unit.
- **Weld Cell:** An automated fabrication center in which programmed robotic welding units perform a specified series of welds on successive sets of components.

## DEFINITION AND BASIC PRINCIPLES

An automated process is any set of tasks that has been combined to be carried out in a sequential order automatically and on command. The tasks are not necessarily physical in nature, although this is the most common circumstance. The execution of the instructions in a computer program represents an automated process, as does the repeated execution of a series of specific welds in a robotic weld cell. The two are often inextricably linked, as the control of the physical process has been given to such digital devices as programmable logic controllers (PLCs) and computers in modern facilities.

Physical regulation and monitoring of mechanical devices such as industrial robots is normally achieved through the incorporation of servomechanisms. A servomechanism is a device that accepts information from the system itself and then uses that information to adjust the system to maintain specific operating conditions. A servomechanism that controls the opening and closing of a valve in a process stream, for example, may use the pressure of the process stream to regulate the degree to which the valve is opened.

The stepper motor is another example of a servomechanism. Given a specific voltage input, the stepper motor turns to an angular position that exactly corresponds to that voltage. Stepper motors are essential components of disk drives in computers, moving the read and write heads to precise data locations on the disk surface.

Another essential component in the functioning of automated processes and servomechanisms is the feedback control systems that provide self-regulation and auto-adjustment of the overall system. Feedback control systems may be pneumatic, hydraulic, mechanical, or electrical in nature. Electrical feedback may be analogue in form, although digital electronic feedback methods provide the most versatile method of output sensing for input feedback to digital electronic control systems.

## BACKGROUND AND HISTORY

Automation begins with the first artificial construct made to carry out a repetitive task in the place of a person. Early clock mechanisms, such as the water clock, used the automatic and repetitive dropping of a specific amount of water to accurately measure the passage of time. Water-, animal- or, wind-driven mills and threshing floors automated the repetitive action of processes that had been accomplished by humans. In many underdeveloped areas of the world, this repetitive human work is still a common practice.

With the mechanization that accompanied the Industrial Revolution, other means of automatically controlling machinery were developed, including self-regulating pressure valves on steam engines. Modern automation processes began in North America with the establishment of the assembly line as a standard industrial method by Henry Ford. In this method, each worker in his or her position along the assembly line performs a limited set of functions, using only the parts and tools appropriate to that task.

Servomechanism theory was further developed during World War II. The development of the transistor in 1951, and hence, digital electronics, enabled the development of electronic control and feedback devices. The field grew rapidly, especially following the development of the microcomputer in 1969. Digital logic and machine control can now be interfaced in an effective manner, such that today's automated systems function with an unprecedented degree of precision and dependability.

## HOW IT WORKS

An automated process is a series of repeated, identical operations under the control of a master operation or program. While simple in concept, it is complex in practice and difficult in implementation and execution. The process control operation must be designed in a logical, step-by-step manner that will provide the desired outcome each time the process is cycled. The sequential order of operations must be set so that the outcome of any one step does not prevent or interfere with the successful outcome of any other step in the process. In addition, the physical parameters of the desired outcome must be established and made subject to a monitoring protocol that can then act to correct any variation in the outcome of the process.

A plain analogy is found in the writing and structuring of a simple computer programming function. The definition of the steps involved in the function must be exact and logical, because the computer, like any other machine, can do only exactly what it is instructed to do. Once the order of instructions and the statement of variables and parameters have been finalized, they will be carried out in exactly the same manner each time the function is called in a program. The function is thus an automated process.

The same holds true for any physical process that has been automated. In a typical weld cell, for example, a set of individual parts are placed in a fixture that holds them in their proper relative orientations. Robotic welding machines may then act upon the setup to carry out a series of programmed welds to join the individual pieces into a single assembly. The series of welds is carried out in exactly the same manner each time the weld cell cycles. The robots that carry out the welds are guided under the control of a master program that defines the position of the welding tips, the motion that it must follow, and the duration of current flow in the welding process for each movement, along with many other variables that describe the overall action that will be followed. Any variation from this programmed pattern of movements and functions will result in an incorrect output.

The control of automated processes is carried out through various intermediate servomechanisms. A servomechanism uses input information from both the controlling program and the output of the process to carry out its function. Direct instruction from the controller defines the basic operation of the servomechanism. The output of the process generally includes monitoring functions that are compared to the desired output. They then provide an input signal to the servomechanism that informs how the operation must be adjusted to maintain the desired output. In the example of a robotic welder, the movement of the welding tip is performed through the action of an angular positioning device. The device may turn through a specific angle according to the voltage that is supplied to the mechanism. An input signal may be provided from a proximity sensor such that when the necessary part is not detected, the welding operation is interrupted and the movement of the mechanism ceases.

The variety of processes that may be automated is practically limitless given the interface of digital electronic control units. Similarly, servomechanisms may be designed to fit any needed parameter or to carry out any desired function.

## APPLICATIONS AND PRODUCTS

The applications of process automation and servomechanisms are as varied as modern industry and its products. It is perhaps more productive to think of process automation as a method that can be applied to the performance of repetitive tasks than to dwell on specific applications and products. The commonality of the automation process can be illustrated by examining a number of individual applications, and the products that support them.

"Repetitive tasks" are those tasks that are to be carried out in the same way, in the same circumstances, and for the same purpose a great number of times. The ideal goal of automating such a process is to ensure that the results are consistent each time the process cycle is carried out. In the case of the robotic weld cell described above, the central tasks to be repeated are the formation of welded joints of specified dimensions at the same specific locations over many hundreds or thousands of times. This is a typical operation in the manufacturing of subassemblies in the automobile industry and in other industries in which large numbers of identical fabricated units are produced.

Automation of the process, as described above, requires the identification of a set series of actions to be carried out by industrial robots. In turn, this requires the appropriate industrial robots be designed and constructed in such a way that the actual physical movements necessary for the task can be carried out. Each robot will incorporate a number of servomechanisms that drive the specific movements of parts of the robot according to the control instruction set. They will also incorporate any number of sensors and transducers that will provide input signal information for the self-regulation of the automated process. This input data may be delivered to the control program and compared to specified standards before it is fed back into the process, or it may be delivered directly into the process for immediate use.

Programmable logic controllers (PLCs), first specified by the General Motors Corporation in 1968, have become the standard devices for controlling automated machinery. The PLC is essentially a dedicated computer system that employs a limited-instruction-set programming language. The program of instructions for the automated process is stored in the PLC memory. Execution of the program sends the specified operating parameters to the corresponding machine in such a way that it carries out a set of operations that must otherwise be carried out under the control of a human operator.

A typical use of such methodology is in the various forms of CNC machining. CNC (computer numeric control) refers to the use of reduced-instruction-set computers to control the mechanical operation of machines. CNC lathes and mills are two common applications of the technology. In the traditional use of a lathe, a human operator adjusts all of the working parameters such as spindle rotation speed, feed rate, and depth of cut, through an order of operations that is designed to produce a finished piece

to blueprint dimensions. The consistency of pieces produced over time in this manner tends to vary as operator fatigue and distractions affect human performance. In a CNC lathe, however, the order of operations and all of the operating parameters are specified in the control program, and are thus carried out in exactly the same manner for each piece that is produced. Operator error and fatigue do not affect production, and the machinery produces the desired pieces at the same rate throughout the entire working period. Human intervention is required only to maintain the machinery and is not involved in the actual machining process.

Servomechanisms used in automated systems check and monitor system parameters and adjust operating conditions to maintain the desired system output. The principles upon which they operate can range from crude mechanical levers to sophisticated and highly accurate digital electronic-measurement devices. All employ the principle of feedback to control or regulate the corresponding process that is in operation.

In a simple example of a rudimentary application, units of a specific component moving along a production line may in turn move a lever as they pass by. The movement of the lever activates a switch that prevents a warning light from turning on. If the switch is not triggered, the warning light tells an operator that the component has been missed. The lever, switch, and warning light system constitute a crude servomechanism that carries out a specific function in maintaining the proper operation of the system.

In more advanced applications, the dimensions of the product from a machining operation may be tested by accurately calibrated measuring devices before releasing the object from the lathe, mill, or other device. The measurements taken are then compared to the desired measurements, as stored in the PLC memory. Oversize measurements may trigger an action of the machinery to refine the dimensions of the piece to bring it into specified tolerances, while undersize measurements may trigger the rejection of the piece and a warning to maintenance personnel to adjust the working parameters of the device before continued production.

Two of the most important applications of servomechanisms in industrial operations are control of position and control of rotational speed. Both commonly employ digital measurement. Positional control is generally achieved through the use of servomotors, also known as stepper motors. In these devices, the rotor turns to a specific angular position according to the voltage that is supplied to the motor. Modern electronics, using digital devices constructed with integrated circuits, allows extremely fine and precise control of electrical and electronic factors, such as voltage, amperage, and resistance. This, in turn, facilitates extremely precise positional control. Sequential positional control of different servomotors in a machine, such as an industrial robot, permits precise positioning of operating features. In other robotic applications, the same operating principle allows for extremely delicate microsurgery that would not be possible otherwise.

The control of rotational speed is achieved through the same basic principle as the stroboscope. A strobe light flashing on and off at a fixed rate can be used to measure the rate of rotation of an object. When the strobe rate and the rate of rotation are equal, a specific point on the rotating object will always appear at the same location. If the speeds are not matched, that point will appear to move in one direction or the other according to which rate is the faster rate. By attaching a rotating component to a representation of a digital scale, such as the Gray code, sensors can detect both the rate of rotation of the component and its position when it is functioning as part of a servomechanism. Comparison with a digital statement of the desired parameter can then be used by the controlling device to adjust the speed or position, or both, of the component accordingly.

## IMPACT ON INDUSTRY

Automated processes, and the servomechanisms that apply to them, have had an immeasurable impact on industry. While the term "mass production" does not necessarily imply automation, the presence of automated systems in an operation does indicate a significant enhancement of both production and precision. Mass production revolutionized the intrinsic nature of industry in North America, beginning with the assembly-line methods made standard by Ford. This innovation enabled Ford's industry to manufacture automobiles at a rate measured in units per day rather than days per unit.

While Ford's system represents a great improvement in productions efficiency, that system can only be described as an automatic system rather than an automated process. The automation of automobile production began when machines began to carry out some of the assembly functions in the place of humans. Today, it is entirely possible for the complete assembly process of automobiles, and of other goods, to be fully automated. This is not the most effective means of production, however, as even the most sophisticated of computers pales in comparison to the human brain in regard to intuition and intelligence. While the computer excels at controlling the mechanical function of processes, the human values of aesthetics and quality control and management are still far beyond computer calculation. Thus, production facilities today utilize the cooperative efforts of both machines and humans.

As technology changes and newer methods and materials are developed, and as machines wear out or become obsolete, constant upgrading of production facilities and methods is necessary. Process automation is essential for minimizing costs and for maximizing performance in the delivery of goods and services, with the goal of continuous improvement. It is in this area that process automation has had another profound effect on modern industry, identified by the various total quality management (TQM) programs that have been developed and adopted in many different fields throughout the world.

TQM can be traced to the Toyota method, developed by the Toyota Motor Corporation in Japan following World War II. To raise Japan's economy from the devastating effects of its defeat in that conflict, Toyota executives closely examined and analyzed the assembly-line methods of manufacturers in the United States, then enhanced those methods by stressing the importance of increasing efficiency and minimizing waste. The result was the method that not only turned Toyota into the world's largest and most profitable manufacturer of automobiles and other consumer goods, but revolutionized the manner in which manufacturers around the world conducted their businesses. Today, TQM programs such as Lean, 6Sigma, and other International Organization for Standardization (ISO) designations use process automation as a central feature of their operations; indeed, TQM represents an entirely new field of study and practice.

## CAREERS AND COURSEWORK

Students looking to pursue a career that involves automated processes and servomechanisms can

## Fascinating Facts About Automated Processes

- The ancient Egyptians used a remote hydraulic system to monitor the level of the water in the Nile River to determine the beginning and end of annual religious festivals.
- The water clock designed to measure the passage of a specific amount of time was perhaps the first practical, artificial automated process.
- Grain threshing mills powered by animals, wind, or water to automate the separation of grain kernels from harvested plants are still in use today.
- In relatively recent times, the turning of roasting spits in roadhouse kitchens was automated through the use of dogs called turnspits, which ran inside a treadmill wheel. A morsel of food suspended just beyond the reach of the dog was the "servomechanism" that kept the dog running.
- Control mechanisms of automated processes can use either analogue signal processing or digital signal processing.
- Computer control of physical operations is much more effective than human control, but human intervention is far superior for the subjective functions of management and quality control.
- Servomechanisms that perform some kind of automated control function can be found in unexpected places, including the spring-loaded switch that turns on the light in a refrigerator when its door is opened.
- Modern digital electronics permits extremely fine control of many physical actions and movements through servomechanisms.
- Process automation is as applicable to the fields of business management and accounting as it is to mass production in factories.

expect to take foundational courses in applied mathematics, physics, mechanics, electronics, and engineering. Specialization in feedback and control systems technology will be an option for those in a community college or pursuing an associate's, degree. Industrial electronics, digital technology, and machining and millwrighting also are optional routes at this level. More advanced levels of studies can be pursued through college or university programs in computer programming, mechanical engineering, electrical engineering, and some applied sciences in the biomedical field.

As may be imagined, robotics represents a significant aspect of work in this field, and students should expect that a considerable amount of their coursework will be related to the theory and practice of robotics.

As discussed, process automation is applicable in a variety of fields in which tasks of any kind must be repeated any number of times. The repetition of tasks is particularly appropriate in fields for which computer programming and control have become integral. It is therefore appropriate to speak of automated accounting practices, automated blood-sample testing, and other biomedical testing both in research and in treatment contexts, automated traffic control, and so on. One may note that even the procedure of parallel parking has become an automated process in some modern automobiles. The variety of careers that will accept specialized training geared to automated processes is therefore much broader than might at first be expected through a discussion that focuses on robotics alone.

### SOCIAL CONTEXT AND FUTURE PROSPECTS

While the vision of a utopian society in which all menial labor is automated, leaving humans free to create new ideas in relative leisure, is still far from reality, the vision does become more real each time another process is automated. Paradoxically, since the mid-twentieth century, knowledge and technology have changed so rapidly that what is new becomes obsolete almost as quickly as it is developed, seeming to increase rather than decrease the need for human labor.

New products and methods are continually being developed because of automated control. Similarly, existing automated processes can be re-automated using newer technology, newer materials, and modernized capabilities.

Particular areas of growth in automated processes and servomechanisms are found in the biomedical fields. Automated processes greatly increase the number of tests and analyses that can be performed for genetic research and new drug development. Robotic devices become more essential to the success of delicate surgical procedures each day, partly because of the ability of integrated circuits to amplify or reduce electrical signals by factors of hundreds of thousands. Someday, surgeons will be able to perform the most delicate of operations remotely, as normal actions by the surgeon are translated into the miniscule movements of microscopic surgical equipment manipulated through robotics.

Concerns that automated processes will eliminate the role of human workers are unfounded. The nature of work has repeatedly changed to reflect the capabilities of the technology of the time. The introduction of electric street lights, for example, did eliminate the job of lighting gas-fueled streetlamps, but it also created the need for workers to produce the electric lights and to ensure that they were functioning properly. The same sort of reasoning applies to the automation of processes today. Some traditional jobs will disappear, but new types of jobs will be created in their place through automation.

*Richard M. Renneboog, M.Sc.*

**FURTHER READING**

Bryan, L. A., and E. A. Bryan. *Programmable Controllers: Theory and Implementation.* Atlanta: Industrial Text, 1988. This textbook provides a sound discussion of digital logic principles and the digital electronic devices of PLCs. Proceeds through a detailed discussion of the operation of PLCs, data measurement, and the incorporation of the systems into more complex and centralized computer networks.

James, Hubert M. *Theory of Servomechanisms.* New York: McGraw-Hill, 1947. Discusses the theory and application of the principles of servomechanisms before the development of transistors and digital logic.

Kirchmer, Mathias. *High Performance Through Process Excellence.* Berlin: Springer, 2009. Examines the processes of business management as related to automation and continuous improvement in all areas of business operations.

Seal, A. M. *Practical Process Control.* Oxford, England: Butterworth-Heinemann/Elsevier, 1998. An advanced technical book that includes a good description of analogue and digital process control mechanisms and a discussion of several specific industrial applications.

Seames, Warren S. *Computer Numerical Control Concepts and Programming.* 4th ed. Albany, N.Y.: Delmar, Thomson Learning, 2002. Provides an overview of numerical control systems and servomechanisms, with an extended discussion of the specific functions of the technology.

Smith, Carlos A. *Automated Continuous Process Control.* New York: John Wiley & Sons, 2002. A comprehensive discussion of process control systems in chemical engineering applications.

**WEB SITES**

*Control-Systems-Principles*
http://www.control-systems-principles.co.uk

**See also:** Applied Mathematics; Applied Physics; Computer-Aided Design and Manufacturing; Computer Engineering; Computer Science; Digital Logic; Electronics and Electrical Engineering; Human-Computer Interaction; Hydraulic Engineering; Mechanical Engineering; Pneumatics; Quality Control; Robotics; Surgery.

# AVIONICS AND AIRCRAFT INSTRUMENTATION

## FIELDS OF STUDY

Aerodynamics; aeronautical engineering; computer science; electrical engineering; electronics; hydraulics; mechanical engineering; meteorology; physics; pneumatics.

## SUMMARY

Flight instrumentation refers to the indicators and instruments that inform a pilot of the position of the aircraft and give navigational information. Avionics comprises all the devices that allow a pilot to give and receive communications, such as air traffic control directions and navigational radio and satellite signals. Early in the history of flight, instrumentation and avionics were separate systems, but these systems have been vastly improved and integrated. These systems allow commercial airliners to fly efficiently and safely all around the world. Additionally, the integrated systems are being used in practically all types of vehicles—ships, trains, spacecraft, guided missiles, and unmanned aircraft—both civilian and military.

## KEY TERMS AND CONCEPTS

- **Airspeed Indicator:** Aircraft's speedometer, giving speed based on the difference between ram and static air pressure.
- **Artificial Horizon:** Gyroscopic instrument that displays the airplane's attitude relative to the horizon.
- **Directional Gyro (Vertical Compass):** Gyroscopically controlled compass rose.
- **Global Positioning System (GPS):** Satellite-based navigational system.
- **Horizontal Situation Indicator (HSI):** Indicator that combines an artificial horizon with a directional gyro.
- **Instrument Flight:** Flight by reference to the flight instruments when the pilot cannot see outside the cockpit because of bad weather.
- **Instrument Landing System (ILS):** Method to guide an airplane to the runway using a sensitive localizer to align the aircraft with the runway and a glide slope to provide a descent path to the runway.
- **Radar Altimeter:** Instrument that uses radar to calculate the aircraft's height above the ground.
- **Ram:** Airflow that the aircraft generates as it moves through the air.
- **Turn And Bank Indicator:** Gyroscopic instrument that provides information on the angle of bank of the aircraft.

### DEFINITION AND BASIC PRINCIPLES

Flight instrumentation refers to the instruments that provide information to a pilot about the position of the aircraft in relation to the Earth's horizon. The term "avionics" is a contraction of "aviation" and "electronics" and has come to refer to the combination of communication and navigational devices in an aircraft. This term was coined in the 1970's after the systems were becoming one integral system.

The components of basic flight instrumentation are the magnetic compass, the instruments that rely on air-pressure differentials, and those that are driven by gyroscopes and instruments. Air pressure decreases with an increase in altitude. The altimeter and vertical-speed indicator use this change in pressure to provide information about the height of the aircraft above sea level and the rate that the aircraft is climbing or descending. The airspeed indicator uses ram air pressure to give the speed that the aircraft is traveling through the air.

Other instruments use gyroscopes to detect changes in the position of the aircraft relative to the Earth's surface and horizon. An airplane can move around the three axes of flight. The first is pitch, or the upward and downward position of the nose of the airplane. The second is roll, the position of the wings. They can be level to the horizon or be in a bank position, where one wing is above horizon and the other below the horizon as the aircraft turns. Yaw is the third. When an airplane yaws, the nose of the airplane moves to the right or left while the airplane is in level flight. The instruments that use gyroscopes to show movement along the axes of flight are the turn and bank indicator, which shows the angle of the airplane's wings in a turn and the rate of turn in degrees per second; the artificial horizon, which indicates the airplane's pitch and bank; and the directional gyro, which is a compass card connected to a gyroscope. Output from the flight instruments can be used to operate autopilots. Modern inertial navigation systems (INS's) use gyroscopes, sometimes in conjunction with a Global Positioning System (GPS), as integrated flight instrumentation and avionics systems.

The radios that comprise the avionics of an aircraft include communications radios that pilots use to talk to air traffic control (ATC) and other aircraft and navigation radios. Early navigation radios relied on ground-based radio signals, but many aircraft have come to use GPS receivers that receive their information from satellites. Other components of an aircraft's avionics include a transponder, which sends a discrete code to ATC to identify the aircraft and is used in the military to discern friendly and enemy aircraft, and radar, which is used to locate rain and thunderstorms and to determine the aircraft's height above the ground.

## BACKGROUND AND HISTORY

Flight instruments originally were separated from the avionics of an aircraft. The compass, perhaps the most basic of the flight instruments, was developed by the Chinese in the second century B.C.E. Chinese navy commander Zheng He's voyages from 1405 to 1433 included the first recorded use of a magnetic compass for navigation.

The gyroscope, a major component of many flight instruments, was named by French physicist Jean-Bernard-Léon Foucault in the nineteenth century. In 1909, American businessman Elmer Sperry invented the gyroscopic compass that was first used on U.S. naval ships in 1911. In 1916, the first artificial horizon using a gyroscope was invented. Gyroscopic flight instruments along with radio navigation signals guided American pilot Jimmy Doolittle to the first successful all-instrument flight and landing of an airplane in 1929. Robert Goddard, the father of rocketry, experimented with using gyroscopes in guidance systems for rockets. During World War II, German rocket scientist Wernher von Braun further developed Goddard's work to build a basic guidance system for Germany's V-2 rockets. After the war, von Braun and 118 of his engineers immigrated to the United States, where they worked for the U.S. Army on gyroscopic inertial navigation systems (INSs) for rockets. Massachusetts Institute of Technology engineers continued the development of the INS to use

*Test pilot David Bonifield talks on the phone while making flight preparations in the cockpit of a Cessna Citation Mustang business jet in Wichita, Kansas in 2005. (AP Photo)*

in Atlas rockets and eventually the space shuttle. Boeing was the first aircraft manufacturer to install INSs into its 747 jumbo jets. Later, the Air Force introduced the system to their C-141 aircraft.

Radios form the basis of modern avionics. Although there is some dispute over who actually invented the radio, Italian physicist Guglielmo Marconi first applied the technology to communication. During World War I, in 1916, the Naval Research Laboratory developed the first aircraft radio. In 1920, the first ground-based system for communication with aircraft was developed by General Electric. The earliest navigational system was a series of lights on the ground, and the pilot would fly from beacon to beacon. In the 1930's, the nondirectional radio beacon (NDB) became the major radio navigation system. This was replaced by the very high frequency omnidirectional range (VOR) system in the 1960's. In 1994, the GPS became operational and was quickly adapted to aircraft navigation. The great accuracy that GPS can supply for both location and time was adapted for use in INS.

## HOW IT WORKS

**Flight Instruments.** Flight instruments operate using either gyroscopes or air pressure. The instruments that use air pressure are the altimeter, the vertical speed indicator, and the airspeed indicator. Airplanes are fitted with two pressure sensors: the pitot tube, which is mounted under a wing or the front fuselage, its opening facing the oncoming air; and the static port, which is usually mounted on the side of the airplane out of the slipstream of air flowing past the plane. The pitot tube measures ram air; the faster the aircraft is moving through the air, the more air molecules enter the pitot tube. The static port measures the ambient air pressure, which decreases with increasing altitude. The airspeed indicator is driven by the force of the ram air calibrated to the ambient air pressure to give the speed that the airplane is moving through the sky. The static port's ambient pressure is translated into altitude above sea level by the altimeter. As air pressure can vary from location to location, the altimeter must be set to the local barometric setting in order to receive a correct altimeter reading. The vertical speed indicator also uses the ambient pressure from the static port. This instrument can sense changes in altitude and indicates feet per minute that the airplane is climbing or descending.

Other flight instruments operate with gyroscopes. These instruments are the gyroscopic compass, the turn and bank indicator, and the artificial horizon. The gyroscopic compass is a vertical compass card connected to a gyroscope. It is either set by the pilot or slaved to the heading indicated on the magnetic compass. The magnetic compass floats in a liquid that allows it to rotate freely but also causes it to jiggle in turbulence; the directional gyro is stabilized by its gyroscope. The magnetic compass will also show errors while turning or accelerating, which are eliminated by the gyroscope. The turn and bank indicator is connected to a gyroscope that remains stable when the plane is banking. The indicator shows the angle of bank of the airplane. The artificial horizon has a card attached to it that shows a horizon, sky, and ground and a small indicator in the center that is connected to the gyroscope. When the airplane pitches up or down or rolls, the card moves with the airplane, but the indicator is stable and shows the position of the aircraft relative to the horizon. Pilots use the artificial horizon to fly when they cannot see the natural horizon. The artificial horizon and directional gyro can be combined into one instrument, the horizontal situation indicator (HSI). These instruments can be used to supply information to an autopilot, which can be mechanically connected to the flight surfaces of the aircraft to fly it automatically.

**Ground-Based Avionics.** Ground-based avionics provide communications, navigational information,

and collision avoidance. Communication radios operate on frequencies between 118 and 136.975 megahertz (MHz). Communication uses line of sight. Navigation uses VOR systems. The VOR gives a signal to the aircraft receiver that indicates the direction to or from the VOR station. A more sensitive type of VOR, a localizer, is combined with a glide slope indicator to provide runway direction and a glide path for the aircraft to follow when it is landing in poor weather conditions and the pilot does not have visual contact with the runway. Collision avoidance is provided by ATC using signals from each aircraft's transponders and radar. ATC can identify the aircrafts' positions and advise pilots of traffic in their vicinity.

**Satellite-Based Systems.** The limitation of line of sight for ground-based avionic transmitters is a major problem for navigation over large oceans or in areas of the world that have large mountain ranges or few transmitters. The U.S. military was very concerned about these limitations and the Department of Defense spearheaded the research and implementation of a system that addresses these problems. GPS is the United States's satellite system that provides navigational information. GPS can give location, movement, and time information. The system uses a minimum of twenty-four satellites orbiting the Earth that send signals to monitoring receivers on Earth. The receiver must be able to get signals from a minimum of four satellites in order to calculate an aircraft's position correctly. Although originally designed solely for military use, GPS is widely used by civilians.

**Inertial Navigation Systems (INS).** The INS is a self-contained system that does not rely on outside radio or satellite signals. INS is driven by accelerometers and gyroscopes. The accelerometer houses a small pendulum that will swing in relation to the aircraft's acceleration or deceleration and so can measure the aircraft's speed. The gyroscope provides information about the aircraft's movement about the three axes of flight. Instead of the gimbaled gyros, more precise strap-down laser gyroscopes have come to be used. The strap-down system is attached to the frame of the aircraft. Instead of the rotating wheel in the gimbaled gyroscopes, this system uses light beams that travel in opposite directions around a small, triangular path. When the aircraft rotates, the path traveled by the beam of light moving in the direction of rotation appears shorter than the path of the other beam of light moving in the opposite direction. The length of the path causes a frequency shift that is detected and interpreted as aircraft rotation. INS must be initialized: The system has to be able to detect its initial position or it must be programmed with its initial position before it is used or it will not have a reference point from which to work.

### APPLICATIONS AND PRODUCTS

**Military INS and GPS Uses.** Flight instrumentation and avionics are used by military aircraft as well as civilian aircraft, but the military have many other applications. INS is used in guided missiles and submarines. It can also be used as a stand-alone navigational system in vehicles that do not want to communicate with outside sources for security purposes. INS and GPS are used in bombs, rockets, and, with great success, unmanned aerial vehicles (UAVs) that are used for reconnaissance as well as delivering ordnance without placing a pilot in harm's way. GPS is used in almost all military vehicles such as tanks, ships, armored vehicles, and cars, but not in submarines as the satellite signals will not penetrate deep water. GPS is also used by the United States Nuclear Detonation Detection System as the satellites carry nuclear detonation detectors.

**Navigation.** Besides the use of flight instrumentation and avionics for aircraft navigation, the systems can also be used for almost all forms of navigation. The aerospace industry has used INS for guidance of spacecraft that cannot use earthbound navigation systems, including satellites that orbit the planet. INS systems can be initialized by manually inputting the craft's position using GPS or using celestial fixes to direct rockets, space shuttles, and long-distance satellites and space probes through the reaches of the solar system and beyond. These systems can be synchronized with computers and sensors to control the vehicles by moving flight controls or firing rockets. GPS can be used on Earth by cars, trucks, trains, ships, and handheld units for commercial, personal, and recreational uses. One limitation of GPS is that it cannot work where the signals could be blocked, such as under water or in caves.

**Cellular Phones.** GPS technology is critical for operating cellular phones. GPS provide accurate time that is used in synchronizing signals with base stations. If the phone has GPS capability built into it, as many smart phones do, it can be used to locate a mobile cell phone making an emergency call. The GPS system in cell phones can be used in cars for navigation as well for recreation such as guidance while hiking, biking, boating, or geocaching.

**Tracking Systems.** In the same manner that GPS can be used to locate a cell phone, GPS can be used to find downed aircraft or pilots. GPS can be used by biologists to track wildlife by placing collars on the animals, a major improvement over radio tracking that was line-of-sight and worked only over short ranges. Animals that migrate over great distances can be tracked by using only GPS. Lost pets can be tracked through GPS devices in their collars. Military and law enforcement use GPS to track vehicles.

**Other Civilian Applications.** Surveyors and map makers use GPS to mark boundaries and identify

---

### Fascinating Facts About Avionics and Aircraft Instrumentation

- During the Civil War, surveillance balloons communicated with ground crews using telegraph. The telegraph wires from the balloon could be connected with ground telegraph wires so that the observations could be relayed directly to President Lincoln in the White House.

- Before radios were installed in airplanes, inventors tried to communicate with people on the ground by dropping notes tied to rocks and by smoke signals.

- The technology that tells airplanes how high in the sky they are is used by skydivers to know when to open their parachutes and by scuba divers to know how far under the water they are.

- Some researchers are investigating using robotic dirigibles flying in the stratosphere to replace expensive communication satellites and unsightly cell phone towers.

- GPS, which is commonly used in the family car and is an "app" in smart phones, was originally developed by the military for defense.

- GPS can be used to help lost children and pets get home safely.

- Aircraft flight instrumentation is used to construct flight simulators, including flight simulators that can be used on a home computer.

- Someday unmanned airplanes may be used in commercial transportation, carrying freight and passengers.

locations. GPS units installed at specific locations can detect movements of the Earth to study earthquakes, volcanoes, and plate tectonics.

**Next Generation Air Transportation System (NextGen).** While ground-based navigational systems such as the VOR are still used by pilots and radar is used by ATC to locate airplanes, the Federal Aviation Administration (FAA) is researching and designing NextGen, a new system for navigation and tracking aircraft that will be based on GPS in the National Airspace System (NAS). Using NextGen GPS navigation, aircraft will be able to fly shorter and more direct routes to their destinations, saving time and fuel. ATC will change to a satellite-based system of managing air traffic.

### IMPACT ON INDUSTRY

**Throughout the history of aviation, the United States has been a leader in research and development of flight instrumentation and avionics.** During the Cold War, the Soviet Union and China worked to develop their own systems but lagged behind the United States. China has produced its own avionics but has not found an international market for its products. China has partnered with U.S. companies such as Honeywell and Rockwell Collins to produce avionics.

**Government and University Research.** Most government research is spearheaded by the FAA and conducted in government facilities or in universities and businesses supported by grants. The FAA is concentrating on applying GPS technology to aviation. Ohio University's Avionics Engineering Center is researching differential GPS applications for precision runway approaches and landings, as GPS is not sensitive enough to provide the very precise positioning required. The University of North Dakota is researching GPS applications for NextGen. Embry-Riddle Aeronautical University is working with the FAA's Office of Commercial Space Transportation to develop a space-transportation information system that will tie in with air traffic control. Embry-Riddle is also partnering with Lockheed Martin, Boeing, and other companies on the NextGen system. The FAA has agreements with Honeywell and Aviation Communication & Surveillance Systems to develop and test airfield avionics systems to be used in NextGen. The military also does research on new products, but their efforts are usually classified.

**Commercial and Industry Sectors.** The major consumers of flight instrumentation and avionics are aircraft manufacturers and airlines. In the United States, Boeing is the foremost commercial manufacturer for civilian as well as military aircraft, while Lockheed Martin, Bell Helicopter, and Northrop Grumman produce military aircraft. In Europe, Airbus is the major aircraft manufacturer.

The GPS technology developed for aircraft has also been used by companies such as Verizon and AT&T for cellular phones, Garmin for GPS navigation devices, and by power plants to synchronize power grids.

Spacecraft and space travel sectors have several companies actively manufacturing rocket ships, spacecraft, and navigational systems. The British company Virgin Galactic and the American company Space Adventures operate spacecraft, and private citizens can book flights, although it is extremely expensive.

Flight instrumentation and avionics are being used to construct UAVs for the military. Northrop Grumman is working with the military to develop this technology further, and it has been used very successfully in Iraq and Afghanistan. Other companies and universities are researching the possibility of having unmanned commercial aircraft in the United States, although the problems with coordinating unmanned aircraft with traditional aircraft and the safety concerns of operating an aircraft without a pilot are daunting.

**Major Corporations.** Several companies manufacture avionics, most of which are in the United States. Some of these companies have joint ventures with other countries and have established factories in those countries. Major avionics and flight instrument manufacturers are Honeywell, Rockwell Collins, Bendix/King, ARINC, Kollsman, Narco Avionics, Sigtronics, and Thales.

Many companies continue to develop new flight and avionics systems for aircraft. These products often emphasize integrating existing systems and applying declassified military advances to civilian uses. For example, Rockwell Collins's Pro Line Fusion integrated avionics system uses the military head-up display (HUD) technology for civilian business jets.

### CAREERS AND COURSE WORK

The possible careers associated with flight instruments and avionics include both civilian and military positions ranging from mechanics and technicians to designers and researchers. The education required for these occupations usually requires at least two years of college or technical training, but research and design may require a doctorate.

Maintenance and avionics technicians install and repair flight instruments and avionics. They may work on general aviation airplanes, commercial airliners, or military aircraft. With more and more modes of transportation using INS and GPS, mechanics and technicians may also be employed to install and repair these systems on other types of vehicles—ships, trains, guided missiles, tanks, or UAVs. NASA and private companies employ technicians to work with spacecraft. Most of these positions require an associate's degree with specialization as an aircraft or avionics technician, or the training may be acquired in the military.

As computers are becoming more and more important in these fields, the demand for computer technicians, designers, and programmers will increase. Jobs in these fields range from positions in government agencies such as the FAA or National Aeronautics and Space Administration (NASA) or the military to private-sector research and development. The education required for these occupations varies from high school or vocational computer training to doctorates in computer science or related fields.

Flight instrument and avionics systems are being designed and researched by persons who have been educated in mechanical, electrical, and aeronautical engineering, computer science, and related fields. Some of these occupations require the minimum of a bachelor's degree, but most require a master's or doctorate.

## SOCIAL CONTEXT AND FUTURE PROSPECTS

Aviation, made possible by flight instrumentation and avionics, has revolutionized how people travel and how freight is moved throughout the world. It has also dramatically changed how wars are fought and how countries defend themselves. In the future, flight instrumentation and avionics will continue to affect society not only through aviation but also through applications of the technology in daily life.

Military use of UAVs controlled by advances in flight instrumentation and avionics will continue to change how wars are fought. However, in the not-too-distant future this technology may be used in civilian aviation. UAVs could be used to inspect pipelines and perform surveys in unpopulated areas or rough terrain, but it is unsure whether they will be used for passenger flights. Many people will certainly be fearful of traveling in airplanes with no human operators. The FAA would have to develop systems that would incorporate unmanned aircraft into the airspace. However, the use of unmanned vehicles may be an important part of future space exploration.

Perhaps the avionics system that has had the most impact on society is GPS. As GPS devices are being made more compact and more inexpensively, they are being used more and more in daily life. GPS can permit underdeveloped countries to improve their own air-navigation systems more rapidly without the expensive of buying and installing expensive ground-based navigational equipment or radar systems used by air traffic control facilities.

*Polly D. Steenhagen, B.S., M.S.*

## FURTHER READING

Collinson, R. P. G. *Introduction to Avionics Systems.* 2d ed. Boston: Kluwer Academic, 2003. A comprehensive and well-illustrated review of both civilian and military flight and avionics systems.

Dailey, Franklyn E., Jr. *The Triumph of Instrument Flight: A Retrospective in the Century of U.S. Aviation.* Wilbraham, Mass.: Dailey International, 2004. A history of instrument flight with factual information as well as the author's personal flying experiences.

El-Rabbany, Ahmed. *Introduction to GPS: The Global Positioning System.* 2d ed. Norwood, Mass.: Artech House, 2006. A thorough overview of GPS and a discussion of future applications for the technology.

Federal Aviation Administration. *Instrument Flying Handbook.* New York: Skyhorse, 2008. The FAA's official manual on instrument flying includes discussions of flight instruments and avionics and the two allow a pilot to fly by reference to the instruments.

Johnston, Joe. *Avionics for the Pilot: An Introduction to Navigational and Radio Systems for Aircraft.* Wiltshire, England: Airlife, 2007. A straightforward basic explanation of aircraft avionics, what they do, and how they operate.

Tooley, Mike. *Aircraft Digital Electronic and Computer Systems: Principles, Operation and Maintenance.* Burlington, Mass.: Butterworth-Heinemann, 2007. A more technical look at aircraft avionics and computer systems and how they work.

**WEB SITES**

*Aircraft Electronics Association*
http://www.aea.net

*Aviation Instrument Association*
http://www.aia.net

*Federal Aviation Administration*
http://www.faa.gov

*Official Government Information About the GPS*
http://www.gps.gov

**See also:** Aeronautics and Aviation; Computer Science; Electrical Engineering; Mechanical Engineering; Meteorology; Pneumatics.

# B

## BAROMETRY

### FIELDS OF STUDY

Atmospheric science; physics; chemistry; fluid mechanics; electromagnetics; signal processing, meteorology;

### SUMMARY

The science and engineering of pressure measurement in gases take its practitioners far beyond its original realm of weather prediction. The sensors used in barometry range from those for the near vacuum of space and the small amplitudes of soft music to those for ocean depths and the shock waves of nuclear-fusion explosions.

### KEY TERMS AND CONCEPTS

- **Aneroid Barometer:** Instrument that measures atmospheric pressure without using liquid.
- **Bar:** Unit of pressure equal to 100,000 newtons per square meter.
- **Pascal:** Unit of pressure equal to 1 newton per square meter.
- **Piezoelectric Material:** Any material that generates a voltage when the pressure acting on it changes.
- **Pounds Per Square Inch Absolute (PSIA):** Unit of pressure in pounds per square in absolute.
- **Pounds Per Square Inch Gauge (PSIG):** Unit of pressure in pounds per square inch relative to some reference pressure.
- **Pressure-Sensitive Paint:** Liquid mixture that, when painted and dried on a surface and illuminated with ultraviolet light, emits radiation, usually infrared, the intensity of which changes with the pressure of air acting on the surface.
- **Torr:** Unit of pressure equal to one part in 760 of a standard atmosphere, also equal to the pressure at the bottom of a column of mercury 1 millimeter high with vacuum above it.
- **Torricelli Barometer:** Instrument to measure absolute atmospheric pressure, consisting of a graduated tube closed at the top end, containing a vacuum at the top above a column of mercury, and standing in a reservoir of mercury.
- **U-Tube Manometer:** Instrument consisting of a graduated tube containing liquid and shaped like the letter U with each arm connected to a source of pressure, so that the difference in liquid levels between the two tubes indicates the differential pressure.

### DEFINITION AND BASIC PRINCIPLES

Barometry is the science of measuring the pressure of the atmosphere. Derived from the Greek words for "heavy" or "weight" (*baros*) and "measure" (*metron*), it refers generally to the measurement of gas pressure. In gases, pressure is fundamentally ascribed to the momentum flowing across a given surface per unit time, per unit area of the surface. Pressure is expressed in units of force per unit area. Although pressure is a scalar quantity, the direction of the force due to pressure exerted on a surface is taken to be perpendicular and directed onto the surface. Therefore, methods to measure pressure often measure the force acting per unit area of a sensor or the effects of that force. Pressure is expressed in newtons per square meter (pascals), in pounds per square foot (psf), or in pounds per square inch (psi). The pressure of the atmosphere at standard sea level at a temperature of 288.15 Kelvin (K) is 101,325 pascals, or 14.7 psi. This is called 1 atmosphere. Mercury and water barometers have become such familiar devices that pressure is also expressed in inches of water, inches of mercury, or in torrs (1 torr equals about 133.3 pascals).

The initial weather-forecasting barometer, the Torricelli barometer, measured the height of a liquid column that the pressure of air would support, with a vacuum at the closed top end of a vertical tube. This barometer is an absolute pressure instrument. Atmospheric pressure is obtained as the product of the height, the density of the barometric liquid, and the acceleration because of gravity at the Earth's surface. The aneroid barometer uses a partially evacuated box the spring-loaded sides of which expand or contract depending on the atmospheric pressure, driving a clocklike mechanism to show the pressure on a circular dial. This portable instrument was convenient to carry on mountaineering, ballooning, and mining expeditions to measure altitude by the change in atmospheric pressure. A barograph is an aneroid barometer mechanism adapted to plot a graph of the variation of pressure with time, using a stylus moving on a continuous roll of paper. The rate of change of pressure helps weather forecasters to predict the strength of approaching storms.

The term "manometer" derives from the Greek word *manos*, meaning "sparse," and denotes an instrument used to measure the pressure relative to a known pressure. A U-tube manometer measures the pressure difference from a reference pressure by the difference between the height of a liquid column in the leg of the U-tube that is connected to a known pressure source and the height of the liquid in the other leg, exposed to the pressure of interest. Manometers of various types have been used extensively in aerospace engineering experimental-test facilities such as wind tunnels. The pitot-static tubes used to measure flow velocity in wind tunnels were initially connected to water or mercury manometers. Later, electronic equivalents were developed. Inclined tube manometers were used to increase the sensitivity of the instrument in measuring small pressure differences amounting to fractions of an inch of water.

## BACKGROUND AND HISTORY

In 1643, Italian physicist and mathematician Evangelista Torricelli proved that atmospheric pressure would support the weight of a thirty-five-foot water column leaving a vacuum above that in a closed tube and that this height would change with the weather. Later Torricelli barometers used liquid mercury to reduce the size of the column and make such instruments more practical.

The technology of pressure measurement has evolved gradually since then, with the aneroid barometer demonstrating the reliability of deflecting a diaphragm. This led to electrical means of measuring the amount of deflection. The most obvious method was to place strain gauges on the diaphragm

and directly measure the strain. Later electrical methods used the change in capacitance caused by the changing gap between two charged plates. Such sensors dominated the market until the 1990's at the low end of the measurement range. Piezoresistive materials expanded the ability of miniaturized strain gauge sensors to measure high pressures changing at high frequency. Microelectromechanical system (MEMS) technology enabled miniaturized solid-state sensors to challenge the market dominance of the diaphragm sensors. In the early twenty-first century, pressure-sensitive paints allowed increasingly sensitive and faster-responding measurements of varying pressure with very fine spatial resolution.

## HOW IT WORKS

Barometry measures a broad variety of pressures using an equally broad variety of measurement techniques, including liquid column methods, elastic element methods, and electrical sensors. Electrical sensors include resistance strain gauges, capacitances, piezoresistive instruments, and piezoelectric devices. The technologies range from those developed by French mathematician Blaise Pascal, Greek mathematician Archimedes, and Torricelli to early twenty-first century MEMS sensors and those used to conduct nanoscale materials science.

Pressures can be measured in environments from the near vacuum of space to more than 1,400 megapascals (MPa) and from steady state to frequencies greater than 100,000 cycles per second. Sensors that measure with respect to zero pressure or absolute vacuum are called absolute pressure sensors, whereas those that measure with respect to some other reference pressure are called gauge pressure sensors. Vented gauge sensors have the reference side open to the atmosphere so that the pressure reading is with respect to atmospheric pressure. Sealed gauges report pressure with respect to a constant reference pressure.

Where rapid changes in pressure must be measured, errors due to the variation of sensitivity with the rate of change must be considered. A good sensor is one whose frequency response is constant over the entire range of frequency of fluctuations that might occur. Condenser microphones with electromechanical diaphragms have long been used to measure acoustic pressure in demanding applications such as music recording, with flat frequency response from 0.1 cycles per second (hertz) to well

## Fascinating Facts About Barometry

- Italian physicist and mathematician Evangelista Torricelli built a tall water barometer protruding through the roof of his house in 1643 to display his invention. His neighbors accused him of practicing sorcery.
- The sonic boom generated on the ground by an aircraft flying at supersonic speeds produces a pressure change shaped like the letter N: a sharp increase, a more gradual decrease and a sharp recovery at the end.
- The mean atmospheric pressure at the surface of Mars is roughly 600 pascals, compared with 101,300 pascals at Earth's surface. This is roughly equal to the atmospheric pressure at 34,500 meters above Earth.
- The atmospheric static pressure inside a hurricane may go down to only 87 percent of the normal atmosphere. Pressure in the core of a tornado is believed to be similar to this; however, the changes occur within a few seconds as opposed to hours in the case of a hurricane.
- The pressure at the bottom of the Mariana Trench—at 11,034 meters, the lowest surveyed point of the Pacific Ocean—is roughly 111 megapascals, or 1,099 times the pressure at the surface.
- Solar radiation at Earth's orbit around the Sun exerts a pressure of roughly 4.56 micropascals.
- The threshold of human hearing is a pressure fluctuation of roughly twenty micropascals, while the threshold of pain is a pressure change of 100 pascals.
- The pressure inside the core of the Sun is calculated to be around 250 billion bars, while that occurring during the explosion of an American W80 nuclear-fusion weapon is roughly 64 billion bars.

over 20,000 hertz, covering the range of human hearing. Pressure-sensitive paint in certain special formulations has been shown to achieve excellent frequency response to more than 1,600 hertz but only when the fluctuation amplitude is quite large, near the upper limit of human tolerance. Using digital signal processing, inexpensive sensors can be corrected to produce signals with frequency response quality approaching that of much more expensive sensors.

In the 1970's, devices based on the aneroid barometer principle were developed, in which the deflection of a diaphragm caused changes in electrical capacitance that then were indicated as voltage changes in a circuit. In the 1980's, piezoelectric materials were developed, enabling electrical voltages to be created from changes in pressure. Micro devices based on these largely replaced the more expensive but accurate diaphragm-based electromechanical sensors. Digital signal processing enabled engineers using the new small, inexpensive devices to recover most of the accuracy possessed by the more expensive devices.

## APPLICATIONS AND PRODUCTS

Barometry has ubiquitous applications, measured by a broad variety of sensors. It is key to weather prediction and measuring the altitude of aircraft as well as to measuring blood pressure to monitor health. Pressure-sensitive paints enable measurement of surface pressure as it changes in space and time. The accuracy of measuring and controlling gas pressure is fundamental to manufacturing processes.

**Weather Forecasting.** Scientists learned to relate the rate of change of atmospheric pressure to the possibility of strong winds, usually bringing rain or snow. For example, if the pressure drops by more than three millibar per hour, winds of up to fifty kilometers per hour are likely to follow. Powerful storms may be preceded by drops of more than twenty-four millibar in twenty-four hours. If the pressure starts rising, clear calm weather is expected. However, these rules change with regional conditions. For instance in the Great Lakes region of the United States, rising pressure may indicate an Arctic cold front moving in, causing heavy snowfall. In other regions, a sharply dropping pressure indicates a cold front moving in, followed by a quick rise in pressure as the colder weather is established. As a warm front approaches, the pressure may level out and rise slowly after the front passes. Modern forecasters construct detailed maps showing contours of pressure from sensors distributed over the countryside and use these to predict weather patterns. Aircraft pilots use such maps to identify safe routes and areas to avoid. Using Doppler radar wind measurements, infrared temperature maps and cloud images from satellites, and computational fluid dynamics, modern weather bureaus are able to issue warnings about severe weather several hours in advance for smaller local weather fronts and storms and several days ahead for major storms moving across continents or oceans. However, the number of weather-monitoring pressure sensors available to forecasters is quite inadequate to issue accurate predictions for minor weather changes, particularly when predicting rain or snow.

**Electrical Gauges.** Gauges operating on the electrical changes induced by deflection of a diaphragm are used in industrial process monitoring and control where computer interfacing is required. Unsteady pressure transducers come in many ranges of amplitude and frequency. Piezoresistive sensors are integrated into an electrical-resistance bridge and constructed as miniature self-contained, button-like sensors. These are suitable for high amplitudes and frequencies, such as those encountered in shock waves and explosions, and transonic or supersonic wind-tunnel tests. Condenser microphones are used in acoustic measurements. As computerized data-acquisition systems became common, but pressure sensors remained expensive, pressure switches enabled dozens of pressure-sensing lines connected through the switches to each sensor to be measured one at a time. This required a long time to collect data from all the sensors, spending enough time at each to capture all the fluctuations and construct stable averages, making it unsuitable for rapidly changing conditions. Inexpensive, miniaturized, and highly sensitive solid-state piezoelectric sensors and fast, multichannel analogue-digital converters have made it possible to connect each pressure port to an individual sensor, vastly reducing the time between individual measurements at each sensor.

**Aircraft Testing.** Water and mercury manometers were used extensively in aerospace test facilities such as wind tunnels, where banks of manometers indicated the distribution of pressure around the surfaces of models from pressure-sensing holes in the models. Pressure switches connecting numerous pressure-sensing ports to a single sensor became common in the 1970's. In the 1990's, inexpensive sensors based on microelectromechanical systems technology enabled numerous independent sensing channels to be monitored simultaneously.

**Sphygmomanometers for Blood Pressure.** Other than weather forecasting, the major common application of pressure measurement is in measuring blood pressure. The device used is called a sphygmomanometer. The high and low points of pressure reached in the heartbeat cycle are noted on a mercury manometer tube synchronized with the heartbeat sounds detected through a stethoscope.

**Bourdon Tubes for Household Barometry.** The Bourdon tube is a pressure-measuring device in which a coiled tube stretches and uncoils depending on the difference between pressures inside and outside the tube and drives a levered mechanism connected to an indicator dial. Diaphragm-type pressure gauges and Bourdon-tube gauges are still used in the vast majority of household and urban plumbing. These instruments are highly reliable and robust, but they operate over fairly narrow ranges of pressure.

**Pressure-Sensitive Paints to Map Pressure Over Surfaces.** So-called pressure-sensitive paints (PSPs) offer an indirect technique to measure pressure variations over an entire surface, using the fact that the amount of oxygen felt at a surface is proportional to the density and thus to the pressure if the temperature does not change. These paints are luminescent dyes dispersed in an oxygen-permeable binder. When illuminated at certain ultraviolet wavelengths, the dye molecules absorb the light and move up into higher energy levels. The molecules then release energy in the infrared wavelengths as they relax to equilibrium. If the molecule collides with an oxygen molecule, the energy gets transferred without emission of radiation. Therefore, the emission from a surface becomes less intense if the number of oxygen molecules being encountered increases. This occurs when the pressure of air increases. The observed intensity from a painted surface is inversely proportional to the pressure of oxygen-containing air. Light-intensity values at individual picture element (pixel) are converted to numbers, compared with values at some known reference pressure, and presented graphically as colors. Typically, an accurate pressure sensor using either piezoelectric or other technology is used for reference. As of 2011, pressure-sensitive paints had reached the sensitivity required to quantify pressure distributions over passenger automobiles at moderate highway speeds, given expert signal processing and averaging a large number of images.

**Smart Pressure Transmitters for Automatic Control Systems.** Wireless pressure sensors are used in remote applications such as weather sensing. Modern automobiles incorporate tire pressure transmitters. Manifold pressure sensors send instantaneous readings of the pressure inside automobile engine manifolds so that a control computer can calculate the best rate of fuel flow to achieve the most efficient combustion. Smart pressure transmitters incorporating capacitance-type diaphragm pressure sensors and microprocessors can be configured to adjust their settings remotely, perform automatic temperature compensation of data, and transmit pressure data and self-diagnosis data in digital streams.

**Nuclear Explosion Sensors.** Piezoresistive transducers have been developed to report the extremely high overpressure, as high as sixty-nine megapascals, of an air blast of a nuclear weapon, with the microsecond rise time required to measure the blast wave accurately. One design uses a silicon disk with integral diffused strain-sensitive regions and thermal barriers. Another design uses the principle of Fabry-Perot interferometry, in which laser light reflecting in a cavity changes intensity depending on the shape of the cavity when the diaphragm bounding the cavity flexes because of pressure changes. This sensor has the response speed and ruggedness required to operate in a hostile environment, where there may be very large electromagnetic pulses. In such environments a capacitance-based sensor or piezoelectric sensor may not survive.

**Extreme Applications of Barometry.** The basic origins of pressure can be used to explain the pressure due to radiation as the momentum flux of photons. At Earth's orbit around the Sun, the solar intensity of 1.38 kilowatts per square meter causes a radiation pressure of roughly 4.56 micropascals. Solar sails have been proposed for long-duration missions in space, driven by this pressure. Close to the center of the Earth, the pressure reaches 3.2 to 3.4 million bars. Inside the Sun, pressure as high as 250 billion bars is expected, while the explosion of a nuclear-fusion weapon may produce a quarter of that. Metallic solid hydrogen is projected to form at pressures of 250,000 to 500,000 bars.

## IMPACT ON INDUSTRY

**Government and University Research.** Barometers enabled rapid development of scientific weather forecasting. Weather forecasting, in turn, has had a tremendous effect on emergency preparedness. Research sponsored by the defense research offices provides fertile opportunities and challenges for new pressure-measurement techniques. Any experiment that uses fluids, either flowing or in containers, requires monitoring and often rapid measurement of pressure. The development of pressure-sensitive paint technology is a frontier in research in the early twenty-first century.

Both the sensitivity and the frequency response of such paints need substantial improvement before they can be routinely used in laboratory measurements and transitioned to industrial measurements.

**Industry and Business.** Quantitative knowledge of the detailed surface pressure distribution on wind tunnel models of flight vehicles enables engineers to develop modifications to improve the performance of the vehicle and reduce fuel consumption. The ability to monitor pressure is critical in the nuclear and petroleum industries as well. Submarine and oil-rig crews depend on pressure measurements for their lives. Oil exploration involves several steps in which the pressure must be tracked with extreme care, especially when there is a danger of gas rising through drilling tubes from subterranean reservoirs. Pressure buildup in steam or other gas circuits is critical in the nuclear industry and in most of the chemical industry wherever leaks of gas into the atmosphere must be strictly controlled.

## CAREERS AND COURSE WORK

Because barometry is so important to so many industries and so many branches of scientific research, most students who are planning on a career in engineering, other technological jobs, and the sciences must understand it.

Modern pressure-measurement technology integrates ideas from many branches of science and engineering derived from physics and chemistry. The pressure-measurement industry includes experts in weather forecasting, plumbing, atmospheric sciences, aerospace wind-tunnel experimentation, automobile-engine development, the chemical industry, chemists developing paint formulations, electrical and electronics engineers developing microelectromechanical sensors, software engineers developing smart sensor logic, and the medical community interested in using barometry to monitor patients' health and vital signs.

Pressure measurement therefore comes up as a subject in courses offered in schools of mechanical, chemical, civil, and aerospace engineering. The numerous other related issues come up in specialized courses in materials science, electronics, atmospheric sciences, and computer science.

## SOCIAL CONTEXT AND FUTURE PROSPECTS

Instrumentation for measuring pressure, normal and shear stresses, and flow rate from numerous sensors are becoming integrated into computerized measurement systems. In many applications, such sensors are mass-produced using facilities similar to those for making chips for computers. Very few ideas exist for directly measuring pressure, as it changes rapidly at a point in a flowing fluid, without intrusive probes of some kind. Such nonintrusive measurements, if they become possible, could help us to understand the nature of turbulence and assist in a major breakthrough in fluid dynamics.

Measurement of pressure is difficult to make inside flame environments, where density and temperature fluctuate rapidly. Better methods of measuring pressure in biological systems, such as inside blood vessels, would have major benefits in diagnosing heart disease and improving health.

As of 2011, pressure-measurement systems are still too expensive to allow sufficient numbers to be deployed to report pressure with enough spatial and time resolution to permit development of a real-time three-dimensional representation. Research in this area will doubtless improve the resolution and response and hopefully bring down the cost. With more pressure sensors distributed over the world, weather prediction will become more accurate and reliable.

*Narayanan M. Komerath, Ph.D.*

## FURTHER READING

American Society of Mechanical Engineers. *Pressure Measurement.* New York: American Society of Mechanical Engineers, 2010. Authoritative document with guidance on determining pressure values, according to the American Society of Mechanical Engineers performance test codes. Discusses how to choose and use the best methods, instrumentation, and corrections, as well as the allowable uncertainty.

Avallone, Eugene A., Theodore Baumeister III, and Ali M. Sadegh. *Marks' Standard Handbook for Mechanical Engineers.* 11th ed. New York: McGraw-Hill, 2006. Best reference for solving mechanical engineering problems. Discusses pressure sensors and measurement techniques and their applications in various parts of mechanical engineering.

Benedict, Robert P. *Fundamentals of Temperature, Pressure, and Flow Measurements.* 3d ed. New York: John Wiley & Sons, 1984. Suited for practicing engineers in the process control industry.

Burch, David. *The Barometer Handbook: A Modern Look at Barometers and Applications of Barometric Pressure.* Seattle: Starpath Publications, 2009. Written to assist the practicing weather forecaster, with chapters on weather forecasting on land and sea. Contains an excellent history of the field as well as methods for instrument calibration and maintenance.

Gillum, Donald R. *Industrial Pressure, Level, and Density Measurement.* 2d ed. Research Triangle Park, N.C.: International Society for Automation, 2009. Teaching and learning resource on the issues and methods of pressure measurement, especially related to industrial control systems. Contains assessment questions at the end of each section.

Green, Don W., and Robert H. Perry. *Perry's Chemical Engineers' Handbook.* 8th ed. New York: McGraw-Hill, 2008. Still considered the best source for bringing together knowledge from various parts of the field of chemical engineering, where the student can find the different applications of pressure measurement and many other things. Contains a succinct, illustrated explanation of pressure-measurement techniques.

Ryans, J. L. "Pressure Measurement." In *Kirk-Othmer Encyclopedia of Chemical Technology.* 5th ed. Hoboken, N.J.: John Wiley & Sons, 2000. Describes mechanical and electronic sensors and instrumentation for pressure measurements from 1,380 megapascals to near vacuum.

Taylor, George Frederic. *Elementary Meteorology.* New York: Prentice Hall, 1954. This classic remains an excellent resource for the basic methods of weather prediction, including the methods for measuring temperature, pressure, and humidity, as well as the methods for using these measurements in predicting the weather.

## WEB SITES

*American Meteorological Society*
http://www.ametsoc.org

*American Society of Mechanical Engineers*
http://www.asme.org

*National Oceanic and Atmospheric Administration*
Office of Oceanic and Atmospheric Research
http://www.oar.noaa.gov

**See also:** Atmospheric Sciences; Chemical Engineering; Civil Engineering; Mechanical Engineering; Meteorology.

# BIOCHEMICAL ENGINEERING

## FIELDS OF STUDY

Biochemistry; microbiology; biotechnology; cell biology; biology; chemical engineering; chemistry; genetics; molecular biology; pharmacology; medicine; agriculture; food science; environmental science; petroleum refinement; physiology; waste management.

## SUMMARY

Biochemical engineers are responsible for designing and constructing those manufacturing processes that involve biological organisms or products made by them. Biochemical engineers take commercially valuable biological or biochemical commodities and design the means to produce those commodities effectively, cheaply, safely, and in mass quantities. They do this by optimizing the growth of organisms that produce valuable molecules or perform useful biochemical processes, establishing the most effective way to purify the desired molecules, and designing the operation systems that execute these processes, while adhering to a high standard of quality, purity, worker safety, and environmental cleanliness.

## KEY TERMS AND CONCEPTS

- **Biofuels:** Solid, liquid, or gaseous fuels derived from biomass.
- **Biomass:** Plant materials and animal waste used especially as a source for fuel.
- **Bioreactor:** Device in which industrial biochemical reactions occur with the help of either enzymes or living cells.
- **Enzymes:** Particular proteins or ribonucleic acids (RNAs) that accelerate the rate of chemical reactions without being consumed or changed in the process.
- **Genetically Engineered Organisms:** Biological organisms that have had their endogenous deoxyribonucleic acid (DNA) altered, usually by the introduction of foreign DNA.
- **Hollow-Fiber Membrane Bioreactor (HFMB):** Cylindrical bioreactor that has a series of thin, porous, narrow, hollow tubes inside a plastic cylinder. Cultured cells grow in the spaces between the hollow tubes or fibers (extra-capillary spaces), while oxygen and nutrients continuously flow through the hollow fibers to the cells.
- **Hybridoma:** Cultured cell line that results from the fusion of an antibody-making B lymphocyte and a myeloma (B lymphocyte tumor cell) that secretes a monoclonal antibody.
- **Monoclonal Antibodies:** Proteins secreted by specific cells of the immune system that precisely bind to specific sites on the surface of foreign invaders and facilitate the destruction or neutralization of the foreign invaders.
- **Phage Display:** Test-tube selection technique that genetically fuses a protein to the outer-coat protein of a virus that infects bacteria, resulting in display of the fused protein on the outside of the virus. This allows screening of vast numbers of variants of the protein, each encoded by its corresponding DNA sequence.
- **Photobioreactor:** Translucent container that incorporates a light source and is used to grow small photosynthetic organisms for controlled biomass production.
- **Wave Bioreactor:** Disposable, sterile, plastic bag bioreactor that is mounted on a rocking platform, which creates wave action inside the bag to mix the culture that grows inside it.

### DEFINITION AND BASIC PRINCIPLES

Biochemical engineering involves designing and building those industrial processes that use catalysts, feedstocks, or absorbents of biological origin. Industrial processes used in food, waste-management, pharmaceutical, and agricultural plants are often called unit operations. Those unit operations used in combination with biological organisms or molecules include heat and mass transfer, bioreactor design and operation, filtration, cell isolation, and sterilization.

One of the main tasks of bioengineers is to optimize the production of commercially valuable molecules by genetically engineered microorganisms. Biochemical engineers design culture containers known as bioreactors that accommodate growing cultures and maintain an environment that keeps growth at optimal levels. They also create the protocols that separate the cultured cells and their growth medium from the molecule of interest and purify this molecule from all contaminating components. Biochemical engineers do not make the genetically engineered organisms that produce or do valuable things, but instead they maximize the capacities of such organisms in the safest and most cost-effective ways.

Biochemical engineers also design systems that degrade organic or industrial waste. In these cases, bioreactors house biological organisms that receive and decompose waste. They select the right organism or mix of organisms for the job at hand, establish environments that allow these organisms to thrive, and design systems that feed waste to the organisms and remove the degradation products.

A branch of biochemical engineering called tissue engineering combines cultured cells with synthetic materials and external forces to mold those cells into organs that can serve as a replacement for diseased or damaged organs. Biochemical engineers determine the forces, materials, or biochemical cues that drive cells to form fully functional organs and then design the bioreactor and associated instrumentation to provide the proper environment and cues.

## BACKGROUND AND HISTORY

Biochemical engineering is a subspecialty of chemical engineering. Chemical engineering began in 1901 when George E. Davis, its British pioneer, mathematically described all the physical operations commonly used in chemical plants (distillation, evaporation, filtration, gas absorption, and heat transfer) in his landmark book, *A Handbook of Chemical Engineering*.

Biochemical engineering emerged in the 1940's as advancements in biochemistry, the genetics of microorganisms, and engineering shepherded in the era of antibiotics. World War II created shortages in commonly used industrial agents; therefore, manufacturers turned to microorganisms or enzymes to synthesize many of the chemicals needed for the war effort. Growing large batches of microorganisms presented scaling, mixing, and oxygenation problems that had never been encountered before, and biochemical engineers solved these problems.

During the 1960's, advances in biochemistry, genetics, and engineering drove the creation of biomedical engineering, which is the application of all engineering disciplines to medicine, and separated it

from biochemical engineering. During this decade, biochemical engineers developed new types of bioreactors and new instrumentation and control circuits for them. They also made breakthroughs in kinetics (the science that mathematically describes the rates of reactions) within bioreactors and whole-cell biotransformations.

The 1970's saw the development of enzyme technologies, biomass engineering, single-cell protein production, and advances in bioreactor design and operation. From 1980 to 2000 there was a virtual explosion in biochemical-engineering advances that had never been seen before. The advent of recombinant DNA and hybridoma technologies, cell culture, molecular models, large-scale protein chromatography, protein and DNA sequencing, metabolic engineering, and bioremediation technologies changed biochemical engineering in a drastic and profound way. These technologies also presented new challenges and problems, many of which are still the subject of intense research and development.

## HOW IT WORKS

**Bioreactors.** Bioreactors that utilize living cells are typically called fermenters. There are several different types of bioreactors: mechanically stirred or agitated tanks; bubble columns (cylindrical tanks that are not stirred but through which gas is bubbled); loop reactors, which have forced circulation; packed-bed reactors; membrane reactors; microreactors; and a variety of different types of reactors that are not easily classified (such as gas-liquid reactors and rotating-disk reactors). Biochemical engineers must choose the best bioreactor type for the desired purpose and outfit it with the right instrumentation and other features.

Bioreactor operation is either batch-wise or continuous. Batch-wise operation or batch cultures include all the nutrients required for the growth of cells prior to cultivation of the organisms. After inoculation, cell growth commences and ceases once the organisms have exhausted all the available nutrients in the culture medium. A modification of this type of operation is a fed-batch or semi-batch operation in which the reactants are continuously fed into the bioreactor, and the reaction is allowed to go to completion, after which the products are recovered. Continuously operated bioreactors, use "continuous culture systems" that continuously feed culture medium into

## Chemical Reactor

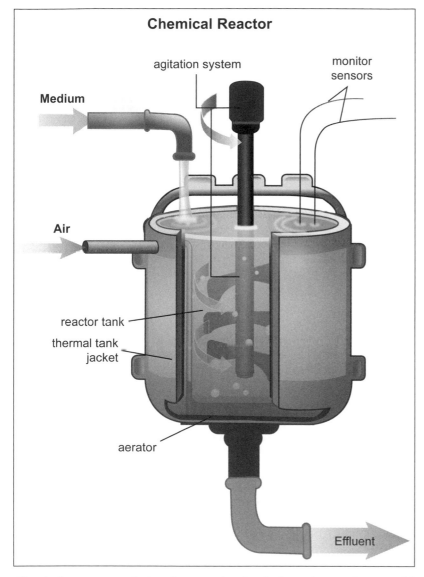

*Chemical reactors are designed to contain chemical reactions such as emulsifying, solid suspension, and gas dispersion.*

the culture with adequate oxygen requires the use of agitators or stirring equipment that must operate at high enough levels to aerate the culture without severely damaging the growing cells. Fourth, the bioreactor must have sensors to measure accurately the physical properties of the culture system, such as temperature, acidity (pH), and ionic strength. Fifth, the bioreactor should also be equipped with the means to adjust these physical properties as needed. Finally, the bioreactor must be integrated into a network of peripheral equipment that allows automated monitoring and adjustment of the culture's physical factors.

**Separation.** Once a bioreactor makes a product, separating this molecule or group of molecules from the remaining contaminants, byproducts, and other components is an integral part of preparing that molecule for market.

There are several different separation techniques. Filtration separates undissolved solids from liquids by passing the solid-liquid mixture through solids perforated by pores of a particular size (like a membrane). If the liquid is viscous or the particle size of the solid is too small for filtration, centrifugation can separate such solids from liquids. The liquid samples are loaded into centrifuges, which spin rotors at very high speeds. This process creates pellets from the solids and separates them from liquids. Neither filtration nor centrifugation can separate dissolved components from liquids.

the bioreactor and simultaneously remove excess medium at the same rate. Batch-culture bioreactors work best for fast-growing biological organisms. Slow-growing organisms usually require continuous-culture bioreactors.

Several factors influence the success of bioreactor-based operations. First, choosing the right strain to make the desired product is essential. Second, the culture medium and growth conditions must optimize the growth of the chosen organism. Third, supplying

Adsorption and chromatography can effectively separate dissolved molecules. Adsorption involves the accumulation of dissolved molecules on the surface of a solid in contact with the liquid. The solid in most cases consists of a resin made of porous charcoal, silica, polysaccharides (complex chains of sugars), or other molecules. Chromatography runs the liquid through a stationary medium packed into a cylindrical column that has particular chemical properties. The interaction between the desired molecules and the stationary medium facilitates their isolation. Other types of separation techniques include crystallization, in which the molecule of interest is driven to form crystals. This effectively removes it from solution and facilitates "salting out," in which gradually increased salt concentrations precipitate the molecules of interest, or contaminating molecules, from a liquid solution.

**Sterilization.** If a culture of genetically engineered organisms is used to produce a commercially useful product, contamination of that culture can decrease the amount of product or cause the production of harmful byproducts. Therefore, all tubes, valves, the bioreactor container, and the air supplied to it during operation must be effectively sterilized before the start of any production run.

Heat, radiation, chemicals, or filtration can sterilize equipment and liquids. One of the most economical means of sterilization is moist steam. Calculating the time it takes to sterilize something depends on the initial number of organisms present, the resilience of those organisms to killing with the chosen agent, the ability of the air or liquid to conduct the sterilizing agent, and length of time the organisms are exposed to the sterilizing agent.

## APPLICATIONS AND PRODUCTS

**Pharmaceuticals.** Hundreds of pharmaceuticals are proteins made by genetically engineered organisms. Because these reagents are intended for clinical use, they must be produced under completely sterile conditions and are usually grown in disposable (plastic), prepackaged, sterile bioreactor systems. A variety of wave bioreactors, hollow-fiber membrane bioreactors, and variations on these devices help grow the cells that make these products.

Some of the proteins made by genetically engineered cells are enzymes. Genentech, for example, makes dornase alfa, an enzyme that degrades DNA. This enzyme is made by genetically engineered Chinese hamster ovary (CHO) cells and is purified by filtration and column chromatography. Dornase alfa is administered as an inhalable aerosol to allay the symptoms of cystic fibrosis. Other therapeutic enzymes include clotting factors such as Helixate FS (native clotting factor VIII made by CSL Behring), NovoSeven (clotting factor VII made by Novo Nordisk) to treat

hemophilia, and Fabrazyme or Replagal (agalsidase alfa) to treat Anderson-Fabry disease.

Other pharmaceuticals are peptide hormones. Serostim and Saizen are commercially available versions of recombinant human growth hormone. Both products are made with cultured mouse C127 cells in bioreactors. Human growth hormone is used to treat children with hypopituitary dwarfism or those who experience the chronic wasting associated with AIDS.

Therapeutic proteins are normally made in the human body under certain conditions, and synthetic versions of these proteins that are made in labs can be used as medicine. For example, human cells make a protein called interferon in response to viral infections, but synthetic interferon can also be used to treat multiple sclerosis. Two synthetic forms of interferon-1$\beta$, Rebif, which is made in CHO cells by EMD Serono, and Avonex, also made in CHO cells by Biogen Idec, serve as treatments for multiple sclerosis. Alefacept (brand name Amevive), which is made by Astellas Pharma, is a fusion protein that blocks the growth of specific T cells (immune cells). No such protein exists in the human body, but alefacept is used to treat psoriasis and various cancers.

These are only a few examples of the hundreds of pharmaceutical compounds made by genetically engineered organisms in bioreactors designed by biochemical engineers.

**Monoclonal Antibodies.** Monoclonal antibodies are Y-shaped proteins secreted by specific cells of the immune system that precisely bind to specific sites (epitopes) on the surface of foreign invaders, and act as guided missiles that facilitate the destruction or neutralization of the foreign invaders.

Immune cells called B lymphocytes secrete antibodies, and the fusion of these antibody-producing cells with myelomas (B-cell tumor cells) produces a hybridoma, an immortal cell that grows indefinitely in culture and secretes large quantities of a particular antibody. Antibodies made by hybridoma cells can bind to one and only one site on a specific target and are known as monoclonal antibodies.

Monoclonal antibodies are powerful clinical and industrial tools, and by growing hybridoma cell lines in bioreactors, biotechnology companies can produce large quantities of them for a variety of applications.

Mouse monoclonal antibodies end with the suffix "-omab." Tositumomab (brand name Bexxar) was approved by the Food and Drug Administration (FDA) for treatment of non-Hodgkin's lymphoma in 2003.

Chimeric or humanized monoclonal antibodies, and have the suffixes "-ximab" (chimeric antibodies that are about 65 percent human) or "-zumab" (humanized antibodies that are about 95 percent human). Cetuximab (Erbitux) is a chimeric antibody that was approved by the FDA in 2004 for the treatment of colorectal, head, and neck cancers. Bevacizumab (Avastin) is a humanized antibody approved by the FDA in 2004 that shrinks tumors by preventing the growth of new blood vessels into them.

Human monoclonal antibodies are made either by hybridomas from transgenic mice that have had their mouse antibody genes replaced with human antibody genes, or by a process called phage display. Human monoclonal antibodies end with the suffix "-mumab." The first human monoclonal antibody developed through phage display technologies was adalimumab (Humira), which was approved by the FDA to treat several immune system diseases.

**Tissue Engineering.** Making artificial organs for transplantation represents a unique challenge. Bioreactors tend to grow cells in two-dimensional cultures, but organs are three-dimensional structures. Thus, biochemical engineers have designed synthetic scaffolds that support the growth of cultured cells and mold them into structures that bear the shape and properties of organs. They have also designed special bioreactors that subject cells to the physical conditions that induce the cells to form the tissues that compose particular organs.

People often need cartilage repair or replacement, but bone and cartilage form only when their progenitor cells are subjected to mechanical stresses and shear forces. Biochemical engineers have grown bone by seeding bone marrow stem cells on a ceramic disc imbued with zirconium oxide and loading these discs into bioreactors with a rotating bed. Cartilage biopsies are taken from the nose or knee and grown in a bioreactor in which the cells are perfused into a complex sugar called glycosaminoglycan (GAG). This engineered cartilage is then used for transplantations. Such experiments have established that nasal cartilage responds to physical forces similarly to knee cartilage and might substitute for knee cartilage.

Heart muscle is grown in bioreactors that pulse the liquid growth medium through the chamber under high-oxygen tension. Blood vessels are grown in two-chambered bioreactors and contain a reservoir of smooth muscle cells and a chamber through which culture medium is repeatedly pulsed.

**Food Engineering.** Companies making foods that require fermentation by microorganisms or digestion of complex molecules by enzymes use bioreactors to optimize the conditions under which these reactions occur. Biochemical engineers design the industrial processes that manufacture, package, and sterilize foods in the most cost-effective manner.

Starch is a polymer of sugar made by plants and is a very cheap source of sugar. To convert starch into glucose, enzymes called amylases are employed. These enzymes are often isolated from bacteria or fungi, and some are even stable at high temperatures. Degrading starch at high temperatures often clarifies it and rids it of contaminating proteins.

Lactic acid fermentation metabolizes simple sugars to lactic acid and is commonly used in the production of yogurt, cheeses, breads, and some soy products. Cheese production begins with curdling milk by adding acids such as vinegar that separate solid curds from liquid whey and an enzyme mixture called rennet that comes from mammalian stomachs and coagulates the milk. Starter bacterial cultures then ferment the milk sugars into lactic acid. Yogurt is made from heat-treated milk to which starter cultures are added. The acidity of the culture is monitored, and when it reaches a particular point, the yogurt is heated to sterilize the culture for packaging.

Ethanol fermentation converts simple sugars to ethyl alcohol and is used in the production of alcoholic beverages. The most common organism utilized for ethanol fermentation is the baker's yeast, *Saccharomyces cerevisiae*. Malted barley is the sugar source in beer production, and grapes are used to make wine. Beer production involves the extraction of wort, a sugar-rich liquid from barley, which is treated with hops to add aroma and flavor and is then fermented by yeast to form beer. For wine production, the juice from crushed grapes is fermented by yeast for from five to twelve days to generate ethanol. For most red wines and some white wines, the mixture is fermented a second time by malolactic bacteria that degrade the malic acid in the wine, which has a rather harsh, bitter taste, to lactic acid. This lowers the acidity of the wine.

**Biofuel Production.** Burning of fossils fuels as an energy source is not sustainable, since the supply of these fuels is finite and their combustion generates greenhouse gases such as carbon dioxide ($CO_2$), sulfur dioxide ($SO_2$), and nitrogen oxides. First-generation biofuels (biodiesel and bioethanol) utilize biomass from cultivated crops such as corn, sugar beets, and sugar cane. This results in the unfortunate consequences of tying up large swathes of farmland for fuel production and raising food prices. Second-generation biofuels come from grasses, rice straw, and bio-ethers, which are economically superior to first-generation biofuels. Third-generation biofuels show the most ecological and economic promise and come from microalgae. The oil content of some microalgae can exceed 80 percent of their dry weight, and since they use sunlight as their energy source and atmospheric $CO_2$ as their carbon source, microalgae can produce substantial amounts of oil with little material investment.

Microalgae can be grown in open ponds, which ties up land, or special bioreactors called photobioreactors. The fast-growing microalgae are harvested and then liquefied by microwave high-pressure reactors. Oils extracted from the algal species *Dunaliella tertiolecta* at 340 degrees Celsius for 60 minutes had physical properties comparable to fossil fuel oil.

**Waste Management.** The removal of pollutants from air and water provides a large global challenge to environmental engineers. While there are nonbiological ways to degrade pollution, biological strategies represent some of the most innovative and potentially effective ways to remediate pollution.

To treat polluted air, it is piped through a biofilter, which consists of an inert substance called a carrier. Nutrients are trickled over the carrier, and consequently the carrier is colonized by biological organisms that can degrade the pollutant. Devices called bioscrubbers eliminate pollutants such as hydrogen sulfide ($H_2S$), which smells like rotten eggs, or $SO_2$, by dissolving the air pollutants in water and running the water into a bioreactor where the pollutants are degraded. For air pollutants that are poorly soluble in water, such as methane ($CH_4$) or nitric oxide (NO), hollow-fiber membrane bioreactor (HFMB) systems that house a robust population of biological organisms that can degrade gas-phase pollutants effectively treat air polluted with such molecules. Many of these same strategies can also treat polluted water.

Bioreactor landfills were designed to accelerate the degradation of municipal solid waste (MSW) in landfills. Bioreactor landfills use microorganisms to degrade solid wastes, but they also drain the water (leachate) that moves through the landfill, clean it, and recycle it back through the landfill in a process called leachate recirculation. The design of a bioreactor landfill requires extensive knowledge of the surroundings, the nature of the MSWs to be treated, and the quality of the water that becomes the leachate.

**IMPACT ON INDUSTRY**

**Government Agencies.** Several government agencies regulate the work of biochemical engineers. The Environmental Protection Agency (EPA) enforces several regulations enacted by the U.S. Congress, which include the Clean Air Act, the Clean Water Act, and the Resource Conservation and Recovery Act (which amended the Solid Waste Disposal Act). In response to this legislation, the EPA established National Ambient Air Quality Standards, which set allowable ceilings for specific pollutants. The Clean Air Act listed almost two hundred chemicals the emissions of which had to be reduced or completely phased out. The EPA also established National Emissions Standards for Hazardous Air Pollutants, which specifies emission standards for pollutants such as asbestos, organic compounds, and heavy metals.

The National Institute for Occupational Safety and Health (NIOSH), part of the Centers for Disease Control, is responsible for conducting research and helping companies prevent work-related illness and injury. The Occupational Safety and Health Administration (OSHA) enforces the Hazardous Waste Operations and Emergency Response Standard, which specifies safe work conditions, safety training, and effective emergency-response plans.

All companies that produce medicines or food products for human consumption are subject to oversight by the FDA.

**University Research.** Several universities have biochemical engineering departments that engage in high-level research, much of which is supported by government funding bodies such as the National Science Foundation and the National Institutes of Health. Most of the research in biochemical engineering investigates bioreactor design, fermentation application, biotechnology/biomolecular engineering, nanopharmaceutics, pharmaceutical engineering, energy, and process systems. Some of the leading biochemical engineering programs in the United States are found at the following universities and colleges: University of California, Irvine; California Polytechnic State University, San Luis Obispo; Northwestern University; University of Maryland; Dartmouth College; Rutgers University; and Drexel University.

**Industry and Business Sectors.** Several industries make extensive use of biochemical engineers to design and maintain their production plants. Chemical industries use biological organisms to make useful

---

## Fascinating Facts About Biochemical Engineering

- Capromab pendetide (ProstaScint), made by Cytogen Corporation, an engineered monoclonal antibody that specifically labels prostate cancers, was the first recombinant protein produced in hollow-fiber membrane bioreactors to be approved by the Food and Drug Administration in 1996.

- Heart muscle tissue grows in bioreactors only in the presence of high oxygen concentrations, but oxygen is poorly soluble in aqueous environments. To solve this problem, tissue engineers use an artificial oxygen carrier, such as perfluorocarbon, to act as a kind of artificial hemoglobin.

- A common component of household mildew, the fungus *Aspergillus niger*, converts sugars in beets or cane molasses to citric acid, which is used to acidulate foods, beverages, and candies.

- By injecting a gel made from seaweed (calcium alginate) into rabbit leg bones, tissue engineers were able to grow engineered bone that provided viable material for bone-transplant surgeries without affecting normal bone growth.

- Tempeh, a popular fermented food eaten as a meat substitute, is made in a solid-state fermentation bioreactor by cultivating the fungus *Rhizopus oligosporus* on cooked soybeans. The fungus binds the soybeans into compact cakes that are fried and packaged to sell to the public.

- Bioartificial hollow-fiber liver devices have cultured liver cells (hepatocytes) growing in between hollow fibers. The patient's blood is passed through the fibers, where it diffuses into the extra-fiber spaces, interacts with the hepatocytes, and returns to the patient. In laboratory tests, such devices maintained liver function for a few months.

---

chemicals. Since the vast majority of organic compounds are made from petroleum, the price of which keeps increasing, using microorganisms to make organic molecules is economically wise and environmentally sound. Energy industries, particularly those involved in cultivating alternative energy sources, use biochemical engineers to develop first-, second- and third-generation biofuels. Environmental firms also use bioreactors to detoxify and degrade pollutants. Food producers use a variety of enzymes and fermentations to produce their products. Pharmaceutical industries employ about 20 percent of all biochemical engineers. This industry utilizes genetically engineered organisms to make a variety of medicinal compounds ranging from antibiotics to over-the-counter drugs. Finally, the pulp and paper industry employs biochemical engineers to fashion cellulose, a biopolymer, into various types of products, such as stationery, cardboard, and recycled paper.

**Major Corporations.** In the pharmaceutical industry, some of the leading companies include Merck, Pfizer, Novartis, GlaxoSmithKline, Eli Lilly, Roche, AstraZeneca, Bristol-Myers-Squibb, Genentech, and Sanofi-Aventis. In the food industry, major companies include Unilever, Nestlé, PepsiCo, Diageo, and Mars. Some of the companies that use biochemical engineering to mitigate environmental pollution include Environmental Resources Management; Malcolm Pirnie; Versar; Ecology and Environment; EnSafe; First Environment; and Heritage-Crystal Clean. Biofuel and energy companies include Schlumberger, Dominion, E.ON, CenterPointEnergy, GE Power Systems, Sempra Energy, and Edison International. Chemical industries include Dow Chemical and DuPont. Finally, the paper and pulp industries that use biochemical engineering include Iogen Corporation, Eka Chemicals, Kalamazoo Paper Chemicals, and Plasmine Technology.

### CAREERS AND COURSE WORK

Foundational course work for biochemical engineering includes classes in biology, chemistry, physics, advanced mathematics, and computer programming. Additional course work in basic engineering concepts, like statics, dynamics, electronics, and thermodynamics are also necessary to train as a biochemical engineer. These would be followed by advanced courses in chemistry and chemical engineering. To work as a biochemical engineer, a bachelor's degree in biological, biochemical, or chemical engineering is required, as is becoming registered as professional engineer (PEng). Advanced degrees (M.S. or Ph.D.) are required for those who wish to work in academics or professional engineering research. In industry, advanced degrees are typically not required for entry-level jobs but are helpful for promotions to managerial or supervisory positions.

Biochemical engineering is a highly collaborative field, and an engineer must be able to work well with other professionals. Therefore, good communication skills are essential. Biochemical engineering is a highly analytical field that requires skill and an affinity for mathematics and chemistry as well as and problem solving. Since much of the problem solving occurs on a computer, the ability to visualize and design processes on computers is becoming an integral part of the field.

If a company employs biological organisms or enzymes for biological or chemical conversions, that company requires the expertise of a biochemical engineer. The majority of biochemical engineers work in chemical process industries (CPI), which include the chemical, gas and oil, food and beverage, textile, and agricultural sectors. Other enterprises that employ biochemical engineers include biotechnology and pharmaceutical companies, petroleum refining and oil sands extraction, waste management, paper and pulp manufacturing, farm machinery, construction and engineering design companies, and environmental companies and agencies. Government-employed biochemical engineers may work for the Departments of Energy or Agriculture, the EPA, or other such agencies.

## SOCIAL CONTEXT AND FUTURE PROSPECTS

Two aspects of biochemical engineering can be cause for concern to the general public. First, biochemical engineers work with genetically engineered organisms. Many people have never completely made peace with the use of such organisms, despite the fact that many of the items people consume on a regular basis, from seasonal flu vaccines and other medicines to the foods they eat, are made by genetically engineered organisms. Nevertheless, fear of genetically engineered organisms remains. For example, despite repeated tests establishing that genetically engineered foods are as safe as food from nongenetically modified crops, some people still feel the need to label genetically engineered food as Frankenfood.

As long as this fear persists, the work of biochemical engineers will make some people uncomfortable. Second, biochemical engineers tend to work for large industries that are sometimes painted as inveterate polluters by environmental groups or as greedy, unconcerned capitalists by consumer-advocate groups. Since many companies abide by strict environmental standards and engage in humanitarian work, these accusations are somewhat unfair.

The development of new technologies in fields like genetic engineering, biomedicine, bioinstrumentation, biomechanics, waste management, and alternative energy development are driving new employment opportunities for biochemical engineers. According to the U.S. Department of Labor, Bureau of Labor Statistics, biochemical engineers are expected to have 21 percent employment growth during the period of 2006 through 2016, which is faster than the average for all occupations. Greater demands for more sophisticated medical equipment, procedures, and medicines will increase the need for greater cost-effectiveness. The growing interest in biochemical engineering is reflected in the tripling of the annual number of bachelor's degrees earned in biochemical engineering since 1991 and in the doubling of the annual number of graduate degrees earned in this field per year since 1991.

According to the U.S. Department of Labor, Bureau of Labor Statistics, the median annual salary for a biochemical engineer in 2009 was $73,930, with the lowest 10 percent earning $44,930 and highest 10 percent earning $116,330.

*Michael A. Buratovich, B.S., M.A., Ph.D.*

## FURTHER READING

Katoh, Shigeo, and Fumitake Yoshida. *Biochemical Engineering: A Textbook for Engineers, Chemists, and Biologists.* Weinheim, Germany: Wiley-VCH Verlag, 2009. A basic, though rather technical, textbook of biochemical engineering by two prominent Japanese biochemical engineers that contains many tables of mathematical symbols and conversions, graphs that illustrate the application of the equations presented, and problems for interested students to solve.

McNamee, Gregory. *Careers in Renewable Energy: Get a Green Energy Job.* Masonville, Colo.: PixyJack Press, 2008. This highly readable summary of the renewable-energy job market includes more than just engineering jobs and discusses potential future employment opportunities in alternative energy sources.

Mosier, Nathan S., and Michael R. Ladisch. *Modern Biotechnology: Connecting Innovations in Microbiology and Biochemistry to Engineering Fundamentals.* Hoboken, N.J.: Wiley-AIChE, 2009. A very practical and richly illustrated and referenced guide to the advances in molecular biology for aspiring biochemical engineers.

Murphy, Kenneth M., Paul Travers, and Mark Walport. *Janeway's Immunobiology.* 7th ed. Oxford, England: Taylor & Francis, 2007. A standard immunology textbook that has an excellent section on the therapeutic use of antibodies.

Pahl, Greg. *Biodiesel: Growing a New Energy Economy.* 2d ed. White River Junction, Vt.: Chelsea Green, 2008. A popular guide to the advances in biodiesel technology and biofuel industries that also examines the food-for-fuel controversy and the issues surrounding genetically modified crops.

Vasic Racki, Durda. "History of Biotransformations: Dreams and Realities." *Industrial Biotransformations,* edited by Andreas Liese, Karsten Seelbach, and Christian Wandrey. Weinheim, Germany: Wiley-VCH Verlag, 2000. This essay chronicles the history of using microorganisms and enzymes to synthesize commercially valuable products and the rise of biochemical engineering as an inevitable consequence of these developments.

Walker, Sharon. *Biotechnology Demystified.* New York: McGraw-Hill Professional, 2006. This is an introduction to the basics of and latest advances in molecular biology and the latest applications of these concepts to a range of discoveries, including new drugs and gene therapies.

**WEB SITES**

*Biohealthmatics*
http://www.biohealthmatics.com/careers/PID00269.aspx

*National Society for Professional Engineers*
http://www.nspe.org

*Sloan Career Cornerstone Center*
http://www.careercornerstone.org/pdf/bioeng/bioeng.pdf

**See also:** Bioenergy Technologies; Biofuels and Synthetic Fuels; Cell and Tissue Engineering; Chemical Engineering; DNA Analysis; DNA Sequencing; Enzyme Engineering; Food Preservation; Food Science; Genetically Modified Organisms; Genetic Engineering; Industrial Fermentation; Stem Cell Research and Technology.

# BIOENERGY TECHNOLOGIES

## FIELDS OF STUDY

Biology; chemistry; physics; chemical engineering; engineering technology; fuel resources and logistics; combustion technology; mathematical statistics and modeling; agricultural science; genetic engineering.

## SUMMARY

The introduction of green bioenergy technologies may reduce pollution and dependence on finite fossil fuels. Bioenergy derived from biomass has the potential to provide renewable and sustained energy on both a local and a global scale. According to the International Energy Agency Bioenergy, the fundamental objective of bioenergy technology is to increase the use and implementation of ecologically sound, economically viable, and sustainable bioenergy that will help meet the world's increasing energy demands.

## KEY TERMS AND CONCEPTS

- **Bioenergy:** Energy derived from biomass (living or recently living biological organisms); can be either a liquid (ethanol), gas (methane), or solid (biochar).
- **Biofuel:** Liquid fuels that are most commonly derived from plant material, such as sugar and starch crops (ethanol), or vegetable oils and animal fat (biodiesel) and used as a replacement for or additive to traditional fossil fuels (petrol).
- **Biomass:** Any renewable organic material derived from plants (such as crops or agricultural plant residue) or animals (such as human sewage or animal manure) that is used to produce bioenergy.
- **Feedstock:** Primary raw material used in the manufacture of a product or in an industrial process.
- **Fossil Fuels:** Deposits within the Earth's crust of either solid (coal), liquid (oil), or gaseous (natural gas) hydrocarbons produced through the natural decomposition of organic material (plants and animals) over many millions of years. These deposits contain high amounts of carbon, which can be burned with oxygen to provide heat and energy.
- **Peak Oil:** Point at which global oil production reaches its maximum point; it occurs when about

half of the oil deposits in the Earth's crust have been extracted. After this, oil extraction falls into terminal decline. Production is thought to follow a bell curve, starting slowly, peaking, and falling at a relatively steady rate.
- **Pyrolysis:** Thermochemical process in which organic biomass is combusted at high temperatures in the absence of oxygen reagents to achieve decomposition and produce energy.
- **Renewable Energy:** Energy obtained from (usually) natural and sustainable sources, including the Sun (solar), ocean (marine), water (hydro), wind, geothermal sources, and organic matter (biomass); also known as alternative energy.

### DEFINITION AND BASIC PRINCIPLES

Although they are closely related terms, bioenergy should not be confused with biomass. Fundamentally, bioenergy is energy derived from biomass, that is, energy derived from living (or recently living) biological organisms. Although fossil fuels are naturally occurring substances formed through the decomposition of biological organisms, the creation of these types of fuels takes millions of years, which means they are, on a human history scale, nonrenewable and unsustainable. The global demand for fossil fuels such as oil continues to increase. Oil has allowed human society to thrive because it has supplied seemingly endless cheap energy, creating diverse industries and employment. Scientific evidence continues to mount, however, in regard to its environmental impact. Oil is also a finite resource, and research has indicated that peak oil has either already occurred or will occur by the mid-2000's. A growing number of environmental and scientific organizations state that once the point of peak oil is reached, demand will far outstrip supply, and therefore, it is imperative that an alternative and sustainable source of fuel is found and implemented.

### BACKGROUND AND HISTORY

The history of bioenergy is as long as the history of human civilization itself. The most basic form of bioenergy from biomass—the burning of wood for heat and light—has been used for thousands of years. Human society might rely heavily on fossil fuels for its energy needs, but bioenergy has been the world's primary energy source for most of human history and is still in use.

Wood and other combustibles such as corn husks are not the only sources of bioenergy with a long history of human use. One of the most popular biofuels is ethanol. Ethanol was first developed in the early to mid-1800's, before Edwin Drake's 1859 discovery of petroleum. The push for alternative fuels during that time was driven by the need to replace the whale oil used in lamps, as supplies dwindled and prices increased. By the late 1830's, ethanol mixed with naturally derived turpentine was the preferred and cheaper alternative.

In 1826, Samuel Morey invented the internal combustion engine. His engine was powered by a composite fuel of ethanol and turpentine, and although he was unable to find a suitable investor, his engine is considered to be his greatest and most progressive invention. It was not until 1860, however, that the German inventor Nicholas Otto independently invented the internal combustion engine (again fueled by ethanol) and achieved financial backing to develop the engine.

The next considerable leap in interest in bioenergy and in technology development occurred with the invention of the automobile in the early 1900's and Henry Ford's vision of an ethanol-fueled vehicle. As early as 1917, scientists knew that ethanol and other alcohols could be used as fuels. Because ethanol and other alcohols could be derived easily from any vegetable matter that undergoes fermentation, they could be cheaply and easily produced.

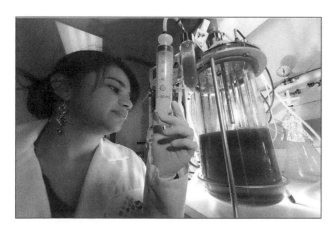

*Researcher with algae fermentation bioreactor.* (Volker Steger/Photo Researchers, Inc.)

Despite this, however, ethanol fuels were not widely embraced as the fuel of choice for automobiles, particularly in counties such as the United States, where an ethanol tax made the alternative fuel more expensive than petrol. By 1906 when the tax was removed, gasoline fuel had developed an extensive infrastructure, and ethanol could not compete.

Because of rationing and shortages of petrol during World War II, ethanol and other vegetable-based fuels were used extensively. In particular, the use of vegetable-based fuels such as palm oil became common in European colonies in Africa so as to increase fuel self-sufficiency. During the oil crisis of the 1970's, oil prices skyrocketed and oil-dependent countries, such as the United States, became desperate to find a replacement for fossil fuels and to reduce their dependence on oil-producing countries. The 1974 oil embargo was influential in renewing interest in alternative fuels, such as ethanol, particularly in the United States.

Despite this, however, fossil fuels have continued to be used to a much greater degree than ethanol fuels. Although many automotive fuels are mixed with ethanol, there is still significant room for growth and technological and infrastructure development of bioenergy and biofuels. The potential of biofuels as a sustainable alternative to fossil fuels has once again increased interest in bioenergy. However, there are arguments against the use of bioenergy, particularly the fuel-versus-food debate. Some people believe that as long as large numbers of people worldwide do not get enough to eat, using crops for fuel instead of food is at best misguided and at worst highly unethical, and it contributes to the rates of starvation and malnutrition seen in many developing countries.

### HOW IT WORKS

Although scientific debate still surrounds human-influenced climate change and the timing of peak oil, many experts believe that developing sustainable fuels from renewable sources and implementing technology that helps reduce pollution is important and necessary. Bioenergy technologies play an important role in accelerating the adoption of environmentally sound bioenergy at a reasonable cost and in a sustainable manner, thereby helping meet future energy demands.

Bioenergy can be produced from many different biological materials, including wood and various crops, as well as human and animal waste. All these materials can be used to produce electricity and heat, and after coal, oil, and natural gas, biomass is the fourth largest energy resource in the world. About 14 percent of the world's electricity is produced from biomass sources. This percentage differs from country to country and is greater in developing countries, accounting for between 70 and 90 percent of the energy produced in some countries, while in developed countries such as the United States, bioenergy accounts for only between 3 and 4 percent. The United States is, however, the world's largest biopower generator and, according to the U.S. Energy Information Administration, it possesses more than half of the world's installed bioenergy capacity.

Since the 1990's, bioenergy technologies have experienced continuous development. Generally, these technologies can be divided into two main groups in relation to bioenergy production: energy crops and waste energy. Energy crops, which include trees, sugarcane, and rapeseed, are either combusted or fermented to produce high-energy alcohols such as ethanol and biodiesel that can be used as a replacement for petrol and liquid fuels. Certain waste, including organic waste from agriculture, human and animal effluent, food and plant waste, and industrial residue, can be used to produce methane gas. This methane gas can be combusted to produce steam, which can then be used to turn turbine generators and produce heat and electricity.

Most forms of bioenergy require combustion and thus the release of carbon dioxide into the atmosphere at some stage during their production. This release, however, is offset by the initial absorption of carbon dioxide by the fuel crops during the growing process. According to research, even accounting for all carbon dioxide released as a result of the planting, harvesting, producing, and transporting of bioenergy, net carbon emissions are significantly reduced (up to 90 percent).

## APPLICATIONS AND PRODUCTS

A number of different types of domestic biomass resources (also referred to as feedstocks) are used to produce bioenergy. These include biomass processing residues such as paper and pulp, agricultural and forestry wastes, urban landfill waste and gas, animal (including human) sewage and manure waste, and land and aquatic crops. There are basically two very important and useful applications of bioenergy: the production of electricity and the replacement of liquid fuels such as petrol.

**Electricity Production.** Biomass is capable of producing electricity in many different ways. The most commonly used methods include pyrolysis, cofiring, direct-fired/conventional stream method, biomass gasification, anaerobic digestion, and landfill gas collection.

The term "pyrolysis" is derived from the Greek words *pyro* (meaning fire) and *lysys* (meaning decomposition). Pyrolysis is a thermochemical conversion technology that involves the combustion of biomass at very high temperatures and its decomposition without oxygen. Although this process is energy consumptive and expensive, it can be used to produce electricity through the creation of pyrolysis oil, biochar, and syngas (oil, coke, and gas). These three products can be used for electricity production, as soil fertilizer, and for carbon storage. There are two types of pyrolysis—fast and slow. Fast, or flash, pyrolysis, which uses any organic material as a biomass source, takes place within seconds at temperatures of 300 to 550 degrees Celsius with rapid accumulation of biochar. In slow, or vacuum, pyrolysis, which uses any organic material as a biomass source, the combustion of the biomass occurs within a vacuum to reduce the boiling point and adverse chemical reactions.

Cofiring basically involves the combustion of solid biomass such as wood or agricultural waste with a traditional fossil fuel such as coal to produce energy. Many consider this form of bioenergy to be the most efficient in terms of the existing fossil fuel infrastructure, with its high dependence on coal. In addition, because the combustion of biomass is carbon neutral (the carbon absorbed during growth is equal to the amount released during combustion), mixing biomass with coal can assist in reducing net carbon emissions and other pollutants such as sulfur. The production of electricity by this method is considered advantageous because it is inexpensive and makes use of already existing power plants.

The direct-fired/conventional stream method is the most commonly used method of producing bioenergy. This process involves the direct combustion of biomass to produce steam, which then turns turbines that drive generators to produce electricity.

Biomass gasification is another type of thermochemical conversion technology, in which any organic biomass is converted into its gaseous form (known as syngas) and used to produce energy. The process relies on biomass gasifiers that heat solid biomass until it forms a combustible gas, which is then used in power production systems that merge gas and steam turbines to generate electricity. Although this technology is still in its infancy, it is hoped that gasification of biomass may lead to more efficient bioenergy production. Biomass gasification technologies basically can be categorized as fixed bed gasification, fluidized bed gasification, and novel design gasification.

Anaerobic digestion, a natural biological process, has a long history and involves the decomposition of organic biomass material such as manure and urban solid wastes in an air-deficient environment. Fundamentally, anaerobic digestion involves the production of methane gas through bacteria and archaea activity. This methane gas, also known as a type of biogas, is captured and then used to power turbines and produce electric and heat energy, as well as a soil-enhancement material called digestate. The advantage of this method is that it uses waste, such as wastewater sludge, to produce renewable energy and reduces the amount of greenhouse gas being released into the atmosphere.

The process of landfill gas collection is closely related to anaerobic digestion and involves the capture of gas from the decomposition of landfill urban wastes. This gas, which is about 50 percent methane, 45 percent carbon dioxide, 4 percent nitrogen, and 1 percent other gases, is then used to produce energy.

**Fuel Production.** Liquid biofuels are a significant alternative to petroleum-based vehicle and transportation fuels. Biofuels are attractive because they can be used in already existing vehicles with little modification required and also in the production of electricity. These fuels are estimated to account for almost 2 percent of the transportation fuels used in the world's vehicles.

Bioenergy production is generally divided into first-generation and second-generation fuels. The main distinction between these two types of fuels relates to the feedstock used. First-generation fuels are already in commercial production in many countries and are primarily made from edible grains, sugars, or seeds. The two most common biofuels are ethanol and biodiesel, both of which are already used in large quantities in many countries. Second-generation

fuels, although not yet in commercial production, are considered superior to first-generation fuels because they are primarily made from nonedible whole plants or waste from food crops such as husks and stalks.

Ethanol, also known as ethyl alcohol, is an alcohol that can be used as a vehicle or transportation fuel. It is a renewable energy produced from sustainable agricultural feedstocks, particularly sugar and starch crops such as sugarcane, potatoes, and maize. Although it can be used as a direct replacement fuel, it is more commonly used as a fuel additive. Many vehicles use ethanol-petrol blends of 10 percent ethanol, which improve octane and decrease emissions. Ethanol is particularly popular in Brazil and the United States, which together produce almost 90 percent of the world's ethanol. Brazil has been at the forefront of ethanol use and as early as 1976 mandated the use of ethanol with fuels, eventually requiring a blend containing 25 percent ethanol (the most of any country). Although using arable land to grow crops to produce ethanol is somewhat controversial, the development of ethanol from cellulosic biomass (obtained from the cellulose of trees and grasses) is considered promising and may play a large role in the future of ethanol as a biofuel.

Biodiesel is a renewable energy produced from sustainable animal fat and vegetable oil feedstocks, such as soy, rapeseed, sunflowers, palm oil, hemp, and algae, and can be used as a vehicle or transportation fuel. As with ethanol, however, biodiesel is more often used as a diesel additive to reduce the levels of pollution emitted by traditional diesel engines. It is primarily produced through a process known as transesterification, which is the exchange or conversion of an organic acid ester into another ester.

Biobutanol can be used as fuel in internal combustion engines. It is usually produced from the fermentation of biomass, and because of its chemical properties, it is actually more similar to petrol than ethanol is. It can be produced from the same feedstocks used for ethanol production, such as corn, sugarcane, potatoes, and wheat. Despite its possible applications, however, this type of biofuel has not been produced commercially.

## IMPACT ON INDUSTRY

Many governments, universities, and organizations have been investigating bioenergy because of

## Fascinating Facts About Bioenergy Technologies

- Organic material, such as plant and animal waste, can be used to produce both liquid fuel and electricity.
- Bioenergy can be produced from biomass, which is sustainable and renewable and offers a viable alternative to fossil fuels.
- The burning of biomass for fuel is carbon neutral–that is, the carbon absorbed during growth is equal to the amount released during combustion.
- Some experts are concerned that the use of biomass for fuel will create both social and environmental problems because it uses food for fuel and reduces biodiversity.
- The political climate in the twenty-first century has created renewed interest in bioenergy for the world's growing human population. The percentage of bioenergy used is much higher in less-developed countries than in developed nations.

its potential for becoming one of the most important fuels of the future. Its ability to replace fossil fuels and its sustainability have pushed bioenergy into the forefront of fuel research.

**Major Organizations.** Many agencies and nongovernmental organizations (NGOs) are moving forward with bioenergy technology research and applications. A number of agencies and organizations are committed to increasing research and information exchange in an attempt to promote the use of bioenergy.

One of the earliest organizations established was the International Energy Agency (IEA). This autonomous body of the Organisation for Economic Co-operation and Development was founded in 1974 in response to the oil crisis. In 1978, the IEA established IEA Bioenergy to promote the development of new green energy and improve international bioenergy research and information exchange, with a particular focus on economic and social development and environmental protection.

Given the global impact of fossil fuel consumption in terms of particulates and atmospheric pollution, many international organizations are taking an active interest in bioenergy technology and have released a number of statements regarding the adoption of

bioenergy as an alternative fuel, including the Food and Agriculture Organization of the United Nations, which has stated that bioenergy may play a role in eradicating hunger and poverty and in ensuring environmental sustainability.

A number of organizations have been working toward the adoption of specific bioenergies. The International Biochar Initiative, formed in 2006 at the World Soil Science Congress, aims to encourage the development and use of biochar for carbon capture, soil enhancement, and energy production. The goal of the volunteer-led Biofuelwatch is to ensure that biofuels used within the European Union are obtained only from sustainable sources, thereby lessening the possibility of loss of biodiversity. Global support for organizations involved in bioenergy technology research and applications is provided by organizations such as the World Bioenergy Association, formed in 2008.

**Government and University Research.** Many countries have been expressing greater interest in bioenergy technology and conducting research. Some of the most significant research that has been undertaken on bioenergy has taken place in countries such as the United States, particularly within the Midwest. The oil embargo of 1973-1974 was influential in renewing interest in alternative fuels, particularly in countries such as the United States, which began intensive research and development of the dual fuel project at Ohio State University. Iowa State University has led the way in terms of academic research and education, with its Biorenewable Resources and Technology graduate program, the first of its kind in the United States. It not only offers advanced study in the use of plant and crop biomass for the creation of fuels and energy but also is home to the Bioeconomy Institute, one of the largest bioenergy institutes in the United States investigating the use of biorenewable resources as sustainable feedstocks for bioenergy.

The U.S. Department of Energy's Biomass Program, run by the office of Energy Efficiency and Renewable Energy, also aims to expand and enhance biomass power technology and increase production and use of biofuels such as ethanol and biodiesel. Specifically, the program seeks to produce diverse bioenergy products including electricity, liquid/solid/gaseous fuels, heat, and other bio-based materials. Bioenergy has become the second largest source of renewable energy used in the United States, and the president's Biofuels Initiative aims to make cellulosic ethanol economically competitive.

Many other countries, including those in the European Union, Australia, and the United Kingdom, have been increasing research into bioenergy technology and have contributed significantly to international knowledge. Australia, for example, established Bioenergy Australia in 1997 as part of its IEA Bioenergy collaborative agreement and to provide a forum to encourage and facilitate the expansion of domestic biomass energy development and use. The European Union has set goals for meeting 5.75 percent of transportation fuel requirements with biofuels. The United Kingdom Biomass Strategy, initially published in 2007, provides guidelines for the sustainable development of biomass for heat, fuel, and electricity applications. The Biomass Strategy, the Renewable Energy Strategy, and the Bio-energy Capital Grants Scheme are among the United Kingdom's initiatives to achieve its 15 percent share of the 2020 renewable energy target established by the European Union.

Of most significance, perhaps, are the bioenergy initiatives of China and India. Although both of these rapidly developing countries use less energy per capita than most developed countries, such as the United States, Australia, and Canada (three of the top energy-consuming countries per capita), the sheer size of their populations means that they are among the world's highest energy consumers on a per country basis. Both China and India have been pursuing and researching bioenergy technology alternatives. China, the third-largest ethanol producer in the world, set a ten-year bioenergy target of generating 1 percent of its alternative energy from biomass sources. India has announced the development of bioenergy power plants to help meet growing energy demands and supply adequate energy to rural areas.

## CAREERS AND COURSE WORK

Undergraduate and graduate courses in bioenergy technology are offered at many universities. Most students who follow this path have a strong background in science, agriculture, and fuel production technology. Following graduation, students studying bioenergy technology will understand methods for improving the efficiency of new energy technologies and have a solid understanding of environmentally friendly bioenergy concepts, theories, processes, and

practices. These include biology and plant production, bioenergy electricity and fuel production from crops and waste, combustion science, sustainability in energy production, agricultural engineering, power-engine design, and emission-reduction techniques. They will also know how to integrate environmental issues and global economics into decision making. The primary purpose in bioenergy course work and research is to provide students with an understanding of energy as it relates to both economics and the environment. In addition, as with many modern sciences, students of bioenergy technology will require an understanding of computer modeling.

Bioenergy technology responds to the ever-changing needs of human society in relation to renewable and sustainable energy requirements, while focusing on environmental engineering in an internationally, socially, and ecologically responsible way. Students involved in bioenergy technology research and application can pursue various careers in environmental auditing and consulting, bioenergy education, agricultural and combustion engineering, sustainable agriculture, farm energy specializations, power plant and steam boiler engineering, pollution prevention and emissions reduction, waste treatment, renewable energy program management, and project and resource management. These careers span a wide range of industries and sectors, including, most prominently, the energy industry, private sector, nongovernmental organizations, specialized government organizations and agencies, and universities and institutions undertaking teaching and research.

## SOCIAL CONTEXT AND FUTURE PROSPECTS

The concept of bioenergy relies on the fact that such energy is sustainable. Although many believe bioenergy is synonymous with green energy, this is not always the case. Individual types of bioenergy produced in different ways and from various biomasses can have very diverse environmental impacts. Although the goal of bioenergy is to reduce the world's dependence on nonrenewable (to all intents and purposes) fossil fuels and thereby reduce greenhouse gas emissions and pollution, some forms of bioenergy can be equally harmful in terms of pollution or the energy expended to produce the bioenergy. As such, there have been significant movements since the 1980's to develop cleaner, greener, and move advanced biofuels and technologies. Many countries are investigating the potential of bioenergy, and many researchers believe that bioenergy will become a key contributor to sustainable global energy use.

Biofuels are, however, controversial, and some are calling for a moratorium on their use and advancement because of environmental and social concerns, particularly the loss of biodiversity and habitat destruction and the use of crops for fuel rather than food. First-generation fuels, already in commercial production in many countries, have been criticized because they are obtained from edible seeds and plants. However, because of the increasing problems of fossil fuel dependence, many of the world's governments and international organizations are stepping up research into bioenergy as a viable and sustainable alternative fuel. Many researchers believe that investigation and implementation of second-generation fuels, which are made from nonedible whole plants or waste from food crops, is the way of the future in terms of both efficiency and social responsibility. Biomass does offer many countries the opportunity to use fuels that are both sustainable and domestically sourced.

*Christine Watts, Ph.D., B.App.Sc., B.Sc.*

## Further Reading

Geller, Howard. *Energy Revolution: Policies for a Sustainable Future.* Washington, D.C.: Island Press, 2003. Examines renewable energy, concentrating on energy patterns, trends and consequences, the barriers to sustainable energy use, and case studies on effective energy from different parts of the world.

Rosillo-Calle, Frank, et al., eds. *The Biomass Assessment Handbook: Bioenergy for a Sustainable Environment.* Sterling, Va.: Earthscan, 2008. Provides information on the supply and consumption of biomass and the skills and tools needed to understand biomass resource assessment and identify the effects and benefits of exploitation.

Scragg, Alan. *Biofuels: Production, Application, and Development.* Cambridge, Mass.: CAB International, 2009. Looks at biofuel production, concentrating on technological issues, benefits and problems, and existing and future forms of biofuels.

Silveira, Semida, ed. *Bioenergy: Realizing the Potential.* San Diego, Calif.: Elsevier, 2005. Investigates and integrates the fundamental technical, policy, and economic issues as they relate to bioenergy projects in both developed and developing countries. Focuses on biomass availability and potential and covers market development and technical and economic improvements.

Sims, Ralph. *The Brilliance of Bioenergy: In Business and Practice.* London: James and James, 2002. Examines the main biomass resources, technologies, processes, and principles of bioenergy production, with a focus on social, economic, and environmental issues in the form of small- and large-scale case studies in developed and developing countries.

Singh, Om V., and Steven P. Harvey, eds. *Sustainable Biotechnology: Sources of Renewable Energy.* London: Springer, 2009. An extensive collection of research reports and reviews, focuing on the progress and challenges involved in the use of sustainable resources for the production of renewable biofuels.

## Web Sites

*International Energy Agency Bioenergy*
http://www.ieabioenergy.com

*U.S. Department of Agriculture, Economic Research Service*
Featuring Bioenergy
http://www.ers.usda.gov/features/bioenergy

*U.S. Department of Energy*
Bioenergy
http://www.energy.gov/energysources/bioenergy.htm

*U.S. Energy Information Administration*
Renewable and Alternative Fuels
http://www.eia.doe.gov/fuelrenewable.html

**See also:** Agricultural Science; Biofuels and Synthetic Fuels; Climate Modeling; Environmental Biotechnology; Fossil Fuel Power Plants; Fuel Cell Technologies; Gasoline Processing and Production; Hydroelectric Power Plants; Land-Use Management; Solar Energy; Wind Power Technologies.

# BIOENGINEERING

## FIELDS OF STUDY

Cell biology; molecular biology; biochemistry; physiology; ecology; microbiology; pharmacology; genetics; medicine; immunology; neurobiology; biotechnology; biomechanics; bioinformatics; physics; mechanical engineering; electrical engineering; materials science; buildings science; architecture; chemical engineering; genetic engineering; thermodynamics; robotics; mathematics; computer science; biomedical engineering; tissue engineering; bioinstrumentation; bionics; agricultural engineering; human factors engineering; environmental health engineering; biodefense; nanotechnology; nanoengineering.

## SUMMARY

Bioengineering is the field in which techniques drawn from engineering are used to tackle biological problems. For example, bioengineers may use mechanics principles—knowledge about how to design and construct mechanical objects using the most ideal materials—to create drug delivery systems. They may work on developing efficient ways to irrigate and drain land for growing crops, or they may be involved in building artificial environments that can support life even in the harsh climate of outer space. A highly interdisciplinary, collaborative field that synthesizes expertise from multiple research areas, bioengineering has had a significant impact on many fields of study, including the health sciences, technology, and agriculture.

## KEY TERMS AND CONCEPTS

- **Biocompatible Material:** Material used to replace or repair tissues in the body, or to perform a biological function in a living organism.
- **Bioinformatics:** Application of data processing, retrieval, and storage techniques to biological research, especially in genomics.
- **Biomechanics:** Application of mechanical principles to questions of motor control in biological systems.
- **Bioreactor:** Tool or device that generates chemical reactions to create a product.

- **Bioremediation:** Use of bacteria and other microorganisms to solve environmental problems, such as neutralizing hazardous waste.
- **Geoengineering:** Use of engineering techniques to modify environmental or geological processes, such as the weather, on a global scale.
- **Prosthetic Device:** Artificial part or implant designed to replace the function of a lost or damaged part of the body.
- **Regenerative Medicine:** Therapies that aim to restore the function of tissues that have been damaged or lost through injury or disease by using tissue or cells grown in laboratories or compounds created in laboratories.
- **Systems Biology:** Theoretical branch of bioengineering that creates models of complex biological processes or systems, using them to predict future behavior.
- **Transgenic Organism:** Plant or animal containing genetic information taken from another species.

## DEFINITION AND BASIC PRINCIPLES

Bioengineering is an interdisciplinary field of applied science that deals with the application of engineering methods, techniques, design approaches, and fundamental knowledge to solve practical problems in the life sciences, including biology, geology, environmental studies, and agriculture. In many contexts, the term bioengineering is used to refer solely to biomedical engineering. This is the application of engineering principles to medicine, such as in the development of artificial limbs or organs. However, the field of bioengineering has many applications beyond the field of health care. For example, genetically modified crops that are resistant to pests, suits that protect astronauts from the ultra-low pressures in space, and brain-computer interfaces that may allow soldiers to exercise remote control over military vehicles all fall under the wide umbrella of bioengineering.

Each of the subdisciplines within bioengineering relies on different sets of basic engineering principles, but a few fundamental approaches can be said to apply broadly across the entire field. From an engineering perspective, three basic steps are involved in solving any problem: an analysis of how the system in question works, an attempt to synthesize the information gathered from this analysis and generate potential solutions, and finally an attempt to design and test a useful product. Bioengineers apply this three-stage problem-solving process to problems in the life sciences. What is somewhat novel about this approach is that it is a holistic one. In other words, it treats biological entities as systems—sets of parts that work together and form an integrated whole—rather than looking at individual parts in isolation. For example, to develop an artificial heart, bioengineers need to consider not just the structure of the heart itself on a cellular or tissue level but also the complex dynamics of the organ's interactions with the rest of the body through the circulatory system and the immune system. They must build a device whose parts can mimic the functionality of a healthy heart and whose materials can be easily integrated into the body without triggering a harmful immune response.

## BACKGROUND AND HISTORY

Principles of chemical and mechanical engineering have been applied to problems in specific biological systems for centuries. For example, bioengineering applications include the fermention of alcoholic beverages, the use of artificial limbs (which are documented as far back as 500 B.C.E.), and the building of heating and cooling systems that regulate human environments.

Bioengineering did not emerge as a formal scientific discipline, however, until the middle of the twentieth century. During this period, more and more scientists began to be interested in applying new technologies from electronic and mechanical engineering to the life sciences. As the United States, Japan, and Europe began to enter a period of economic recovery and growth following World War II, governments increased funding for bioengineering efforts. The cardiac pacemaker and the defibrillator, both developed during this postwar period, were two of the earliest and most significant inventions to come out of the quickly developing field. In 1966, the Engineers Joint Council Committee on Engineering Interaction with Biology and Medicine first used the term "bioengineering." At about the same time, academic institutions began to form specialized departments and programs of study to train professionals in the application of engineering principles to biological problems. In the twenty-first century, rapid technological advances continue to produce growth in the field of bioengineering.

## HOW IT WORKS

Because bioengineering is such a large and diverse field, it would be impossible to enumerate all the processes involved in creating the totality of its applications. The following are a few of the most significant examples of the types of technological tools used in bioengineering.

**Materials Science.** One of the most important areas of bioengineering is the intersection of materials science and biology. Scientists working in this field are charged with developing materials that, although synthetic, are able to successfully interact with living tissues or other natural biological systems without impeding them. (For example, it is vital that biocompatible materials not allow blood platelets to adhere to them and form clots, which can be fatal.) Depending on the specific application in question, other properties, such as tensile strength, resistance to wear, and permeability to water, gases, and small biological molecules, are also important. To manipulate these properties to achieve a desired end, engineers must carefully control both the chemical structure and the molecular organization of the materials. For this reason, biocompatible materials are generally made out of some kind of synthetic polymer—substances with simple and extremely regular molecular structures that repeat again and again. In addition, additives may be incorporated into the materials, such as inorganic fillers that allow for greater mechanical flexibility or stabilizers and antioxidants that keep the material from becoming degraded over time.

**Biochemical Engineering.** Since living cells are essentially chemical systems, the tools of chemical engineering are especially applicable to biology. Biochemical engineers study and manipulate the behavior of living cells. Their basic tool for doing this is a fermenter, a large reactor within which chemical processes can be carried out under carefully controlled conditions. For example, the modern production of virtually all antibiotics, such as penicillin and tetracycline, takes place inside a fermenter. A central vessel, sealed tight to prevent contamination and surrounded by jackets filled with coolants to control its temperature, contains propellers that stir around the nutrients, culture ingredients, and catalysts that are associated with the reaction at hand.

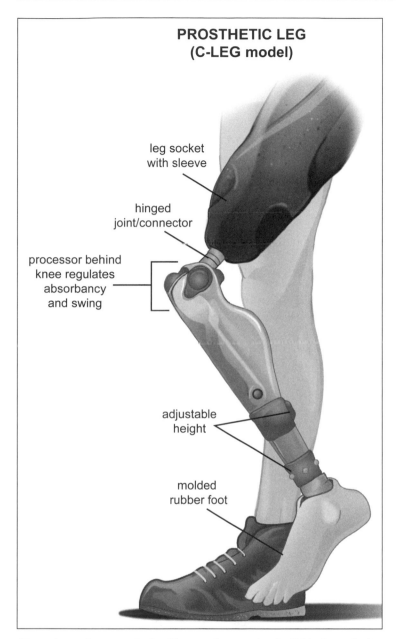

**PROSTHETIC LEG**
**(C-LEG model)**

leg socket
with sleeve

hinged
joint/connector

processor behind
knee regulates
absorbancy
and swing

adjustable
height

molded
rubber foot

*Bioengineers have made significant advances in the field of prosthetics; the C-Leg prosthesis was introduced in 1997.*

Genetic engineering is a subfield of biochemical engineering that is growing increasingly significant. Scientists alter the genetic information in one cell by inserting into it a gene from another organism. To do this, a vector such as a virus or a plasmid (a small strand of DNA) is placed into the cell nucleus and combines with the existing genes to form a new genetic code. The technology that enables scientists to alter the genetic information of an organism is called gene splicing. The new genetic information created by this process is known as recombinant DNA. Genetic engineering can be divided into two types: somatic and germ line. Somatic genetic engineering is a process by which gene splicing is carried out within specific organs or tissues of a fully formed organism; germ-line genetic engineering is a process by which gene splicing is carried out within sex cells or embryos, causing the recombinant DNA to exist in every cell of the organism as it grows.

**Electrical Engineering.** Electrical engineering technologies are an essential part of the bioengineering tool kit. In many cases, what is required is for the bioengineer to find some way to convert sensory data into electric signals, and then to produce these electric signals in such a way as to enable them to have a physiological effect on a living organism.

The cochlear implant is an example of one such development. The cochlea is the part of the brain that interprets sounds, and a cochlear implant is designed for people who are profoundly deaf. A cochlear implant uses electronic devices that capture sounds and relay them to the cochlea. The implant has four parts: a microphone, a tiny computer processor, a radio transmitter, and a receiver, which surgeons implant in the user's skull. The microphone picks up nearby sounds, such as human speech or music emerging from a pair of stereo speakers. Then the processor converts the sounds into digital information that can be sent through a wire to the radio transmitter. The software used by the processor separates sounds into different channels, each representing a range of frequencies. In turn, the radio transmitter translates the digital information into radio signals, which it relays through the skull to the receiver. The receiver then turns the radio signals into electric impulses, which directly stimulate the nerve endings in the cochlea. It is these electric signals that the brain is able to interpret as sounds, allowing even profoundly deaf people to hear.

Another example of how electric signals can be used to direct biological systems can be found in brain-computer interfaces (BCIs). BCIs are direct channels of communication between a computer and the neurons in the human brain. They work because activity in the brain, such as that produced by thoughts or sensory processing, can be detected by bioinstruments designed to record electrophysiological signals. These signals can then be transmitted to a computer and used to generate commands. For example, BCIs allow stroke victims who have lost the use of a limb to regain mobility; a patient's thoughts about movement are transmitted to an external machine, which in turn transmits electric signals that precisely control the movements of a cradle holding his or her paralyzed arm.

### APPLICATIONS AND PRODUCTS

**Biomedical Applications.** Biomedical engineering is a vast subdiscipline of bioengineering, which itself encompasses multiple fields of interest. The many clinical areas in which applications are being developed by biomedical engineers include medical imaging, cell and tissue engineering, bioinstrumentation, the development of biocompatible materials and devices, biomechanics, and the emerging field of bionanotechnology.

Medical imaging applications collect data about patients' bodies and turn that data into useful images that physicians can interpret for diagnostic purposes. For example, ultrasound scans, which map the reflection and reduction in force of sounds as they bounce off an object, are used to monitor the development of fetuses in the wombs of pregnant women. Magnetic resonance imaging (MRI), which measures the response of body tissues to high-frequency radio waves, is often used to detect structural abnormalities in the brain or other body parts.

Cell and tissue engineering is the attempt to exploit the natural characteristics of living cells to regenerate lost or damaged tissue. For example, bioengineers are working on creating viable replacement heart cells for people who have suffered cardiac arrests, as well as trying to discover ways to regenerate brain cells lost by patients with neurodegenerative disorders such as Alzheimers disease. Genetic engineering is a closely related area of biomedicine in which DNA from a foreign organism is introduced into a cell so as to create a new genetic code with desired characteristics.

Bioinstrumentation is the application of electrical engineering principles to develop machines that can sense and respond to biological or physiological signals, such as portable devices for diabetics that measure and report the level of glucose in their blood. Other common examples of bioinstrumentation include electroencephalogram (EEG) machines that continuously monitor brain waves in real time, and electrocardiograph (ECG) machines that perform the same task with heartbeats.

Many biomedical engineers work on developing materials and devices that are biocompatible, meaning that they can replace or come into direct contact with living tissues, perform a biological function, and refrain from triggering an immune system response. Pacemakers, small artificial devices that are implanted within the body and used to stimulate heart muscles to produce steady, reliable contractions, are a good example of a biocompatible device that has emerged from the collaboration of engineers and clinicians.

Biomechanics is the study of how the muscles and skeletal structure of living organisms are affected by and exert mechanical forces. Biomechanics applications include the development of orthotics (braces or supports), such as spinal, leg, and foot braces for patients with disabling disorders such as cerebral palsy, multiple sclerosis, or stroke. Prostheses (artificial limbs) also fall under the field of biomechanics; the sockets, joints, brakes, and pneumatic or hydraulic controls of an artificial leg, for example, are manufactured and then combined in a modular fashion, in much the same way as are the parts of an automobile in a factory.

**Bionanotechnology.** Nanotechnology is a fairly young field of applied science concerned with the manipulation of objects at the nanoscale (about 1-100 nanometers, or about one-thousandth the width of a strand of human hair) to produce machinery. Bionanotechnological applications within medicine include microscopic biosensors installed on small chips; these can be specialized to recognize and flag specific proteins or antibodies, helping physicians conduct extremely fast and inexpensive diagnostic tests. Bioengineers are also developing microelectrodes on a nanoscale; these arrays of tiny electrodes can be implanted into the brain and used to stimulate specific nerve cells to treat movement disorders and other diseases.

**Military Applications.** Bioengineering applications are making themselves felt as a powerful presence on the front lines of the military. For example, bioengineering students at the University of Virginia designed lighter, more flexible, and stronger bullet-proof body armor using specially created ceramic tiles that are inserted into protective vests. The armor is able to withstand multiple impacts and distributes shock more evenly across the wearer's body, preventing damaging compression to the chest. Others working in the field are creating sophisticated biosensors that soldiers can use to detect the presence of potential pathogens or biological weapons that have been released into the air.

One of the most significant contributions of bioengineering to the military is in the development of treatments for severe traumas sustained during warfare. For example, stem cell research may one day enable military physicians to regenerate functional tissues such as nerves, bone, cartilage, skin, and muscle—an invaluable tool for helping those who have lost limbs or other body parts as a result of explosives. The United States military was responsible for much of the early research done in creating safe, effective artificial blood substitutes that could be easily stored and relied on to be free of contamination on the battlefield.

**Agriculture.** Agricultural engineering involves the application of both engineering technologies and knowledge from animal and plant biology to problems in agriculture, such as soil and water conservation, food processing, and animal husbandry. For example, agricultural engineers can help farmers maximize crop yields from a defined area of land. This technique, known as precision farming, involves analyzing the properties of the soil (factors such as drainage, electrical conductivity, pH [acidity] level, and levels of chemicals such as nitrogen) and carefully calibrating the type and amount of seeds, insecticides, and fertilizers to be used.

Farm machinery and implements represent another area of agriculture in which engineering principles have made a big impact. Tractors, harvesters, combines, and grain-processing equipment, for example, have to be designed with mechanical and electrical principles in mind and also must take into account the characteristics of the land, the needs of the human operators, and the demands of working with particular agricultural products. For example, many crops require specialized equipment to be

## Fascinating Facts About Bioengineering

- Bioengineering has enabled scientists to grow replacement human skin, tracheas, bladders, cartilage, and other tissues and organs in the laboratory.
- Materials scientists and clinical researchers are working together to develop contact lenses that can deliver precise doses of drugs directly into the eye.
- By genetically engineering crops that are naturally resistant to insects, bioengineers have helped reduce the need to use harmful pesticides in industrial farming.
- Bacteria whose genetic information has been carefully reengineered may eventually provide an endless supply of crude oil, helping meet the world's energy needs without engaging in damaging drilling.
- In 2009, an MIT bioengineer invented a new way to pressurize space suits that does not use gas, making them far sleeker and less bulky than conventional astronaut gear.
- One military application of bioengineering is a robotic system that seeks out and identifies tiny pieces of shrapnel lodged within tissue, then guides a needle to those precise spots so that the shrapnel can be removed.
- Bionic men and women are not just the stuff of television and motion-picture fantasy. In fact, anyone who has an artificial body part, such as a prosthetic leg, a pacemaker, or an implanted hearing aid, can be considered bionic.
- Some bioengineers are working on developing artificial noses that can detect and diagnose disease by smell–literally sniffing out infections and cancer, for example.
- One day, it may be possible to "print out" artificial organs using a three-dimensional printer. Layer by layer, cells would be deposited onto a glass slide, building up specialized tissues that could be used to replace damaged kidneys, livers, and other organs.
- Each year, more women choose to enter the field of biomedical engineering than any other specialty within engineering.

successfully mechanically harvested. Thus a pea harvester may have several components—one that lifts the vines and cuts them from the plant, one that strips pea pods from the stalk, and one that threshes the pods, causing them to open and release the peas inside them. Another example of an agricultural engineering application is the development of automatic milking machines that attach to the udders of a cow and enable dairy farmers to dispense with the arduous task of milking each animal by hand.

The management of soil and water is also an important priority for bioengineers working in agricultural settings. They may design structures to control the flow of water, such as dams or reservoirs. They may develop water-treatment systems to purify wastewater coming out of industrial agricultural production centers. Alternatively, they may use soil walls or cover crops to reduce the amount of pesticides and nutrients that run off from the soil, as well as the amount of erosion that takes place as a result of watering or rainfall.

**Environmental and Ecological Applications.** Environmental and ecological engineers study the impact of human activity on the environment, as well as the ways in which humans respond to different features of their environments. They use engineering principles to clean, control, and improve the quality of natural spaces, and find ways to make human interactions with environmental resources more sustainable. For example, the reduction and remediation of pollution is an important area of concern. Therefore, an environmental engineer may study the pathways and rates at which volatile organic compounds (such as those found in many paints, adhesives, tiles, wall coverings, and furniture) react with other gases in the air, causing smog and other forms of air pollution. They may design and build sound walls in residential areas to cut down on the amount of noise pollution caused by airplanes taking off and landing or cars racing up and down highways.

The life-support systems designed by bioengineers to enable astronauts to survive in the harsh conditions of outer space are also a form of environmental engineering. For example, temperatures around a space shuttle can vary wildly, depending on which side of the vehicle is facing the Sun at any given moment. A complex system of heating, insulation, and ventilation helps regulate the temperature inside the cabin. Because space is a vacuum, the shuttle itself must be filled with pressurized gas. In addition, levels of oxygen, carbon dioxide, and nitrogen within the cabin must be controlled so that they resemble the atmosphere on Earth. Oxygen is stored on board in tanks, and additional supplies of the essential gas are produced from electrolyzed water; in turn, carbon dioxide is channeled out of the shuttle through vents.

**Geoengineering.** Geoengineering is an emerging subfield of bioengineering that is still largely theoretical. It would involve the large-scale modification of environmental processes in an attempt to counteract the effects of human activity leading to climate change. One proposed geoengineering project involves depositing a fine dust of iron particles into the ocean in an attempt to increase the rate at which algae grows in the water. Since algae absorbs carbon dioxide as it photosynthesizes, essentially trapping and containing it, this would be a means of reducing the amount of this greenhouse gas in the atmosphere. Other geoengineering proposals include the suggestion that it might be possible to spray sulfur dust into the high atmosphere to reflect some of the Sun's light and heat back into space, or to spray drops of seawater high up into the air so that the salt particles they contain would be absorbed into the clouds, making them thicker and more able to reflect sunlight.

## IMPACT ON INDUSTRY

Bioengineering is a global industry that boasts consistent revenues. With an aging public, biomedical engineering looks poised for significant growth. The global focus on reversing climate change, ensuring an adequate supply of food and clean water, and improving and maintaining health means that bioengineering is likely to be an ever-expanding field. Those nations that embrace its possibilities will find themselves reaping the rewards in a better quality of life for their citizens.

**Government and University Research.** The United States is generally considered to be the world leader in bioengineering research, especially within the field of biomedical engineering. However, significant strides are being made in many European countries, including France and Germany, as well as by many growing economies in Asia, such as China, Singapore, and Taiwan. In the United States, the main governmental organization funding studies in this area is the National Institute of Biomedical Imaging and

Bioengineering, a branch of the National Institutes of Health. Among the United States universities whose faculty and students are recognized as conducting the most leading-edge research in bioengineering are The Johns Hopkins School of Medicine, the Massachusetts Institute of Technology (MIT), and the University of California, San Diego—all of which were ranked at the top of the 2009 *U.S. News and World Report*'s list of the best biomedical/bioengineering schools in the country.

**Major Corporations.** Major biomedical engineering corporations include Medtronic, Abbot, Merck, and Glaxo-Smith Kline, all international producers of products such as pharmaceuticals and medical devices. Medtronic, for example, manufactures items such as defibrillators, pacemakers, and heart valves, while Merck produces drugs to treat cancer, heart disease, diabetes, and infections. Within the field of agricultural biotechnology, Monsanto and DuPont are industry leaders. Both corporations produce, patent, and market seeds for transgenic crops. The plants grown from these seeds possess traits attractive to industrial farmers, such as resistance to pesticides, longer ripening times, and higher yield.

**Industry and Business.** Bioengineering is considered a strong growth industry with a great deal of potential for expansion in the twenty-first century. The production of biomedical devices and biocompatible materials alone, for example, is a market worth an estimated $170 billion per year. In the United States, Department of Labor statistics indicate that in 2006, engineers working in the biomedical, agricultural, health and safety, and environmental engineering fields—all of which fall under the bioengineering umbrella—held a total of nearly 100,000 jobs nationwide. In the following years, the department estimated that each of these sectors would add jobs at a rate that either keeps pace with or far exceeds the national average for all occupations.

## CAREERS AND COURSE WORK

Although bioengineering is a field that exists at the intersection between biology and engineering, the most common path for professionals in the field is to first become trained as engineers and later apply their technical knowledge to problems in the life sciences. (A less common path is to pursue a medical degree and become a clinical researcher.) At the high school level, it is important to cover a broad range of mathematical topics, including geometry, calculus, trigonometry, and algebra. Biology, chemistry, and physics should also be among an aspiring bioengineer's course work. At the college level, a student should pursue a bachelor of science in engineering. At many institutions, it is possible to further concentrate in a subfield of engineering: Appropriate subfields include biomedical engineering, electrical engineering, mechanical engineering, and chemical engineering. Students should continue to take electives in biology, geology, and other life sciences wherever possible. In addition, English and humanities courses, especially writing classes, can provide the aspiring bioengineer with strong communication skills—important for working collaboratively with colleagues from many different disciplines.

Many, though not all, choose to pursue graduate-level degrees in biomedical engineering, agricultural engineering, environmental engineering, or another subfields of bioengineering. Others go through master's of business administration programs and combine this training with their engineering background to become entrepreneurs in the bioengineering industry. Additional academic training beyond the undergraduate level is required for careers in academia and higher-level positions in private research and development laboratories, but entry-level technical jobs in bioengineering may require only a bachelor's degree. Internships (such as at biomedical companies) or evidence of experience conducting original research will be helpful in obtaining one's first job.

A variety of career options exist for bioengineers; many work as researchers in academic settings, private industry, government institutions, or research hospitals. Some are faculty members, and some are administrators, managers, supervisors, or marketing consultants for these same organizations. Others are engaged in designing, developing, and conducting safety and performance testing for bioengineering instruments and devices.

## SOCIAL CONTEXT AND FUTURE PROSPECTS

Bioengineering is a field with the capacity to exert a powerful impact on many aspects of social life. Perhaps most profound are the transformations it has made in health care and medicine. By treating the body as a complex system—looking at it almost as if it were a machine—bioengineers and physicians working together have enabled countless patients to overcome what once might have seemed to be insurmountable damage. After all, if the body is a machine, its parts might be reengineered or replaced entirely with new ones—as when the damaged cilia of individuals with hearing impairments are replaced with electro-mechanical devices. Some aspects of bioengineering, however, have drawn concern from observers who worry that there may be no limit to the scientific ability to interfere with biological processes. Transgenic foods are one area in which a contentious debate has sprung up. Some are convinced that the ecological and health ramifications of growing and ingesting crops that contain genetic information from more than one species have not yet been fully explored. Stem cell research is another area of controversy; some critics are uncomfortable with the fact that human embryonic stem cells are being obtained from aborted fetuses or fertilized eggs that are left over from assisted reproductive technology procedures.

One aspect of bioengineering that has been the subject of both fear and hope in the twenty-first century is the question of whether it might be possible to stop or even reverse the harmful effects of climate change by carefully and deliberately interfering with certain geological processes. Some believe that geoengineering could help the international community avoid the devastating effects of global warming predicted by scientists, such as widespread flooding, droughts, and crop failure. Others, however, warn that any attempt to interfere with complex environmental systems on a global scale could have wildly unpredictable results. Geoengineering is especially controversial because such projects could potentially be carried out unilaterally by countries acting without international agreement and yet have repercussions that could be felt all across the world.

*M. Lee, B.A., M.A.*

## FURTHER READING

Artmann, Gerhard M., and Shu Chien, eds. *Bioengineering in Cell and Tissue Research.* New York: Springer, 2008. Examines bioengineering's role in cell research. Heavily illustrated with diagrams and figures; includes a comprehensive index and references after each section.

Enderle, John D., Susan M. Blanchard, and Joseph D. Bronzino, eds. *Introduction to Biomedical Engineering.* 2d ed. Boston: Elsevier Academic Press, 2005. A broad introductory textbook designed for undergraduates. Each chapter contains an outline, objectives, exercises, and suggested reading.

Huffman, Wallace E., and Robert E. Evenson. *Science for Agriculture: A Long-Term Perspective.* 2d ed. Ames, Iowa: Blackwell, 2006. A history of agricultural engineering research within the United States. Includes a glossary and list of relevant acronyms.

Madhavan, Guruprasad, Barbara Oakley, and Luis G. Kun, eds. *Career Development in Bioengineering and Biotechnology.* New York: Springer, 2008. An extensive guide to careers in bioengineering, biotechnology, and related fields, written by active practitioners. Covers both traditional and alternative job opportunities.

Nemerow, Nelson Leonard, et al., eds. *Environmental Engineering.* 3 vols. 6th ed. Hoboken, N.J.: John Wiley & Sons, 2009. Discusses topics such as food protection, soil management, waste management, water supply, and disease control. Each section includes references and a bibliography.

## WEB SITES

*Biomedical Engineering Society*
http://www.bmes.org

*National Institutes of Health*
National Institute of Biomedical Imaging and Bioengineering
http://www.nibib.nih.gov

*Society for Biological Engineering*
http://www.aiche.org/sbe

**See also:** Agricultural Science; Artificial Organs; Audiology and Hearing Aids; Biochemical Engineering; Biomathematics; Biomechanical Engineering; Bionics and Biomedical Engineering; Bioprocess Engineering; Cell and Tissue Engineering; Climate Engineering; Electrical Engineering; Genetically Modified Organisms; Genetic Engineering; Human Genetic Engineering; Military Sciences and Combat Engineering; Rehabilitation Engineering.

# BIOFUELS AND SYNTHETIC FUELS

## FIELDS OF STUDY

Biology; microbiology; plant biology; chemistry; organic chemistry; biochemistry; agriculture; biotechnology; bioprocess engineering; chemical engineering.

## SUMMARY

The study of biofuels and synthetic fuels is an interdisciplinary science that focuses on development of clean, renewable fuels that can be used as alternatives to fossil fuels. Biofuels include ethanol, biodiesel, methane, biogas, and hydrogen; synthetic fuels include syngas and synfuel. These fuels can be used as gasoline and diesel substitutes for transportation, as fuels for electric generators to produce electricity, and as fuels to heat houses (their traditional use). Both governmental agencies and private companies have invested heavily in research in this area of applied science.

## KEY TERMS AND CONCEPTS

- **Biodiesel:** Biofuel with the chemical structure of fatty acid alkyl esters.
- **Biogas:** Biofuel that contains a mixture of methane (50-75 percent), carbon dioxide, hydrogen, and carbon monoxide.
- **Biomass:** Mass of organisms that can be used as an energy source; plants and algae convert the energy of the sun and carbon dioxide into energy that is stored in their biomass.
- **Ethanol:** Colorless liquid with the chemical formula $C_2H_5OH$ that is used as a biofuel; also known as ethyl alcohol, grain alcohol, or just alcohol.
- **Fischer-Tropsch Process:** Process that indirectly converts coal, natural gas, or biomass through syngas into synthetic oil or synfuel (liquid hydrocarbons).
- **Fuel:** Any substance that is burned to provide heat or energy.
- **Gasification:** Conversion of coal, petroleum, or biomass into syngas.
- **Methane:** Colorless, odorless, nontoxic gas, with the molecular formula $CH_4$, that is the main chemical component of natural gas (70-90 percent) and is used as a biofuel.
- **Molecular Hydrogen:** Also known by its chemical symbol $H_2$, a flammable, colorless, odorless gas; hydrogen produced by microorganisms is used as a biofuel and is called a biohydrogen.
- **Synfuel:** A synthetic liquid fuel (synthetic oil) obtained via the Fischer-Tropsch process or methanol-to-gasoline conversion.
- **Synthesis Gas:** Synthetic fuel that is a mixture of carbon monoxide and $H_2$; also known as syngas.

## DEFINITION AND BASIC PRINCIPLES

The science of biofuels and synthetic fuels deals with the development of renewable energy sources, alternatives to nonrenewable fossil fuels such as petroleum. Biofuels are fuels generated from organisms or by organisms. Living organisms can be used to generate a number of biofuels, including ethanol (bioethanol), biodiesel, biomass, butanol, biohydrogen, methane, and biogas. Synthetic fuels (synfuel and syngas) are a class of fuels derived from coal or biomass. Synthetic fuels are produced by a combination of chemical and physical means that convert carbon from coal or biomass into liquid or gaseous fuels.

Around the world, concerns about climate change and possible global warming due to the emission of greenhouse gases from human use of fossil fuels, as well as concerns over energy security, have ignited interest in biofuels and synthetic fuels. A large-scale biofuel and synthetic fuel industry has developed in many countries, including the United States. A number of companies in the United States have conducted research and development projects on synthetic fuels with the intent to begin commercial production of synthetic fuels. Although biofuels and synthetic fuels still require long-term scientific, economic, and political investments, investment in these alternatives to fossil fuels is expected to mitigate global warming, to help protect the global climate, and to reduce U.S. reliance on foreign oil.

## BACKGROUND AND HISTORY

People have been using biofuels such as wood or dried manure to heat their houses for thousands of years. The use of biogas was mentioned in Chinese literature more than 2,000 years ago. The first biogas plant was built in a leper colony in Bombay, India, in the middle of the nineteenth century. In Europe, the first apparatus for biogas production was built in Exeter, England, in 1895. Biogas from this digester was used to fuel street lamps. Rudolf Diesel, the inventor of the diesel engine, used biofuel (peanut oil) for his engine during the World Exhibition in Paris in 1900. The version of the Model T Ford built by Henry Ford in 1908 ran on pure ethanol. In the 1920's, 25 percent of the fuels used for automobiles in the United States were biofuels rather than petroleum-based fuels. In the 1940's, biofuels were replaced by inexpensive petroleum-based fuels.

Gasification of wood and coal for production of syngas has been done since the nineteenth century. Syngas was used mainly for lighting purposes. During World War II, because of shortages of petroleum, internal combustion engines were modified to run on syngas and automobiles in the United States and the United Kingdom were powered by syngas. The United Kingdom continued to use syngas until the discovery in the 1960's of oil and natural gas in the North Sea.

The process of converting coal into synthetic liquid fuel, known as the Fischer-Tropsch process, was developed in Germany at the Kaiser Wilhelm Institute by Franz Fischer and Hans Tropsch in 1923. This process was used by Nazi Germany during World War II to produce synthetic fuels for aviation.

During the 1970's oil embargo, research on biofuels and synthetic fuels resumed in the United States and Europe. However, as petroleum prices fell in the 1980's, interest in alternative fuels diminished. In the twenty-first century, concerns about global warming and increasing oil prices reignited interest in biofuels and synthetic fuels.

### How It Works

Biofuels and synthetic fuels are energy sources. People have been using firewood to heat houses since prehistorical time. During the Industrial Revolution, firewood was used in steam engines. In a steam engine, heat from burning wood is used to boil water; the steam produced pushes pistons, which turn the wheels of the machinery.

Biofuels and synthetic fuels such as ethanol, biodiesel, butanol, biohydrogen, and synthetic oil can be used in internal combustion engines, in which the combustion of fuel expands gases that move pistons or turbine blades. Other biofuels such as methane,

*Corn ethanol processing plant.* (David Nunuk/Photo Researchers, Inc.)

biogas, or syngas are used in electric generators. Burning of these fuels in electric generators rotates a coil of wire in a magnetic field, which induces electric current (electricity) in the wire.

Hydrogen is used in fuel cells. Fuel cells generate electricity through a chemical reaction between molecular hydrogen ($H_2$) and oxygen ($O_2$). Ethanol, the most common biofuel, is produced by yeast fermentation of sugars derived from sugarcane, corn starch, or grain. Ethanol is separated from its fermentation broth by distillation. In the United States, most ethanol is produced from corn starch. Biodiesel, another commonly used biofuel, is made mainly by trans-esterification of plant vegetative oils such as soybean, canola, or rapeseed oil. Biodiesel may also be produced from waste cooking oils, restaurant grease, soap stocks, animal fats, and even from algae. Methane and biogas are produced by metabolism of microorganisms. Methane is produced by microorganisms called *Archaea* and is an integral part of their metabolism. Biogas produces by a mixture of bacteria and archaea.

Industrial production of biofuels is achieved mainly in bioreactors or fermenters of some hundreds gallons in volume. Bioreactors or fermenters are closed systems that are made of an array of tanks or tubes in which biofuel-producing microorganisms are cultivated and monitored under controlled conditions.

Syngas is produced by the process of gasification in gasifiers, which burn wood, coal, or charcoal. Syngas can be used in modified internal combustion engines. Synfuel can be generated from syngas through Fischer-Tropsch conversion or through methanol to gasoline conversion process.

## APPLICATIONS AND PRODUCTS

**Transportation.** Biofuels are mainly used in transportation as gasoline and diesel substitutes. As of the early twenty-first century, two biofuels—ethanol and biodiesel—were being used in vehicles. In 2005, the U.S. Congress passed an energy bill that required that ethanol sold in the United States for transportation be mixed with gasoline. By 2010, almost every fuel station in the United States was selling gasoline with a 10 percent ethanol content. The U.S. ethanol industry has lobbied the federal government to raise the ethanol content in gasoline from 10 to 15 percent. Most cars in Brazil can use an 85 percent/15 percent ethanol-gasoline mix (E85 blend). These cars must have a modified engine known as a flex engine. In the United States, only a small fraction of all cars have a flex engine.

Biodiesel performs similarly to diesel and is used in unmodified diesel engines of trucks, tractors, and other vehicles and is better for the environment. Biodiesel is often blended with petroleum diesel in ratios of 2, 5, or 20 percent. The most common blend is B20, or 20 percent biodiesel to 80 percent diesel fuel. Biodiesel can be used as a pure fuel (100 percent or B100), but pure fuel is a solvent that degrades the rubber hoses and gaskets of engines and cannot be used in winter because it thickens in cold temperatures. The energy content of biodiesel is less than that of diesel. In general, biodiesel is not used as widely as ethanol, and its users are mainly governmental and

state bodies such as the U.S. Postal Service; the U.S. Departments of Defense, Energy, and Agriculture; national parks; school districts; transit authorities; public utilities; and waste-management facilities. Several companies across the United States (such as recycling companies) use biodiesel because of tax incentives.

Hydrogen power ran the rockets of the National Aeronautics and Space Administration for many years. A growing number of automobile manufactures around the world are making prototype hydrogen-powered vehicles. These vehicles emit only water, no greenhouse gases, from their tailpipes. These automobiles are powered by electricity generated in the fuel cell through a chemical reaction between $H_2$ and $O_2$. Hydrogen vehicles offer quiet operation, rapid acceleration, and low maintenance costs because of fewer moving parts. During peak time, when electricity is expensive, fuel-cell hydrogen automobiles could provide power for homes and offices. Hydrogen for these applications is obtained mainly from natural gas (methane and propane), through steam reforming, or by water electrolysis. As of 2010, hydrogen was used only in experimental applications. Many problems need to be overcome before hydrogen becomes widely used and readily available. The slow acceptance of biohydrogen is partly caused by the difficulty in producing it on a cost-effective basis. For hydrogen power to become a reality, a great deal of research and investment must take place.

Methane was used as a fuel for vehicles for a number of years. Several Volvo automobile models with Bi-Fuel engines were made to run on compressed methane with gasoline as a backup. Biogas can also be used, like methane, to power motor vehicles.

**Electricity Generation.** Biogas and methane are mainly used to generate electricity in electric generators. In the 1985 film *Mad Max Beyond Thunderdome*, starring Mel Gibson, a futuristic city ran on methane generated by pig manure. While the use of methane has not reached this stage, methane is a very good alternative fuel that has a number of advantages over biofuels produced by microorganisms. First, it is easy to make and can be generated locally, eliminating the need for an extensive distribution channel. Second, the use of methane as a fuel is a very attractive way to reduce wastes such as manure, wastewater, or municipal and industrial wastes. In farms, manure is fed into digesters (bioreactors), where microorganisms metabolize it into methane. There are several landfill gas facilities in the United States that generate electricity using methane. San Francisco has extended its recycling program to include conversion of dog waste into methane to produce electricity and to heat homes. With a dog population of 120,000, this initiative promises to generate a significant amount of fuel and reduce waste at the same time.

**Heat Generation.** Some examples of biomass being used as an alternative energy source include the burning of wood or agricultural residues to heat homes. This is a very inefficient use of energy, because typically only 5 to 15 percent of the biomass energy is actually used. Burning biomass also produces harmful indoor air pollutants such as carbon monoxide. On the positive side, biomass is an inexpensive resource whose costs are only the labor to collect it. Biomass supplies more than 15 percent of the energy consumed worldwide. Biomass is the number-one source of energy in developing countries; in some countries, it provides more than 90 percent of the energy used.

In many countries, millions of small farmers maintain a simple digester for biogas production to generate heat energy. More than 5 million household digesters are being used in China, mainly for cooking and lighting, and India has more than 1 million biogas plants of various capacities.

## IMPACT ON INDUSTRY

In 2009, the annual revenue of the global biofuels industry was $46.5 billion, with revenue in the United States alone reaching $20 billion. The United States is leading the world in research on biofuels and synthetic fuels. Significant biofuels and synthetic fuels research has also been taking place in many European countries, Russia, Japan, Israel, Canada, Australia, and China.

**Government and University Research.** Many governmental agencies such as the U.S. Department of Energy (DOE), the National Science Foundation (NSF), and the U.S. Department of Agriculture provide funding for research in biofuels and synthetic fuels. The DOE has several national laboratories (such as the National Renewable Energy Laboratory in Golden, Colorado) where cutting-edge research on biofuels and synthetic fuels is performed. In addition, three DOE research centers are concentrated entirely on biofuels. These centers are the BioEnergy

Science Center, led by Oak Ridge National Laboratory; the Great Lakes Bioenergy Research Center, led by the University of Wisconsin, Madison; and the Joint BioEnergy Institute, led by Lawrence Berkeley National Laboratory.

In 2007, DOE established the Advanced Research Projects Agency-Energy (ARPA-E) to fund the development and deployment of transformational energy technologies in the United States. Several projects funded by this agency are related to biofuels, such as the development of advanced or second-generation biofuels. Traditional biofuels such as biomass (wood material) and ethanol and biodiesel from crops are sometimes called first-generation biofuels. Second-generation biofuels such as cellulosic ethanol are produced from agricultural and forestry residues and do not take away from food production. Another second-generation biofuel is biohydrogen.

Biofuels such as butanol are referred to as third-generation biofuels. Butanol ($C_4H_9OH$) is an alcohol fuel, but compared with ethanol, it has a higher energy content (roughly 80 percent of gasoline energy content). It does not absorb water as ethanol does, is not as corrosive as ethanol is, and is more suitable for distribution through existing gasoline pipelines.

Scientists are trying to create "super-bugs" for superior biofuel yields and studying chemical processes and enzymes to improve existing bioprocesses for biofuels production. They also are working to improve the efficiency of the existing production process and to make it more environmentally friendly. Engineers and scientists are designing and developing new apparatuses (bioreactors or fermenters) for fuel generation and new applications for by-products of fuel production.

**Industry and Business Sectors.** The major products of the biofuel industry are ethanol, biodiesel, and biogas; therefore, research in industry has concentrated mainly on these biofuels. Some small businesses in the biofuel industry include startup research and development companies that study feedstocks (such as cellulose) and approaches for production of biofuels at competitive prices. These companies, many of which are funded by investment firms or government agencies, analyze biofuel feedstocks, looking for new feedstocks or modifying existing ones (corn, sugarcane, or rapeseed).

Big corporations such as Poet Energy, ExxonMobil, and BP are spending a significant part of their revenues

## Fascinating Facts About Biofuels and Synthetic Fuels

- Scientists have discovered that *Gliocladium roseum*, a tree fungus, is able to convert cellulose directly into biodiesel, thus making transesterification unnecessary. The fungus eats the tree and, interestingly, keeps other fungi away from the tree by producing an antibiotic. Scientists are studying enzymes that will help this fungus eat cellulose.

- Termites can produce two liters of biofuel, molecular hydrogen, or $H_2$, by fermenting just one sheet of paper with microbes that live in their guts. Study of the biochemical pathways involved in $H_2$ production in termite guts may lead to application of this process industrially.

- In 2009, Continental Airlines successfully powered a Boeing 737-800 using a biodiesel fuel mixture partly produced from algae.

- Firewood can power an automobile, but its engine must be modified slightly and a trailer with a syngas generator (gasifier) must be attached to the automobile.

- An automobile that runs on diesel fuel can easily be modified to run on used cooking oil, which can be obtained for little or no charge from local restaurants.

- By attaching a water electrolyzer, an automobile can be modified to run partly on water. Electricity from the car battery splits water in the electrolyzer into hydrogen and oxygen. The hydrogen can be burned in the internal combustion engine and power the automobile.

- Modifying an automobile to run partly on methane is simple and definitely saves gasoline.

on biofuel or synthetic fuels research. One area of biofuel research examines using algae to generate biofuels, especially biodiesel. More than fifty research companies worldwide, including GreenFuel Technologies, Solazyme, and Solix Biofuels, are conducting research in this area. Research conducted by the U.S. Department of Energy Aquatic Species Program from the 1970's to the 1990's demonstrated that many species of algae produce sufficient quantities of oil to become economical feedstock for biodiesel production. The oil productivity of many algae greatly exceeds the productivity of the best-producing oil crops. Algal oil content can exceed 80 percent per cell dry weight, with oil levels commonly at about 20 to 50 percent. In addition, crop land and potable water are not required to cultivate algae, because algae can grow in wastewater. Although development of biodiesel from algae is a very promising approach, the technology needs further research before it can be implemented commercially.

### CAREERS AND COURSE WORK

The alternative fuels industry is growing, and research in the area of biofuels and synthetic fuels is increasing. Growth in these areas is likely to produce many jobs. The basic courses for students interested in a career in biofuels and synthetic fuels are microbiology, plant biology, organic chemistry, biochemistry, agriculture, bioprocess engineering, and chemical engineering. Many educational institutions are offering courses in biofuels and synthetic fuels, although actual degrees or concentrations in these disciplines are still rare. Several community colleges offer associate degrees and certificate programs that prepare students to work in the biofuel and synthetic fuel industry. Some universities offer undergraduate courses in biofuels and synthetic fuels or concentrations in these areas. Almost all these programs are interdisciplinary. Graduates of these programs will have the knowledge and internship experience to enter directly into the biofuel and synthetic fuel workforce. Advanced degrees such as a master's degree or doctorate are necessary to obtain top positions in academia and industry related to biofuels and synthetic fuels. Some universities such as Colorado State University offer graduate programs in biofuels.

Careers in the fields of biofuels and synthetic fuels can take different paths. Ethanol, biodiesel, or biogas industries are the biggest employers. The available jobs are in sales, consulting, research, engineering, and installation and maintenance. People who are interested in research in biofuels and synthetic fuels can find jobs in governmental laboratories and in universities. In academic settings, fuel professionals may share their time between research and teaching.

### SOCIAL CONTEXT AND FUTURE PROSPECTS

The field of biofuels and synthetic fuels is undergoing expansion. Demands for biofuels and synthetic fuels are driven by environmental, social, and economic factors and governmental support for alternative fuels.

The use of biofuels and synthetic fuels reduces the U.S. dependence on foreign oil and helps mitigate the devastating impact of increases in the price of oil, which reached a record $140 per barrel in 2008. The production and use of biofuels and synthetic fuels reduces the need for oil and has helped hold world oil prices 15 percent lower than they would have been otherwise. Many experts believe that biofuels and synthetic fuels will replace oil in the future.

Pollution from oil use affects public health and causes global climate change because of the release of carbon dioxide. Using biofuels and synthetic fuels as an energy source generates fewer pollutants and little or no carbon dioxide.

The biofuel and synthetic fuel industry in the United States was affected by the economic crisis in 2008 and 2009. Several ethanol plants were closed, some plants were forced to work below capacity, and other companies filed for Chapter 11 bankruptcy protection. Such events led to layoffs and hiring freezes. Nevertheless, overall, the industry was growing and saw a return to profitability in the second half of 2009. Worldwide production of ethanol and biodiesel is expected to grow to $113 billion by 2019; this is more than 60 percent growth of annual earnings. One segment of the biofuel and synthetic fuel industry, the biogas industry, was not affected by recession at all. More than 8,900 new biogas plants were built worldwide in 2009. According to market analysts, the biogas industry has reached a turning point and may grow at a rate of 24 percent from 2010 to 2016. Research and development efforts in biofuels and synthetic fuels actually increased during the economic crisis. In general, the future of biofuels and synthetic fuels is bright and optimistic.

*Sergei A. Markov, Ph.D.*

**FURTHER READING**

Bart, Jan C. J., and Natale Palmeri. *Biodiesel Science and Technology: From Soil to Oil.* Cambridge, England: Woodhead, 2010. A comprehensive book on biodiesel fuels.

Bourne, Joel K. "Green Dreams." *National Geographic* 212, no. 4 (October, 2007): 38-59. An interesting discussion about ethanol and biodiesel fuels and their future.

Glazer, Alexander N., and Hiroshi Nikaido. *Microbial Biotechnology: Fundamentals of Applied Microbiology.* New York: Cambridge University Press, 2007. Provides an in-depth analysis of ethanol and biomass for fuel applications.

Mikityuk, Andrey. "Mr. Ethanol Fights Back." *Forbes,* November 24, 2008, 52-57. Excellent discussion about the problems and the hopes of the ethanol industry. Examines how the ethanol industry fought the economic crisis in 2008 and returned to profitability in 2009.

Probstein, Ronald F., and Edwin R. Hicks. *Synthetic Fuels.* Mineola, N.Y.: Dover Publications, 2006. A comprehensive work on synthetic fuels. Contains references and sources for further information.

Service, Robert F. "The Hydrogen Backlash." *Science* 305, no. 5686 (August 13, 2004): 958-961. Discusses the future of hydrogen power, including its maturity. Written in an easy-to-understand manner.

Wall, Judy, ed. *Bioenergy.* Washington, D.C.: ASM Press, 2008. Provides the information on generation of biofuels by microorganisms and points out future areas for research. Ten chapters focus on ethanol production from cellulosic material.

**WEB SITES**

*Advanced BioFuels USA*
http://advancedbiofuelsusa.info

*International Energy Agency Bioenergy*
http://www.ieabioenergy.com

*U.S. Department of Agriculture, Economic Research Service*
Featuring Bioenergy
http://www.ers.usda.gov/features/bioenergy

*U.S. Department of Energy*
Energy Efficiency and Renewable Energy
http://www.energy.gov/energysources/index.htm

*U.S. Energy Information Administration*
Renewable and Alternative Fuels
http://www.eia.doe.gov/fuelrenewable.html

**See also:** Agricultural Science; Bioenergy Technologies; Chemical Engineering; Hybrid Vehicle Technologies.

# BIOINFORMATICS

## FIELDS OF STUDY

Molecular biology; genetics; molecular genetics; phylogenetics; cell biology; physics; biochemistry; biophysics; biostatistics; computational biology; computer science; evolutionary biology; structural biology; systems biology; mathematics.

## SUMMARY

Bioinformatics is simultaneously a new type of re search practice and a rapidly emerging new discipline. It has shifted the practice of scientific research from traditional laboratory bench research to computer-based data analysis and experimentation with massive datasets available on the Internet. To support this research, bioinformaticians develop the software required for data analyses, design the biodatabases, organize and manage the data within them, create the online computing environment, and develop the highly specialized mathematical algorithms and statistical packages to search, retrieve, and analyze biodata by bioresearchers. Bioinformatics has accelerated the pace and understanding of biological systems and molecules exponentially, with its greatest promise indicated in medical and environmental applications.

## KEY TERMS AND CONCEPTS

- **Genome-Wide Association Studies:** Scanning of a human genome biocomputationally to associate genetic variations to specific diseases, drug reactions, and other relevant issues.
- **Genomics:** Study of the structure, function, and interaction of the complete set of genetic material found in an organism.
- **Multiple Sequence Alignment:** Algorithm that aligns three or more sequences to one another according to their best sequence similarities.
- **Pharmacogenomics:** Identification and study of the genes involved in an organism's response to drugs.
- **Phylogenetics:** Bioinformatically, the evolutionary relationships between biosequences to trace sequence ancestries of genes or organisms.
- **Protein Modeling:** Manipulation of three-dimensional structures of biosequence molecules computationally and through visualization software to correlate structure with function.
- **Proteomics:** Study and comparison of the entire complement of proteins (proteome) under a given condition: a cell's proteome, comparison of a normal liver proteome before and after drug treatment, comparison of proteomes from normal tissues with cancerous tissues.
- **Single Nucleotide Polymorphisms (SNPs):** Normal variations found in individual genomes at a single nucleotide position; rare variations are often disease-causing mutations.

## DEFINITION AND BASIC PRINCIPLES

Bioinformatics is simultaneously a relatively new type of research practice and a rapidly emerging discipline. As research, bioinformatics is defined as the manipulation and the varied analyses performed by laboratory-based researchers on massive biological datasets residing in thousands of Internet-based databases, each with a distinct set of data and a specific purpose. Originating from molecular biology, bioinformatics has rapidly spread to cell biology, chemistry, statistics, computer sciences, physics, biomedical engineering, psychology, and even anthropology.

As a discipline, bioinformatics draws on those professionals with advanced skills from the computer sciences, information sciences, and mathematics disciplines to bear on biological problems posed by laboratory-based bioresearchers. Collectively, these specialists are referred to as "bioinformaticians" to distinguish them from the laboratory-based science researchers carrying out experiments with the products bioinformaticians have created.

Researchers often define bioinformatics from within the perspective of their specific discipline or individual research efforts. Some biologists view bioinformatics as only involving DNA or protein sequencing. Chemists and physicists tend to view bioinformatics as involving protein molecular structures. Computer scientists describe bioinformatics from a programming or information infrastructure perspective. Pharmacologists often define bioinformatics from the viewpoint of drug-protein interactions. All of these variations share the concept of applying computational analyses to biological processes.

A definition that encompasses bioinformatics both as a profession and a research practice and also takes into account the multitude of disciplines involved is still very much a work in progress as bioinformatics continues to evolve. A unified definition views bioinformatics as the convergence of the biological sciences and computer technologies and the integration of statistics and probability mathematics to understand biological processes of molecules on a very large scale. In turn, collecting, cataloging, classification, storage, organization, management, and retrieval of these massive biodatasets requires information theory and practice (informatics) from the information sciences disciplines to make them available for problem solving.

## BACKGROUND AND HISTORY

Bioinformatics originates from within the fields of genetics and molecular biology. The computational, mathematical, and biodatabase origins of bioinformatics arose not from within the biological, computer, or mathematical sciences but rather from two individuals who had a fascination with the computer technologies being introduced in the 1960's: Robert Ledley, a dentist turned theoretical physicist, and Margaret Dayhoff, a quantum chemist. Ledley, the inventor of the whole-body computerized tomography machine, founded the National Biomedical Research Foundation (NBRF) in 1960 to research and discover possible uses of computers in biomedical research. He recruited Dayhoff to apply her knowledge and skills at data entry and processing toward protein sequencing, which, at that time was taking more than a year to sequence a single protein by traditional laboratory methods.

Using computational analyses, Dayhoff discovered sequence patterns that identified similar proteins and predicted possible functions. She created a series of mathematical scoring matrices and defined a set of mathematical expressions that accurately reflected these similarities across evolutionary distances. In so doing, she created the first bioinformatics algorithms. Her sequence similarity matrices and rules still provide the basis for contemporary sequence similarity searching algorithms, most notably the suite of Basic Local Alignment Sequence Tools (BLAST) created by the National Center for Biotechnology Information (NCBI) in 1997.

In 1963, Dayhoff began compiling protein sequences into a series of books titled *Atlas of Protein Structure and Function*. By 1978, the *Atlas of Protein Structure and Function* had grown too large to make comparisons and perform analyses. Her second major contribution was to create a database infra-

---

### Fascinating Facts About Bioinformatics

- During the 2009 H1N1 swine flu epidemics, more than 24,000 individual virus genomes were sequenced immediately from infected patients worldwide. These data played a major role in the autumn 2009 vaccine development.

- GenBank has become a major sequence resource containing 150 million sequence records in 2009 and giving rise to hundreds of secondary, specialized databases worldwide. More than 500 million records in thirty-plus secondary databases exist at the National Center for Biotechnology Information alone. Worldwide, the number of bioinformatics records based on GenBank is in the billions.

- The genomes between different humans are 99.9 percent similar. The 0.1 percent difference is due to single DNA nucleotide variations at very specific points within the human genome, making each person different from all others physically, behaviorally, and physiologically.

- Before the Human Genome Project, scientists thought that the human genome contained up to 100,000 genes because of the large number of proteins that are known to exist in humans. Scientists now know that there are fewer than 30,000 genes in humans, with each gene estimated to give rise to 3 to 8 different proteins. There are exceptions. The *dscam* gene is involved in the development of neural circuits. In humans, this gene codes for more than 16,000 variations of the dscam protein, and in the fruit fly, 38,016 isoforms of the dscam protein have been proven to exist.

- Human DNA is 98 percent similar in sequence to chimpanzees. However, the genetic difference between any two chimpanzees is four to five times greater than the difference between any two humans.

- Of the greater than 3 billion nucleotides in the human genome, less than 3 percent actually codes for protein molecules. The functions of the vast majority of the human genome and what it does or does not do remain unknown.

structure to convert the atlas to the first online biological database accessible to researchers who could use it to sort, manipulate, and align multiple protein sequences. The database created, the Protein Information Resource (PIR), has become the major Internet UniProt protein bioinformatics resource at the European Bioinformatics Institute (EBI).

Although the National Institutes of Health (NIH) founded the DNA bioinformatics database, Gen-Bank, in 1982 to specifically accelerate nucleic acid sequence experimentation, progress in DNA sequencing and gene cloning technologies lagged behind protein sequencing. In 1985, then Chancellor of the University of California, Santa Cruz and molecular biologist Robert Sinsheimer convened a workshop of prominent scientists and made what was considered a radical and controversial proposal. He proposed to sequence the entire human genome and then use computational analyses to discover unknown genes and their functions and interrelationships. Thus, the Human Genome Project was initiated in 1990 by the National Institutes of Health. It soon became obvious that existing computational power and hardware were insufficient to process or hold the data being generated. Major engineering innovations were needed to process larger sample numbers, faster. The computer sciences and engineering disciplines responded. Within a few years, specialized robotics, miniaturization of samples, faster computers and processors, larger data storage capacity, and new kinds of software engineering tools were in use, greatly accelerating DNA sequencing.

## HOW IT WORKS

How bioinformatics is practiced depends on whether the research is conducted on small data as typified by individual research laboratories or on a mega-scale. Small-scale data handling is often called low throughput, while large-scale is always called high throughput.

**Low-Throughput Bioinformatics.** In a simple sequencing scenario, researchers working to identify a protein or a gene perform "wet research" experiments ultimately yielding DNA or protein candidates. The candidates are sequenced. The researcher accesses the appropriate bioinformatics databases over the Internet and searches for similar sequences using sequence similarity algorithms, analyzing the results to provide clues to the function and identity

of the candidate sequences. Once the researcher has clues to possible functions, additional bioinformatics databases are searched to aid in the development of the next experiment to be performed. In the process, many different bioinformatics databases and tools are used. Learning what databases and tools exist and are best is part of the process of learning bioinformatics research. There are times when the tool may not exist or existing databases are not sufficient. The bench researcher may ask the local bioinformatician to help design a more specific programming tool. If this becomes a critical problem for this area of research in general, bioinformaticians develop new tools and/or databases. These are published in the peer-reviewed literature and tried out by the scientific community. Those that work eventually become established as bioinformatics resources.

Sequences recovered by laboratory researchers with federally funded grants must be uploaded to a sequence repository, along with any information discovered. In the United States, this is NCBI GenBank. Data uploaded in the United States, Europe, and Asia are shared among the countries daily, permitting rapid access to the biodata generated worldwide. NCBI curators then work on the uploaded sequences to integrate and incorporate them into any of the thirty-plus databases at the National Center for Biotechnology Information. When new types of data are being uploaded as research progresses into new areas, the center or its European and Asian counterparts design new kinds of databases and algorithms or fund others to do so.

**High-Throughput Bioinformatics.** This research typically involves massive generation of data, such as large-scale genome sequencing efforts, or the simultaneous analyses of very large datasets. An example of the latter would be clinical data arising from the identification of proteins unique to a specific cancer isolated from many patients. In these scenarios, millions of sequences need to be processed daily. This kind of bioinformatics requires robotic bioinstrumentation and different algorithms to process. It typically is carried out by supercomputing facilities supported by bioinformaticians with experience in parallel computing, networking, grid computing, advanced algorithms, statistical programming skills, and advanced database modeling and design. Any sequence data recovered from research supported from federal funds must be uploaded to the National

Center for Biotechnology Information. In this case, since the functions of the sequences are unknown, the center computationally processes these data to different databases than GenBank, making them available for others to search and identify the function of the sequences.

### APPLICATIONS AND PRODUCTS

**Biological Databases.** Biosequence databases are at the very foundation of bioinformatics research and discovery. The National Center for Biotechnology Information, the European Bioinformatics Institute, and the DNA Database of Japan are the major biosequence spaces; each has an extensive suite of hyperlinked protein, genomes, nucleotide, genes, gene expressions, disease, and chromosome databases. Scientific organizations, government agencies, and research institutes have collaborated to create other databases.

The major protein databases are UniProt of EBI and the three-dimensional structural protein resource at Protein DataBank. Online Mendelian Inheritance of Man (OMIM) and Animal (OMIA) correlate mutations and their inheritance patterns with disease phenotypes. PharmGKB is a major pharmacogenetics database that monitors human genetic variations to specific patient drug reactions and their symptoms. Biological pathway databases, including BRENDA, Reactome, and KEGG, enable researchers to locate proteins that interact with each other and determine how protein sequence alterations could give rise to abnormal biological processes.

Genomes of many different organisms have been sequenced, each representing biodata that detail a model biological system or disease process. Finally, there are thousands of smaller "boutique" biodatabases for specific diseases, the different functional or structural components of genes or proteins, and similar topics.

**Algorithms.** Although there are many mechanisms to search biodatabases, the most critical and extensively used is sequence similarity searching. Needleman-Wunsch, Smith-Waterman, FASTA, and BLAST represent the major similarity algorithms. They differ in algorithmic mechanism and computational speed, with Needleman-Wunsch being the most accurate but also the most computationally intense. At the time of its publication in 1970, it took days to return results. BLAST is the least accurate of

the set but computationally the fastest, taking only minutes to return results. BLAST supported laboratory bench research in real time and is the major sequence similarity algorithm in use. However, as personal computers have advanced to faster processors, the Needleman-Wunsch and Smith-Waterman algorithms have been reengineered and made available at the National Center for Biotechnology Information and the European Bioinformatics Institute.

**Molecular Visualization and Modeling.** A combination of software engineering and sequence algorithm, three-dimensional molecular viewers enable researchers to manipulate and computationally model proteins. They are particularly important in drug design and analyses of mutant proteins involved in disease, as researchers can introduce changes *in silico* and view how they alter drug interaction or structure or compare a mutant protein directly with a normal protein superimposed in three dimensions. The two most important molecular viewers are Cn3D at NCBI and RasMol for the other protein and nucleic acid databases.

**Biodiversity.** Microbes (bacteria, fungi, protozoans, and viruses) represent half of the Earth's biomass. It is estimated that there are at least 10 million bacterial species alone, with only a few thousand described. Since the 1990's, it has become clear that most microbes live in mixed communities with other microbes, with any given species present in a small number, none of which can be cultured in a laboratory.

Metagenomics is that part of bioinformatics that determines and then studies what organismal communities are present in various environmental samples such as soil or oceans. It can also detect the organisms present in animal organs or tissues such as the digestive tract or skin. The present-day state of bioinformatics technology and data acquisition and storage permits the identification of microorganismal communities only. This includes mixed communities containing bacteria, fungi, and viruses. The samples are collected and all the organisms present in the sample are recovered. Without an attempt to isolate, culture, or identify any of the organisms present, all the DNA from all the organisms is extracted in mass, sequenced, and reassembled into the original genomes, thereby identifying what organisms are present.

Initiated in 2008, the Human Microbiome Project aims to identify all the microorganisms present in five areas of the human body—the digestive tract, the mouth, the skin, the nose, and the vagina—from samples taken from healthy human volunteers. Once the healthy human microbiome has been characterized, the human microbiome will be studied in different disease, nutritional, or treatment states. The aim is to use the human microbiome to identify particular diseases and to study the effectiveness of probiotics, pharmaceutical drug treatments, and other therapies. In 2010, 900 microbial genomes had been identified as components of the human microbiome, of which 178 have been fully sequenced. The data indicate that the human microbiome is massive and at least one hundred times larger than the human genome itself. It contains nearly twice the microbial diversity already identified in public domain databases.

Understanding the oceanic microbiome and how it responds to climate and human impact is an important step to oceanic conservation. In addition, adding to the catalog of known proteins enhances the ability to discover new proteins that could be reengineered or repurposed for medicinal or bioremediation uses. In a metagenomics approach similar to that taken by the Human Microbiome Project, oceanic samples have been collected, all the microorganisms recovered, DNA extracted, sequenced, and genomically reassembled. This Global Ocean Sampling expedition has identified at least 400 new microbial species and 6 million predicted proteins, doubling the total number of proteins previously identified.

Bioinformatics' contribution to biodiversity is not limited to the present day. Museums worldwide contain unique specimens (both plant and animal) that can be sequenced, genomically cataloged, and characterized. Ancient DNA, the DNA recovered from fossil organisms trapped in underground ancient lake beds or water droplets trapped in various geological samples, is also available for genomic analyses, adding to the publicly available Neanderthal and *Mastodon* genomes.

**Personal Genomes.** In less than seven years, the cost of sequencing a human genome dropped from almost $3 billion for the Human Genome Project to less than $30,000 in 2010 because of rapid advances in computational and engineering technologies in bioinformatics. By the end of 2011, it is estimated that more 30,000 different human genomes will have been sequenced by various genomic centers and institutes worldwide. As costs continue to drop, sequencing a human genome will be within the reach

of individual research laboratories in several years and affordable by many private citizens in possibly five more years. Several private companies are already advertising (at a cost ranging from $400 to $1,500) to scan people's genome for common DNA sequence variations that are associated with specific diseases or conditions such as diabetes or high cholesterol or are known to reduce or enhance the metabolism of pharmaceutical drugs. Some of these companies offer services that trace an individual's ancestry through his or her inheritance of specific DNA patterns now known to be specific to particular ethnicities or to have originated in distinct geographical areas around the world. This new branch of genomics, in which individual human genomes are sequenced and analyzed, is called personal genomics and carries with it evolving ethical issues that are themselves undergoing rapid debate and analyses.

At the academic research level, large consortiums are being formed to analyze vast numbers of individual human genomes to first catalog and then study all the known genetic differences among both individuals and different kinds of populations. The 1000 Genomes project is a consortium of more than seventy-five universities, institutes, and companies worldwide. Regardless of its name, it aims to sequence the genomes of 2,300 individuals with ancestry from Europe, east Asia, south Asia, West Africa, and the Americas. Each genome will be independently analyzed as well as compared to genomes within the same populations. Early studies indicate that each person may carry 250 to 300 mutations in genes known to cause disease, as well as 50 to 100 sequence variations known to be implicated in inherited disorders. Not all genetic variations give rise to disease. The 1000 Genomes catalog has already identified several candidate genetic differences in two genes inherited within one family group that may be responsible for this family having very low cholesterol levels. The hope is that the study of these genes can lead to new cholesterol-lowering strategies.

## IMPACT ON INDUSTRY

Bioinformatics is creating new industries and services. Industrial applications are very much in their infancy and in the research and development phase for the most part.

**Government and University Research.** The National Institutes of Health is a major source of research funding for basic science, biomedical, and clinical research. The National Science Foundation funds more in the environmental and educational sectors. Both are active policy setters and enforcers of data sharing and integration, cyberinfrastructure, supercomputing, and bioethical issues. Because of a federal mandate, the National Institutes of Health focuses on issues related to health and disease, which is the reason that the majority of bioinformatics research in the United States is related to medical and clinical research. The Department of Energy, through the Office of Biological and Environmental Research (BER), funds research on selected organisms, as well as on environmental genomics and proteomics projects related to bioremediation. BER also studies the ethical, legal, and social ramifications of genome projects through the Ethical, Legal, and Social Issues Program. The Department of Energy's Advanced Scientific Computing Research office funds computer science, networking, and mathematics research.

**Industry and Business Sectors.** The Howard Hughes Medical Institute (HHMI) is a nonprofit independent research institution that both funds and performs bioinformatics research. Its funding program entails appointing scientists as Hughes Investigators, providing them with long-term funding at their home institutions with the freedom to explore research projects as they choose. It is influential in recommending policies and standards for undergraduate science and medical education, including incorporating bioinformatics into the curriculum.

Many independent research institutions (including the Broad Institute, the J. Craig Venter Institute, and the Sanger Institute) carry out bioinformatics research. Several are large biodata producers, performing only high-throughput genomic and metagenomic computationally intense bioinformatics research.

In the for-profit sector, the biotechnology and pharmaceutical companies are significantly invested in protein engineering and modeling for pharmaceutical drug and diagnostic kit development. Bioproduct companies produce the enzymes, reagents, and kits needed to support the laboratory-based molecular biology research related to bioinformatics. Bioinstrumentation companies research, develop, and provide the highly specialized automated genomics and proteomics sequence analyzers and processors needed by both for-profit and nonprofit research efforts. They are an important source of innovation in sequencing methodologies that is largely responsible for continuing to advance bioinformatics into practical applications including metagenomics and personalized medicine initiatives.

## CAREERS AND COURSE WORK

Optimum bioinformatics practice needs bioinformaticians trained or experienced in biocomputational research and development. As bioinformatics continues to evolve and expand, the employment outlook is excellent. Jobs are available as algorithmic developers, programmers, data analysts/integrators, software engineers, or database designers in private industry, academic environments, government agencies, and nonprofit research institutions. Positions exist at all levels, from entry programmers to senior-level scientists to research directors.

Formal undergraduate and graduate academic bioinformatics or computational biology educational programs are just becoming available. Curricula try to provide a broad foundational core of understanding of the mathematics, computer, molecular, cellular, and genetics sciences. Course work in PERL programming languages, data structures, database design, algebra, probability and statistics, calculus, and discrete mathematics are designed to fill the existing gap in the marketplace. Laboratory courses provide exposure to the basic data and tools used, including DNA and protein sequence similarity and alignment, protein structure, phylogenetic analyses, and finding and cloning genes. Bioethics, justice courses, or a senior research project/thesis are not uncommon requirements. Graduate level curricula add courses in bioinstrumentation, protein engineering, population genetics, molecular diagnostics and prediction, the emerging field of genetic association studies, statistical computing packages such as R, and topically specialized seminars.

## SOCIAL CONTEXT AND FUTURE PROSPECTS

Metagenomics, personalized medicine, and future bioinformatics initiatives yet to be discovered will rapidly affect nearly all individuals. It is not surprising that with these major advances in bioinformatics technologies comes a caution by many for careful introspection and debate of their implications. Interest in bioethics is on the rise and has been added to the curricula of not only bioinformatics educational programs but those of many other disciplines as well.

*Diane C. Rein, Ph.D., M.L.S.*

**FURTHER READING**

Baxevanis, Andreas D., and B. F. Francis Ouellette, eds. *Bioinformatics: A Practical Guide to the Analysis of Genes and Proteins.* 3d ed. Hoboken, N.J.: John Wiley & Sons, 2005. Covers bioinformatics from the database and searching perspective. Contains chapters on various biological databases and their search interfaces.

Gu, Jenny, and Phillip E. Bourne, eds. *Structural Bioinformatics.* 2d ed. Hoboken, N.J.: John Wiley & Sons, 2009. Combination textbook and manual covering all aspects of protein bioinformatics, including the major protein databases, visualization, mass spectrometry, and protein modeling.

Lesk, Arthur M. *Introduction to Bioinformatics.* 3d ed. New York: Oxford University Press, 2008. Comprehensive overview of genomes, proteomics, protein structure, databases, phylogenetics, programming languages, and more.

Zvelebil, Marketa, and Jeremy Baum. *Understanding Bioinformatics.* New York: Garland Science, 2008.

Intermediate text with detailed descriptions on sequence alignments, phylogenetics, genomics, proteomics, and protein structure and modeling.

**WEB SITES**

*American Medical Informatics Association*
http://www.amia.org

*Bioinformatics Organization*
http://www.bioinformatics.org

*European Bioinformatics Institute*
http://www.ebi.ac.uk

*National Center for Biotechnology Information*
http://www.ncbi.nlm.nih.gov

**See also:** Bioengineering; Biomathematics; DNA Analysis; DNA Sequencing; Genomics; Human Genetic Engineering; Pattern Recognition; Probability and Statistics; Proteomics and Protein Engineering.

# BIOMATHEMATICS

## FIELDS OF STUDY

Algebra; geometry; calculus; probability; statistics; cellular biology; genetics; differential equations; molecular biology; oncology; immunology; epidemiology; prokaryotic biology; eukaryotic biology

## SUMMARY

Biomathematics is a field that applies mathematical techniques to analyze and model biological phenomena. Often a collaborative effort, mathematicians and biologists work together using mathematical tools such as algorithms and differential equations in order to understand and illustrate a specific biological function. Biomathematics is used in a wide variety of applications from medicine to agriculture. As new technologies lead to a rise in the amount of biological data available, biomathematics will become a discipline that is increasingly in demand to help analyze and effectively utilize the data.

## KEY TERMS AND CONCEPTS

- **Algorithm:** Use of symbols and a set of rules for solving problems.
- **Biology:** Study of living things.
- **Cell:** Unit that is the basis of an organism and encompasses genetic material, as well as other molecules, and is defined by a cell membrane.
- **Deoxyribonucleic Acid (DNA):** Nucleic acid that forms the molecular basis for heredity.
- **Differential Equation:** Mathematic expression that uses variables to express changes over time. Differential equations can be linear or nonlinear.
- **Genetics:** Study of an organism's traits, including how they are passed down through generations.
- **Matrix:** Mathematical structure for arranging numbers or symbols that have particular mathematical rules for use.
- **Oncology:** Field of science that studies the cause and treatment of cancer.

## DEFINITION AND BASIC PRINCIPLES

Biomathematics is a discipline that quantifies biological occurrences using mathematical tools.

Biomathematics is related to and may be a part of other disciplines including bioinformatics, biophysics, bioengineering, and computational biology, as these disciplines include the use of mathematical tools in the study of biology.

Biologists have used different ways to explain biological functions, often employing words or pictures. Biomathematics allows biologists to illustrate these functions using techniques such as algorithms and differential equations. Biological phenomena vary in both scale and complexity, encompassing everything from molecules to ecosystems. Therefore, the creation of a model requires the scientist to make some assumptions in order to simplify the process. Biomathematical models vary in length and complexity and several different models may be tested.

The use of biomathematics is not limited to modeling a biological function and includes other techniques, such as structuring and analyzing data. Scientists may use biomathematics to organize data or analyze data sets, and statistics are often considered an integral tool.

## BACKGROUND AND HISTORY

As early as the 1600's, mathematics was used to explain biological phenomena, although the mathematical tools used date back even farther. In 1628, British physician William Harvey used mathematics to prove that blood circulates in the body. His model changed the belief at that time that there were two kinds of blood. In the mid-1800's, Gregor Mendel, an Augustinian monk, used mathematics to analyze the data he obtained from his experiments with pea plants. His experiments would become the basis for genetics. In the early 1900's, British mathematician R. A. Fisher applied statistical methods to population biology, providing a better framework for studying the field. In 1947, theoretical physicist Nicolas Rashevsky argued that mathematical tools should be applied to biological processes and created a group dedicated to mathematical biology. Despite the fact that some dismiss Rashevsky's work as being too theoretical, many view him as one of the founders of mathematical biology. In the 1950's, the Hodgkin-Huxley equations were developed to describe a cellular function known as ion channels. These equations are still used. In the

1980's, the Smith-Waterman algorithm was created to aid scientists in comparing DNA sequences. While the algorithm was not particularly efficient, it paved the way for the BLAST (Basic Local Alignment Search Tool) software, a program that has allowed scientists to compare DNA sequences since 1990. Despite the fact that mathematical tools have been applied to some biological problems during the second half of the twentieth century, the practice has not been all-inclusive. In the twenty-first century, there has been a renewed interest in biology becoming more quantitative, due in part to an increase in new data.

## How It Works

**Basic Mathematical Tools.** Biomathematicians may use mathematical tools at different points during the investigation of a biological function. Mathematical tools may be used to organize data, analyze data, or even to generate data. Algorithms, which use symbols and procedures for solving problems, are employed in biomathematics in several ways. They may be used to analyze data, as in sequence analysis. Sequence analysis uses specifically developed algorithms to detect similarity in pieces of DNA. Specifically developed algorithms are also used to predict the structure of different biological molecules, such as proteins. Algorithms have led to the development of more useful biological instruments such as specific types of microscopy. Statistics are another common way of analyzing biological data. Statistics may be used to analyze data, and this data may help create an equation to describe a theory: Statistics was used to analyze the movement of single cells. The data taken from the analysis was then used to create partial equations describing cell movement.

Differential equations, which use variables to express changes over time, are another common technique in biomathematics. There are two kinds of differential equations: linear and nonlinear. Nonlinear equations are commonly used in biomathematics. Differential equations, along with other tools, have been used to model the functions of intercellular processes. Differential equations are utilized in several of the important systems used in biomathematics for modeling, including mean field approaches. Other modeling systems include: patch models, reaction-diffusion equations, stochastic models, and interacting particle systems. Each modeling system provides a different approach based on different assumptions. Computers have helped in this area by providing an easier way to apply and solve complex equations. Computer modeling of dynamic systems, such as the motion of proteins, is also a work in progress.

New methods and technology have increased the amount of data being obtained from biological experimentation. The data gained through experimentation and analysis may be structured in different ways. Mathematics may be used to determine the structure. For example, phylogenetic trees (treelike graphs that illustrate how pieces of data relate to one another) use different mathematical tools, including matrices, to determine their structure. Phlyogenic trees also provide a model for how a particular piece of data evolved. Another way to organize data is a site graph, or hidden Markov model, which uses probability to illustrate relationships between the data.

**Modeling a Biological Function.** The scientist may be at different starting points when considering a mathematical model. He or she may be starting with data already analyzed or organized by a mathematical technique or already described by a visual depiction or written theory. However, there are several considerations that scientists must take into account when creating a mathematical model. As biology covers a wide range of matter, from molecules to ecosystems, when creating a model the scale of phenomena must be considered. The time scale and complexity must also be considered, as many biological systems are dynamic or interact with their environment. The scientist must make assumptions about the biological phenomena in order to reduce the parameters used in the model. The scientist may then define important variables and the relationships between them. Often, more than one model may be created and tested.

## Applications and Products

The field of biomathematics is applicable to every area of biology. For example, biomathematics has been used to study population growth, evolution, and genetic variation and inheritance. Mathematical models have also been created for communities, modeling competition or predators, often using differential equations. Whether the scale is large or small, biomathematics allows scientists a greater understanding of biological phenomena.

**Molecules and Cells.** Biomathematics has been applied to various biological molecules, including DNA, ribonucleic acid (RNA), and proteins. Biomathematics may be used to help predict the structure of these molecules or help determine how certain molecules are related to one another. Scientists have used biomathematics to model how bacteria can obtain new, important traits by transferring genetic material between different strains. This information is important because bacteria may, through sharing genetic material, acquire a trait such as a resistance to an antibiotic. To model the sharing of a trait, scientists have combined two of the ways to structure data: the phylogenetic tree and the site graph. The phylogenetic tree illustrates how the types of bacteria are related to one another. The site graph illustrates how pieces of genetic material interact. Then, scientists use a particular algorithm to determine the parameters of the model. By using such tools, scientists can predict which areas of genetic material are most likely to transfer between the bacterium.

Biomathematics has been used in cellular biology to model various cellular functions, including cellular division. The models can then be used to help scientists organize information and gain a deeper understanding about cellular functions. Cellular movement is one example of an application of biomathematics to cellular biology. Cellular movements can be seen as a set of steps. The scientists first considered certain cellular steps or functions, including how a cell senses a signal and how this signal is used within the cell to start movement. Scientists also considered the environment surrounding the cell, how the signal was provided, and the processes that occurred within the cell to read the signal and start movement. The scientists were then able to build a mathematical function that takes these steps into account. Depending on the particular question, the scientists may chose to focus on any of these steps. Therefore, more than one model may be used.

**Organisms and Agriculture.** Biomathematics has been used to create mathematical models for different functions of organisms. One popular area has been organism movement, where models have been created for bacteria movement and insect flight. A more complete understanding of organisms through mathematical models supports new technologies in agriculture. Biomathematics may also be used to help protect harvests. For example, biomathematics has been used to model a type of algae bloom known as brown tide. In the late 1980's, brown tide appeared in the waters near Long Island, New York, badly affecting the shellfish population by blocking sunlight and

---

## Fascinating Facts About Biomathematics

- Scientists are using biomathematics to create a virtual patient. First, biomathematics is used to model the human body. Using the virtual patient, scientists can test cancer-prevention drugs. The result is a quicker and cheaper way to develop drugs.
- Biomathematics has applications in nature. By using biomathematics, the pigment patterns of a leopard's spots or the patterns of seashells can be modeled.
- Scientists and mathematicians have found that by using biomathematics and computers they can simulate kidney functions. This kidney simulation helps doctors understand kidney disease better and can help them provide more effective treatments.
- Biomathematics is being used to model the workings of a heart in order to improve artificial heart valves. With biomathematics models, designs can be tested more quickly and efficiently.
- Mathematical models of biofilms systems, which are layers of usually nonresistant microorganisms such as bacteria that attach to a surface, are critical to the medical and technical industries. Biofilm systems can cause infections in humans and corrosion and deterioration in technical systems. Biomathematics can be used to model biofilms systems to help understand and prevent their formation.
- Biomathematics was used to sort and analyze data from the Human Genome Project. Completed in 2003, the thirteen-year-long project identified the entire human genome.

---

depleting oxygen. Four years later, the algae blooms receded. Both mathematicians and scientists collaborated in order to create a model of the brown tide in order to understand why it bloomed and whether it will bloom in the future. To create a model, the collaborators used differential equations. They focused on the population density, which included factors such as temperature and nutrients. The collaborators had to consider many variables and remove the ones they did not consider important. For instance, they hypothesized that a period of drought followed by rain may have affected growth. They also considered fertilizers and pesticides that were used in the area. A better understanding of the brown tide may help protect the shellfish harvest in future years.

**Medical Uses.** Biomathematical models have been developed to illustrate various functions within the human body, including the heart, kidneys, and cardiac and neural tissue. Biomathematics is useful in modeling cancer, enabling scientists to learn more about the type of cancer, thereby allowing them to study the efficacy of different types of treatment. One project has focused on modeling colon cancer on a genetic and molecular level. Not only did scientists gain information about the genetic mutations that are present during colon cancer, but they also developed a model that predicted when tumor cells would be sensitive to radiation, which is the most common way to treat colon cancer. Studies such as this can be built on in future experimentations, the results of which may someday be used by doctors to create more effective cancer treatments.

Biomathematics has also been used to organize and analyze data from experiments dealing with drug efficacy and gene expression in cancer cells. Using matrices, statistics, and algorithms, scientists have been able to understand if a particular drug is more likely to work based on the patient's cancer cell's gene expression. Biomathematics has also been integral in epidemiology, the field that studies diseases within a population. Biomathematics may be used to model various aspects of a disease such as human immunodeficiency virus (HIV), allowing for more comprehensive planning and treatment.

## IMPACT ON INDUSTRY

**Government and University Research.** Biomathematics is often developed through a collaborative effort between biologists and mathematicians. In the United States, the National Science Foundation has a mathematical biology program that provides grants for research to develop mathematical applications related to biology. In addition, many U.S. universities offer biomathematics programs. Some universities have biomathematics departments, such as Ohio State University's Mathematical Biosciences Institute, which focuses on creating mathematical applications to help solve biological problems. This institute provides research and education opportunities, including workshops and public lectures.

Each biomathematics department or program may emphasize a different aspect of biomathematics: Some focus on medical applications, others focus on the need to quantify biological phenomena using mathematical tools. UCLA's department of biomathematics conducts research in areas such as statistical genetics, evolutionary biology, molecular imaging, and neuroscience.

Biomathematics is also being developed internationally. Many universities in the United Kingdom have biomathematics programs. The University of Oxford has a Centre for Mathematical Biology. There are also independent research institutions and organizations that focus on biomathematics. The Institute for Medical BioMathematics, in Israel, works on developing analytical approaches to treating cancer and infectious diseases. The Society for Mathematical Biology, in Boulder, Colorado, has provided an international forum for biomathematics for more than twenty years. The International Biometric Society in Washington, D.C., also addresses the application of mathematical tools to biological phenomena. The European Society for Mathematical and Theoretical Biology, founded in 1991, promotes biomathematics, and the Society for Industrial and Applied Mathematics in Philadelphia has an activity group on the life sciences that provides a platform for researchers working in the area of biomathematics.

**Industry.** Biomathematics may lead to developments in medical treatments or technology, which may then be marketed. The pharmaceutical industry is a good example of this development. Biomathematics is often used to create models of diseases that can lead to a deeper understanding of the disease and new medicines. In addition, biomathematics provides tools that may be used throughout the drug-creation process and can be used to predict how well a drug will work or how safe a drug will be for a group of patients with a particular genetic makeup. The engineering of microorganisms has also benefited from biomathematics. Biomathematics is being used to create models to understand the fundamentals of microorganisms better. The end result of this understanding may be the changed metabolism or structure of an organism to produce more milk or a sweeter wine. Finally, some companies are targeting software to aid with biological modeling, and others provide consulting in the field of biomathematics. BioMath Solutions in Austin, Texas, is a company that provides analytical software in the area of molecular biology.

## CAREERS AND COURSE WORK

Degree programs in biomathematics are gaining popularity in universities. Some schools have biomathematics departments, and others have biomathematics programs within the mathematics or biology departments. Undergraduate course work for a B.S. in biomathematics encompasses classes in mathematics, biology, and computer science, including: calculus, chemistry, genetics, physics, software development, probability, statistics, organic chemistry, epidemiology, population biology, molecular biology, and physiology. A student may also choose to receive a B.S. in biology or mathematics. In addition, students may seek additional opportunities outside their program. Ohio State's Mathematical Biosciences Institute offers summer programs for undergraduate and graduate students.

In the field of biomathematics, a doctorate is required for many careers. Doctoral programs in biomathematics include course work in statistics, biology, probability, differential equations, linear algebra, cellular modeling, genetics modeling, computer programming, pharmacology, and clinical research methods. In addition, doctoral candidates often will perform biomathematics research with support from departmental faculty. As with the undergraduate degrees, a student may also choose to pursue a doctorate in biology or mathematics.

With the influx of biological data from new technologies and tools, a degree in biomathematics is imperative. Those who receive a Ph.D. may choose to enter a postdoctoral program, such as the one at Ohio State's Mathematical Biosciences Institute, which offers postdoctoral fellowships as well as mentorship and research opportunities. Other career paths include research in medicine, biology, and mathematics with universities and private research institutions; work with software development and computer modeling; teaching; or collaborating with other professionals and consulting in an industry such as pharmaceuticals or bioengineering.

## SOCIAL CONTEXT AND FUTURE PROSPECTS

While mathematical tools have been applied to biology for some time, many scientists believe there is still a need for increased quantitative analysis of biology. Some call for more emphasis on mathematics in high school and undergraduate biology classes. They believe that this will advance biomathematics. As more universities develop biomathematics departments and degrees, more mathematics classes will be added to the curriculum. A concern has been raised in the biomathematics field about the assumptions used to create simplified mathematical models. More complex and accurate models will likely be developed.

Important future applications for biomathematics will be in the bioengineering and medical industries. The development of mathematical models for complex biological phenomena will aid scientists in a deeper understanding that can lead to more effective treatments in such areas such as tumor therapy. As new tools and methods continue to develop, biomathematics will be a field that expands to sort and analyze the large influx of data.

*Carly L. Huth, B.S., J.D.*

## FURTHER READING

Hochberg, Robert, and Kathleen Gabric. "A Provably Necessary Symbiosis." *The American Biology Teacher* 72, No. 5 (2010): 296-300. This article describes some mathematics that can be taught in biology classrooms.

Misra, J. C., ed. *Biomathematics: Modelling and Simulation.* Hackensack, N.J.: World Scientific, 2006. This book provides an in-depth guide to several modern applications of biomathematics and includes many helpful illustrations.

Schnell, Santiago, Ramon Grima, and Philip Maini. "Multiscale Modeling in Biology." *American Scientist* 95 (March-April, 2007): 134-142. This article gives an overview of how biological models are created and provides several modern examples of biomathical applications.

## WEB SITES

*International Biometric Society*
http://www.tibs.org

*National Science Foundation*
http://www.nsf.gov

*Ohio State University Mathematical Biosciences Institute*
http://mbi.osu.edu

*Society for Industrial and Applied Mathematics*
http://www.siam.org

*Society for Mathematical Biology*
http://www.smb.org

**See also:** Bioengineering; Bioinformatics; Biomechanical Engineering; Bionics and Biomedical Engineering; Biophysics; Calculus; Computer Science; Geometry; Probability and Statistics; Software Engineering; Trigonometry.

# BIOMECHANICAL ENGINEERING

## FIELDS OF STUDY

Biomedical engineering; biomechanics; physiology; nanotechnology; implanted devices; modeling; bioengineering; bioinstrumentation; computational biomechanics; cellular and molecular biomechanics; forensic biomechanics; tissue engineering; mechanobiology; micromechanics; anthropometics; imaging; biofluidics.

## SUMMARY

Biomechanical engineering is a branch of science that applies mechanical engineering principles such as physics and mathematics to biology and medicine. It can be described as the connection between structure and function in living things. Researchers in this field investigate the mechanics and mechanobiology of cells and tissues, tissue engineering, and the physiological systems they comprise. The work also examines the pathogenesis and treatment of diseases using cells and cultures, tissue mechanics, imaging, microscale biosensor fabrication, biofluidics, human motion capture, and computational methods. Real-world applications include the design and evaluation of medical implants, instrumentation, devices, products, and procedures. Biomechanical engineering is a multidisciplinary science, often fostering collaborations and interactions with medical research, surgery, radiology, physics, computer modeling, and other areas of engineering.

## KEY TERMS AND CONCEPTS

- **Angular Motion:** Motion involving rotation around a central line or point known as the axis of rotation.
- **Biofluidics:** Field of study that combines the characterization of fluids focused on flows in the body as well as environmental flows involved in disease process.
- **Dynamics:** Branch of mechanics that studies systems that are in motion, subject to acceleration or deceleration.
- **Kinematics:** Study of movement of segments of a body without regard for the forces causing the movement.

- **Kinesiology:** Study of human movement.
- **Kinetics:** Study of forces associated with motion.
- **Linear Motion:** Motion involving all the parts of a body or system moving in the same direction, at the same speed, following a straight (rectilinear) or curved (curvilinear) line.
- **Mechanics:** Branch of physics analyzing the resulting actions of forces on particles or systems.
- **Mechanobiology:** Emerging scientific field that studies the effect of physical force on tissue development, physiology, and disease.
- **Modeling:** Computerized analytical representation of a structure or process.
- **Statics:** Branch of mechanics that studies systems that are in a constant state of motion or constant state of rest.

## DEFINITION AND BASIC PRINCIPLES

Biomechanical engineering applies mechanical engineering principles to biology and medicine. Elements from biology, physiology, chemistry, physics, anatomy, and mathematics are used to describe the impact of physical forces on living organisms. The forces studied can originate from the outside environment or generate within a body or single structure. Forces on a body or structure can influence how it grows, develops, or moves. Better understanding of how a biological organism copes with forces and stresses can lead to improved treatment, advanced diagnosis, and prevention of disease. This integration of multidisciplinary philosophies has lead to significant advances in clinical medicine and device design. Improved understanding guides the creation of artificial organs, joints, implants, and tissues. Biomechanical engineering also has a tremendous influence on the retail industry, as the results of laboratory research guide product design toward more comfortable and efficient merchandise.

## BACKGROUND AND HISTORY

The history of biomechanical engineering, as a distinct and defined field of study, is relatively short. However, applying the principles of physics and engineering to biological systems has been developed over centuries. Many overlaps and parallels to complementary areas of biomedical engineering and biomechanics exist, and the terms are often interchangeably with biomechanical engineering. The mechanical analysis of living organisms was not internationally accepted and recognized until the definition provided by Austrian mathematician Herbert Hatze in 1974: "Biomechanics is the study of the structure and function of biological systems by means of the methods of mechanics." Aristotle introduced the term "mechanics" and discussed the movement of living beings around 322 B.C.E in the first book about biomechanics, *On the Motion of Animals.* Leonardo da Vinci proposed that the human body is subject to the law of mechanics in the 1500's. Italian physicist and mathematician Giovanni Alfonso Borelli, a student of Galileo's, is considered the "father of biomechanics" and developed mathematical models to describe anatomy and human movement mechanically. In the 1890's German zoologist Wilhelm Roux and German surgeon Julius Wolff determined the effects of loading and stress on stem cells in the development of bone architecture and healing. British physiologist Archibald V. Hill and German physiologist Otto Fritz Meyerhof shared the 1922 Nobel Prize for Physiology or Medicine. The prize was divided between them: Hill won "for his discovery relating to the production of heat in the muscle"; Meyerhof won "for his discovery of the fixed relationship between the consumption of oxygen and the metabolism of lactic acid in the muscle."

The first joint replacement was performed on a hip in 1960 and a knee in 1968. The development of imaging, modeling, and computer simulation in the latter half of the 1900's provided insight into the smallest structures of the body. The relationships between these structures, functions, and the impact of internal and external forces accelerated new research opportunities into diagnostic procedures and effective solutions to disease. In the 1990's, biomechanical engineering programs began to emerge in academic and research institutions around the world.

## HOW IT WORKS

Biomechanical engineering science is extremely diverse. However, the basic principle of studying the relationship between biological structures and forces, as well as the important associated reactions of biological structures to technological and environmental materials, exists throughout all disciplines. The biological structures described include all life forms and may include an entire body or organism or even the microstructures of specific tissues or systems. Characterization and quantification of the response of these structures to forces can provide insight into disease process, resulting in better treatments and diagnoses. Research in this field extends beyond the laboratory and can involve observations of mechanics in nature, such as the aerodynamics of bird flight, hydrodynamics of fish, or strength of plant root systems, and how these findings can be modified and applied to human performance and interaction with external forces.

As in biomechanics, biomechanical engineering has basic principles. Equilibrium, as defined by British physicist Sir Isaac Newton, results when the sum of all forces is zero and no change occurs and energy cannot be created or destroyed, only converted from one form to another.

The seven basic principles of biomechanics can be applied or modified to describe the reaction of forces to any living organism.

1. The lower the center of mass, the larger the base of support; the closer the center of mass to the base of support, and the greater the mass, the more stability increases.
2. The production of maximum force requires the use of all possible joint movements that contribute to the task's objective.
3. The production of maximum velocity requires the use of joints in order—from largest to smallest.
4. The greater the applied impulse, the greater increase in velocity.
5. Movement usually occurs in the direction opposite that of the applied force.
6. Angular motion is produced by the application of force acting at some distance from an axis, that is, by torque.
7. Angular momentum is constant when a body or object is free in the air.

The forces studied can be combinations of internal, external, static, or dynamic, and all are important in the analysis of complex biochemical and biophysical processes. Even the mechanics of a single cell, including growth, cell division, active motion, and contractile mechanisms, can provide insight into mechanisms of stress, damage of structures, and disease processes at the microscopic level. Imaging and computer simulation allow precise measurements and observations to be made of the forces impacting the smallest cells.

## APPLICATIONS AND PRODUCTS

Biomechanical engineering advances in modeling and simulation have tremendous potential research and application uses across many health care disciplines. Modeling has resulted in the development of designs for implantable devices to assist with organs or areas of the body that are malfunctioning. The biomechanical relationships between organs and supporting structures allow for improved device design and can assist with planning of surgical and treatment interventions. The materials used for medical and surgical procedures in humans and animals are being evaluated and some redesigned, as biomechanical science is showing that different materials, procedures, and techniques may be better for reducing complications and improving long-term patient health. Evaluating the physical relationship between the cells and structures of the body and foreign implements and interventions can quantify the stresses and forces on the system, which provides more accurate prediction of patient outcomes.

Biomechanical engineering professionals apply their knowledge to develop implantable medical devices that can diagnose, treat, or monitor disease and health conditions and improve the daily living of patients. Devices that are used within the human body are highly regulated by the U.S. Food and Drug Administration (FDA) and other agencies internationally. Pacemakers and defibrillators, also called cardiac resynchronization therapy (CRT) devices, can constantly evaluate a patient's heart and respond to changes in heart rate with electrical stimulation. These devices greatly improve therapeutic outcomes in patients afflicted with congestive heart failure. Patients with arrhythmias experience greater advantages with implantable devices than with pharmaceutical options. Cochlear implants have been designed to be attached to a patient's auditory nerve and can detect sound waves and process them in order to be interpreted by the brain as sound for deaf or hard-of-hearing patients. Patients who have had cataract surgery used to have to wear thick corrective lenses to restore any standard of vision but with the development of intraocular lenses that can be implanted into

---

### Fascinating Facts About Biomechanical Engineering

- Biomechanical engineers design and develop many items used and worn by astronauts.
- Of all the engineering specialties, biomechanical engineering has one of the highest percentages of female students.
- Synovial fluid in joints has such low friction that engineers are trying to duplicate it synthetically to lubricate machines.
- Many biomechanical engineering graduates continue on to medical school.
- On October 8, 1958, in Sweden, forty-three-year-old Arne Larsson became the first person to receive an implanted cardiac pacemaker. He lived to the age of eighty-six.
- Many biomechanical engineers have conducted weightlessness experiments aboard the National Aeronautics and Space Administration's (NASA) C-9 microgravity aircraft.
- Cardiac muscles rest only between beats. Based on 72 beats a minute, they will pump an average of 3.0 trillion times during an eighty-year life span.

---

the eye, their vision can be restored, often to a better degree than before the cataract developed.

Artificial replacement joints comprise a large portion of medical-implant technology. Patients receive joint replacement when their existing joints no longer function properly or cause significant pain because of arthritis or degeneration. More than 220,000 total hip replacements were performed in the United States in 2003, and this number is expected to grow significantly as the baby boomer portion of the population ages. Artificial joints are normally fastened to the existing bone by cement, but advances in biomechanical engineering have lead to a new process called "bone ingrowth," in which the natural bone grows into the porous surface of the replacement joint. Biomechanical engineering contributes considerable knowledge to the design of the artificial joints, the materials from which they are made, the surgical procedure used, fixation techniques, failure mechanisms, and prediction of the lifetime of the replacement joints.

Computer-aided (CAD) design has allowed biomechanical engineers to create complex models of organs and systems that can provide advanced analysis and instant feedback. This information provides insight into the development of designs for artificial organs that align with or improve on the mechanical properties of biological organs.

Biomechanical engineering can provide predictive values to medical professionals, which can help them develop a profile that better forecasts patient outcomes and complications. An example of this is using finite element analysis in the evaluation of aortic-wall stress, which can remove some of the unpredictability of expansion and rupture of an abdominal aortic aneurysm. Biomechanical computational methodology and advances in imaging and processing technology have provided increased predictability for life- threatening events.

Nonmedical applications of biomechanical engineering also exist in any facet of industry that impacts human life. Corporations employ individuals or teams to use engineering principles to translate the scientifically proven principles into commercially viable products or new technological platforms. Biomechanical engineers also design and build experimental testing devices to evaluate a product's performance and safety before it reaches the marketplace, or they suggest more economically efficient design options. Biomechanical engineers also use ergonomic principles to develop new ideas and create new products, such as car seats, backpacks, or even equipment and clothing for elite athletes, military personnel, or astronauts.

## IMPACT ON INDUSTRY

Biomechanical engineering is a dynamic scientific field, and its vast range of applications is having a significant impact and influence on industry. Corporations realize the value of having their designs and products evaluated by biomechanical engineers to optimize the comfort and safety of consumers. Small modifications in the design of a product can influence consumers to select one product over several others, be it clothing, furniture, sporting equipment, or beverage and food containers. With so many products to choose from, it is important that one stand out as safer, more comfortable, or more efficient than another.

Biotechnology and health care are highly competitive, multibillion dollar industries. Biotechnology is an extremely research-intensive industry, so biotech

companies employ teams of biomechanical engineers to research and develop devices, treatments, and diagnostic and monitoring devices to be used in health care. Implantable devices are used in the treatment and management of cardiovascular, orthopedic, neurological, ophthalmic, and various other chronic disorders. Orthopedic implants are the most common and include reconstructive joint replacements, spinal implants, orthobiologics, and trauma implants. In the United States, the demand for implantable medical devices is expected to reach a market value of $48 billion by 2014.

Surgeons, medical personnel, and health care administrators are always looking for new options that will provide their patients with optimal care, earlier diagnosis, pain reduction, and a decreased risk of complications. Aging and disabled persons have more effective options than ever before that allow them to continue vital, independent, and active lives. Industry is aligning with these developments, demands, and discoveries, and there is great competition among companies to be the first to capitalize. The FDA and other international regulatory bodies have strict standards and high levels of control for any device or implement that will be used on patients. The application and approval process can take many years. It is not uncommon for a biotechnology company to invest millions of dollars in the research and development of a medical device before it can be presented to the consumer.

## CAREERS AND COURSE WORK

There are a variety of career choices in biomechanical engineering, and study in this field often evolves into specialized work in related areas. Students who earn a bachelor's degree from an accredited biomechanical engineering program may begin working in areas such as medical device, implant, or product design. Most teaching positions require a master's or doctoral degree. Some students continue to medical school.

Biomechanical engineering programs are composed of a cross section of course work from many disciplines. Students should have a strong aptitude for mathematics as well as biological sciences. Elements from engineering, physics, chemistry, anatomy, biology, and computer science provide core knowledge that is applied to mathematical modeling and computer simulation. Experimental work involving biological, mechanical, and clinical studies are performed to illustrate theoretical models and solve important research problems. The principles of biomechanical engineering can have vast applications, ranging from building artificial organs and tissues to designing products that are more comfortable for consumers.

Biomechanical engineering programs often are included as a subdiscipline of engineering or biomedicine. However, some schools, such as Stanford University, are creating interdisciplinary programs that offer undergraduate and graduate degrees in biomechanical engineering.

## SOCIAL CONTEXT AND FUTURE PROSPECTS

The diversity of studying the relationship between living structure and function has opened up vast opportunities in science, health care, and industry. In addition to conventional implant and replacement devices, the demand is growing for implantable tissues for cosmetic surgery, such as breast and tissue implants, as well as implantable devices to aid in weight loss, such as gastric banding.

Reports of biomechanical engineering triumphs and discoveries are appearing in the mainstream media, making the general public more aware of the scientific work being done and how it impacts daily life. Sports fans learn about the equipment, training, and rehabilitation techniques designed by biomechanical engineers that allow their favorite athletes to break performance records and return to work sooner after being injured or having surgery. The public is accessing more information about their own health options than ever before, and they are becoming knowledgeable about the range of treatments available to them and the pros and cons of each.

Biomechanical engineering and biotechnology is an area that is experiencing accelerated growth, and billions of dollars are being funneled into research and development annually. This growth is expected to continue.

*April D. Ingram, B.Sc.*

## FURTHER READING

Ethier, C. Ross, and Craig A. Simmons. *Introductory Biomechanics: From Cells to Organisms.* Cambridge, England: Cambridge University Press, 2007. Provides an introduction to biomechanics and also discusses clinical specialties, such as cardiovascular, musculoskeletal, and ophthalmology.

Hall, Susan J. *Basic Biomechanics.* 5th ed. New York: McGraw-Hill, 2006. A good introduction to biomechanics, regardless of one's math skills.

Hamill, Joseph, and Kathleen M. Knutzen. *Biomechanical Basis of Human Movement.* 3d ed. Philadelphia: Lippincott, 2009. Integrates anatomy, physiology, calculus, and physics and provides the fundamental concepts of biomechanics.

Hay, James G., and J. Gavin Reid. *Anatomy, Mechanics, and Human Motion.* 2d ed. Englewood Cliffs, N.J.: Prentice Hall, 1988. A good resource for upper high school students, this text covers basic kinesiology.

Peterson, Donald R., and Joseph D. Bronzino, eds. *Biomechanics: Principles and Applications.* 2d ed. Boca Raton, Fla.: CRC Press, 2008. A collection of twenty articles on various aspects of research in biomechanics.

Prendergast, Patrick, ed. *Biomechanical Engineering: From Biosystems to Implant Technology.* London: Elsevier, 2007. One of the first comprehensive books for biomechanical engineers, written with the student in mind.

## WEB SITES

*American Society of Biomechanics*
http://www.asbweb.org

*Biomedical Engineering Society*
http://www.bmes.org

*European Society of Biomechanics*
http://www.esbiomech.org

*International Society of Biomechanics*
http://www.isbweb.org

*World Commission of Science and Sports*
http://www.wcss.org.uk

**See also:** Bioengineering; Bioinformatics; Biomechanics; Bionics and Biomedical Engineering; Biophysics; Calculus; Cell and Tissue Engineering; Computer Science; Nanotechnology.

# BIOMECHANICS

## FIELDS OF STUDY

Kinesiology; physiology; kinetics; kinematics; sports medicine; technique/performance analysis; injury rehabilitation; modeling; orthopedics; prosthetics; bioengineering; bioinstrumentation; computational biomechanics; cellular/molecular biomechanics; veterinary (equine) biomechanics; forensic biomechanics; ergonomics.

## SUMMARY

Biomechanics is the study of the application of mechanical forces to a living organism. It investigates the effects of the relationship between the body and forces applied either from outside or within. In humans, biomechanists study the movements made by the body, how they are performed, and whether the forces produced by the muscles are optimal for the intended result or purpose. Biomechanics integrates the study of anatomy and physiology with physics, mathematics, and engineering principles. It may be considered a subdiscipline of kinesiology as well as a scientific branch of sports medicine.

## KEY TERMS AND CONCEPTS

- **Angular Motion:** Motion involving rotation around a central line or point known as the axis of rotation.
- **Dynamics:** Branch of mechanics that studies systems in motion, subject to acceleration or deceleration.
- **Kinematics:** Study of movement of segments of a body without regard for the forces causing the movement.
- **Kinesiology:** Study of human movement.
- **Kinetics:** Study of forces associated with motion.
- **Lever:** Rigid bars (in the body, bones) that move around an axis of rotation (joint) and have the ability to magnify or alter the direction of a force.
- **Linear Motion:** Motion involving all the parts of a body or system moving in the same direction, at the same speed, following a straight (rectilinear) or curved (curvilinear) line.
- **Mechanics:** Branch of physics analyzing the resulting actions of forces on particles or systems.

- **Qualitative Movement:** Description of the quality of movement without the use of numbers.
- **Quantitative Movement:** Description or analysis of movement using numbers or measurement.
- **Sports Medicine:** Branch of medicine studying the clinical and scientific characteristics of exercise and sport, as well as any resulting injuries.
- **Statics:** Branch of mechanics that studies systems that are in a constant state of motion or constant state of rest.
- **Torque:** Turning effect of a force applied in a direction not in line with the center of rotation of a nonmoving axis (eccentric).

### DEFINITION AND BASIC PRINCIPLES

Biomechanics is a science that closely examines the forces acting on a living system, such as a body, and the effects that are produced by these forces. External forces can be quantified using sophisticated measuring tools and devices. Internal forces can be measured using implanted devices or from model calculations. Forces on a body can result in movement or biological changes to the anatomical tissue. Biomechanical research quantifies the movement of different body parts and the factors that may influence the movement, such as equipment, body alignment, or weight distribution. Research also studies the biological effects of the forces that may affect growth and development or lead to injury. Two distinct branches of mechanics are statics and dynamics. Statics studies systems that are in a constant state of motion or constant state of rest, and dynamics studies systems that are in motion, subject to acceleration or deceleration. A moving body may be described using kinematics or kinetics. Kinematics studies and describes the motion of a body with respect to a specific pattern and speed, which translate into coordination of a display. Kinetics studies the forces associated with a motion, those causing it and resulting from it. Biomechanics combines kinetics and kinematics as they apply to the theory of mechanics and physiology to study the structure and function of living organisms.

### BACKGROUND AND HISTORY

Biomechanics has a long history even though the actual term and field of study concerned with mechanical analysis of living organisms was not internationally accepted and recognized until the early 1970's. Definitions provided by early biomechanics specialists James G. Hay in 1971 and Herbert Hatze in 1974 are still accepted. Hatze stated, "Biomechanics is the science which studies structures and functions of biological systems using the knowledge and methods of mechanics."

Highlights throughout history have provided insight into the development of this scientific discipline. The ancient Greek philosopher Aristotle was the first to introduce the term "mechanics," writing about the movement of living beings around 322 B.C.E. He developed a theory of running techniques and suggested that people could run faster by swinging their arms. In the 1500's, Leonardo da Vinci proposed that the human body is subject to the law of mechanics, and he contributed significantly to the development of anatomy as a modern science. Italian scientist Giovanni Alfonso Borelli, a student of Galileo, is often considered the father of biomechanics. In the mid-1600's, he developed mathematical models to describe anatomy and human movement mechanically. In the late 1600's, English physician and mathematician Sir Isaac Newton formulated mechanical principles and Newtonian laws of motion (inertia, acceleration, and reaction) that became the foundation of biomechanics.

British physiologist A. V. Hill, the 1923 winner of the Nobel Prize in Physiology or Medicine, conducted research to formulate mechanical and structural theories for muscle action. In the 1930's, American anatomy professor Herbert Elftman was able to quantify the internal forces in muscles and joints and developed the force plate to quantify ground reaction. A significant breakthrough in the understanding of muscle action was made by British physiologist Andrew F. Huxley in 1953, when he described his filament theory to explain muscle shortening. Russian physiologist Nicolas Bernstein published a paper in 1967 describing theories for motor coordination and control following his work studying locomotion patterns of children and adults in the Soviet Union.

## HOW IT WORKS

The study of human movement is multifaceted, and biomechanics applies mechanical principles to the study of the structure and function of living things. Biomechanics is considered a relatively new field of applied science, and the research being done is of considerable interest to many other disciplines, including zoology, orthopedics, dentistry, physical education, forensics, cardiology, and a host of other medical specialties. Biomechanical analysis for each particular application is very specific; however, the basic principles are the same.

**Newton's Laws of Motion.** The development of scientific models reduces all things to their basic level to provide an understanding of how things work. This also allows scientists to predict how things will behave in response to forces and stimuli and ultimately to influence this behavior.

Newton's laws describe the conservation of energy and the state of equilibrium. Equilibrium results when the sum of forces is zero and no change occurs, and conservation of energy explains that energy cannot be created or destroyed, only converted from one form to another. Motion occurs in two ways, linear motion in a particular direction or rotational movement around an axis. Biomechanics explores and quantifies the movement and production of force used or required to produce a desired objective.

**Seven Principles.** Seven basic principles of biomechanics serve as the building blocks for analysis. These can be applied or modified to describe the reaction of forces to any living organism.

1. The lower the center of mass, the larger the base of support; the closer the center of mass to the base of support and the greater the mass, the more stability increases.
2. The production of maximum force requires the use of all possible joint movements that contribute to the task's objective.
3. The production of maximum velocity requires the use of joints in order, from largest to smallest.
4. The greater the applied impulse, the greater increase in velocity.
5. Movement usually occurs in the direction opposite that of the applied force.
6. Angular motion is produced by the application of force acting at some distance from an axis, that is, by torque.
7. Angular momentum is constant when an athlete or object is free in the air.

Static and dynamic forces play key roles in the complex biochemical and biophysical processes that underlie cell function. The mechanical behavior of individual cells is of interest for many different biologic processes. Single-cell mechanics, including growth, cell division, active motion, and contractile mechanisms, can be quite dynamic and provide insight into mechanisms of stress and damage of structures. Cell mechanics can be involved in processes that lie at the root of many diseases and may provide opportunities as focal points for therapeutic interventions.

## APPLICATIONS AND PRODUCTS

Biomechanics studies and quantifies the movement of all living things, from the cellular level to body systems and entire bodies, human and animal. There are many scientific and health disciplines, as well as industries that have applications developed from this knowledge. Research is ongoing in many areas to effectively develop treatment options for clinicians and better products and applications for industry.

**Dentistry.** Biomechanical principles are relevant in orthodontic and dental science to provide solutions to restore dental health, resolve jaw pain, and manage cosmetic and orthodontic issues. The design of dental implants must incorporate an analysis of load bearing and stress transfer while maintaining the integrity of surrounding tissue and comfortable function for the patient. This work has lead to the development of new materials in dental practices such as reinforced composites rather than metal frameworks.

**Forensics.** The field of forensic biomechanical analysis has been used to determine mechanisms of injury after traumatic events such as explosions in military situations. This understanding of how parts of the body behave in these events can be used to develop mitigation strategies that will reduce injuries. Accident and injury reconstruction using biomechanics is an emerging field with industrial and legal applications.

**Biomechanical Modeling.** Biomechanical modeling is a tremendous research field, and it has potential uses across many health care applications. Modeling has resulted in recommendations for prosthetic design and modifications of existing devices. Deformable breast models have demonstrated

capabilities for breast cancer diagnosis and treatment. Tremendous growth is occurring in many medical fields that are exploring the biomechanical relationships between organs and supporting structures. These models can assist with planning surgical and treatment interventions and reconstruction and determining optimal loading and boundary constraints during clinical procedures.

**Materials.** Materials used for medical and surgical procedures in humans and animals are being evaluated and some are being changed as biomechanical science is demonstrating that different materials, procedures, and techniques may be better for reducing complications and improving long-term patient health. Evaluation of the physical relationship between the body and foreign implements can quantify the stresses and forces on the body, allowing for more accurate prediction of patient outcomes and determination of which treatments should be redesigned.

**Predictability.** Medical professionals are particularly interested in the predictive value that biomechanical profiling can provide for their patients. An example is the unpredictability of expansion and rupture of an abdominal aortic aneurysm. Major progress has been made in determining aortic wall stress using finite element analysis. Improvements in biomechanical computational methodology and advances in imaging and processing technology have provided increased predictive ability for this life-threatening event.

As the need for accurate and efficient evaluation grows, so does the research and development of effective biomechanical tools. Capturing real-time, real-world data, such as with gait analysis and range of motion features, provides immediate opportunities for applications. This real-time data can quantify an injury and over time provide information about the extent that the injury has improved. High-tech devices can translate real-world situations and two-dimensional images into a three-dimensional framework for analysis. Devices, imaging, and modeling tools and software are making tremendous strides and becoming the heart of a highly competitive industry aimed at simplifying the process of analysis and making it less invasive.

## IMPACT ON INDUSTRY

Many companies have discovered the benefit of biomechanics in various facets of their operations,

## Fascinating Facts About Biomechanics

- Italian artist and scientist Leonardo da Vinci, born in the fifteenth century, called the foot "a masterpiece of engineering and a work of art."
- To take a step requires about two hundred muscles.
- When walking, one's feet bear a force of one and one-half times one's body weight; when running, the force on one's feet increases to three to four times one's body weight.
- During a one-mile walk, a person's feet strike the ground an average of 1,800 times.
- Just to overcome air resistance, a person running a 5,000-meter race uses 9 percent of his or her total energy expenditure, while a sprinter in a 100-meter race uses 20 percent.
- Pound for pound, bone is six times stronger than steel.
- The muscles that power the fingers are strong enough in some people to allow them to climb vertical surfaces by supporting their entire weight on a few fingertips.
- The muscles that bend the finger joints are located in the palm and in the mid-forearm and are connected to the finger bones by tendons, which pull on and move the fingers like the strings of a marionette.
- It is physically impossible for a person to lick his or her elbow.

including the development of products and of workplace procedures and practices. Most products that are made to assist or interact with people or any living being have probably been designed with the input of a biomechanical professional. Corporations want to protect their investment and profits by ensuring that their products will effectively meet the needs of the consumer and comply with strict safety standards. Biomechanics personnel work with product development engineers and designers to create new products and improve existing ones. Athletic equipment has been redesigned to produce better results with the same exertion of force. Two major sports products that have received international attention and led to world-record-breaking performances have been clap skates (used in speed skating) and the LZR Racer swimsuit, designed to reduce drag on swimmers.

Sporting equipment for athletes and the general public is constantly being redesigned to enhance performance and reduce the chance of injury. Sport-specific footwear is designed to maximize comfort and make it easier for athletes to perform the movements necessary for their sport. Equipment such as bicycles and golf clubs are designed using lighter and stronger materials, using optimal angles and maximizing strength to provide athletes with the best experience possible. Sports equipment goes a step further by analyzing the individual athlete and adjusting a piece of equipment specifically to his or her body. A small change to an angle or lever can produce dramatic results for an individual. This customization goes beyond sporting equipment to rehabilitation implements, wheelchairs, and prosthetic devices.

Biomechanics has a profound influence on many industries that are outside of sports. Most products used or handled on a daily basis have undergone biomechanical evaluation. Medicine bottle tops have been redesigned for easy opening by those with arthritic hands, and products from kitchen gadgets to automobiles have all been altered to improve comfort, safety, and effectiveness and to reduce the need for physical exertion.

Corporations are facing stricter regulations regarding the workplace environment. Safety and injury prevention are key to keeping productivity optimal. In a process called ergonomic assessment, equipment and workstations are biomechanically assessed, and adjustments are made that will limit acute or chronic injuries. Employees also receive instruction on the proper procedures to follow, such as lifting techniques. Industry procedures need to be in place and diligently followed to protect both employees and the company.

Accident litigation is becoming more common, and biomechanical science is playing a large role in accident re-creation and law-enforcement training. Investigations at accident or crime scenes can reveal more evidence, with greater accuracy than ever before, leaving less room for speculation when reconstructing the event.

### CAREERS AND COURSE WORK

Careers in biomechanics can be dynamic and can take many paths. Graduates with accredited degrees may pursue careers in laboratories in universities or in private corporations researching and developing ways of improving and maximizing human performance. Beyond research, careers in biomechanics can involve working in a medical capacity in sport medicine and rehabilitation. Biomechanics experts may also seek careers in coaching, athlete development, and education.

Consulting and legal practices are increasingly seeking individuals with biomechanics expertise who are able to analyze injuries and reconstruct accidents involving vehicles, consumer products, and the environment.

Biomechanical engineers commonly work in industry, developing new products and prototypes and evaluating their performance. Positions normally require a biomechanics degree in addition to mechanical or biomedical engineering degrees.

Private corporations are employing individuals with biomechanical knowledge to perform employee fitness evaluations and to provide analyses of work environments and positions. Using these assessments, the biomechanics experts advise employers of any ergonomic changes or job modifications that will reduce the risk of workplace injury.

Individuals with a biomechanics background may chose to work in rehabilitation and prosthetic design. This is very challenging work, devising and modifying existing implements to maximize people's abilities and mobility. Most prosthetic devices are customized to meet the needs of the patient and to maximize the recipient's abilities. This is an ongoing process because over time the body and needs of a patient may change. This is particularly challenging in pediatrics, where adjustments become necessary as a child grows and develops.

## SOCIAL CONTEXT AND FUTURE PROSPECTS

Biomechanics has gone from a narrow focus on athletic performance to become a broad-based science, driving multibillion dollar industries to satisfy the needs of consumers who have become more knowledgeable about the relationship between science, health, and athletic performance. Funding for biomechanical research is increasingly available from national health promotion and injury prevention programs, governing bodies for sport, and business and industry. National athletic programs want to ensure that their athletes have the most advanced training methods, performance analysis methods,

and equipment to maximize their athletes' performance at global competitions.

Much of the existing and developing technology is focused on increasingly automated and digitized systems to monitor and analyze movement and force. The physiological aspect of movement can be examined at a microscopic level, and instrumented athletic implements such as paddles or bicycle cranks allow real-time data to be collected during an event or performance. Force platforms are being reconfigured as starting blocks and diving platforms to measure reaction forces. These techniques for biomechanical performance analysis have led to revolutionary technique changes in many sports programs and rehabilitation methods.

Advances in biomechanical engineering have led to the development of innovations in equipment, playing surfaces, footwear, and clothing, allowing people to reduce injury and perform beyond previous expectations and records.

Computer modeling and virtual simulation training can provide athletes with realistic training opportunities, while their performance is analyzed and measured for improvement and injury prevention.

*April D. Ingram, B.Sc.*

## FURTHER READING

Hamill, Joseph, and Kathleen Knutzen. *Biomechanical Basis of Human Movement.* 3d ed. Philadelphia: Lippincott, Williams & Wilkins, 2009. This introductory text integrates basic anatomy, physics, and physiology as it relates to human movement. It also includes real-life examples and clinically relevant material.

Hatze, H. "The Meaning of the Term 'Biomechanics.'" *Journal of Biomechanics* 7, no. 2 (March, 1974): 89-90. Contains Hatze's definition of biomechanics, then an emerging field.

Hay, James G. *The Biomechanics of Sports Techniques.* 4th ed. Englewood Cliffs, N.J.: Prentice Hall, 1993. A seminal work by an early biomechanics expert, first published in 1973.

Kerr, Andrew. *Introductory Biomechanics.* London: Elsevier, 2010. Provides a clear, basic understanding of major biomechanical principles in a workbook style interactive text.

Peterson, Donald, and Joseph Bronzino. *Biomechanics: Principles and Applications.* Boca Raton,

Fla.: CRC Press, 2008. A broad collection of twenty articles on various aspects of research in biomechanics.

Watkins, James. *Introduction to Biomechanics of Sport and Exercise*. London: Elsevier, 2007. An introduction to the fundamental concepts of biomechanics that develops knowledge from the basics. Many applied examples, illustrations, and solutions are included.

**WEB SITES**

*American Society of Biomechanics*
http://www.asbweb.org

*European Society of Biomechanics*
http://www.esbiomech.org

*International Society of Biomechanics*
http://www.isbweb.org

*World Commission of Science and Sports*
http://www.wcss.org.uk

**See also:** Bioengineering; Biophysics; Dentistry; Ergonomics; Forensic Science; Kinesiology; Orthopedics; Prosthetics; Rehabilitation Engineering; Sports Engineering.

# BIONICS AND BIOMEDICAL ENGINEERING

## FIELDS OF STUDY

Biology; physiology; biochemistry; engineering; orthopedic bioengineering; physics; bionanotechnology; biomechanics; biomaterials; neural engineering; genetic engineering; tissue engineering; prosthetics.

## SUMMARY

Bionics combines natural biologic systems with engineered devices and electrical mechanisms. An example of bionics is an artificial arm controlled by impulses from the human mind. Construction of bionic arms or similar devices requires the integrative use of medical equipment such as electroencephalograms (EEGs) and magnetic resonance imaging (MRI) machines with mechanically engineered prosthetic arms and legs. Biomedical engineering further melds biomedical and engineering sciences by producing medical equipment, tissue growth, and new pharmaceuticals. An example of biomedical engineering is human insulin production through genetic engineering to treat diabetes.

## KEY TERMS AND CONCEPTS

- **Biologics:** Medicines produced from genes by manipulating genes and using genetic technology.
- **Biomaterials:** Substances, including metal alloys, plastic polymers, and living tissues, used to replace body tissues or as implants.
- **Bionanotechnology:** Construction of materials on a very small scale, enabling the use of microscopic machinery in living tissues.
- **Bionic:** Integrating biological function and mechanical devices.
- **Clone:** Genetically engineered organism with genetic composition identical to the original organism.
- **Human Genetic Engineering:** Genetic engineering focused on altering or changing visible human characteristics through gene manipulations.
- **Prosthesis:** Artificial or biomechanically engineered body part.
- **Recombinant DNA:** DNA created by the combination of two or more DNA sequences that do not normally occur together.

## DEFINITION AND BASIC PRINCIPLES

The fields of biomedical engineering and bionics focus on improving health, particularly after injury or illness, with better rehabilitation, medications, innovative treatments, enhanced diagnostic tools, and preventive medicine.

Bionics has moved nineteenth-century prostheses, such as the wooden leg, into the twenty-first century by using plastic polymers and levers. Bionics integrates circuit boards and wires connecting the nervous system to the modular prosthetic limb. Controlling artificial limb movements with thoughts provides more lifelike function and ability. This mind and prosthetic limb integration is the "bio" portion of bionics; the "nic" portion, taken from the word "electronic," concerns the mechanical engineering that makes it possible for the person using a bionic limb to increase the number and range of limb activity, approaching the function of a real limb.

Biomedical engineering encompasses many medical fields. The principle of adapting engineering techniques and knowledge to human structure and function is a key unifying concept of biomedical engineering. Advances in genetic engineering have produced remarkable bioengineered medications. Recombinant DNA techniques (genetic engineering) have produced synthetic hormones, such as insulin. Bacteria are used as a host for this process; once human-insulin-producing genes are implanted in the bacteria, the bacteria's DNA produce human insulin, and the human insulin is harvested to treat diabetics. Before this genetic technique was developed in 1982 to produce human insulin, insulin-dependent diabetics relied on insulin from pigs or cows. Although this insulin was life saving for diabetics, diabetics often developed problems from the pig or cow insulin because they would produce antibodies against the foreign insulin. This problem disappeared with the ability to engineer human insulin using recombinant DNA technology.

## BACKGROUND AND HISTORY

In the broad sense, biomedical engineering has existed for millennia. Human beings have always envisioned the integration of humans and technology to increase and enhance human abilities. Prosthetic devices go back many thousands of years: A three-thousand-year-old Egyptian mummy, for example, was found with a wooden big toe tied to its foot. In the fifteenth century, during the Italian Renaissance, Leonardo da Vinci's elegant drawings demonstrated some early ideas on bioengineering, including his helicopter and flying machines, which melded human and machine into one functional unit capable of flight. Other early examples of biomedical engineering include wooden teeth, crutches, and medical equipment, such as stethoscopes.

Electrophysiological studies in the early 1800's produced biomedical engineering information used to better understand human physiology. Engineering principles related to electricity combined with human physiology resulted in better knowledge of the electrical properties of nerves and muscles.

X rays, discovered by Wilhelm Conrad Röntgen in 1895, were an unknown type of radiation (thus the "X" name). When it was accidentally discovered that they could penetrate and destroy tissue, experiments were developed that led to a range of imaging technologies that evolved over the next century. The first formal biomedical engineering training program, established in 1921 at Germany's Oswalt Institute for Physics in Medicine, focused on three main areas: the effects of ionizing radiation, tissue electrical characteristics, and X-ray properties.

In 1948, the Institute of Radio Engineers (later the Institute of Electrical and Electronics Engineers), the American Institute for Electrical Engineering, and the Instrument Society of America held a conference on engineering in biology and medicine. The 1940's and 1950's saw the formation of professional societies related to biomedical engineering, such as the Biophysics Society, and of interest groups within engineering societies. However, research at the time focused on the study of radiation. Electronics and the budding computer era broadened interest and activities toward the end of the 1950's.

James D. Watson and Francis Crick identified the DNA double-helix structure in 1953. This important discovery fostered subsequent experimentation in molecular biology that yielded important information about how DNA and genes code for the expression of traits in all living organisms. The genetic code in DNA was deciphered in 1968, arming researchers with enough information to discover ways that DNA could be recombined to introduce genes from one organism into a different organism, thereby allowing the host to produce a variety of useful products. DNA recombination became one of the most important tools in the field of biomedical engineering, leading to tissue growth as well as new pharmaceuticals.

In 1962, the National Institutes of Health created the National Institute of General Medical Sciences, fostering the development of biomedical engineering programs. This institute funds research in the diagnosis, treatment, and prevention of disease.

Bionics and biomedical engineering span a wide variety of beneficial health-related fields. The common

## Fascinating Facts About Bionics and Biomedical Engineering

- In 2010, scientists at the University of California, San Diego, developed biosensor cells that can be implanted in the brain to help monitor receptors and chemical signals that allow cells in the brain to communicate with one another. These cells may help scientists understand drug addiction.

- Vanderbilt engineers in 2010 began testing a knowledge repository and interactive software that will help surgeons more accurately and rapidly place electrodes in the brains of people with Parkinson's disease in a procedure called deep brain stimulation. The data allow for faster surgery and the implementation of best practices.

- In 2010, scientists at Vanderbilt University developed a robotic prosthesis for the lower leg that has powered knee and ankle joints. Intent recognizer software takes information from sensors, determines what the user wants to do, and provides power to the leg.

- The Wadsworth Center at the New York State Department of Health in 2009 developed a brain-computer interface that translates brain waves into action. It allowed a patient with amyotrophic lateral sclerosis who was no longer able to communicate with others because of failing muscles to write e-mails and convey his thoughts to others.

- In 2008, a research team at the Johns Hopkins University modified chondroitin sulfate, a natural sugar, so it could glue a hydrogel (like the material used in soft contact lenses) to cartilage tissue. It is hoped that this technique may help those experiencing joint pain from osteoarthritis, in which the natural cartilage in a joint disappears.

thread is the combination of technology with human applications. Dolly the sheep was cloned in 1996. Cloning produces a genetically identical copy of an existing life-form. Human embryonic cloning presents the potential of therapeutic reproduction of needed organs and tissues, such as kidney replacement for patients with renal failure.

In the twenty-first century, the linking of machines with the mind and sensory perception has provided hearing for deaf people, some sight for the blind, and willful control of prostheses for amputees.

## HOW IT WORKS

Restorative bionics integrates prosthetic limbs with electrical connections to neurons, allowing an individual's thoughts to control the artificial limb. Tiny arrays of electrodes attached to the eye's retina connect to the optic nerve, enabling some visual perception for previously blind people. Deaf people hear with electric devices that send signals to auditory nerves, using antennas, magnets, receivers, and electrodes. Researchers are considering bionic skin development using nanotechnology to connect with nerves, enabling skin sensations for burn victims requiring extensive grafting.

Many biomedical devices work inside the human body. Pacemakers, artificial heart valves, stents, and even artificial hearts are some of the bionic devices correcting problems with the cardiovascular system. Pacemakers generate electric signals that improve abnormal heart rates and abnormal heart rhythms. When pulse generators located in the pacemakers sense an abnormal heart rate or rhythm, they produce shocks to restore the normal rate. Stents are inserted into an artery to widen it and open clogged blood vessels. Stents and pacemakers are examples of specialized bionic devices made up of bionic materials compatible with human structure and function.

**Cloning.** Cloning is a significant area of genetic engineering that allows the replication of a complete living organism by manipulating genes. Dolly the sheep, an all-white Finn Dorset ewe, was cloned from a surrogate mother blackface ewe, which was used as an egg donor and carried the cloned Dolly during gestation (pregnancy). An egg cell from the surrogate was removed and its nucleus (which contains DNA) was replaced with one from a Finn Dorset ewe; the resulting new egg was placed in the blackface ewe's uterus after stimulation with an electric pulse.

*ELegs are a wearable artificially intelligent bionic device that powers paraplegics to get them standing and walking.* (AP Photo)

The electrical pulse stimulated growth and cell duplication. The blackface ewe subsequently gave birth to the all-white Dolly. The newborn all-white Finn Dorset ewe was an identical genetic twin of the Finn Dorset that contributed the new nucleus.

**Recombinant DNA.** Another significant genetic engineering technique involves recombinant DNA. Human genes transferred to host organisms, such as bacteria, produce products coded for by the transferred genes. Human insulin and human growth hormone can be produced using this technique. Desired genes are removed from human cells and placed in circular bacterial DNA strips called plasmids. Scientists use enzymes to prepare these DNA formulations, ultimately splicing human genes into bacterial plasmids. These plasmids are used as vectors, taken up and reproduced by bacteria. This type of genetic adaptation results in insulin production if the spliced genes were taken from the part of the human genome producing insulin; other cells and substances, coded for by different human genes, can be produced this way. Many biologic medicines are produced using recombinant DNA technology.

## APPLICATIONS AND PRODUCTS

**Medical Devices.** Biomedical engineers produce life-saving medical equipment, including pacemakers, kidney dialysis machines, and artificial hearts. Synthetic limbs, artificial cochleas, and bionic sight chips are among the prosthetic devices that biomedical engineers have developed to enhance mobility, hearing, and vision. Medical monitoring devices, developed by biomedical engineers for use in intensive care units and surgery or by space and deep-sea explorers, monitor vital signs such as heart rate and rhythm, body temperature, and breathing rate.

**Equipment and Machinery.** Biomedical engineers produce a wide variety of other medical machinery, including laboratory equipment and therapeutic equipment. Therapeutic equipment includes laser devices for eye surgery and insulin pumps (sometimes called artificial pancreases) that both monitor blood sugar levels and deliver the appropriate amount of insulin when it is needed.

**Imaging Systems.** Medical imaging provides important machinery devised by biomedical engineers. This specialty incorporates sophisticated computers and imaging systems to produce computed tomography (CT), magnetic resonance imaging (MRI), and positron emission tomography (PET) scans. In naming its National Institute of Biomedical Imaging and Bioengineering (NIBIB), the U.S. Department of Health and Human Services emphasized the equal importance and close relatedness of these subspecialties by using both terms in the department's name.

Computer programming provides important circuitry for many biomedical engineering applications, including systems for differential disease diagnosis. Advances in bionics, moreover, rely heavily on computer systems to enhance vision, hearing, and body movements.

**Biomaterials.** Biomaterials, such as artificial skin and other genetically engineered body tissues, are areas promising dramatic improvements in the treatment of burn victims and individuals needing organ transplants. Bionanotechnology, another subfield of biomedical engineering, promises to enhance the surface of artificial skin by creating microscopic messengers that can create the sensations of touch and pain. Bioengineers interface with the fields of physical therapy, orthopedic surgery, and rehabilitative medicine in the fields of splint development, biomechanics, and wound healing.

**Medications.** Medicines have long been synthesized artificially in laboratories, but chemically synthesized medicines do not use human genes in their production. Medicines produced by using human genes in recombinant DNA procedures are called biologics and include antibodies, hormones, and cell receptor proteins. Some of these products include human insulin, the hepatitis B vaccine, and human growth hormone.

Bacteria and viruses invading a body are attacked and sometimes neutralized by antibodies produced by the immune system. Diseases such as Crohn's disease, an inflammatory bowel condition, and psoriatic arthritis are conditions exacerbated by inflammatory antibody responses mounted by the affected person's immune system. Genetic antibody production in the form of biologic medications interferes with or attacks mediators associated with Crohn's and arthritis and improves these illnesses by decreasing the severity of attacks or decreasing the frequency of flare-ups.

**Cloning and Stem Cells.** Cloned human embryos could provide embryonic stem cells. Embryonic stem cells have the potential to grow into a variety of cells, tissues, and organs, such as skin, kidneys, livers, or heart cells. Organ transplantation from genetically identical clones would not encounter the recipient's natural rejection process, which transplantations must overcome. As a result, recipients of genetically identical cells, tissues, and organs would enjoy more successful replacements of key organs and a better quality of life. Human cloning is subject to future research and development, but the promise of genetically identical replacement organs for people with failed hearts, kidneys, livers, or other organs provides hope for enhanced future treatments.

## IMPACT ON INDUSTRY

**Government and University Research.** The National Institute of Biomedical Imaging and Bioengineering was established in 2001. It is dedicated to health improvement by developing and applying biomedical technologies. The institute supports research with grants, contracts, and career development awards. Many major universities around the world actively research and develop bionics and biomedical engineering. Top engineering schools, such as the University of Michigan, have undergraduate and graduate degree programs for biomedical engineering. Biomedical engineers work in industries producing medical equipment and pharmaceutical companies. They also conduct research and develop genetic therapies useful for treating a wide variety of illnesses.

**Industry and Business Sectors.** The pharmaceutical industry, along with the medical equipment industry, employs biomedical engineers. The medical industry, including the suppliers of laboratory equipment, imaging technology, bionics, and pharmaceuticals, drives much of the employment demand for biomedical engineers. Medical hardware, such as artificial hearts, pacemakers, and renal dialysis machines, remains an important part of biomedical engineering employment.

## CAREERS AND COURSE WORK

According to the 2010-2011 edition of the *Occupational Outlook Handbook*, issued by the U.S. Department of Labor's Bureau of Labor Statistics, biomedical engineers are projected to have a 72 percent employment growth from 2010 to 2020, a rate of increase much greater than the average for all occupations. Biomedical engineers are employed as researchers and scientists, interfacing with a wide variety of disciplines and specialties.

Many career paths exist in biomedical engineering and bionics. Jobs exist for those holding bachelor's degrees through doctorates. Research scientists, usually with Ph.D. or M.D. degrees, can work in a variety of environments, from private companies to universities to government agencies, including the National Institutes of Health.

A typical undergraduate biomedical engineering curriculum for University of Michigan students, for example, includes three divisions of course study: subjects required by all engineering programs, advanced science and engineering mathematics, and biomedical engineering courses. All engineering students take courses in calculus, basic engineering concepts, computing, chemistry, physics, and humanities or social sciences. Required courses in advanced sciences in the second division include biology, chemistry, and biological chemistry, along with engineering mathematics courses. Biomedical engineering courses in the third division cover circuits and electrical systems, biomechanics, engineering materials, cell biology, physiology, and biomedical design. Students can concentrate in areas such as biomechanical, biochemical, and bioelectric engineering, with some modifications in course selection. The breadth and depth of the course work emphasize the link between engineering and biologic sciences—the essence of bionics and biomedical engineering.

A biomedical engineer is an engineer developing and advancing biomedical products and systems. This type of activity spans many specialties and presents many career opportunities. Biomedical engineering interfaces with almost every engineering discipline because bioengineered products require the breadth and depth of engineering knowledge. Major areas include the production of biomaterials, such as the type of plastic used for prosthetic devices. Bioinstrumentation involves computer integration in diagnostic machines and the control of devices. Genetic engineering presents many opportunities for life modifications. Clinical engineers help integrate new technologies, such as electronic medical records, into existing hospital systems. Medical imaging relieves the need for exploratory surgery and greatly enhances diagnostic capabilities. Orthopedic bioengineering plays important roles in prosthesis development and use, along with assisting rehabilitative medicine. Bionanotechnology offers the hope of using microscopic messengers to treat illness and advanced capabilities for artificial bionic devices. Systems physiology organizes the multidisciplinary approach necessary to complete complex bionic projects such as a functioning artificial eye.

## SOCIAL CONTEXT AND FUTURE PROSPECTS

Bionics technologies include artificial hearing, sight, and limbs that respond to nerve impulses. Bionics offers partial vision to the blind and prototype prosthetic arm devices that offer several movements through nerve impulses. The goal of bionics is to better integrate the materials in these artificial devices with human physiology to improve the lives of those with limb loss, blindness, or decreased hearing.

Cloned animals exist but cloning is not a yet a routine process. Technological advances offer rapid DNA analysis along with significantly lower cost genetic analysis. Genetic databases are filled with information on many life-forms, and new DNA sequencing information is added frequently. This basic information that has been collected is like a dictionary, full of words that can be used to form sentences, paragraphs, articles, and books, in that it can be used to create new or modified life-forms.

Biomedical engineering enables human genetic engineering. The stuff of life, genes, can be modified or manipulated with existing genetic techniques. The power to change life raises significant societal concerns and ethical issues. Beneficial results such as optimal organ transplantations and effective medications are the potential of human genetic engineering.

*Richard P. Capriccioso, B.S., M.D.,*
*and Christina Capriccioso*

## FURTHER READING

Braga, Newton C. *Bionics for the Evil Genius: Twenty-five Build-It-Yourself Projects.* New York: McGraw-Hill, 2006. Step-by-step projects that introduce basic concepts in bionics.

Fischman, Josh. "Merging Man and Machine: The Bionic Age." *National Geographic* 217, no. 1 (January, 2010): 34-53. A well-illustrated consideration of the latest advances in bionics, with specific examples of people aided by the most modern prosthetic technologies.

Hung, George K. *Biomedical Engineering: Principles of the Bionic Man.* Hackensack, N.J.: World Scientific, 2010. Examines scientific bioengineering principles as they apply to humans.

Richards-Kortum, Rebecca. *Biomedical Engineering for Global Health.* New York: Cambridge University Press, 2010. Examines the potential of biomedical engineering to treat diseases and conditions throughout the world. Examines health care systems and social issues.

Smith, Marquard, and Joanne Morra, eds. *The Prosthetic Impulse: From a Posthuman Present to a Biocultural Future.* Cambridge, Mass.: MIT Press, 2007. Examines the developments in prosthetic devices and addresses the social aspects, including what it means to be human.

## WEB SITES

*American Society for Artificial Internal Organs*
http://www.asaio.com

*Biomedical Engineering Society*
http://www.bmes.org

*National Institute of Biomedical Imaging and Bioengineering*
http://www.nibib.nih.gov/HomePage

*Rehabilitation Engineering and Assistive Technology Association of North America*
http://resna.org

*Society for Biomaterials*
http://www.biomaterials.org

**See also:** Artificial Organs; Biosynthetics; Cloning; DNA Sequencing; Genetic Engineering; Human Genetic Engineering; Orthopedics; Prosthetics; Rehabilitation Engineering; Stem Cell Research and Technology.

# BIOPHYSICS

## FIELDS OF STUDY

Physics; physical sciences; chemistry; mathematics; biology; molecular biology; chemical biology; engineering; biochemistry; classical genetics; molecular genetics; cell biology.

## SUMMARY

Biophysics is the branch of science that uses the principles of physics to study biological concepts. It examines how life systems function, especially at the cellular and molecular level. It plays an important role in understanding the structure and function of proteins and membranes and in developing new pharmaceuticals. Biophysics is the foundation for molecular biology, a field that combines physics, biology, and chemistry.

## KEY TERMS AND CONCEPTS

- **Circular Dichroism (CD):** Differential absorption of left- and right-handed circularly polarized light.
- **Electromagnetic Waves:** Waves that can transmit their energy through a vacuum.
- **Molecular Genetics:** Branch of genetics that analyzes the structure and function of genes at the molecular level.
- **Polarized Light:** Light waves that vibrate in a single plane.
- **Quantum Mechanics:** Physical analysis at the level of atoms or subatomic fundamental particles.
- **Thermodynamics:** Branch of physics that studies energy conversions.
- **Vector:** Quantity that has both magnitude and direction.

### DEFINITION AND BASIC PRINCIPLES

The word "biophysics" means the physics of life. Biophysics studies the functioning of life systems, especially at the cellular and molecular level, using the principles of physics. It is known that atoms make up molecules, molecules make up cells, and cells in turn make up tissues and organs that are part of an organism, or a living machine. Biophysicists use this knowledge to understand how the living machine works.

In photosynthesis, for instance, the absorption of sunlight by green plants initiates a process that culminates with synthesis of high-energy sugars such as glucose. To fully understand this process, one needs to look at how it begins—light absorption by the photosystems. Photosystems are groups of energy-absorbing pigments such as chlorophyll and carotenoids that are located on the thylakoid membranes inside the chloroplast, the photosynthetic organelle in the plant cell. Biophysical studies have shown that once a chlorophyll molecule captures solar energy, it gets excited and transfers the energy to a neighboring unexcited chlorophyll molecule. The process repeats itself, and thus, packets of energy jump from one chlorophyll molecule to the next. The energy eventually reaches the reaction center, where it begins a chain of high-energy electron-transfer reactions that lead to the storage of the light energy in the form of adenosine triphosphate (ATP) and nicotinamide adenine dinucleotide phosphate (NADPH). In the second half of photosynthesis, ATP and NADPH provide the energy to make glucose from carbon dioxide.

Biophysics is often confused with medical physics. Medical physics is the science devoted to studying the relationship between human health and radiation exposure. For example, a medical physicist often works closely with a radiation oncologist to set up radiotherapy treatment plans for cancer patients.

### BACKGROUND AND HISTORY

In comparison with other branches of biology and physics, biophysics is relatively new and, therefore, still evolving. Even though the use of physical concepts and instrumentation to explain the workings of life systems had begun as early as the 1840's, biophysics did not emerge as an independent field until the 1920's. Some of the earliest studies in biophysics were conducted in the 1840's by a group known as the Berlin school of physiologists. Among its members were pioneers such as Hermann von Helmholtz, Ernst Heinrich Weber, Carl F. W. Ludwig, and Johannes Peter Müller. This group used well-known physical methods to investigate physiological issues, such as the mechanics of muscular contraction and the electrical changes in a nerve cell during impulse transmission. The first biophysics textbook was written in 1856 by Adolf Fick, a student of Ludwig. Although these early biophysicists made significant advances, subsequent research focused on other areas.

In the 1920's, the first biophysical institutes were established in Germany and the first textbook with the word "biophysics" in its title was published. However, through the 1940's, biophysics research was primarily aimed at understanding the biophysical impact of ionizing radiation. In 1944, Austrian physicist Erwin Schrödinger published *What Is Life ? The Physical Aspect of the Living Cell*, based on a series of lectures that addressed biology from the viewpoint of a classical physicist. This cross-disciplinary work motivated several physicists to become interested in biology and thus laid the foundation for the field of molecular biology. From 1950 to 1970, the field of biophysics experienced rapid growth, tremendously accelerated by the discovery in 1953 of the double helix structure of DNA by James D. Watson and Francis Crick. Both Watson and Crick have stated that they were inspired by Schrödinger's work.

## HOW IT WORKS

Biophysicists study life at all levels, from atoms and molecules to cells, organisms, and environments. They attempt to describe complex living systems with the simple laws of physics. Often, biophysicists work at the molecular level to understand cells and their processes.

The work of Gregor Mendel in the late nineteenth century laid the foundation for genetics, the science of heredity. His studies, rediscovered in the twentieth century, led to the understanding that the inheritance of certain traits is governed by genes and that the alleles of the genes are separated during gamete formation. Experiments in the 1940's revealed that genes are made of DNA, but the mechanisms by which genes function remained a mystery. Watson and Crick's discovery of the double helix structure of DNA in 1953 revealed how genes could be translated into proteins.

Biophysicists use a number of physical tools and techniques to understand how cellular processes work, especially at the molecular level. Some of the important tools are electron microscopy, nuclear magnetic resonance (NMR) spectroscopy, circular dichroism (CD) spectroscopy, and X-ray crystallography. For example, the discovery of Watson and Crick's double helix model was possible in part because of the X-ray images of DNA that were taken by Rosalind Franklin and Maurice H. F. Wilkins. Franklin and Wilkins, both biophysicists, made DNA crystals and then used X-ray crystallography to analyze the structure of DNA. The array of black dots arranged in an X-shaped pattern on the X-ray photograph of wet DNA suggested to Franklin that DNA was helical.

**Electron Microscopy.** Electron microscopes use beams of electrons to study objects in detail. Electron microscopy can be used to analyze an object's surface texture (topography) and constituent elements and compounds (composition), as well as the shape and size (morphology) and atomic arrangements (crystallographic details) of those elements and compounds. Electron microscopes were invented to overcome the limitations posed by light microscopes, which have maximum magnifications of 500x or 1000x and a maximum resolution of 0.2 millimeter. To see and study subcellular structures and processes required magnification capabilities of greater than 10,000x. The first electron microscope was a transmission electron microscope (TEM) built by Max Knoll and Ernst Ruska in 1931. The invention of the scanning electron microscope (SEM) was somewhat delayed (the first was built in 1937 by Manfred von Ardenne) because the field had to figure out how to make the electron beam scan the sample.

**NMR Spectroscopy.** Nuclear magnetic resonance (NMR) spectroscopy is an extremely useful tool for the biophysicist to study the molecular structure of organic compounds. The underlying principle of NMR spectroscopy is identical to that of magnetic resonance imaging (MRI), a common tool in medical diagnostics. The nuclei of several elements, including the isotopes carbon-12 and oxygen-16, have a characteristic spin when placed in an external magnetic field. NMR focuses on studying the transitions between these spin states. In comparison with mass spectroscopy, NMR requires a larger amount of sample, but it does not destroy the sample.

**CD Spectroscopy.** Circular dichroism (CD) spectroscopy measures differences in how left-handed and right-handed polarized light is absorbed. These differences are caused by structural asymmetry. CD spectroscopy can determine the secondary and tertiary structure of proteins as well as their thermal stability. It is usually used to study proteins in solution.

**X-Ray Crystallography.** X rays are electromagnetic waves with wavelengths ranging from 0.02 to 100

## Fascinating Facts About Biophysics

- The concept of biophysics was first developed by ancient Greeks and Romans who were trying to analyze the basis of consciousness and perception.
- Human sight begins when the protein rhodopsin absorbs a unit of light called the quanta. This energy absorption triggers an enzymatic cascade that culminates in an amplified electric signal to the brain and enables vision.
- Crystallin, the lens protein, is made only in the lens of the human eye, and melanin, the skin pigment, is made only in skin cells, or melanocytes, even though all the cells in the human body have the genes to make crystallin and melanin.
- A technique called footprinting allows scientists to determine exactly where a protein binds on DNA and how much of the protein actually binds.
- Genes in human cells are selectively turned on and off by proteins called regulators. A defect in this regulatory mechanism can cause diseases such as cancer.

angstroms (Å). Even before X rays were discovered in 1895 by Wilhelm Conrad Röntgen, scientists knew that atoms in crystals were arranged in definite patterns and that a study of the angles therein could provide clues to the crystal structure. As is true of all forms of radiation, the wavelength of X rays is inversely proportional to its energy. Because the wavelength of X rays is smaller than that of visible light, X rays are powerful enough to penetrate most matter. As X rays travel through an object, they are diffracted by the atomic arrangements inside and thus provide a guideline for the electron densities inside the object. Analysis of this electron density data offers a glimpse into the internal structure of the crystal. As of 2010, about 90 percent of the structures in the Worldwide Protein Data Bank had been elucidated through X-ray crystallography.

### APPLICATIONS AND PRODUCTS

Biophysical tools and techniques have become extremely useful in many areas and fields. They have furthered research in protein crystallography, synthetic biology, and nanobiology, and allowed scientists to discover new pharmaceuticals and to study biomolecular structures and interactions and membrane structure

and transport. Biophysics and its related fields, molecular biology and genetics, are rapidly developing and are at the center of biomedical research.

**Biomolecular Structures.** Because structure dictates function in the world of biomolecules, understanding the structure of the biomolecule (with tools such as X-ray crystallography and NMR and CD spectroscopy), whether it is a protein or a nucleic acid, is critical to understanding its individual function in the cell. Proteins function as catalysts and bind to and regulate other downstream biomolecules. Their functional basis lies in their tertiary structure, or their three-dimensional form, and this function cannot be predicted from the gene sequence. The sequence of nucleotides in a gene can be used only to predict the primary structure, which is the amino acid sequence in the polypeptide. Once the structure-function relationship has been analyzed, the next step is to make mutants or knock out the gene via techniques such as ribonucleic acid interference (RNAi) and confirm loss of function. Subsequently, a literature search is performed to see if there are any known genetic disorders that are caused by a defect in the gene being studied. If so, the structure-function relationship can be examined for a possible cure.

**Membrane Structure and Transport.** In 1972, biologist S. J. Singer and his student Garth Nicolson conceived the fluid mosaic model of the plasma membrane. According to this model, the plasma membrane is a fluid lipid bilayer largely made up of phospholipids arranged in an amphipathic pattern, with the hydrophobic lipid tails buried inside and the hydrophilic phosphate groups on the exterior. The bilayer is interspersed with proteins, which help in cross-membrane transport. Because membranes control the import and export of materials into the cell, understanding membrane structure is key to coming up with ways to block transport of potentially harmful pathogens across the membrane.

Electron microscopes were used in the early days of membrane biology, but fluorescence and confocal microscopes have come to be used more frequently. The development of organelle-specific vital stains has rejuvenated interest in evanescent field (EF) microscopy because it permits the study of even the smallest of vesicles and the tracking of the movements of individual protein molecules.

**Synthetic Biology.** In the 2000's, the term "systems biology" became part of the field of life science, followed by the term "synthetic biology." To many people, these terms appear to refer to the same thing, but they do not, even though they are indeed closely related. While systems biology focuses on using a quantitative approach to study existing biological systems, synthetic biology concentrates on applying engineering principles to biology and constructing novel systems heretofore unseen in nature. Clearly, synthetic biology benefits immensely from research in systems and molecular biology. In essence, synthetic biology could be described as an engineering discipline that uses known, tested functional components (parts) such as genes, proteins, and various regulatory circuits in conjunction with modeling software to design new functional biological systems, such as bacteria that make ethanol from water, carbon dioxide, and light. The biggest challenge to synthetic biologists is the complexity of life-forms, especially higher eukaryotes such as humans, and the possible existence of unknown processes that can affect the synthetic biological systems.

**Drug Discovery.** In the pharmaceutical world, the initial task is to identify the aberrant protein, the one responsible for generating the symptoms in any disease or disorder. Once that is done, a series of biophysical tools are used to ensure that the target is the correct one. First, the identity of the protein is confirmed using techniques such as N-terminal sequencing and tandem mass spectroscopy (MS-MS). Second, the protein sample is tested for purity (which typically should be more than 95 percent) using methods such as denaturing sodium dodecyl sulfate polyacrylamide gel electrophoresis (SDS-PAGE). Third, the concentration of the protein sample is determined by chromogenic assays such as the Bradford or Lowry assay. The fourth and probably the most important test is that of protein functionality. This is typically carried out by either checking the ligand binding capacity of the protein (with biacore ligand binding assays) or by testing the ability of the protein to carry out its biological function. All these thermodynamic parameters need to be tested to develop a putative drug, one that could somehow correct or restrain the ramifications of the protein's malfunction.

**Nanobiology.** With the aid of biophysical tools and techniques, the field of biology has moved from organismic biology to molecular biology to nanobiology. To get a feel for the size of a nanometer, picture a strand of hair, then visualize a width that is 100,000 times thinner. Typically nanoparticles are about the size of either a protein molecule or a short DNA segment. Nanomedicine, or the application of technology that relies on nanoparticles to medicine, has become a popular area for research. In particular, the search for appropriate vectors to deliver drugs into the cells is an endless pursuit, especially in emerging therapeutic approaches such as RNA interference. Because lipid and polymer-based nanoparticles are extremely small, they are easily taken up by cells instead of being cleared by the body.

### IMPACT ON INDUSTRY

As one would expect, biophysics research worldwide has progressed faster in countries that traditionally have had a large base of physicists. The Max Planck Institute of Biophysics, one of the earliest pioneers in this field, was established in 1921 in Frankfurt, Germany, as the Institut für Physikalische Grundlagen der Medizin. The aim of the first director

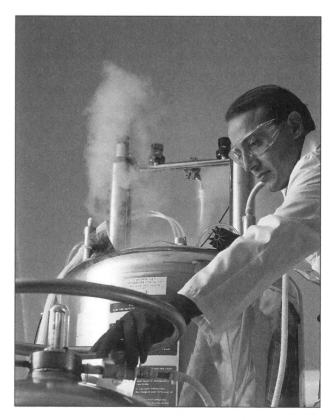

*A nuclear magnetic resonance spectrometer.* (Will and Deni McIntyre/Photo Researchers, Inc.)

director, Friedrich Dessauer, was to look for ways to apply the knowledge of radiation physics to medicine. He was followed by Boris Rajewsky, who coined the term "biophysics." In 1937, Rajewsky established the Kaiser Wilhelm Institute for Biophysics, which incorporated the institute led by Dessauer. The Max Planck Institute of Biophysics has become one of the world's foremost biophysics research institutes, with scientists and students analyzing a wide array of topics in biophysics such as membrane biology, molecular neurogenetics, and structural biology. In addition to Germany, countries active in biophysics research include the United States, Japan, France, Great Britain, Russia, China, India, and Sweden.

The International Union of Pure and Applied Biophysics (IUPAB) was created to provide a platform for biophysicists worldwide to exchange ideas and set up collaborations. The IUPAB in turn is a part of the International Council for Science (ICSU). The primary goal of IUPAB is to encourage and support students and researchers so that the field continues to grow and flourish. By 2010, the national biophysics societies of about fifty countries had become affiliated with the IUPAB. In the United States, the Biophysical Society was created in 1957 to facilitate propagation of biophysics concepts and ideas.

**Government and University Research.** To further broaden its mission, the Biophysical Society includes members from universities as well those in government research agencies such as the National Institutes for Health (NIH) and National Institute of Standards and Technology (NIST), many of whom are also part of the American Institute for Advancement of Sciences (AAAS) and the National Science Foundation (NSF). These members provide useful feedback to federal agencies such as the National Science and Technology Council, National Science Board, and the White House's Office of Science and Technology Policy, which are responsible for formulating national policies and initiatives.

**Industry and Business.** Most countries at the forefront of pharmaceutical breakthroughs have industries that are heavily invested in biophysical and biochemical research. The pharmaceutical industry has been spending billions of dollars to find treatments and cures for diseases and disorders that affect millions of people, including stroke, arthritis, cancer, heart disease, and neurological disorders. Biophysics is at the forefront of the field of drug discovery

because it provides the tools for conducting research in proteomics and genomics and allows the scientific community to identify opportunities for drug design. The next step after drug design is to plan the method of drug delivery, and biophysical research can help provide suitable vectors, including nanovectors. The pharmaceutical industry in the United States—companies such as Novartis, Eli Lilly, Bristol-Myers Squibb, and Pfizer—and the National Institutes of Health have a combined budget of about $60 billion per year. However, neither the industry nor the government support basic research at the interface of life sciences with physics and mathematics. Without this support, biophysics is unlikely to produce new tools to revolutionize or accelerate the ten-to-twelve-year drug development cycle. If this impediment can be overcome, the number of new drugs added to the market every year is likely to grow.

## CAREERS AND COURSE WORK

Few universities offer undergraduate majors in biophysics, but several universities offer graduate programs in biophysics. Students interested in pursuing a career in biophysics can major in molecular biology, physics, mathematics, or chemistry and supplement that with courses outside their major but relevant to biophysics. For example, a mathematics major would take supplementary courses in biology and vice versa. An ideal undergraduate curriculum for the future biophysicist would include courses in biology (genetics, cell biology, molecular biology), physics (thermodynamics, radiation physics, quantum mechanics), chemistry (organic, inorganic, and analytical), and mathematics (calculus, differential equations, computer programming, and statistics). The student should also have hands-on research experience, preferably as a research intern in a laboratory. To become an independent biophysicist, a graduate degree is required, usually a doctorate in biophysics, although some combine that with a medical degree. While in graduate school, the student should determine an area of research to pursue and engage in postdoctoral research for several years. Typically, this is the last step before one becomes an independent biophysicist running his or her own laboratory.

## SOCIAL CONTEXT AND FUTURE PROSPECTS

The discovery of the structure of DNA set off a revolution in molecular biology that has continued into the twenty-first century. In addition, modern scientific equipment has made study at the molecular level possible and productive. Many biophysicists, especially those who have also had course work in genetics and biochemistry, are working in molecular biology, which promises to be an active and exciting area for the foreseeable future.

Organisms are believed to be complex machines made of many simpler machines, such as proteins and nucleic acids. To understand why an organism behaves or reacts a certain way, one must determine how proteins and nucleic acids function. Biophysicists examine the structure of proteins and nucleic acids, seeking a correlation between structure and function. Once proper function is understood, scientists can prevent or treat diseases or disorders that result from malfunctions. This understanding of how proteins function enables scientists to develop pharmaceuticals and to find better means of delivering drugs to patients, and someday, this knowledge may allow scientists to design drugs specifically for a patient, thus avoiding many side effects. In addition, the scientific equipment developed by biophysicists in their research has been adapted for use in medical imaging for diagnosis and treatment. This transformation of laboratory equipment to medical equipment is likely to continue.

Biophysics applications have played and will continue to play a large role in medicine and health care, but future biophysicists may be environmental scientists. Biophysics is providing ways to improve the environment. For example, scientists are modifying microorganisms so that they produce electricity and biofuels that may lessen the need for fossil fuels. They are also using microorganisms to clean polluted water. As biophysics research continues, its applications are likely to cover an even broader range.

*Sibani Sengupta, B.S., M.S., Ph.D.*

## FURTHER READING

Bischof, Marco. "Some Remarks on the History of Biophysics and Its Future." In *Current Development of Biophysics*, edited by Changlin Zhang, Fritz Albert Popp, and Marco Bischof. Hangzhou, China: Hangzhou University Press, 1996. This paper delivered at a 1995 symposium on biophysics in Neuss, Germany, examines how the field of biophysics got its start and predicts future developments.

Claycomb, James R., and Jonathan Quoc P. Tran. *Introductory Biophysics: Perspectives on the Living State.* Sudbury, Mass.: Jones and Bartlett, 2011. This textbook considers life in relation to the universe. Contains a compact disc that allows computer simulation of biophysical phenomena. Relates biophysics to many other fields and subjects, including fractal geometry, chaos systems, biomagnetism, bioenergetics, and nerve conduction.

Glaser, Roland. *Biophysics.* 5th ed. New York: Springer, 2005. Contains numerous chapters on the molecular structure, kinetics, energetics, and dynamics of biological systems. Also looks at the physical environment, with chapters on the biophysics of hearing and on the biological effects of electromagnetic fields.

Goldfarb, Daniel. *Biophysics Demystified.* Maidenhead, England: McGraw-Hill, 2010. Examines anatomical, cellular, and subcellular biophysics as well as tools and techniques used in the field. Designed as a self-teaching tool, this work contains ample examples, illustrations, and quizzes.

Herman, Irving P. *Physics of the Human Body.* New York: Springer, 2007. Analyzes how physical concepts apply to human body functions.

Kaneko, K. *Life: An Introduction to Complex Systems Biology.* New York: Springer, 2006. Provides an introduction to the field of systems biology, focusing on complex systems.

## WEB SITES

*Biophysical Society*
http://www.biophysics.org

*International Union of Pure and Applied Biophysics*
http://iupab.org

*Worldwide Protein Data Bank*
http://www.wwpdb.org

**See also:** Applied Physics; Biofuels and Synthetic Fuels; Biosynthetics; DNA Analysis; DNA Sequencing; Genetic Engineering; Genomics; Human Genetic Engineering; Magnetic Resonance Imaging; Pharmacology; Radiology and Medical Imaging.

# BIOPROCESS ENGINEERING

## FIELDS OF STUDY

Biology; engineering; bioengineering; medicine; genetic engineering; molecular biology.

## SUMMARY

Bioprocess engineering is an interdisciplinary science that combines the disciplines of biology and engineering. It is associated primarily with the commercial exploitation of living things on a large scale. The objective of bioprocess engineering is to optimize either growth of organisms or the generation of target products. This is achieved mainly by the construction of controllable apparatuses. Both government agencies and private companies invest heavily in research within this area of applied science. Many traditional bioprocess engineering approaches (such as antibiotic production by microorganisms) have been advanced by techniques of genetic engineering and molecular biology.

## KEY TERMS AND CONCEPTS

- **Biomass:** Mass of organisms or organic material; traditionally refers to the biomass of plants and microorganisms.
- **Bioreactor:** Apparatus for growing microbial, plant, or animal cells, with a practical purpose under controlled conditions; these closed systems range from small (5- to 10-milliliter), laboratory-scale devices to larger, industrial-scale devices of more than 500,000 liters.
- **Bioremediation:** Use of living organisms to clean up the environment.
- **Enzymes:** Biological catalysts made of proteins.
- **Fermentation:** Metabolic reaction that is necessary to generate energy in microbial cells; used to produce many important compounds, such as alcohol and acetone.
- **Fermenter:** Type of traditional bioreactor (involving either stirred or nonstirred tanks) in which cell fermentation takes place; in continuous-culture fermenters, nutrients are continuously fed into the fermentation vessel so that cells can ferment indefinitely, whereas in batch fermenters, nutrients are added in batches.

## DEFINITION AND BASIC PRINCIPLES

Bioprocess engineering is the use of engineering devices (such as bioreactors) in biological processes carried out by microbial, plant, and animal cells in order to improve or analyze these processes. Large-scale manufacturing involving biological processes requires substantial engineering work. Throughout history, engineering has helped develop many bioprocesses, such as the production of antibiotics, biofuels, vaccines, and enzymes on an industrial scale. Bioprocess engineering plays a role in many industries, including the food, microbiological, pharmaceutical, biotechnological, and chemical industries.

## BACKGROUND AND HISTORY

People have been using bioprocessing for making bread, cheese, beer, and wine—all fermented foods—for thousands of years. Brewing was one of the first applications of bioprocess engineering. However, it was not until the nineteenth century that the scientific basis of fermentation was established, with the studies of French scientist Louis Pasteur, who discovered the microbial nature of beer brewing and wine making.

During the early part of the twentieth century, large-scale methods for treating wastewater were developed. Considerable growth in this field occurred toward the middle of the century, when the bioprocess for large-scale production of the antibiotic penicillin was developed. The World War II goal of industrial-scale production of penicillin led to the development of fermenters by engineers working together with biologists from the pharmaceutical company Pfizer. The fungus *Penicillium* grows and produces antibiotics much more effectively under controlled conditions inside a fermenter.

Later progress in bioprocess engineering has followed the development of genetic engineering, which raises the possibility of making new products from genetically modified microorganisms and plants grown in bioreactors. Just as past developments in bioprocess engineering have required contributions from a wide range of disciplines, including microbiology, genetics, biochemistry, chemistry, engineering, mathematics, and computer science, future developments are likely to require cooperation among scientists in multiple specialties.

## HOW IT WORKS

Living cells may be used to generate a number of useful products: food and food ingredients (such as cheese, bread, and wine), antibiotics, biofuels, chemicals (enzymes), and human health care products such as insulin. Organisms are also used to destroy or break down harmful wastes, such as those created by the 2010 oil spill in the Gulf of Mexico, or to reduce pollution.

A good example of how bioprocess engineering works is the development of a bioprocess using bacteria for industrial production of the human hormone insulin. Without insulin, which regulates blood sugar levels, the body cannot use or store glucose properly. The inability of the body to make sufficient insulin causes diabetes. In the 1970's, the U.S. company Genentech developed a bioprocess for insulin production using genetically modified bacterial cells.

The initial stages involve genetic manipulation (in this case, transferring a human gene into bacterial DNA). Genetic manipulation is done in laboratories by scientists trained in molecular biology or biochemistry. After creating a genetically engineered bacterium, scientists grow it in a small tubes or flasks and study its growth characteristics and insulin production.

Once the bacterial growth and insulin production characteristics have been identified, scientists increase the scale of the bioprocess. They use or build small bioreactors (1-10 liters) that can monitor temperature, pH (acidity-alkalinity), oxygen concentration, and other process characteristics. The goal of this scale-up is to optimize bacterial growth and insulin production.

The next step is another scale-up, this time to a pilot-scale bioreactor. These bioreactors can be as large as 1,000 liters and are designed and built by engineers to study the response of bacterial cells to large-scale production. During a scale-up, decreased product yields are often experienced because the conditions in the large-scale bioreactors (temperature, pH, aeration, and nutrient supply) differ from those in small, laboratory-scale systems. If the pilot-scale bioreactors work efficiently, engineers will design industrial-scale bioreactors and supporting facilities (air supply, sterilization, and process-control equipment).

All these stages are part of upstream processing. An important part of bioprocess engineering is the product recovery process, or so-called downstream processing. Product recovery from cells often can be very difficult. It involves laboratory procedures such as mechanical breakage, centrifugation, filtration, chromatography, crystallization, and drying. The final step in bioprocess engineering is testing of the recovered product, in which animals are often used.

## APPLICATIONS AND PRODUCTS

A wide range of products and applications of bioprocess engineering are familiar, everyday items.

**Foods, Beverages, Food Additives, and Supplements.** Living organisms play a major role in the production of food. Foods, beverages, additives, and supplements traditionally made by bioprocess engineering include dairy products (cheeses, sour cream, yogurt, and kefir), alcoholic beverages (beer, wines, and distilled spirits), plant products (soy sauce, tofu, sauerkraut), and food additives and supplements (flavors, proteins, vitamins, and carotenoids).

Traditional fermenters with microorganisms are used to obtain products in most of these applications. A typical industrial fermenter is constructed from stainless steel. Mixing of the microbial culture in fermenters is achieved by mechanical stirring, often with baffles. Airlift bioreactors have also been applied in the manufacturing of food products such as crude proteins synthesized by microorganisms. Mixing and liquid circulation in these bioreactors are induced by movement of an injected gas (such as air).

**Biofuels.** Bioprocess engineering is used in the production of biofuels, including ethanol (bioethanol), oil (biodiesel), butanol, biohydrogen, and biogas (methane). These biofuels are produced by the action of microorganisms in bioreactors, some of which use attached (immobilized) microorganisms. Cells, when immobilized in matrices such as agar, polyurethane, or glass beads, stabilize their growth and increase their physiological functions. Many microorganisms exist naturally in a state similar to immobilization, either on the surface of soil particles or in symbiosis with other organisms.

**Environmental Applications.** Bioprocess engineering plays an important role in removing pollution from the environment. It is used in treatment of wastewater and solid wastes, soil bioremediation, and mineral recovery. Environmental applications are based on the ability of organisms to use pollutants or other compounds as their food sources. One of the most important and widely used environmental applications is the treatment of wastewater by microorganisms. Microbes eat organic and inorganic compounds in wastewater and clean it at the same time. In this application, microorganisms are placed inside bioreactors (known as digesters) specifically designed by engineers. Engineers have also developed biofilters, bioreactors for removing pollutants from the air. Biofilters are used to remove pollutants, odors, and dust from air by the action of microorganisms. In addition, the mining industry uses bioprocess engineering for extracting minerals such as copper and uranium through the use of bacteria. Microbial leaching uses leaching dumps or tank bioreactors designed by engineers.

**Enzymes.** Enzymes are used in the health, food, laundry, pulp and paper, and textile industries. They are produced mainly from fungi and bacteria using bioprocess engineering. One of these enzymes is glucose isomerase, important in the production of fructose syrup. Genetic manipulation provides the means to produce many different enzymes, including those not normally synthesized by microorganisms. Fermenters for enzyme production are usually up to 100,000 liters in volume, although very expensive enzymes may be produced in smaller bioreactors, usually with immobilized cells.

**Antibiotics and Other Health Care Products.** Most antibiotics are produced by fungi and bacteria. Industrial production of antibiotics usually occurs in fermenters (stirred tanks) of 40,000- to 200,000-liter capacity. The bioprocess for antibiotics was developed by engineers during World War II, although it has undergone some changes since the 1980's. Various food sources, including glucose and sucrose, have been adopted for antibiotic production by microorganisms. The modern bioprocess is highly efficient (90 percent). Process variables such as pH and aeration are controlled by computer, and nutrients are fed continuously to sustain maximum antibiotic production. Product recovery is also based on continuous extraction.

The other major health care products produced with the help of bioprocess engineering are steroids, bacterial vaccines, gene therapy vectors, and therapeutic proteins such as interferon, growth hormone, and insulin. Steroids are important hormones that are manufactured by the process of biotransformation, in which microorganisms are used to chemically modify an inexpensive material to create a desired product. Health care products are produced in traditional fermenters.

**Biomass Production.** Biomass is used as a fuel source, as a source of protein for human food or animal feed, and as a component in agricultural pesticides or fertilizer. Baker's yeast biomass is a major product of bioprocess engineering. It is required for making bread and other baked goods, beer, wine, and ethanol. Yeast is produced in large aerated fermenters of up to 200,000 liters. Molasses is used as a nutrient source for the cells. Yeast is recovered from the fermentation liquid by centrifugation and then is dried. People also use the biomass of algae. Algae are a source of animal feed, plant fertilizer, chemicals, and biofuels. Algal biomass is produced in open ponds, in tubular glass, or in plastic bioreactors.

**Animal and Plant Cell Cultures.** Bioprocess engineering incorporating animal cell culture is used primarily for the production of health care products such as viral vaccines or antibodies in traditional fermenters or bioreactors with immobilized cells. Antibodies, for example, are produced in bioreactors with hollow-fiber immobilized animal cells. Plant cell culture is also an important target of bioprocess engineering. However, only a few processes have been successfully developed. One successful process is the production of the pigment shikonin in Japan. Shikonin is used as a dye for coloring food and has applications as an anti-inflammatory agent.

**Chemicals.** There is an on-going trend in the chemical industry to use bioprocess engineering instead of pure chemistry for production of a variety of chemicals such as amino acids, polymers, and organic acids (citric, acetic, and lactic). Some of these chemicals (citric and lactic acids) are used as food preservatives. Many chemicals are produced in traditional fermenters by the action of microbes.

### IMPACT ON INDUSTRY

Bioprocess engineering plays a major role in many multibillion-dollar industries, including biotechnological, microbiological, food, chemical, and biofuel industries. Because bioprocess engineering is part of several industries, it is difficult to estimate its worldwide revenues. However, global revenues from enzyme production are more than $3 billion, and biofuel industry annual revenues are $46.5 billion.

The United States maintains a dominant global position in a number of industries because of advances created by bioprocess engineering research. The same is true for many European countries, as well as Japan, Israel, Canada, and Australia. Bioprocess engineering has affected industry in many developing countries as well. A good example is Brazil's use of bioprocessing in ethanol production. Brazil, the world's largest ethanol producer for 2010, ferments sugarcane in bioreactors to generate ethanol.

Much of the success of bioprocess engineering is because of the hard work of scientists, engineers, and technicians who have spent countless hours working to improve biological processes to increase the yields of desired products. Scientists are trying to create powerful microorganisms (superbugs) and also working to improve the efficiency of existing production processes and to make them more environmentally friendly. Scientists are studying biochemical processes and enzymes to develop new bioprocesses. In addition, engineers and scientists are designing and developing new types of bioreactors and fermenters.

**Government and University Research.** Many government agencies such as the U.S. Department of Energy, the National Science Foundation, and the U.S. Department of Agriculture provide funding for research in bioprocess engineering. The Department of Energy has several national laboratories that are involved in bioprocess engineering research. The vast majority of research is on biofuel generation by microorganisms and on environmental applications.

**Industry and Business Sectors.** Scientists employed by industry traditionally perform most of the research in bioprocess engineering. A significant proportion of the research in industry has been directed to health care or medical products (such as vaccines) and biofuels.

### CAREERS AND COURSE WORK

There is an increasing demand for students trained in bioprocess engineering who can convert new discoveries in biology into industrial applications. There are many career options for young specialists in bioprocess engineering. Their work may be in the areas of biological process development, manufacturing operations, environmental bioremediation, food technology, or therapeutic stem cell

---

## Fascinating Facts About Bioprocess Engineering

- Citric acid, a common supplement of soft drinks, is a major product of bioprocess engineering. It is produced in fermenters by the common mold *Aspergillus niger.*
- Bioprocess engineering is used to recover gold from gold ores. The bioprocess uses bacteria in bioreactors to attack ores, releasing the trapped gold.
- Most insecticides are produced by the genetically modified bacterium *Bacillus thuringiensis* in bioreactors.
- One of the bacteria commonly used to produce antibodies is *Escherichia coli,* largely because so much is known about *E. coli* protein expression and because gene manipulation is relatively easy in this bacterium.
- Lysine, an amino acid added to animal feed, is produced in fermenters using the bacterium *Corynebacterium glutamicum.* About 700,000 tons are produced this way each year.
- Global Cell Solutions and Hamilton has developed a benchtop incubator-bioreactor for high-density three-dimensional cell cultures. This type of bioprocess engineering may enable pharmaceutical companies to test the toxicity of their drugs without using animals.
- Metabolomics, a technique from functional genetics in which all the metabolites in a cell are analyzed and compared, allows scientists to optimize bioprocesses by improving the strains of bacteria and the medium used in fermentation.

---

research. They may engage in the development and manufacture of gene therapy vectors, vaccines, or renewable biofuels.

Bioprocess engineering is widely used in industry. Many educational institutions offer bioprocess courses for undergraduates and degrees or concentrations in bioengineering or bioprocess engineering. Several community colleges offer associate degrees and certificate programs that typically prepare students to work in industry. Most of these programs are interdisciplinary. Graduates of these programs will have the knowledge and internship experience to enter directly into the bioprocess engineering workforce. Advanced degrees such as a master's degree or doctorate are necessary to obtain top positions in academia and industry in the bioprocess engineering area. Some universities such as Cornell University offer graduate programs in bioprocess engineering.

The basic courses for students interested in a career in bioprocess engineering are microbiology, plant biology, organic chemistry, biochemistry, agriculture, bioprocess engineering, and chemical engineering. Students must master basic engineering calculations and principles and understand physical and chemical processes including material and energy balances, reactor engineering, fluid flow and mixing, heat and mass transfer, filtration and centrifugation, and chromatography.

Careers in the bioprocess engineering field can take different paths. Biotechnological, microbiological, chemical, and biofuel companies are the biggest employers. People who are interested in research in bioprocess engineering can find jobs in government laboratories and universities. In universities, bioprocess engineers may divide their time between research and teaching.

## Social Context and Future Prospects

The role of bioprocess engineering in industry is likely to expand because scientists are increasingly able to manipulate organisms to expand the range and yields of products and processes. Developments in this field continue rapidly.

Bioprocess engineering can potentially be the answer to several problems faced by humankind. One such problem is global warming, which is caused by rising levels of carbon dioxide and other greenhouse gases. A suggested method of addressing this issue is carbon dioxide removal, or sequestration, based on bioprocess engineering. This bioprocess uses microalgae (microscopic algae) in photobioreactors to capture the carbon dioxide that is discharged into the atmosphere by power plants and other industrial facilities. Photobioreactors are various types of closed systems made of an array of transparent tubes in which microalgae are cultivated and monitored under illumination.

The health care industry is another area where bioprocess engineers are likely to be active. For example, if pharmaceutical applications are found for stem cells, a bioprocess must be developed to produce a reliable, plentiful source of stem cells so that these drugs can be produced on a large scale. The process for growing and harvesting cells must be standardized so that the cells have the same characteristics and behave in a predictable manner. Bioprocess engineers must take these processes from laboratory procedures to industrial protocols.

In general, the future of bioprocess engineering is bright, although questions and concerns, primarily about using genetically modified organisms, have arisen. Public education in such a complex area of science is very important to avoid public mistrust of bioprocess engineering, which is very beneficial in most applications.

*Sergei A. Markov, Ph.D.*

## Further Reading

Bailey, James E., and David F. Ollis. *Biochemical Engineering Fundamentals.* 2d ed. New York: McGraw-Hill, 2006. Covers all aspects of biochemical engineering in an understandable manner.

Bougaze, David, Thomas R. Jewell, and Rodolfo G. Buiser. *Biotechnology. Demystifying the Concepts.* San Francisco: Benjamin/Cummings, 2000. Classical book on biotechnology and bioprocessing.

Doran, Pauline M. *Bioprocess Engineering Principles.* London: Academic Press, 2009. A solid, basic textbook for students entering the field.

Glazer, Alexander N., and Hiroshi Nikaido. *Microbial Biotechnology: Fundamentals of Applied Microbiology.* New York: Cambridge University Press, 2007. In-depth analysis of the application of microorganisms in bioprocessing.

Heinzle, Elmar, Arno P. Biwer, and Charles L. Cooney. *Development of Sustainable Bioprocesses: Modeling and Assessment.* Hoboken, N.J.: John Wiley & Sons, 2007. Looks at making bioprocesses sustainable by improving them. Includes case studies on citric acid, biopolymers, antibiotics, and biopharmaceuticals.

Nebel, Bernard J., and Richard T. Wright. *Environmental Science: Towards a Sustainable Future.* 10th ed. Englewood Cliffs: Prentice Hall, 2008. Describes several bioprocesses used in waste treatment and pollution control.

Yang, Shang-Tian. *Bioprocessing for Value-Added Products from Renewable Resources: New Technologies and Applications.* Amsterdam: Elsevier, 2007. Reviews the techniques for producing products through bioprocesses and lists suitable organisms, including bacteria and algae, and describes their characteristics.

**WEB SITES**

*Biotechnology Industry Association*
http://www.bio.org

*International Society for BioProcess Technology*
http://www.isbiotech.org

*Society for Industrial Microbiology*
http://www.simhq.org/index.aspx

*U.S. Department of Agriculture*
http://usda.gov

*U.S. Department of Energy*
Bioenergy
http://www.energy.gov/energysources/bioenergy.htm

**See also:** Agricultural Science; Biochemical Engineering; Bioengineering; Food Science; Genetic Engineering; Proteomics and Protein Engineering.

# BIOSYNTHETICS

## FIELDS OF STUDY

Organic chemistry; biochemistry; bio-organic chemistry; bioinorganic chemistry; medicinal chemistry; pharmaceutical chemistry; pharmacology; analytical chemistry; nanotechnology; biomedical engineering; genetic engineering; genetics; synthetic biology; biology; molecular biology.

## SUMMARY

Biosynthesis is the process of using small, simple molecules to make larger, more complex molecules, either inside the body or in the laboratory. Numerous applications for drug development and medicine include the synthesis of proteins, hormones, dietary supplements, blood products, and surgical dressings for wounds. Additional techniques to facilitate the diagnosis and treatment of disease include protein biomarkers for immune assays, the development of proteomics to analyze changes in proteins in response to a drug, the development of polyclonal and monoclonal antibodies, immunizations, and various drug delivery systems.

## KEY TERMS AND CONCEPTS

- **Amino Acid:** Building block of proteins.
- **Antibody:** Glycoprotein that binds to and immobilizes a substance that the cell recognizes as foreign.
- **Antigen:** Substance that triggers an immune response.
- **Binding Assay:** Experimental method for selecting one molecule out of a number of possibilities by specific binding.
- **DNA (Deoxyribonucleic Acid):** Molecule that contains the genetic code.
- **Enzyme:** Biological catalyst, usually a globular protein.
- **Gene:** Individual unit of inheritance that consists of a sequence of DNA.
- **Hormone:** Substance produced by endocrine glands and delivered by the bloodstream to target cells, producing a desired effect.
- **Hydrophilic:** Property of tending to dissolve in water.
- **Insulin:** Hormone released from the pancreas.
- **Monoclonal Antibody:** Antibody produced from the progeny of a single cell and specific for a single antigen.
- **Peptide:** Molecule formed by linking two to several dozen amino acids.
- **Protein:** Macromolecule formed by polymerization of amino acids.

## DEFINITION AND BASIC PRINCIPLES

In general, the term "biosynthetic" refers to any type of material produced via a biosynthetic process. A biosynthetic process uses enzymes and energetic molecules to transform small molecules into larger molecules within the cells of organisms. The two types of metabolites produced from cellular biosynthetic pathways include the primary metabolites of fatty acids and DNA needed by cells and the secondary metabolites of pheromones, antibiotics, and vitamins that assist the entire organism. Additional small molecules, such as adenosine triphosphate (ATP), provide the energetic driving force for the biosynthetic pathways, and other small molecules, including enzymes, further facilitate the reactions in these pathways. Thus, there have been many possibilities for numerous types of scientists, including chemists, biochemists, biologists, and geneticists, to create innovations.

The term "biosynthetic" differs from the term "chemosynthetic," because chemosynthetic indicates the production of materials that cannot take place within a living organism. Scientists generally begin the process of developing a new medical application or dietary supplement by first isolating and characterizing the DNA of the proteins or other small molecules directly involved in the biological process. They then try to duplicate this naturally occurring biological process to produce massive quantities of the desired material, and ultimately they combine these naturally occurring processes with chemicals that can mimic the process during laboratory manufacturing processes.

## BACKGROUND AND HISTORY

The biochemical pharmacologist Hermann Karl Felix "Hugh" Blaschko was a trailblazer whose discoveries in the 1930's initiated the field of biosynthetics.

His work elucidated the biosynthetic pathway for adrenaline, which is often called the fight-and-flight hormone, and encompassed the study of the enzymes important for regulation of this hormone. This work led the way toward the development of syntheses using amino acids for therapeutic applications.

Another key development in biosynthetics was the discovery of the role of the amino acid L-arginine in the synthesis of creatine, an important biomolecule, by G. L. Foster, Rudolf Schoenheimer, and D. Rittenberg in 1939. Since that time, L-arginine has also been shown to be a precursor to nitrous oxide and nitric oxide, as well as a component of the urea cycle, which is important for ammonia regulation and thus influences the operation of the kidneys and other organs. Nitric oxide is important in the regulation of blood flow to muscles. These discoveries involving L-arginine have led to dietary supplements useful to bodybuilders who wish to enhance their weight-lifting performance.

Throughout the 1940's, 1950's, and 1960's, progress was made toward understanding the genetic composition of organisms, enzymes, and biosynthetic pathways. Researchers made contributions to understanding pyrimidine, galactosidase, *Escherichia coli*, and chlorophyll. Practical biosynthetic applications that were made possible by these fundamental discoveries began to manifest themselves throughout the 1970's, 1980's, and 1990's, with the development of surgical dressings, therapeutic hormones, and plant supplements for increased nutritional value.

## HOW IT WORKS

**General Process.** Often the isolation and characterization of a specific gene responsible for producing an important enzyme or other small molecule is the first step in a lengthy process toward synthesis of a product that undergoes lengthy clinical trials before the final, approved product is ready for manufacture. Once the gene has been characterized, its DNA is further characterized to facilitate the process of peptide synthesis (the process of producing long peptides is known as protein biosynthesis).

The process of peptide synthesis involves the general concepts of antigenicity, hydrophilicity, and surface probability, as well as flexibility indexes. The process involves an analysis of the peptide's characteristics, the use of software and databases to determine hydrophilicity (affinity for water), study of the antigenicity (capacity to stimulate the production of antibodies) to assist with antibody production, the study of surface probability (which determines the likelihood of inducing the formation of antibodies), the determination of the protein sequence, phosphorylation (process that activates or deactivates many protein enzymes), and then selection of two to three peptides, followed by comparison of their homology (similarity of structure).

In a general process called screening, the efficacy of an antibiotic is first tested using bacterial cultures, followed by injection of the antibiotic into laboratory animals, such as rats, rabbits, or guinea pigs; then clinical trials are conducted according to protocols established by the Food and Drug Administration (FDA). Combinatorial chemistry, a faster screening method, is often used instead. FDA-approved products are then manufactured on a larger scale.

**Antibody Production.** The application of a binding assay is used for isolation of the purified protein that is to be the source of an antigen. This antigen is then used as a conjugate to a carrier protein, such as kehole limpet hemocyanin (KLH), to produce a target peptide with a length of thirteen to twenty amino acids to stimulate the immune system. A carrier protein is a membrane protein that can bind to a substance to facilitate the substance's passive transport into a cell. Injection into a laboratory animal occurs next, and then the animals undergo a series of four to six immunizations separated by about twenty days. Enzyme-linked immunosorbent assay (ELISA) is used to detect antibodies. ELISA is based on the antibody-antigen binding interaction and often uses color to visually indicate the concentration of antibodies. Purification of antibodies obtained from the antiserum for specific antigen binding completes the antibody production process.

**Antigen Preparation.** This process is facilitated through bioinformatics analysis to choose the appropriate two to three peptides based on the protein sequence provided by a customer. KLH conjugation used for immunization, and bovine serum albumin (BSA) conjugation is carried out for screening. After immunization protocols and specific antibodies have been selected during fusion and screening, a cell can be cryopreserved.

**Combinatorial Chemistry.** In combinatorial chemistry synthesis, a high-throughput screening method, the starting small molecule is attached to a type of polymeric resin, followed by different permutations of reagents, to produce large libraries containing hundreds of unique products that can be rapidly screened for enzymatic activity, specific antigen-binding, or protein-protein interactions. Often the process is controlled by a computer and completed through the application of robotics. A customer can specify antigen details, and a pharmaceutical company can design a protocol involving the general phases of preparation of antigen, immunization, fusion and screening of assays, and finally selection, purification, and production of antibodies.

## APPLICATIONS AND PRODUCTS

**Biosensors.** Biosensors are microelectronic devices that use antibodies, enzymes, or other biological molecules to interact with an optical device or electrode to record data electronically. These devices can be operated by home health care providers to transmit data obtained from blood or urine samples, for example, to a clinical laboratory some distance away.

**Therapeutic Proteins.** Plasmids are used to transfer human genes that provide the code for proteins important for growth hormones, blood clotting, and insulin production to bacterial cells.

**Disposable Micropumps for Drug Delivery.** Disposable micropumps manufactured by Acuros in Germany are capable of delivering a preset amount of liquid hormones, proteins, antibodies, or other medications. An osmotic microactuator, based on osmotic pressure, is used to regulate the amount of drug delivered, and there are no moving parts or power supply components.

**High-Throughput Screening.** High-throughput screening can assay more than twenty thousand potentially useful drugs per week by using multiwell plates, standard binding assay methodologies, and robotics.

**Protein Biomarker Assays.** NextGen Sciences has developed a mass spectrometry method for protein biomarker assays that does not depend on antibodies but instead uses surrogate proteins to facilitate development of assays. The mass spectrometer measures the amount of surrogate peptides and applies statistical evaluation to assess each biomarker. This first stage requires that a protein be confirmed; then only these selected proteins are used for the second stage of validation of these protein biomarkers. The mass spectrometry data are used along with carbon-13 or

nitrogen-15 isotopically labeled standards to calculate protein concentrations. Reporting the protein biomarkers in terms of concentration is important to allow batches containing hundreds of samples to be analyzed and validated. This technique uses proteomics (the quantitative analysis of proteins based on a physiological response) to allow for much faster development of assays than immunoassays. A wide range of at least 500 plasma proteins and 3,000 tissue proteins can be analyzed at once.

**Gene Expression Databases.** Gene Logic's Bio-Express System is a comprehensive genome-wide gene expression database. The BioExpress System allows cells from a patient to be collected and analyzed to develop a useful biomarker profile for comparison with a database sample to indicate a therapeutic target. This process is made possible by the use of high-throughput gene expression profiling of the mononuclear cell fractions present in a blood sample. The software is capable of mining a database that has access to more than 18,000 samples containing biomarkers for the expression of the gene associated with ovarian cancer. This system is also capable of developing biomarker profiles to help diagnosis autoimmune diseases. Autoimmune diseases include rheumatoid arthritis, Crohn's disease, multiple sclerosis, systemic lupus erythematosus, and psoriasis, which affect about 20 million people in the United States.

**Biosynthetic Temporary Skin Substitute.** A biosynthetic skin substitute is a useful treatment for partial-thickness wounds, including skin tears, burns, and abrasions. After applying a gel to the surface of the wound, a semipermeable membrane of biosynthetic skin is used to cover the wound for protection from infection. Before the development of biosynthetic skin grafts, a physician had to choose between an allograft, which uses cadaver skin, and a xenograft, which uses tissue from another species. Biosynthetic dressings have also been developed. The dressing called Hydrofiber contains ionic silver and has been shown to prevent the spread of bacteria.

**Needle-Free Drug Delivery Systems.** The three types of needle-free drug delivery systems are liquid, powder, and depot injections. Each of these types uses some form of mechanical compression to create enough pressure to force the medication into the skin. Although these needle-free delivery systems cost more initially and require more technical expertise because of their complexity, they also have many advantages. In addition to eliminating pain from needle injections and reducing physician visits, these needle-free delivery systems decrease the frequency of incorrect doses. They are being used to deliver anesthetics, chemotherapy injections, vaccines, and hormones.

**Nanoparticles.** DNA nanotechnology uses discoveries involving nanoparticles and nanomaterials to manipulate DNA's molecular recognition abilities to build tiny medical robots that mimic bond parts or function within cells.

## IMPACT ON INDUSTRY

The United States has been the world leader in biosynthetic innovation, followed closely by India. Many other countries, including Germany, France, the United Kingdom, the Czech Republic, Switzerland, Belgium, Denmark, Australia, Canada, Italy, Spain, Japan, and China, also make significant contributions.

**Government and University Research.** Because the fundamental starting point for the identification and synthesis of a biosynthetic target involves the isolation and characterization of a gene, the Human Genome Project has led to an explosion of developments in this field. The U.S. Department of Energy along with the National Institutes of Health financially supported the Human Genome Project, which was formally proposed in 1985 and officially initiated in 1990. The goal of this project was to identify all the genes in human DNA and organize this information into databases. Among the most significant contributors to the project were the U.S. Department of Energy Joint Genome Institute in Walnut Creek, California; the Baylor College of Medicine Human Genome Sequencing Center in Houston, Texas; the Washington University School of Medicine Genome Sequencing Center in St. Louis, Missouri; and the Whitehead Institute/MIT Center for Genome Research in Cambridge, Massachusetts. This project quickly became international in scope with the establishment of the International Human Genome Sequencing Consortium. Participating research institutes were in the United States—New York, Tennessee, California, Oklahoma, Texas, and Washington—and the United Kingdom, China, France, Germany, and Japan. In addition to identifying the genes in the human genome, these government and university research centers led the way in developing better methods for analyzing and

manipulating genes, important for the initial stages of any biosynthetic project. Although the Human Genome Project was completed in 2003, the data obtained continue to be analyzed and used as the starting point for numerous innovations involving biosynthetic targets.

**Industry and Major Corporations.** Although companies based in the United States, such as Amgen, Genzyme, Gilead Sciences, Eli Lilly, Johnson & Johnson, Abbott Laboratories, Wyeth, Bristol-Myers Squibb, and Pfizer, have traditionally led the way in biosynthetic developments, they have been expanding worldwide, contributing to the rapid growth taking place in other countries. For example, Johnson & Johnson has subsidiaries in hundreds of countries and is responsible for a wide range of products, including surgical dressings, bandages, contact lenses, medications, and various hygiene products. Abbott Laboratories has subsidiaries in more than one hundred different countries and generated more than $29 billion in revenue in 2008. Abbott Laboratories has been a leader in the development of medical devices and medical diagnostic systems, including the first blood screening test for the human immunodeficiency virus (HIV) in 1985. In the 1950's, Wyeth became the first company to apply biosynthetic principles to manufacture the vaccine for polio, a freeze-dried smallpox vaccine. It has evolved to produce more than $3 billion in annual revenue, primarily from the sale of oral contraceptives and over-the-counter medications. Eli Lilly had revenues of more than $20 billion in 2008 and was the first company to apply biosynthetic principles to manufacture penicillin. It has expanded its range of manufactured products to include several successful psychiatric drugs, as well as pacemakers for the heart and systems to monitor intravenous fluid infusions. Pfizer became famous for its application of biosynthetic principles to develop fermentation technology, and it later manufactured penicillin, several other antibiotics, several anti-inflammatory medications, and kinase inhibitors to block the metabolic pathways of several cancers.

## CAREERS AND COURSE WORK

A bachelor of science degree is adequate training for an entry-level position in the biosynthetics field, but an advanced degree, such as a master's of science or a doctorate, is required to have the opportunity to lead research project teams in research and development, whether in academia, industry, or government. Because the field of biosynthetics involves several disciplines, college courses in chemistry, biology, genetics, microbiology, biochemistry, biomedical engineering, molecular biology, or biochemical engineering are the most helpful. Degrees in any of these disciplines would be appropriate preparation for entry into the biosynthetic field.

Many researchers with a doctorate in one of the appropriate fields work in academia and teach in addition to pursuing research. A significantly larger number of employment opportunities exist in the pharmaceutical industry and require either a bachelor's of science or master's of science degree. These positions are in research and development and various areas of manufacturing, including quality control, quality assurance, and process development. There are also opportunities for technicians who do not have a bachelor's degree. Technicians primarily record and analyze data while monitoring experiments and are often responsible for the maintenance of laboratory equipment.

According to the U.S. Bureau of Labor Statistics, employment in pharmaceutical and medical product manufacturing is expected to be one of the fastest growing manufacturing areas, with a growth rate of about 6 percent expected through 2018. The majority of these jobs within the United States are projected to be found in New Jersey, New York, Pennsylvania, Indiana, California, North Carolina, and Illinois.

## SOCIAL CONTEXT AND FUTURE PROSPECTS

The Human Genome Project has facilitated the mapping of genes, which has been instrumental to the development of vaccines to treat influenza, cervical cancer, and malaria, as well as the creation of new diagnostic tools for analysis. As a result, the pharmaceutical industry in the United States has become a multibillion-dollar industry. The generation of biosynthetic products has enhanced the lives of thousands of people through the development of treatments for many types of cancer, pneumonia, cardiovascular diseases, diabetes, tuberculosis, neurological disorders, strokes, blood disorders, and many other diseases.

Combinatorial chemistry has allowed for rapid screening of potentially successful medications that may enhance and extend the lives of many people. Normally, only one out of every 5,000 to 10,000 compounds

screened makes it through the multiyear process of clinical trials to become an FDA-approved drug. However, the desire to recoup the money spent during the years of research required to bring a drug to market has caused some pharmaceutical companies to launch a product as early as possible, which has resulted in serious litigation because some drugs

### Fascinating Facts About Biosynthetics

- The generation of biosynthetic products has led to the development of successful treatments for many types of cancer, pneumonia, cardiovascular diseases, diabetes, tuberculosis, neurological disorders, strokes, blood disorders, and many other diseases and conditions.
- Biosynthetic corneas were used to restore vision in people with keratoconus, a condition that causes corneal scarring. These biosynthetic corneas replaced rejection-prone, scarce cadaver corneas.
- The J. Craig Venter Institute synthesized the first self-replicating synthetic bacteria cell in 2010. Synthesis of such cells may aid researchers and help develop new drugs.
- Synthetic genomics has made it possible to design and assemble chromosomes and genes and gene pathways, which may be used in creating green biofuels, pharmaceuticals, and vaccines.
- Scientists at the University of Sheffield are mapping the metabolism of the *Nostic* bacterium, which fixes nitrogen and releases hydrogen, which could be used as fuel. Once they understand the metabolic process thoroughly, they hope to be able to genetically engineer an organism that can produce hydrogen more efficiently.
- Scientists have identified biosynthetic gene clusters for many aminoglycoside antibiotics, including streptomycin, kanamycin, butirosin, neomycin and gentamicin. A full understanding of how these antibiotics work may enable scientists to get around the problem of antibiotic-resistant bacteria.
- Mass-produced biosynthetic bovine growth hormone, which when injected into dairy cows raises milk production, has been used in many developing countries. However, its use is controversial as questions have arisen regarding its effects on the health of the cows and the people who drink the milk.

proved to have harmful side effects. The application of biosynthetic growth hormones for nonmedical applications, such as bodybuilding, has also caused ethical and medical controversy. However, as the global population continues to grow and the percentage of elderly persons increases, the need for the products of biosynthetic research will continue to grow.

*Jeanne L. Kuhler, B.S., M.S., Ph.D.*

## FURTHER READING

Arya, Dev. *Aminoglycoside Antibiotics: From Chemical Biology to Drug Discovery.* New York: Wiley-Interscience, 2007. Describes the design and synthesis of antibiotics and the process of antibiotic resistance.

Dewick, Paul. *Medicinal Natural Products: A Biosynthetic Approach.* New York: John Wiley & Sons, 2009. Comprehensive textbook describing biosynthetic methods and processes, including new techniques in genetic engineering and isolation of genes.

Lazo, John, and Peter Wipf. "Combinatorial Chemistry and Contemporary Pharmacology." *The Journal of Pharmacology and Experimental Therapeutics* 293, no. 3 (February, 2000): 705-709. Describes the process of combinatorial chemistry. Includes experimental strategies and flow charts describing the screening of compounds.

Pettit, George. *Biosynthetic Products for Cancer Chemotherapy.* Vol. 5 London: Elsevier Science, 1985. A discussion of the fundamental processes involved with screening for antitumor agents.

Savageau, Michael. *Biochemical Systems Analysis: A Study of Function and Design in Molecular Biology.* New York: CreateSpace, 2010. Detailed textbook describing the immune system and gene regulation.

Spentzos, Dimitri. "Gene Expression Signature with Independent Prognostic Significance in Epithelial Ovarian Cancer." *Journal of Clinical Oncology* 22, no. 23 (December, 2004): 4648-4658. The research article describes the diagnosis of ovarian cancer and the use of biomarkers for detection.

Stanforth, Stephen. *Natural Product Chemistry at a Glance.* New York: Wiley-Blackwell, 2006. An introductory textbook that describes much of the organic chemistry involved in biosynthesis.

## WEB SITES

*American Chemical Society*
http://acs.org

*Society for Industrial Microbiology*
http://www.simhq.org

**See also:** Bioengineering; Bioprocess Engineering; DNA Sequencing; Enzyme Engineering; Genetically Modified Organisms; Genetic Engineering; Pharmacology; Proteomics and Protein Engineering; Xenotransplantation.

# BRIDGE DESIGN AND BARODYNAMICS

## FIELDS OF STUDY

Chemistry, civil engineering, construction, material engineering, mechanics, physics, structural engineering.

## SUMMARY

Barodynamics is the study of the mechanics of heavy structures that may collapse under their own weight. In bridge building, barodynamics is the science of the support and mechanics of the methods and types of materials used in bridge design to ensure the stability of the structure. Concepts to consider in avoiding the collapse of a bridge are the materials available for use, what type of terrain will hold the bridge, the obstacle to be crossed (such as river or chasm), how long the bridge needs to be to cross the obstacle, what types of natural obstacles or disasters are likely to occur in the area (high winds, earthquakes), the purpose of the bridge (foot traffic, cars, railway), and what type of vehicles will need to cross the bridge.

## KEY TERMS AND CONCEPTS

- **Arch Bridge:** Type of bridge where weight is distributed outward along two paths that curve toward the ground in the shape of an arch.
- **Beam Bridge:** Simplest type of bridge consisting of two or more supports holding up a beam; ranges from a complex structure to a plank of wood.
- **Cantilever Bridge:** Bridge type where two beams, well anchored, support another beam.
- **Cofferdam:** A temporary watertight structure that is pumped dry to enclose an area underwater and allow construction work on a bridge in the underwater area.
- **Keystone:** The wedge-shape stone at the highest point of an arch; function is to keep the other stones locked into place through gravity.
- **Pontoon Bridge:** A floating bridge supported by floating objects that contain buoyancy sufficient to support the bridge and any load it must carry; often a temporary structure.
- **Suspension Bridge:** A bridge hung by cables, which, in turn, hang from towers; weight is transferred from the cables to the towers to the ground.
- **Truss Bridge:** A type of bridge supported by a network of beams in triangular sections.

## DEFINITION AND BASIC PRINCIPLES

Barodynamics is a key component of any bridge design. Bridges are made of heavy materials, and many concepts, such as tension and compression of building materials, and other factors, such as wind shear, torsion, and water pressure, come into play in bridge building.

Bridge designers and constructors must keep in mind the efficiency (the least amount of material for the highest-level performance) and economy (lowest possible costs while still retaining efficiency) of bridge building. In addition, some aesthetic principles must be followed; public outcry can occur when a bridge is thought to be "ugly." Conversely, a beautiful bridge can become a landmark symbol for an area, such as the Golden Gate Bridge has become for San Francisco.

The four main construction materials for bridges are wood, stone, concrete (including prestressed concrete), and iron (from which steel is made). Wood is nearly always available and inexpensive but is comparatively weak in compression and tension. Stone, another often available material, is strong in compression but weak in tension. Concrete, or "artificial stone," is, like stone, strong in tension and weak in compression.

The first type of iron used in bridges, cast iron, is strong in compression but weak in tension. The use of wrought iron helped in bridge building, as it is strong in compression but still has tensile strength. Steel, a further refinement of iron, is superior in compression and tensile strength, making it a preferred material for bridge building. Reinforced concrete or prestressed concrete (types of concrete with steel bars running through concrete beams) are also popular bridge-building materials because of their strength and lighter-weight design.

## BACKGROUND AND HISTORY

From the beginning of their existence, humans have constructed bridges to cross obstacles such as rivers and chasms using the materials at hand, such as trees, stones, or vines. In China during the third century B.C.E., the emperor of the Qin Dynasty built canals to transport goods, but when these canals interfered with existing roads, his engineers built bridges of stone or wood over these canals.

However, the history of barodynamics in bridge building truly begins in Roman times, when Roman

engineers perfected the keystone arch. In addition to the keystone and arch concepts, the Romans improved bridge-building materials such as cement and concrete and invented the cofferdam so that underwater pilings could be made for bridges. These engineers built a network of bridges throughout the Roman empire to keep communication with and transportation to and from Rome intact. The Romans made bridges of stone because of its durability, and many of these bridges are still intact today.

In the Middle Ages, bridges were an important part of travel and transportation of goods, and many bridges were constructed during this period to support heavy traffic. This is also the period when people began to live in houses built on bridges, in part because in walled cities, places to build homes were limited. Possibly the most famous inhabited bridge was London Bridge, the world's first stone bridge to be built over a tidal waterway where the water rose and fell considerably every twelve hours. However, in Paris in the sixteenth century, there were at least five inhabited bridges over the Seine to the Île de la Cité.

The first iron bridge was built in 1779. Using this material changed the entire bridge-building industry because of the size and strength of structure that became possible. In the 1870's, a fall in the price of steel made bridges made of this material even more popular, and in 1884, Alexandre Gustave Eiffel, of Eiffel Tower fame, designed a steel arch bridge that let wind pass through it, overcoming many of the structural problems with iron and steel that had previously existed. Iron and steel are still the most common materials to use in bridge building.

Suspension bridges began to be quite popular as they are the most inexpensive way to span a longer distance. In the early 1800's, American engineer John Roebling designed a new method of placing cables on suspension bridges. Famous examples of suspension bridges include the Golden Gate Bridge (completed in 1937) and Roebling's Brooklyn Bridge (completed in 1883).

Girder bridges were often built to carry trains in the early twentieth century. Though capable of carrying heavyweight railroad cars, this type of bridge is usually only built for short distances as is typical with beam-type bridges. In the 1950's, the box girder was designed, allowing air to pass through this type of bridge and making longer girder bridges possible.

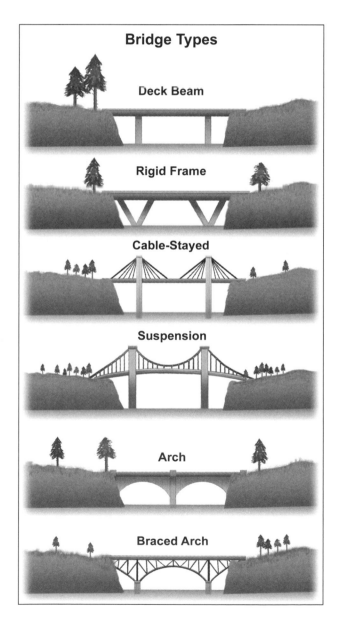

**Bridge Types**

Deck Beam

Rigid Frame

Cable-Stayed

Suspension

Arch

Braced Arch

*In describing the design of a bridge, there are four primary factors: span, placement, material, and form.*

## Fascinating Facts About Bridge Design and Barodynamics

- The Tacoma Narrows Bridge across Puget Sound tore apart and fell into the water below in 1940 because of engineering miscalculations concerning vertical and torsional motion. Viewing motion pictures of the disaster helped engineers rethink the aerodynamics of building bridges.
- The I-35W bridge in Minneapolis collapsed in 2007. The deaths and injuries led to a renewed interest in maintenance and retrofitting of public bridges across the United States.
- London Bridge is possibly the most famous bridge in the world. It was the only bridge across the River Thames from the tenth century until the eighteenth century, when the original stone bridge was torn down and rebuilt with stone. It was rebuilt again in 1973 of steel and concrete to accommodate the heavy traffic using the bridge. Robert McCulloch bought the old stone bridge, numbering each stone, and had it rebuilt in Lake Havasu City, Arizona, where it remains a tourist attraction.
- The origin of the word "bridge" goes back to an Old English word "brycg," which is believed to be derived from the German word root "brugj." The name of the Belgian city Brugge can be translated as "a bridge or a place of bridges."
- The Ponte Vecchio in Florence has an added covered walk forming a top story above the shops that line the bridge. This addition was constructed in 1565 so members of the Medici family could walk across the river from the Uffizi to the Palazzo Pitti without descending to the street level and mingling with commoners.
- The world's first iron bridge was built in 1779 over the River Severn in an English industrial area called Coalbrookdale.
- In *Ramayana*, a mythological Indian epic, tales are told of bridges constructed by the army of Sri Rama, a mythological king of Ayodhya, from India to the island of Sri Lanka, a feat thought to have been impossible in that time period. However, space images taken by the National Aeronautics and Space Administration reveal an ancient bridge, called Adam's Bridge, in the Palk Strait with a unique curvature and composition that indicate that it is man-made.
- The Chinese made suspension bridges as early as 200 B.C.E.
- The oldest surviving stone bridge in China, the Zhaozhou Bridge, is thought to have been built during the Sui Dynasty, around the year 600.

### HOW IT WORKS

The engineering principles that must be used to construct even a simple beam bridge are staggering. Supports must be engineered to hold the weight of the entire structure correctly as well as any traffic that will cross the bridge. The bridge itself, or "span," must be strong enough to bear the weight of traffic and stable enough to keep that traffic safe. Spans must be kept as short as reasonably possible but must sometimes be built across long distances, for example, over deep water.

**Arch.** The Roman arch concept uses the pressure of gravity on the material forming the arch to hold the bridge together with the outward thrust contained by buttresses. It carries loads by compressing and exerting pressure on the foundation, which must be prevented from settling and sliding. This concept allowed bridges to be built that were longer than ever before. For example, a surviving bridge over the Tagus River in Spain has two central arches that are 110 feet wide and 210 feet above the water level.

These arches are made of uncemented granite and each keystone weighs eight tons. This type of bridge is constructed by building a huge timber structure to support the bridge during the building phase, then winching blocks into place with a pulley system. After the keystone to the arch is put into place, the scaffolding is removed, leaving the bridge to stand alone.

**Beam.** This is the most common form of bridge and may be as simple as a log across a stream. This type of bridge carries a load by bending, which horizontally compresses the top of the beam and simultaneously causes horizontal tension on the bottom of the beam.

**Truss.** A truss bridge is popular because it requires a relatively small amount of construction material to carry a heavy load. It works like a beam bridge, carrying loads by bending and causing compression and tension in the vertical and diagonal supports.

**Suspension.** Suspension bridges are essentially steel-rope bridges: Thick steel cables are entwined like ropes into a larger and stronger steel cable or rope. These thick, strong cables then suspend the bridge itself between pylons that support the weight of the bridge. A suspension bridge can be thought of as an upside-down arch, as the curved cables use tension and compression to support the load.

**Cantilever.** Cantilevered means something that projects outward and is supported at only one end (similar to a diving board). This type of bridge is generally made with three spans with the outside spans supported on the shore and the middle span supported by the outside spans. This is a type of beam bridge that uses tension in the lower spans and compression in the upper span to carry a load.

**Pontoon.** A pontoon bridge is built across water with materials that float. Each pontoon, or floating object, can support a maximum load equal to the amount of water it displaces. If the load placed on one pontoon-supported section exceeds the water displaced, the pontoon will submerge and cause the entire bridge to sink.

## APPLICATIONS AND PRODUCTS

Bridges are continuously being built to cross physical obstacles, and as the nature of materials changes, the ability to cross even larger obstacles becomes reality. Nature is the defining force on a bridge; most bridges fail because of flooding or other natural disasters.

Improvements in building materials are ongoing. For example, the Jakway Park Bridge in Buchanan County, Iowa, was the first bridge in North America to be built with ultrahigh performance concrete (UHPC) with pi-girders. This moldable material combines high compressive strength and flexibility and offers a wide range of design possibilities. It is very durable and has low permeability.

Bridge-building products may even be developed that help the environment. For example, the rebuilt I-35W bridge in Minnesota uses a concrete that is said to "eat smog." The concrete contains photo-catalytic titanium dioxide, which accelerates the decomposition of organic material. Other materials like this may change the future of bridge building.

## IMPACT ON INDUSTRY

Bridges have a significant impact on travel and transportation of goods. They are, generally, made for the public, for transportation or travel, and are often built with public funds. Therefore, bridge building has its greatest impact in the governmental or public-transportation areas. Bridge engineers often find jobs with government agencies, such as the U.S. Department of Transportation, or with private companies that are subcontractors on government projects.

## CAREERS AND COURSE WORK

Those who engineer and design bridges may have backgrounds in a variety of fields, including architecture and design. However, those who are involved in the barodynamic aspects of bridge building are engineers, usually either civil, materials, or mechanical engineers. Earning a degree in one of these fields is required to get the training needed in geology, math, and physics to learn about the physical limitations and considerations in bridge building. Many bridge engineers have advanced degrees in a specific related field. After earning a degree, a candidate for this type of job usually works for a few years as an assistant engineer in a sort of apprenticeship, learning the specifics of bridge building such as drafting, blueprint reading, surveying, and stabilization of materials. To become a professional engineer (PE), one must then take a series of written exams to get his or her license.

## SOCIAL CONTEXT AND FUTURE PROSPECTS

Barodynamics is a rapidly changing field. As new materials are created and existing materials change, the possibilities for future improvements in this field increase. Development of future lightweight materials may change the way bridges are engineered and designed to make the structure stable and avoid collapse. Just as innovations in iron refinement changed the face of bridge building in the late 1800's, new materials may refine and improve bridge building even further, bringing more efficient and economical bridges.

Materials engineers are usually the people who provide the technological innovation to create these kinds of new materials. They examine materials on the molecular level to understand how materials can be improved and strengthened in order to provide better building materials for structures such as bridges. Possible future bridge-building materials include ceramics, polymers, and other composites. Two other rapidly growing materials fields that may affect bridge barodynamics are biomaterials and nanomaterials.

*Marianne M. Madsen, M.S.*

## FURTHER READING

Blockley, David. *Bridges: The Science and Art of the World's Most Inspiring Structures.* New York: Oxford University Press, 2010. Written by a professor of engineering with a lay reader in mind; discusses basic forces such as tension, compression, and shear, and bridge failures. Includes a comprehensive history of bridge building with fifty illustrations.

Chen, Wai-Fah, and Lian Duan, eds. *Bridge Engineering Handbook.* Boca Raton, Fla.: CRC Press, 1999. Contains more than 1,600 tables, charts, and illustrations with step-by-step design procedures. Covers fundamentals, superstructure design, substructure design, seismic design, construction and maintenance, and worldwide practice; includes a special topics section.

Haw, Richard. *Art of the Brooklyn Bridge: A Visual History.* New York: Routledge, 2007. A visually interesting compilation of artists' renderings of the Brooklyn Bridge, contributing to the idea that bridges are artful as well as functional.

Tonias, Demetrios E., and Jim J. Zhao. *Bridge Engineering: Design, Rehabilitation, and Maintenance of Modern Highway Bridges.* 2d ed. New York: McGraw-Hill, 2007. Details the entire highway-bridge design process; includes information on design codes.

Unsworth, John F. *Design of Modern Steel Railway Bridges.* Boca Raton, Fla.: CRC Press, 2010. Focuses on new steel superstructures for railway bridges but also contains information on maintenance and rehabilitation and a history of existing steel railway bridges.

Van Uffelen, Chris. *Masterpieces: Bridge Architecture and Design.* Salenstein, Switzerland: Braun Publishing, 2010. Includes photos of sixty-nine bridges from around the world, displaying a variety of structures and materials.

Yanev, Bojidar. *Bridge Management.* Hoboken, N.J.: John Wiley & Sons, 2007. Contains case studies of bridge building and design; discusses bridge design, maintenance, and construction with topics such as objectives, tools, practices, and vulnerabilities.

## WEB SITES

*American Society of Civil Engineers*
http://www.asce.org

*Design-Build Institute of America*
http://www.dbia.org

*National Society of Professional Engineers*
http://www.nspe.org/index.html

**See also:** Civil Engineering; Earthquake Engineering; Structural Composites.

# C

# CALCULUS

## FIELDS OF STUDY

Algebra; geometry; trigonometry.

## SUMMARY

Calculus is the study of functions and change. It is the bridge between the elementary mathematics of algebra, geometry, and trigonometry, and advanced mathematics. Knowledge of calculus is essential for those pursuing study in fields such as chemistry, engineering, medicine, and physics. Calculus is employed to solve a large variety of optimization problems; one example is the so-called least squares solution method commonly used in statistics and elsewhere. The least squares function best fits a set of data points and can then be used to generalize or predict results based on that set.

## KEY TERMS AND CONCEPTS

- **Antiderivative:** Function whose derivative is equal to that of a given function.
- **Continuity:** Characteristic manifested by a function when its output values are equal to the values of its limits.
- **Converge:** Action of an improper integral or series with a finite value.
- **Definite Integral:** Limit of a Riemann sum as the number of terms approaches infinity.
- **Derivative:** Function derived from a given function by means of the limit process, whose output equals the instantaneous rates of change of the given function.
- **Diverge:** Action of an improper integral or series with no finite value.
- **Gradient:** Vector whose components are each of the partial derivatives of a function of several variables.
- **Indefinite Integral:** Antiderivative of a given function.
- **Limit:** Number that the output values of a function approach the closer the input values of the function get to a specified target.
- **Riemann Sum:** Sum of the products of functional values and the lengths of the subintervals over which the function is defined.
- **Series:** Formal sum of an infinite number of terms; it may be convergent or divergent.
- **Taylor Series:** Series whose sums equal the output values of a given function, at least along some interval of input values.

## DEFINITION AND BASIC PRINCIPLES

Calculus is the study of functions and their properties. Calculus takes a function and investigates it according to two essential ideas: rate of change and total change. These concepts are linked by their common use of calculus's most important tool, the limit. It is the use of this tool that distinguishes calculus from the branches of elementary mathematics: algebra, geometry, and trigonometry. In elementary mathematics, one studies problems such as "What is the slope of a line?" or "What is the area of a parallelogram?" or "What is the average speed of a trip that covers three hundred miles in five and a half hours?" Elementary mathematics provides methods or formulas that can be applied to find the answer to these and many other problems. However, if the line becomes a curve, how is the slope calculated? What if the parallelogram becomes a shape with an irregularly curved boundary? What if one needs to know the speed at an instant, and not as an average over a longer time period?

Calculus answers these harder questions by using the limit. The limit is found by making an approximation to the answer and then refining that approximation by improving it more and more. If there is a pattern leading to a single value in those improved approximations, the result of that pattern is called the limit. Note that the limit may not exist in some cases. The limit process is used throughout calculus to provide answers to questions that elementary mathematics cannot handle.

The derivative of a function is the limit of average slope values within an interval as the length of the interval approaches zero. The integral calculates the total change in a function based on its rate of change function.

## BACKGROUND AND HISTORY

Calculus is usually considered to have come into being in the seventeenth century, but its roots were formed much earlier. In the sixteenth century, Pierre de Fermat did work that was very closely related to calculus's differentiation (the taking of derivatives) and integration. In the seventeenth century, René Descartes founded analytic geometry, a key tool for developing calculus.

However, it is Sir Isaac Newton and Gottfried Wilhelm Leibniz who share the credit as the (independent) creators of calculus. Newton's work came first but was not published until 1736, nine years after his death. Leibniz's work came second but was published first, in 1684. Some accused him of plagiarizing Newton's work, although Leibniz arrived at his results by using different, more formal methods than Newton employed.

Both men found common rules for differentiation, but Leibniz's notation for both the derivative and the integral are still in use. In the eighteenth century, the work of Jean le Rond d'Alembert and Leonhard Euler on functions and limits helped place the methods of Newton and Leibniz on a firm foundation. In the nineteenth century, Augustin-Louis Cauchy used a definition of limit to express calculus concepts in a form still familiar more than two hundred years later. German mathematician Georg Riemann defined the integral as a limit of a sum, the same definition learned by calculus students in the twenty-first century. At this point, calculus as it is taught in the first two years of college reached its finished form.

## HOW IT WORKS

Calculus is used to solve a wide variety of problems using a common approach. First, one recognizes that the problem at hand is one that cannot be solved using elementary mathematics alone. This recognition is followed by an acknowledgment: There are some things known about this situation, even if they do not provide a complete basis for solution. Those known properties are then used to approximate a solution to the problem. This approximation may not be very good, so it is refined by taking a succession of better and better approximations. Finally, the limit

is taken, and if the limit exists, it provides the exact answer to the original problem.

One speaks of taking a limit of a function $f(x)$ as $x$ approaches a particular value, for example, $x = a$. This means that the function is examined on an interval around, but not including $x = a$. Values of $f(x)$ are taken on that interval as the varying $x$ values get closer and closer to the target value of $x = a$. There is no requirement that $f(a)$ exists, and many times it does not. Instead the pattern of functional values is examined as $x$ approaches $a$. If those values continue to approach a single target value, it is that value that is said to be equal to the limit of $f(x)$ as $x$ approaches $a$. Otherwise, the limit is said not to exist. This method is used in both differential calculus and integral calculus.

**Differential Calculus.** Differentiation is a term used to mean the process of finding the derivative of a function $f(x)$. This new function, denoted $f'(x)$, is said to be "derived" from $f(x)$. If it exists, $f'(x)$ provides the instantaneous rate of change of $f(x)$ at $x$. For curves (any line other than a straight line), the calculation of this rate of change is not possible with elementary mathematics. Algebra is used to calculate that rate between two points on the graph, then those two points are brought closer and closer together until the limit determines the final value.

Shortcut methods were discovered that could speed up this limit process for functions of certain types, including products, quotients, powers, and trigonometric functions. Many of these methods go back as far as Newton and Leibniz. Using these formulas allows one to avoid the more tedious limit calculations. For example, the derivative function of sine $x$ is proven to be cosine $x$. If the slope of sine $x$ is needed at $x = 4$, the answer is known to be cosine 4, and much time is saved.

**Integral Calculus.** A natural question arises: If $f'(x)$ can be derived from $f(x)$, can this process be reversed? In other words, suppose an $f(x)$ is given. Can an $F(x)$ be determined whose derivative is equal to $f(x)$? If so, the $F(x)$ is called an antiderivative of $f(x)$; the process of finding $F(x)$ is called integration. In general, finding antiderivatives is a harder task than finding derivatives. One difficulty is that constant functions all have derivatives equal to zero, which means that without further information, it is impossible to determine which constant is the correct one. A bigger problem is that there are functions, such as sine $(x^2)$, whose derivatives are reasonably easy to calculate but for which no elementary function serves as an antiderivative.

The definite integral can be thought of as an attempt to determine the amount of area between the graph of $f(x)$ and the $x$-axis, usually between a left and right endpoint. This cannot typically be answered using elementary mathematics because the shape of the graph can vary widely. Riemann proposed approximating the area with rectangles and then improving the approximation by having the width of the rectangles used in the approximation get smaller and smaller. The limit of the total area of all rectangles would equal the area being sought. It is this notion that gives integral calculus its name: By summing the areas of many rectangles, the many small areas are integrated into one whole area.

As with derivatives, these limit calculations can be quite tedious. Methods have been discovered and proven that allow the limit process to be bypassed. The crowning achievement of the development of calculus is its fundamental theorem: The derivative of a definite integral with respect to its upper limit is the integrand evaluated at the upper limit; the value of a definite integral is the difference between the values of an antiderivative evaluated at the limits. If one is looking for the definite integral of a continuous $f(x)$ between $x = a$ and $x = b$, one need only find any antiderivative $F(x)$ and calculate $F(b) - F(a)$.

## APPLICATIONS AND PRODUCTS

**Optimization.** A prominent application of the field of differential calculus is in the area of optimization, either maximization or minimization. Examples of optimization problems include What is the surface area of a can that minimizes cost while containing a specified volume? What is the closest that a passing asteroid will come to Earth? What is the optimal height at which paintings should be hung in an art gallery? (This corresponds to maximizing the viewing angle of the patrons.) How shall a business minimize its costs or maximize its profits?

All of these can be answered by means of the derivative of the function in question. Fermat proved that if $f(x)$ has a maximum or minimum value within some interval, and if the derivative function exists on that interval, then the derivative value must be zero. This is because the graph must be hitting either a peak or the bottom of a valley and has a slope of zero at its highest or lowest points. The search for optimal values then becomes the process of finding the correct function modeling the situation in question, finding its derivative, setting that derivative equal to zero, and solving. Those solutions are the only candidates for optimal values. However, they are only candidates because derivatives can sometimes equal zero even if no optimal value exists. What is certain is that if the derivative value is not zero, the value is not optimal.

The procedure discussed here can be applied in two dimensions (where there is one input variable) or three dimensions (where there are two input variables).

**Surface Area and Volume.** If a three-dimensional object can be expressed as a curve that has been rotated about an axis, then the surface area and volume of the object can be calculated using integrals. For example, both Newton and Johannes Kepler studied the problem of calculating the volume of a wine barrel. If a function can be found that represents the curvature of the outside of the barrel, that curve can be rotated about an axis and pi (p) times the function squared can be integrated over the length of the barrel to find its volume.

**Hydrostatic Pressure and Force.** The pressure exerted on, for example, the bottom of a swimming pool of uniform depth is easily calculated. The force on a dam due to hydrostatic pressure is not so easily computed because the water pushes against it at varying depths. Calculus discovers the answer by integrating a function found as a product of the pressure at any depth of the water and the area of the dam at that depth. Because the depth varies, this function involves a variable representing that depth.

**Arc Length.** Algebra is able to determine the length of a line segment. If that path is curved, whether in two or three dimensions, calculus is applied to determine its length. This is typically done by expressing the path in parametric form and integrating the function representing the length of the vector that is tangent to the path. The length of a path winding through three-dimensional space, for example, can be determined by first expressing the path in the parametric form $x = f(t)$, $y = g(t)$, and $z = h(t)$, in which $f$, $g$, and $h$ are continuous functions defined for some interval of values of $t$. Then the square root of the sum of the squares of the three derivatives is integrated to find the length.

**Kepler's Laws.** In the early seventeenth century,

Kepler formulated his three laws of planetary motion based on his analysis of the observations kept by Tycho Brahe. Later, calculus was used to prove that these laws are correct. Kepler's laws state that any planet's orbit around the Sun is elliptical, with the Sun at one focus of the ellipse; that the line joining the Sun to the planet sweeps out equal areas in equal times; and that the square of the period of revolution is proportional to the cube of the length of the major axis of the orbit.

**Probability.** Accurate counting methods can be sufficient to determine many probabilities of a discrete random variable. This would be a variable whose values could be, for example, counting numbers such as 1, 2, 3, and so on, but not numbers inbetween, such as 2.4571. If the random variable is continuous, so that it can take on any real number within an interval, then its probability density function must be integrated over the relevant interval to determine the probability. This can occur in two or three dimensions.

One common example is determining the likelihood that a customer's wait time is longer than a specified target, such as ten minutes. If the manager knows the average wait time that a customer experiences at an establishment is, for example, six minutes, then this time can be used to determine a probability density function. This function is integrated to determine the probability that a person's wait time will be longer than ten minutes, less than three minutes, between five and thirteen minutes, or within any range of times that is desired.

## IMPACT ON INDUSTRY

The study of calculus is the foundation of more advanced work in mathematics, engineering, economics, and many areas of science. In some cases, it is these other disciplines that affect industry, but much of the time, the effect of calculus can be seen directly. Businesses, government, and industry throughout the developed world continue to apply calculus in a wide variety of settings.

**Government and University Research.** Knowing the importance of calculus in many fields, the National Science Foundation funded many projects in the 1990's that were designed to renew and refresh calculus education. The foundation's Division of Mathematical Sciences continues to fund projects related to both education and research. Projects funded in 2009 included research on the suspension of aerosol

---

## Fascinating Facts About Calculus

- Archimedes, who lived in the third century B.C.E., derived the formula for the volume of a sphere by using a method that foreshadowed integration. This was understood only when Archimedes's explanation of his method was discovered on a palimpsest in 1906.

- The controversies as to whether Sir Isaac Newton or Gottfried Wilhelm Leibniz (or both) should be credited as the founders of calculus and whether Leibniz stole Newton's ideas led to Leibniz's disgrace. In fact, his secretary was the only mourner to attend Leibniz's funeral.

- The geometric figure known as Gabriel's horn or Toricelli's trumpet is found by revolving the curve $1/x$ about the $x$-axis, beginning at $x = 1$. Calculus shows that this object, an infinitely long horn shape, has an infinite surface area but only finite volume.

- The Bernoulli brothers, Jakob and Johann, proved that a chain hanging from two points has the shape of a catenary, not a parabola, and that of all possible shapes, the catenary has the lowest center of gravity and thus the minimal potential energy.

- The logical foundations of calculus were not established until well after the methods themselves had been employed. One of the critics who spurred this development was George Berkeley, a bishop of the Church of Ireland, who accused mathematicians of accepting calculus as a matter of faith, not of science.

- Calculus, combined with probability, is used in the pricing, construction, and hedging of derivative securities for the financial market.

---

particles in the atmosphere, including relating these concepts to the teaching of calculus, and research in the area of partial differential equations and their applications.

**Industry and Business.** Many branches of engineering use calculus methods and results. In chemical engineering, knowledge of vector calculus and Taylor polynomials is particularly important. In electrical engineering, these same topics, together with an understanding of integration techniques, are emphasized. Mechanical engineering makes significant use of vector calculus and the solving of differential equations. The latter are often solved by numerical procedures such as Euler's method or the Runge-Kutta method when exact solutions are either impossible or impractical to obtain.

In finance, series are used to find the present value of revenue streams with an unlimited number of perpetuities. Calculus is also used to model and calculate levels of risk and benefit for investment schemes. In economics, calculus methods find optimal levels of production based on cost and revenue functions.

## CAREERS AND COURSE WORK

A person preparing for a career involving the use of calculus will most likely graduate from a university with a degree in mathematics, physics, actuarial science, statistics, or engineering. In most cases, engineers and actuaries are able to join the profession after earning their bachelor's degree. For actuaries, passing one or more of the exams given by the Society of Actuaries or the Casualty Actuarial Society is also expected, which requires a thorough understanding of calculus.

In statistics, a master's degree is typically preferred, and to work as a physicist or mathematician, a doctorate is the standard. In terms of calculus-related course work, in addition to the calculus sequence, students will almost always take a course in differential equations and perhaps one or two in advanced calculus or mathematical analysis.

In 2008, about 428,000 people were working as either chemical, electrical, or mechanical engineers in the United States, most in either the manufacturing or service industries. Private industry and the government employed 23,000 statisticians. Insurance carriers, brokers, agents, or other offices provided jobs for 20,000 actuaries. About 15,000 physicists worked as researchers for private industry or the government. The Bureau of Labor Statistics counted only 3,000 individuals as simply mathematicians but tallied 55,000 mathematicians who teach at the college or university level. As of 2010, all these careers were projected to have a job growth rate at or above the national average.

## SOCIAL CONTEXT AND FUTURE PROSPECTS

Calculus itself is not an industry, but it forms the foundation of other industries. In this role, it continues to power research and development in diverse fields, including those that depend on physics. Physics derives its results by way of calculus techniques. These results in turn enable developments in small- and large-scale areas. An example of a small-scale application is the ongoing development of semiconductor chips in the field of electronics. Large-scale applications are in the solar and space physics critical for ongoing efforts to explore the solar system. These are just two examples of calculus-based fields that will continue to have significant impact in the twenty-first century.

*Michael J. Caulfield, B.S., M.S., Ph.D.*

## FURTHER READING

Bardi, Jason Socrates. *The Calculus Wars: Newton, Leibniz, and the Greatest Mathematical Clash of All Time.* New York: Thunder's Mouth Press, 2006. Examines the controversy over who should be considered the originator of calculus.

Dunham, William. *The Calculus Gallery: Masterpieces from Newton to Lebesgue.* Princeton, N.J.: Princeton University Press, 2005. Focuses on thirteen individuals and their notable contributions to the development of calculus.

Kelley, W. Michael. *The Complete Idiot's Guide to Calculus.* 2d ed. Indianapolis: Alpha, 2006. Begins by examining what calculus is, then leads the reader through calculus basics.

Simmons, George F. *Calculus Gems: Brief Lives and Memorable Mathematics.* 1992. Reprint. Washington, D.C.: The Mathematical Association of America, 2007. Includes dozens of biographies from ancient times to the nineteenth century, together with twenty-six examples of the remarkable achievements of these people.

Stewart, James. *Essential Calculus.* Belmont, Calif.: Thomson Brooks/Cole, 2007. A standard text that relates all of the concepts and methods of calculus, including examples and applications.

## WEB SITES

*American Mathematical Society*
http://www.ams.org

*Mathematical Association of America*
http://www.maa.org

*Society for Industrial and Applied Mathematics*
http://www.siam.org

**See also:** Applied Mathematics; Engineering Mathematics; Numerical Analysis; Probability and Statistics.

# CAMERA TECHNOLOGIES

## FIELDS OF STUDY

Cinematography; photography; mechanical and fluid engineering; electromechanical engineering; electrical engineering; computer programming; design and media management; electronics; cartography; computer science; mathematics; material science; digital photography; video production; film and digital production; design and media management; astrophotography.

## SUMMARY

Camera technologies are concerned with the design, development, operation, and assessment of film and digital cameras. The field includes traditional photographic imaging as well as measuring and recording instruments in the fields of science, medicine, engineering, surveillance, and cartography. Camera technologies have widespread applications in almost every area of modern life, from digital video cameras used at banks to record information about customers to long-range space voyagers transmitting images back to scientists on Earth for study. Medical applications also have been developed to help surgeons "see" inside the human body and perform minimally invasive laparoscopic surgeries. Camera technologies continue to evolve and change rapidly as they integrate smaller digital technology for use in medicine, entertainment, industry, and science.

## KEY TERMS AND CONCEPTS

- **Aperture:** Opening in the lens through which light passes.
- **Camera Obscura:** Darkened enclosure, such as a room, with a small aperture in one of the walls that allows light in and casts an inverted image on the opposite wall.
- **Exposure:** Total amount of light allowed to fall on the photographic medium (photographic film or image sensor) during the process of taking a photograph.
- **Focus:** Process of adjusting the camera lens in order to capture an image clearly.
- **Image Capture:** Process of taking a picture and storing it on film or an image sensor.

- **Lens:** Part of the camera that captures the light from the subject and brings it to photographic memory.
- **Negative Image:** Camera image with all the tones of the original scene reversed.
- **Photographic Film:** Sheet of plastic coated with light-sensitive pigments or chemicals used to capture a visible image.
- **Photographic Medium:** Photographic film or image sensor used to capture an image.
- **Pixel Count:** Resolution of an image captured by a digital camera.
- **Positive Images:** Standard image capture by a camera.
- **Processing:** Chemical means by which photographic film and paper are treated after photographic exposure to produce a negative or positive image.
- **Shutter:** Device that allows light to pass for a determined period of time for the purpose of exposing light to photographic film or a light-sensitive electronic sensor to capture a permanent image of a scene.
- **Stereoscopic:** Technique for enhancing a three-dimensional image that creates the illusion of depth by presenting two offset images separately to the left and right eye of the viewer.

### DEFINITION AND BASIC PRINCIPLES

Camera technologies revolve around the science of capturing images. The field includes the design, development, operation, and assessment of both film and digital camera technologies. Film cameras use traditional photosensitive films to capture images by exposing the films to the right amount of light for the right length of time. Digital and video cameras use an array of photosensitive electronic devices or image sensors to capture images electronically and store them as a series of digital data.

Whether the camera is film, digital, or video, it typically consists of five basic parts: the body, the shutter and shutter-release button, the viewfinder, the camera lens complex, and the film or device that captures the image. The camera's body or casing is made of metal or high-grade plastic that holds the camera's other parts together and provides protection for the camera parts. The shutter captures the image, and the shutter-release button tells the camera to take the picture. The viewfinder is the window that shows the object or objects that compose the image that will be captured by the camera. The camera lens complex is made of several smaller pieces. Camera lens complexes control and create different image effects, such as the ability to zoom, focus, and distortion correction, which combine to control the appearance of the image being captured. The optical lens is the curved piece of glass located on the outside of the lenses that focuses light into the camera and onto the film or digital sensor. The aperture is controlled by an aperture ring, which controls the size of the opening and determines how much light goes into the lens. The aperture size is measured by an f-stop number; the smaller the f-stop, the more light is let in. The final component of the camera is the photographic medium used to capture and store the image, film or digital.

The series of scientific discoveries that bring us to the current array of camera technologies began with primitive temporary imaging devices, such as the camera obscura, and evolved to the various cameras using photographic film and digital image capturing. More advanced camera technologies, such as the Hubble Space Telescope and the laparoscope, have greatly furthered medical treatment and scientific knowledge. In both cases, the application of camera technology on a large and small scale allows human operators to direct actions and observe phenomena from a vantage point they cannot achieve with their eyes. The camera-technologies field is integrated in several other fields of science, medicine, engineering, surveillance, and cartography.

## BACKGROUND AND HISTORY

The term camera comes from *camera obscura* (Latin for "dark chamber"), an ancient mechanism for projecting images. Camera obscura was a pinhole device used to produce temporary images on flat surfaces in dark rooms. Camera obscura was improved during the sixteenth and seventeenth centuries, but it was not until 1727, when German physician Johann Heinrich Schulze accidentally created the first photosensitive compound, that the next leap forward in camera technologies occurred. The early nineteenth century saw the application of Schulze's discovery to create development processes that translated temporarily viewed images to permanent images. By

the mid-nineteenth century, direct positive images could be permanently affixed to glass (ambrotypes) or metal (tintypes or ferrotypes). In 1861, another development in camera technologies occurred when Scottish physicist James Clerk Maxwell demonstrated the first color photographs using a system of filters and slide projection called the color separation method. During this same era, the application of photos to record current events and new areas of the country occurred as photographers captured images taken during the American Civil War and of the West. The next innovation, by English physician Richard Leach Maddox in 1871, paved the way for a revolution in camera technologies with the development of the dry plate process—using an emulsion of gelatin and silver bromide on a glass plate. In the late nineteenth century, companies such as Kodak created commercially available cameras that used film to capture pictures. In 1907, the first commercial color film became available, and by 1914, the standard modern 24-by-36-millimeter (mm) frame and sprocketed 35 mm movie film were developed. In 1932, the idea of Technicolor for movies arose. In the Technicolor process, three black-and-white negatives were made in the same camera under different filters to create a colorful finished product. The rest of the twentieth century saw continued technical improvements and optimization of the quality and abilities of cameras and films, including development of multilayer color film, "instant" Polaroid film, underwater cameras, auto-focus cameras, the automatic diaphragm as well as introduction of the single-lens reflex (SLR) camera. The SLR camera changed photography by using a semiautomatic moving mirror system that permitted the photographer to see the exact image that would be captured by the film or digital imaging system. In 1972, Texas Instruments was granted the patent for an all-electric camera. In 1982, camera technologies moved forward again with the introduction of Sony's still video camera. The ability to manipulate and transform images was then revolutionized with the 1990 release of Adobe Photoshop, which allowed nonprofessionals to transform photo images from their home computers. Also in the early 1990's, Kodak brought digital cameras to the general population by developing the photo CD system. The camera phone was introduced in 2000 in Japan. By 2003, more affordable digital cameras were made available and quickly became the dominant image-capturing format, so much so that Kodak ceased production of film cameras in 2004, with Nikon following suit with many of its film cameras as well, in 2006.

## How It Works

The core element in camera technology is the capture of a desired image. The mechanism of capture varies depending on the type of camera used: film or digital. Film cameras capture images on a sheet of plastic coated with light-sensitive pigments (silver halide salts bonded by gelatin). The pigments contain variable crystal sizes that determine the sensitivity, contrast, and resolution of the film. When the film is sufficiently exposed to light or other electromagnetic radiation, it forms an image. The film is then developed using specialized chemical processes to create a visible image. Digital cameras use a special sensor or pixilated metal oxide semiconductors (photodiodes) made from silicon to convert the light that falls onto them into electrons in order to capture a desired image and store it in a memory device. The two most common types of sensors used in digital cameras are the charge-coupled device (CCD) and the complementary metal-oxide semiconductor (CMOS). A film recorded with a digital camcorder is captured in a similar way—saved as a series of frames rather than a single snapshot. In both traditional film and digital cameras, the image itself can be modified using different filters and lenses.

## Applications and Products

**Astronomy and Physics.** Camera technologies have expanded the ability of astronomers to explore the universe far beyond manned space exploration or early telescopes. High-powered telescopes and cameras placed in space have provided photographic evidence of rare astronomical events and features that have resulted in revisions of scientific theory. For example, photographs from the Hubble Space Telescope led to the discovery of dark energy, the hypothetical and unexplained force that seems to be drawing galaxies away from each other. Other unique types of cameras, such as the near-infrared camera Lucifer 1, are powerful tools used to gain spectacular insights into universe phenomena such as star formation.

**Medicine.** Use of camera technologies has revolutionized medicine. Physicians can use cameras and robots to evaluate and treat distant patients as far

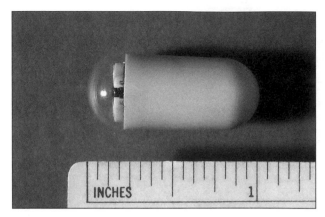

*A digital camera the size of a vitamin is used to transmit images of the gastro-intestinal tract when swallowed.* (Leonard Lessin/Photo Researchers, Inc.)

away as the astronauts on the space station. Closer to home, medical professionals can use tiny cameras and image-capturing devices in pill form to evaluate the inner workings of their patients' intestines. Many surgeries can be performed laparoscopically, where surgeons use tiny cameras on medical instruments that project images of their patients' bodies. The surgeons then use images and controls attached to consoles to guide their surgical implements within the patients' bodies precisely. A major benefit of surgeries using camera technologies is that they are less invasive—smaller incisions need to be made because manipulators using cameras can be extremely narrow. Surgeries can also be performed remotely using imaging-incorporated technology such as the da Vinci Surgical System to conduct operations such as prostatectomy, cardiac surgery, bariatric surgery, and various forms of neurosurgery.

**Journalism.** The adage that "a picture is worth a thousand words" summarizes the incredible impact of images in conveying current events, news, and history. Photojournalists using camera technologies add depth to news stories that cannot always be conveyed through words alone. Examples of the impact of photography in journalism include photos of the famine in Ethiopia and the devastation wrought by Hurricane Katrina and other natural disasters.

**Cartography.** The field of cartography has been revolutionized through the use of camera technologies. The ability to take images of the geologic features of the Earth from the ground or from space satellites has increased the accuracy and speed of mapmaking. In fact, companies such as Google have traveled the United States to take street-level images of addresses that are then incorporated into maps and directions.

**Engineering.** Modern camera development has expanded the capabilities of engineering. Using computer technology and images, engineers can create models to predict a variety of outcomes from the impact of a head-on car collision to the minute changes in an electrical circuit. Modern images and specialized image-analysis systems can be used to enhance an observer's ability to make measurements from a large or complex set of images by improving accuracy, objectivity, or processing speed.

**Entertainment.** The development of camera technologies over time has provided entertainment options from the early moving pictures to specialized gaming systems. The use of advances in cameras can provide a more realistic or personalized experience. For example, a particular Wii fitness game comes with a motion-tracking camera that ensures the user is exercising in the optimal manner.

**Surveillance and Security.** The ability to use camera technologies to track and identify the movements of individuals or intruders often relies heavily on images. Photographic evidence can be used in court as proof of wrongdoing.

**Military Operations.** Satellite imaging using high-powered cameras has transformed war and peacetime military operations, as the images can provide real-time information on the movement of individuals, troops, and resources. Unmanned drone airplanes with cameras allow the military to explore locations that may be dangerous for humans to enter and provide needed information for military operations. The development of image-based technologies such as thermal and infrared cameras to detect changes in geologic features and topography are vital for planning military missions.

**Exploration and Rescue.** Sending robots and other unmanned machines with camera technologies to places too remote or dangerous for human beings to work in—outer space, great ocean depths, and disaster zones—is an important application. The space exploration robots Voyager 1 and 2 have been traveling through the solar system for nearly thirty-three years sending back images of distant planets, their moons, and the Earth itself. Images and cameras were also invaluable during the 2010 Deepwater

Horizon oil spill, as underwater explorer robots equipped with cameras were able to send images of the damaged oil well as well as reports on how repairs were proceeding. Thermal-imaging cameras can also identify regions on fire even as they allow rescue robots to seek out injured humans trapped in fires or under rubble.

## IMPACT ON INDUSTRY

The total value of the camera-technologies industry is difficult to estimate as camera technologies are presently a startling cross-section of industries from photofinishing and retail camera industry to video surveillance to the motion picture and video industries. The camera technologies industry is, however, an expanding market. The video-surveillance market is expected to grow from $11.5 billion in 2008 to $37.7 billion in 2015 in the United States. According to market research company First Research, the U.S. photofinishing and retail camera industry is expanding and currently includes about 4,000 locations with combined annual revenue of about $6 billion. In another sector of the camera technologies, wage and salary employment in the motion picture and video industries is projected to grow 14 percent between 2008 and 2018, compared with 11 percent growth

---

### Fascinating Facts About Camera Technologies

- One of the main missions of the Hubble Space Telescope was to look at the images and discover when the universe was born.
- The first photographic film was highly flammable
- The earliest cameras produced in significant numbers used glass plates to capture images.
- Barack Obama's official portrait was taken with a digital camera, a presidential first.
- The word photography is derived from the Greek words *photo* ("light") and *graphein* ("to write").
- American electrical engineer Steven Sasson, the inventor of the digital camera, was awarded the National Medal of Technology and Innovation in 2010. This is the highest honor bestowed by the U.S. government on scientists and inventors.
- As of 2011, about 80 percent of households in the United States had a digital camera.

projected for wage and salary employment in all industries combined.

**Medicine.** The impact of cameras and camera technologies continues to revolutionize the medical field. From improvements in laparoscopic surgery and noninvasive imaging using tiny cameras, the use of camera technologies in medicine is expanding and improving the safety of medical procedures through knowledge of an individual's specific anatomy and medical issues. In addition, tiny, ingestible cameras can provide real-time imaging of body structure and function.

**United States Military.** The United States government, particularly the military, relies heavily on camera technologies to gather intelligence, perform safety inspections, and conduct military operations. Accordingly, they are one of the biggest sources of funding for camera research and development. The government has funded grants for projects such the improvement of optical communications and imaging systems and the use of magneto-optic imaging technology to detect hard-to-access corrosion on Air Force fighter planes. The purpose of these grants is to improve information gathering, military operations, security, and troop safety.

**Industry and Business.** The manufacturing industry is a large user of camera technologies, which help it to monitor industrial areas, cut costs, and increase efficiency. For example, many manufacturers use video-surveillance cameras to monitor areas that are intolerable to humans, as the cameras can be designed to have a high resistance to weather and corrosion. Video cameras can also be used near heavy-duty machinery in an industrial environment to check the efficiency of the automated processes and safeguard the employees working there.

## CAREERS AND COURSE WORK

There are many careers in a variety of industries that directly correlate to camera technologies, and entry-level requirements vary significantly by position. Given the wide spectrum of difference between the careers, a sampling of careers and course work follows.

The motion picture and video industries provide career options such as cinematographers, camera operators, and gaffers who work together to capture the scripted scenes on film and perform the actual shooting. Formal training can be an asset to workers in film and television production, but experience, talent, creativity, and professionalism are usually the most important factors in getting a job. In addition to colleges and technical schools, many independent centers offer training programs on various aspects of filmmaking, such as screen writing, editing, directing, and acting. For example, the American Film Institute offers training in directing, production, cinematography, screen writing, editing, and production design.

Another camera-technologies career option is photographer. Photographers produce and preserve images that paint a picture, tell a story, or record an event. They often specialize in areas such as portrait, commercial and industrial, scientific, news, or fine arts photography. Employers usually seek applicants with a "good eye," imagination, and creativity, as well as a good technical understanding of photography. Photojournalists or industrial or scientific photographers generally need a college degree. Freelance and portrait photographers need technical proficiency, gained through a degree, training program, or experience. Photography courses are offered by many universities, community colleges, and vocational and technical institutes. Basic courses in photography cover equipment, processes, and techniques. Learning good business and marketing skills is important and some bachelor's degree programs offer courses that focus on those. Art schools offer useful training in photographic design and composition. A good way for a photographer to start out is to work as an assistant to an experienced photographer.

Still another option is a mapmaker or geographer who works with cameras and images to create maps. Photogrammetrists interpret the more detailed data from aircraft to produce maps. Most mapmakers work for architectural and engineering services companies, governments, and consulting firms. Remote-sensing specialists and photogrammetrists often have at least a bachelor's degree in geography or a related subject, such as surveying or civil engineering. Classes in statistics, geometry, and matrix algebra also are useful. Many remote-sensing specialists have degrees in the natural sciences, including forestry, biology, and geology. They often take courses in remote sensing or mapping while earning these degrees. Not everyone working in this field has a bachelor's degree, however. People who have an associate's degree or a certificate in remote sensing or photogrammetry usually begin as assistants and gain additional skills on the job. Taking high school or college-level classes in mapping, drafting, and science can also lead to assistant jobs. Some employers hire entry-level workers who do not have college training but do have an aptitude for math and visualizing in three dimensions.

## SOCIAL CONTEXT AND FUTURE PROSPECTS

The ability to record and view the world will continue to change as new camera technologies allow individuals and companies to explore further. Future technological innovations in image capturing, size of cameras, storage of images, and viewing of images will change as the industry progresses.

Since the early days of camera technology, cameras have been used to bring the reality of different lives and social issues to the general population through books, exhibits, and newspapers. Early examples of this include Jacob Riis's 1890 publication *How the Other Half Lives*, which presented images of tenement life in New York City, and Lewis Hine's commission by the United States National Child Labor Committee to photograph children working in mills in 1909. A more recent example is the interest in the plight of Afghan women, which was stimulated in 1985 by photojournalist Steve McCurry's haunting photo of a young Afghan refugee named Sharbat Gula that appeared on the cover of *National Geographic*.

Employment in camera technologies is increasing for many careers. As an example, employment of photographers is expected to grow 12 percent over the period from 2008 to 2018, about as fast as the average for all occupations.

*Dawn A. Laney, B.A., M.S., C.G.C., C.C.R.C.*

## FURTHER READING

Allan, Roger. "Robotics Give Doctors a Helping Hand." *Electronic Design*, June 19, 2008. Discusses the use of cameras and robots in medicine.

Ang, Tom. *Photography*. New York: Dorling Kindersley, 2005. Covers a number of aspects of photography, both historical and technical.

Chamberlain, P. "From Screen to Monitor." *Engineering & Technology* 3, issue 15 (2008): 18-21. A discussion of movie technology, images, and changes in viewing options.

Freeman, Michael. *The Photographer's Eye: Composition and Design for Better Digital Photos*. Burlington, Mass.: Focal Press, 2007. Discusses how to develop one's photographic eye and compose striking, effective photographs; with illustrations.

Gustavson, Todd. *Camera: A History of Photography from Daguerreotype to Digital*. New York: Sterling Innovation, 2009. An excellent history of the camera and photography.

Heron, Michal. *Creative Careers in Photography: Making a Living With or Without a Camera*. New York: Allworth Press, 2007. Provides information on numerous camera- and photography-related careers and includes a self-assessment tool.

## WEB SITES

*American Photographic Artists*
http://www.apanational.com

*National Press Photographers Association*
http://www.nppa.org

*Professional Photographers of America*
http://www.ppa.com

*Women in Photography International*
http://www.womeninphotography.org

**See also:** Cinematography; Computer Science; Electrical Engineering; Mechanical Engineering; Optics; Photography.

# CARDIOLOGY

## FIELDS OF STUDY

Biology; chemistry; anatomy; physiology; biochemistry; neurology; pharmacology; pathology.

## SUMMARY

Cardiology is the study of the heart. By understanding the heart's normal functional state, applied science can address its abnormalities, designing stents and bypasses for blocked coronary vessels, replacements for faulty valves, and pacemakers to compensate for electrical abnormalities. In addition, cardiopulmonary bypass machines allow the patient to undergo open-heart surgery, and artificial hearts briefly extend the life of a patient waiting for a transplant. Automated external defibrillators for applying an electrical shock to reset a heart are becoming increasingly available in public places such as airplanes and shopping malls.

## KEY TERMS AND CONCEPTS

- **Angiogram:** X ray of blood vessels using radioactive dye.
- **Angioplasty:** Surgical procedure that unblocks blood vessels.
- **Aorta:** Largest artery that carries oxygenated blood to the body.
- **Arrhythmia:** Irregular heartbeat.
- **Atrium:** One of two upper chambers of the heart.
- **Bradycardia:** Lower than normal heart rate.
- **Catheter:** Thin tube that fits within blood vessels and other body tunnels.
- **Defibrillator:** Machine that delivers an electrical shock to the heart.
- **Echocardiogram:** Image of the heart using ultrasound.
- **Electrocardiogram (EKG):** Recording over time of the heart's electrical activity.
- **Pacemaker:** Device implanted in the chest to regulate the heartbeat.
- **Stent:** Implanted device that keeps arteries open for smooth blood flow.
- **Stethoscope:** Medical device for listening to sounds inside the body such as blood flow.
- **Tachycardia:** Higher than normal heart rate.
- **Vena Cava:** Largest vein that delivers deoxygenated blood from the body to the heart.
- **Ventricle:** One of two lower chambers of the heart.

## DEFINITION AND BASIC PRINCIPLES

Cardiology is the branch of medicine that deals with the heart and the cardiovascular system. The heart is a muscular organ in the middle of the chest and consists of four chambers with valves that send blood to the lungs for oxygenation and then send the oxygenated blood throughout the body. It is essentially a pump; the muscle contracts in response to electrical signals that arise from the sinoatrial node, a patch of specialized tissue located at the top of the right atrium. Without oxygenated blood circulating to the body's tissues, a person would die.

Blood that is poor in oxygen enters the right atrium from the vena cava. The right atrium collects this oxygen-poor blood and sends it through the tricuspid valve into the right ventricle. The right ventricle sends this oxygen-poor blood through the pulmonary valve and pulmonary artery to the lungs for oxygenation. Oxygen-rich blood returns to the heart and enters the left atrium through the pulmonary vein. The left atrium collects this oxygen-rich blood and sends it though the mitral valve into the left ventricle. The left ventricle sends this oxygen-rich blood through the aortic valve to the rest of the body. A wall of muscle, called the septum, separates the left and right halves of the heart.

## BACKGROUND AND HISTORY

A great deal of applied technology related to the heart began to evolve in the twentieth century. In 1903, Dutch physiologist Willem Einthoven invented the electrocardiograph; for this, he was awarded the Nobel Prize in Physiology or Medicine in 1924. Werner Forssmann performed the first human cardiac catheterization in Eberswalde, Germany, in 1929. Twelve years later, in New York, André F. Cournand and Dickinson W. Richards first used cardiac catheter techniques to measure cardiac output for diagnostic purposes. In 1956, these three men shared the Nobel Prize in Physiology or Medicine for their work.

In 1952, at the University of Minnesota, American surgeons F. John Lewis and C. Walton Lillehei successfully performed the first open-heart surgery on a human patient. After reducing the body temperature of a five-year-old girl, they repaired a hole in her heart in a ten-minute operation. In 1967, the first whole-heart transplant was performed by South African surgeon Christiaan Barnard. The patient, Louis Washkansky, died eighteen days later of pneumonia.

In 1982, American surgeon William C. DeVries implanted the first permanent artificial heart, the Jarvik-7, invented by American physician Robert K. Jarvik, into Seattle dentist Barney Clark. The dentist lived for another 112 days.

## How It Works

**Blockage of Coronary Vessels.** Blockage in the blood vessels leading to the heart may result from blood clots or fatty deposits called plaques. Blockages may be found on an electrocardiogram, echocardiogram, or angiogram. Small blockages may be managed by lifestyle changes such as increased exercise, smoking cessation, and a diet low in cholesterol. Medications are also available that lower cholesterol levels, lower blood pressure, or thin the blood, minimizing blood clot formation. Applied science has contributed to the development of surgical procedures: An artery can be kept open for adequate blood flow by the insertion of a stent, blood vessels can be expanded by balloon angioplasty, and oxygenated blood can bypass a blocked coronary artery and reach the heart through an artery from another part of the body that is used to connect the blocked artery and the aorta.

**Faulty Heart Valves.** A heart valve may function improperly as the result of a birth defect, an infection, or age-related changes. It may also become deformed as the result of damage and scar tissue. Properly functioning valves permit only the one-way flow of blood. One malfunction, called regurgitation, occurs when the valve does not close tightly, allowing blood to reenter the chamber from which it came. Another malfunction, called stenosis, occurs when the valve becomes thick and stiff so it does not open completely and insufficient blood passes through with each heartbeat. The third malfunction, called atresia, occurs when the valve is fused shut, not allowing blood to pass. Untreated valve malfunctions may progress to cause sudden cardiac arrest and

death. Less severe valve malfunctions may be managed by lifestyle changes and medications to relieve the symptoms. More severe valve malfunctions may be surgically corrected by repair or replacement. Replacement tissue valves may come from cows, pigs, or human cadavers. They last ten to fifteen years in older, less-active patients but wear out and must be replaced sooner in younger, more active patients. Applied science has developed mechanical replacement valves that are intended to last beyond a patient's lifetime, so that only one surgery is needed. However, the risk of blood clot formation is higher, so the patient must remain on anticoagulant drugs.

**Electrical Irregularities.** Electrical irregularities of the heart may result in a heartbeat that is too fast (tachycardia), too slow (bradycardia), or erratic (arrhythmia). Such dysfunction may be detected using a stethoscope, feeling peripheral pulses, or generating an electrocardiogram. Treatment depends on the stability of the patient's condition. In some cases, physical maneuvers may be employed to regulate occasional palpitations. In other cases, medications that prevent arrhythmia may be prescribed; these must be taken in conjunction with anticoagulant drugs to reduce the risk of blood clot formation. In cases of chronic bradycardia, an electrical pacemaker may be implanted to deliver a shock to the heart when the heart rate falls too low.

## Applications and Products

**Stethoscopes and Electrocardiograms.** Electrical irregularities such as an erratic heartbeat may be detected through a stethoscope, felt in a peripheral pulse, or seen as an abnormal tracing on electrocardiography. The electrocardiography invented by Dutch physiologist Einthoven in 1903 was a large tablelike contraption. The patient put his or her hands and feet in buckets of saltwater to facilitate electrical conduction. Four decades later, the machine had become smaller and received input from wire leads on metal disks attached to the patient's wrists and ankles. Modern machines are compact and easily transported on a wheeled cart from one exam room to another. The wire leads clip to disposable, self-adhesive disks for easy placement and removal.

**Pacemakers.** Heartbeats that flutter instead of beating strongly and regularly may be corrected by electrical stimulation from a cardiac pacemaker. In 1950, Canadian John A. Hopps invented the first

## Fascinating Facts About Cardiology

- The average adult heart weighs 7 to 15 ounces and is slightly smaller than two clenched fists.
- Each day, the heart beats 100,000 times and pumps 2,000 gallons of blood. Most adults have a total blood volume of 10 pints, which is 1.25 gallons, in their body.
- A healthy heart beats with enough pressure to shoot blood 30 feet.
- In 1949, a crude prototype of an artificial heart was built by two Yale doctors from an erector set, small cheap toys, and mismatched household items. It kept a dog alive for more than an hour.
- A modern version of the Jarvik-7 total artificial heart has been implanted in more than eight hundred people since 1982 but each device was removed when a donor heart became available.
- In the 1980's, the external pneumatic power sources that drove artificial hearts were large and based on milking machines.
- The prevalence of heart disease increased so dramatically between 1940 and 1967 that the World Health Organization called it the world's most serious epidemic.
- Surgeons got the idea to lower the body temperatures of patients to slow their metabolism and heart rate during surgery from observing hibernating groundhogs.

cardiac pacemaker; it was external because it was simply too heavy to be implanted into the chest. His background in electrical engineering led him to be called the father of biomedical engineering. When transistors replaced vacuum tubes, the pacemaker became less cumbersome, and in 1958, Colombian scientist Jorge Reynolds Pombo designed the first internal cardiac pacemaker.

**Defibrillators.** A heart that has just stopped beating or that is pumping with no apparent rhythm may be restarted with an electrical charge delivered by a defibrillator. The shock disrupts the chaotic heart action and allows the sinoatrial node to resume its regulatory function. Automated external defibrillators are becoming increasingly available for emergency situations in public places. When activated, these machines deliver audible instructions so that untrained bystanders may use them effectively.

**Stents.** A coronary stent is a wire-mesh tube that is inserted into an artery to improve blood flow. It is placed as part of an angioplasty to remove a blood clot or plaque deposit and remains permanently. Charles Dotter invented the first coronary stent in 1969 and implanted it in a dog. Stents were implanted in humans in Europe as early as 1986. The U.S. Food and Drug Administration (FDA) approved the Palmaz-Schatz stent, a balloon-expandable coronary stent, for human use in 1994.

**Angiography, Angioplasty, and Catheters.** A cardiac catheter is a long, small-diameter tube that is threaded to the heart through the femoral artery. When contrast dye is injected through the tube, the coronary arteries may be visualized on X rays, indicating the location of any blockage; this process is called coronary angiography. These catheters and specific techniques for using them were developed by American radiologist Melvin Judkins in the 1960's.

One such catheter, a balloon-tipped catheter, is used to open a blocked artery. When the catheter with a balloon fitted on its tip reaches the site of a blockage, the balloon is inflated to enlarge the interior of the artery by flattening plaque deposits against the vessel wall. In 1977, German cardiologist Andreas Gruentzig performed the first balloon angioplasty in a human in Zurich, Switzerland.

When angioplasty cannot sufficiently open a closed artery, coronary artery bypass surgery may be performed. In this procedure, an artery from another part of the body is grafted to the heart to reroute blood flow around the blockage. This procedure requires open-heart surgery, for which the patient must be put on a heart-lung machine (cardiopulmonary bypass machine) to remain alive while the heart is not beating. The patient's blood is pumped from the body through the vena cava into the machine, where it is filtered, cooled, diluted with a specific solution to lower its viscosity, and oxygenated. The blood is then returned to the body through the ascending aorta. The patient is also given an anticoagulant to prevent the formation of blood clots.

**Heart Valves.** The cardiopulmonary bypass machine, which allows blood flow to bypass the heart and lungs, is also used in other surgical procedures, such as valve repair and replacement, repairs of septal defects and congenital heart defects, and heart transplantation. Surgeons repair or replace heart valves in 99,000 operations per year in the United States. The valves most commonly affected are those on the left side of the heart, namely the mitral and aortic valves, because they are exposed to higher blood pressure than those on the right side of the heart to pump oxygenated blood into the body.

When any one of the four heart valves requires replacement, the transplanted valve may be from a cow, pig, or deceased human, or it may be mechanical. Biological valves of animal tissue last only about ten years and are better suited for use in older, less-active patients. Mechanical valves are typically fashioned from plastic, carbon, or metal. Although they last longer than tissue valves and seldom require replacement, blood clots may form on their surface, so the recipient must take an anticoagulant for the rest of his or her life.

**Artificial Hearts.** An artificial heart is a machine that substitutes for a heart in which both halves no longer function properly. It is generally used as a temporary measure until a healthy human heart becomes available for transplantation. However, the goal of ongoing development is to create a lightweight, durable, functional machine that does not need to be replaced and will not be rejected by the body. Various models of artificial hearts are available; some have tubes for pumping pressure and wires for electrical charging that attach to equipment outside of the body; however, later models are designed to be completely enclosed within the chest. Novel biomaterials lessen the risk of foreign-body rejection and complete enclosure reduces the risk of infection.

## IMPACT ON INDUSTRY

**Private Industry.** Private industry has created a market known as worldwide cardiac rhythm management, which covers the research and development, manufacturing, and marketing of cardiac-related medical devices such as internal and external pacemakers, defibrillators, and monitoring leads. Regulatory issues involve countries that trade parts and processes, including the United States, Japan, China, Korea, India, Australia, and France. Companies are also developing technology such as wireless heart monitors so that physicians may receive a patient's cardiac data over the Internet and stents coated with drugs that discourage scarring and collapsing of arteries that have been unblocked with angioplasty. Because the demand for transplantable organs consistently exceeds the available supply, companies are continuing to develop artificial heart valves and artificial hearts from biomaterials that are compatible with magnetic resonance imaging as well as the body's immune system. Biometric studies are being conducted to expand product lines to accommodate the smaller sizes of women, children, and infants.

The countries in the European Union and the European Free Trade Area have an average of fifty-eight cardiologists per million inhabitants. In the United States, there are seventy cardiologists per million inhabitants. A density below fifty is considered insufficient and a density greater than eighty is excessive. However, shortages are predicted in the future. Cardiologists and cardiology researchers are needed in various settings.

**The Human Heart**

aortic arch
pulmonary artery
superior vena cava
pulmonary veins
right atrium
right ventricle
inferior vena cava
right bundle branch
aorta
pulmonary artery
left atrium
pulmonary veins
left ventricle
left bundle branch

*A tirelessly-working organ, the human heart, composed of muscle, pumps nourishment, in the form of blood, to the human body's every cell.*

**National Agencies.** Heart disease is the number-one killer in the United States. The American Heart Association is committed to research and education aimed at reducing deaths from heart disease. Funding is available for research in such areas as cardiovascular aging, pediatric cardiomyopathy, and Friedreich's ataxia (a genetic cause of heart disease).

The National Institutes of Health oversees the National Heart Lung and Blood Institute, which in June, 1991, launched the National Heart Attack Alert Program. The goals of this program are to reduce the severity of heart attacks through early detection and treatment and to improve the quality of life for patients and their families.

The Food and Drug Administration, the agency that oversees pharmaceuticals, alerts consumers to heart-related risks associated with newly developed medications such as antidiabetes drugs, attention deficit hyperactivity disorder drugs, and diet pills. The FDA oversees the manufacturing of medical devices and takes action when devices such as stents or pacemakers are suspected of acting in a faulty manner.

**Research Hospitals.** At the Covenant Heart and Vascular Institute, a member of the Covenant Health System in West Texas, research and clinical trials are conducted to develop therapeutics, procedures, and medical devices for the treatment of heart disease. It is an accredited Cycle III Chest Pain Center with Percutaneous Coronary Intervention, one of only fourteen in the state. This certificate is for excellence in providing acute cardiac medicine.

The Heart Institute of Childrens Hospital Los Angeles is part of a fifteen-member consortium of pediatric cardiac institutes called the Pediatric Heart Network. Its research focus is the optimal survival of babies born with hypoplastic left heart syndrome, in which the left ventricle is missing or underdeveloped; without corrective surgery, the condition is fatal.

The Jim Moran Heart and Vascular Research Institute at Holy Cross Hospital in Fort Lauderdale, Florida, conducts cardiac research in collaboration with private corporations, university medical centers, and other research institutes.

**University Medical Centers.** The Dorothy M. Davis Heart & Lung Research Institute at Ohio State University Medical Center engages in research into the basic science of heart disease, including such processes as natural cell death and inflammation.

Researchers are also seeking gene-based information that will allow therapies to be customized.

The Penn Cardiovascular Institute at the University of Pennsylvania provides education, research, and patient care. The researchers there have been awarded grant money to study stem cell therapies.

**Nonprofit Organizations.** The Florida Heart Research Institute is a nonprofit organization dedicated to stopping heart disease through research, education, and prevention. It facilitates collaborative research nationally and offers free community services locally.

The Heart Disease Research Institute is an international nonprofit organization based in Arizona that offers information on the prevention and treatment of heart disease to the general public.

The American College of Cardiology is a national medical society that makes transparent its members' relationships with drug and device manufacturers, especially regarding research funding, publications, and continuing medical education for practitioners.

### CAREERS AND COURSE WORK

According to the U.S. Bureau of Labor Statistics, cardiologists will see a 27 percent growth in their field by 2014. To become a cardiologist, a person must complete an undergraduate college degree, majoring in a life science such as biology, chemistry, or biochemistry. That graduate must then complete a medical degree, studying detailed anatomy, physiology, pathology, and pharmacology. That physician must then complete additional training in cardiology, through residencies and fellowships, and become board certified as a cardiologist. Cardiologists may further specialize in areas such as cardiac diagnosis, pediatric cardiology, and electrophysiology. Cardiologists commonly have a private practice with hospital privileges.

Cardiac researchers must also complete an undergraduate college degree, majoring in a life science. That graduate may follow the same path as a cardiologist but most choose instead to pursue additional education in master's and doctoral programs. They then do postdoctoral work and become published authors before they apply for and receive funding for their own research projects. Most cardiac research laboratories are associated with university medical schools, although some are associated with government agencies or private companies such as pharmaceutical and medical device manufacturers.

Jobs are available for cardiac technicians who set up electrocardiograms, echocardiograms, and other diagnostic procedures in a cardiologist's office or a hospital cardiology department. Educational opportunities vary; some schools offer specific vocational programs with job placement, although some physicians prefer to train a college graduate who majored in a life science.

## SOCIAL CONTEXT AND FUTURE PROSPECTS

Researchers are continuing their efforts to develop an artificial heart that will sustain patients for longer periods while they wait for a transplant, with the ultimate goal of implanting a mechanical heart that will not need replacement. The AbioCor artificial heart, manufactured by Abiomed of Danvers, Massachusetts, was the first self-contained implantable device; previous versions of an artificial heart required patients to remain in bed connected to machines with tubes and electrodes. The AbioCor device is powered by an external battery pack, allowing the patient to be ambulatory. In human trials, this device was implanted in patients whose life expectancy was thought to be less than 30 days. The goal was to extend their lives by an additional 30 days. This device received FDA approval on September 5, 2006. So far, fifteen patients have received an AbioCor artificial heart; the longest a recipient lived was 512 days.

A modern prototype of a fully implantable artificial heart contains cutting-edge electronic sensors, synthetic microporous skins, and other novel biomaterials.

*Bethany Thivierge, B.S., M.P.H.*

## FURTHER READING

Holler, Teresa. *Cardiology Essentials.* Sudbury, Mass.: Jones and Bartlett Learning, 2007. Presents cardiology with a practical clinical orientation for medical personnel working in a cardiology office.

Mueller, Richard L., and Timothy A. Sanborn. "The History of Interventional Cardiology: Cardiac Catheterization, Angioplasty, and Related Interventions." *American Heart Journal* 129, no. 1 (January, 1995): 146-172. Contains plenty of names, dates, and details of interest in cardiology history.

Murphy, Joseph G. *Mayo Clinic Cardiology: Concise Textbook.* 3d ed. London: Informa Healthcare Communications, 2006. This easy-to-read textbook features information on all aspects of cardiology.

Topol, Eric J., ed. *Textbook of Cardiovascular Medicine.* 3d ed. Philadelphia: Lippincott Williams & Wilkins, 2006. A complete, well-organized, user-friendly reference book, complete with audio and visual aids.

## WEB SITES

*Alliance of Cardiovascular Professionals*
http://www.acp-online.org

*American College of Cardiology*
http://www.cardiosource.org

*American Heart Association*
http://www.heart.org/HEARTORG

*Heart Disease Research Institute*
http://heart-research.org

*National Institutes of Health*
National Heart Lung and Blood Institute
http://www.nhlbi.nih.gov

**See also:** Artificial Organs; Bioengineering; Bionics and Biomedical Engineering; Cell and Tissue Engineering; Geriatrics and Gerontology; Pediatric Medicine and Surgery.

# CELL AND TISSUE ENGINEERING

## FIELDS OF STUDY

Cell biology; cardiology; cardiac surgery; biochemistry; organic chemistry; developmental biology; physiology; transplant surgery; stem cell research; biomaterial science; drug and gene delivery; neuroscience; bioinformatics; molecular engineering; orthopedic surgery; mechanical engineering; physical therapy; biophysical tools; medical ethics; public health.

## SUMMARY

Cell and tissue engineering are fields dedicated to discovering the mechanisms that underlie cellular function and organization to develop biological or hybrid biological and nonbiological substitutes to restore or improve cellular tissues. The most immediate goal of cell and tissue engineering is to allow physicians to replace damaged or failing tissues within the body. The field was first recognized as a distinct branch of bioengineering in the 1980's and has since grown to attract participation from numerous medical and biological disciplines.

Engineered cellular materials may be used to grow new tissue within a patient's heart or to replace damaged bone, cartilage, or other tissues. In addition, research into the mechanisms affecting cellular organization and development may aid in the treatment of congenital and developmental disorders. Cell and tissue engineering has developed in conjunction with stem cell research and is therefore subject to debate over the ethics of stem cell research.

## KEY TERMS AND CONCEPTS

- **Bioartificial Device:** Substance, tissue, or organ that combines biological and synthetic components.
- **Bioengineering:** Medical or biological application of engineering principles, including the process of engineering biological tissues and components from raw materials.
- **Biomaterial:** Cellular or synthetic material that can be introduced into living tissues; often part of a medical device.

- **Cardiology:** Branch of medicine concerned with the disorders, diseases, and function of the heart and associated systems.
- **Cell Therapy:** Introduction of cells or tissues to treat disease or other physiological disorders.
- **Differentiation:** Processes by which cells change morphology and function to fill a specific role within an organism or tissue.
- **Drug And Gene Delivery:** Field of study dedicated to the methods involved in introducing medical chemicals and genes to an organism.
- **Extracellular Matrix:** Substance surrounding cellular tissues in which connecting tissues are fixed.
- **Growth Factor:** Substance that stimulates growth of a cell, tissue, or other part of an organism.
- **Heterologous:** Derived from an organism of a different species.
- **In Vitro:** Outside a living body or organism, in an artificial environment.
- **In Vivo:** Within a living body or organism.
- **Orthopedics:** Branch of medicine concerned with disorders, diseases, and injuries to the skeleton and associated tissues.
- **Regenerative Medicine:** Branch of medicine concerned with applying techniques and tools from a variety of disciplines to restore or repair damaged tissues, cells, and organs by stimulating the biological healing and regeneration processes.
- **Rejection:** Immune response in which the host's immune system attempts to defend against cells or tissues introduced from a foreign organism.
- **Stem Cell:** Undifferentiated cell type that gives rise to specialized cells within the body; most are derived from embryonic tissues.
- **Transplant:** Transfer of an organ, tissue, or other cellular material from one individual to another.

### DEFINITION AND BASIC PRINCIPLES
Cell and tissue engineering is a branch of bioengineering concerned with two basic goals: studying and understanding the processes that control and contribute to cell and tissue organization and developing substitutes to replace or improve existing tissues in an organism. Substitute tissues can be composed either of biological materials or of a blend of biological and nonbiological materials.

The basic goal of cell and tissue engineering is to create more effective treatments for tissue degeneration and damage resulting from congenital disorders, disease, and injury. Engineers may, for instance, introduce foreign tissues that have been modified to stimulate healing within the patient's own tissues, or they may implant synthetic structures that help control and stimulate cellular development. Another goal in cell and tissue engineering is to create tissues that are resistant to rejection from the host organism's immune system. Rejection is one of the primary difficulties in organ transplant and limb replacement surgery.

One of the basic principles of cell and tissue engineering is to use and enhance an organism's innate regenerative capacity. Engineers therefore examine the ways that tissues grow and change during development. Using cutting-edge development in genomics and gene therapy, engineers are working to develop ways to stimulate a patient's immune system and enhance healing.

Cell and tissue engineering have a wide variety of potential applications. In addition to creating new therapies, engineering principles can be used to create new methods for delivering drugs and engineered cells to target locations within a patient. The potential applications of cell and tissue engineering depend on the capability to create cultures of cells and tissues to use for experimentation and transplantation. Research on cell growth is a major facet of the bioengineering field.

## BACKGROUND AND HISTORY

Cell and tissue engineering emerged from a field of study known as regenerative medicine, a branch concerned with developing and using methods to enhance the regenerative properties of tissues involved in the healing process. Ultimately, cell and tissue engineering became most closely associated with transplant medicine and surgery.

Medical historians have found documents from as early as 1825 recording the successful transplantation of skin. The first complete organ transplants occurred in the 1950's, and the first heart transplant was completed successfully in 1964.

The science of cell and tissue engineering arose from attempts to combat the problems that affect transplantation, including scarcity of organs and frequent issues involving rejection by the host's immune system. In the 1970's and 1980's, scientists began working on ways to build artificial or semi-artificial substitutes for organ transplants. Most early work in tissue engineering involved the search for a suitable artificial substitute for skin grafts.

By the mid-1980's, physicians were using semi-synthetic compounds to anchor and guide transplanted tissues. The first symposium for tissue engineering was held in 1988, by which time the field had adherents around the world. The rapid advance of research into the human genome and genetic medicine in the mid-1990's had a considerable effect on bioengineering. In the twenty-first century, cell and tissue engineers work closely with genetic engineers in an effort to create new and better tissue substitutes.

## HOW IT WORKS

Broadly speaking, cell and tissue engineering involves creating cell cultures and tissues that are introduced to an organism to repair damaged or degenerated tissues. There are a wide variety of techniques and specific applications for cell and tissue engineering, ranging from cellular manipulation at the chemical or genetic level to the creation of artificial organs for transplant.

Most cell and tissue engineering methods share several common procedures. First, scientists must produce cells or tissues. Next, engineers must tell the cells what to do. This can be done in a variety of ways, from physically manipulating cellular development and tissue formation to altering the genes of cells in such a way as to direct their function. Finally, engineered tissues and cells must be integrated into the body of the host organism under controlled conditions to limit the potential for rejection. Cell and tissue engineering can be divided into two main categories, in vitro engineering and in vivo engineering.

**In Vitro Engineering.** In vitro engineering is the development of cell cultures and tissues outside of the body in a controlled laboratory environment. This method has several advantages. Producing tissues in a laboratory has the potential for growing large amounts of tissue and eventually entire organs. This could help solve a major issue with transplant surgery: the scarcity of viable organs for transplantation. Scientists can more precisely control the growing environment and can therefore exert greater control over developing cells and tissues. In vitro engineering

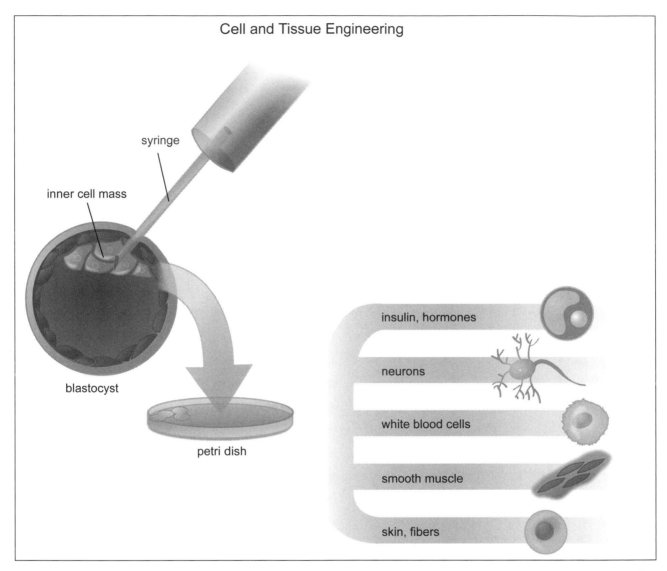

## Cell and Tissue Engineering

*One of the most immediate goals of cell and tissue engineering is to allow physicians to replace damaged or failing tissues within the body.*

allows engineers to modify and adjust cellular properties without the need for surgery or invasive techniques.

In vitro engineering is commonly used in the creation of skin tissues, cartilage, and some bone replacement tissues. Although in vitro techniques have certain advantages, they have serious drawbacks, including a higher rejection rate for cells and tissues created in vitro. In addition, there are physiological advantages to engineering within the host organism's body, including the presence of accessible cellular nutrients.

**In Vivo Engineering.** In vivo engineering is the family of techniques that involves creating engineered cellular cultures or tissues within the host's body. It involves the use of chemicals to alter cellular function and the use of synthetic materials that interact with the host's body to stimulate or direct cellular growth.

In vivo procedures typically involve introducing only minor changes to the host's internal environment, and therefore, these tissues are more likely to be resistant to rejection. In addition, working in vivo allows engineers to take full advantage of the host's existing cellular networks and the physiological environment of the body. The body provides the essential nutrients, exchange of materials, and disposal of waste, helping create healthy tissues.

The primary disadvantages of the in vivo approach are that engineers have less direct control over the development of the cells and tissues and cannot make exact changes to the microenvironment during development. In addition, in vivo engineering does not allow for the production of mass quantities of cells and is therefore not an avenue toward addressing the shortage of available tissues and organs for transplant.

## APPLICATIONS AND PRODUCTS

Hundreds of bioengineers are working around the world, and they have created a wide variety of applications using cell and tissue engineering research. Among the most promising applications are cell matrices and bioartificial organ assistance devices.

**Cell Matrices.** In an effort to improve the success of tissue transplants, bioengineers have developed a method for using artificial matrices, also called "scaffolds," to control and direct the growth of new tissues. Using cutting-edge microengineering techniques and materials, engineers create three-dimensional structures that are implanted into an organism and thereafter serve as a "guide" for developing tissues.

The scaffold acts like an extracellular matrix that anchors growing cells. New cells anchor to the artificial matrix rather than to the organism's own extracellular material, allowing engineers to exert control over the eventual size, shape, and function of the new tissue. In addition, scaffolds can aid in the diffusion of resources within the growing tissue and can help engineers direct the placement of functional cells, as the scaffold can be installed directly at the site of an injury.

Matrices may be constructed from a variety of materials, including entirely synthetic combinations of polymers and other structures that are created from derivatives of the extracellular matrix. Many researchers have been designing scaffolds that dissolve as the tissues form and are then absorbed into the organism. These biodegradable scaffolds allow engineers to avoid further surgical procedures to remove implanted material.

Cellular scaffolds represent a middle ground between in vivo and in vitro engineering. Engineers can create a scaffold in a laboratory environment and can allow tissue to anchor and grow around the matrix before implantation, or they can place a scaffold in their target area within the organism and allow the organism's own cells to populate the matrix.

Scaffolds have been used successfully in cardiac repair, especially in conjunction with stem cells. A scaffold seeded with stem cells may be implanted directly into a heart valve, roughly at the site where a cardiac infarction has occurred. The scaffold then directs the growing cells toward the injured area and facilitates regeneration of damaged tissue.

Artificial matrices have also been successful in treating disorders that affect the kidney, bone, and cartilage. Researchers are hopeful that cellular scaffolds could eventually allow the creation of entire organs by coaxing cells to develop around a scaffold designed as an organ template.

**Bioartificial Organs.** One of the major areas of research in tissue engineering is the creation of machines that assist organs damaged by disease or injury. Made from a combination of synthetic and organic materials, these machines are sometimes called bioartificial devices.

One of the most promising organ assistance devices is the bioartificial liver (BAL), which has been developed to help patients suffering from congenital liver disease, acute liver failure, and other metabolic disorders affecting the liver. The BAL consists of cells incorporated into a bioreactor, which is a small machine that provides an environment conducive to biological processes. Cells growing within the BAL receive optimal nutrients and are exposed to hormones and growth factors to stimulate development. The bioreactor is also designed to facilitate the delivery of any chemicals produced by the developing tissues to surrounding areas.

The BAL performs some of the functions usually performed by the liver: It processes blood, removes impurities, produces proteins, and aids in the synthesis of digestive enzymes. The BAL is not intended to permanently replace the liver but rather to supplement liver function or to allow a patient to survive until a liver transplant can be arranged. The bioartificial liver enables patients to forgo dialysis treatments, and some researchers hope to develop BAL devices that may function as a permanent replacement for patients in need of dialysis.

Researchers are working on bioartificial kidney devices that would aid patients with diabetes and other disorders leading to kidney failure. Again, the bioartificial kidney devices are bioreactors, using stem cells and kidney cells to perform some of the purification and detoxification functions of the kidney. Researchers are also developing bioartificial devices to treat disorders of the pancreas and the heart and to help patients suffering from nervous system or circulatory disorders. Taken as a whole, the development of organ assistance devices may be a step toward the development of bioartificial devices that can function to fully replace a patient's malfunctioning organ.

## IMPACT ON INDUSTRY

Bioengineering has become one of the fastest growing fields in medical and biological research. As of 2008, the global market was estimated to have an approximate value of $1.5 billion, according to a study by Life Science Intelligence.

Analysts have said that the bioengineering market has not reached its full potential, as research organizations are only beginning to use the potential funding and resources available for cell and tissue engineering applications. According to Life Science Intelligence, the global market for tissue and cell engineering may reach $118 billion by 2013.

In 2008, the United States controlled more than 90 percent of the bioengineered products market. Europe and Japan were the next leading providers, and both the European and the Japanese markets were expected to continue to grow. There are more than two hundred companies in the United States working to provide research, equipment, and products in the industry.

**Government Funding and Research.** In the United States, a variety of public funding organizations, including the National Institutes of Health, the National Science Foundation, and the National Institute of Standards and Technology support cell and tissue research.

Within the National Institutes of Health, the government's largest research and funding organization, there are six divisions that provide funding for projects dealing with bioengineering, including the National Institute of Biomedical Imaging and Bioengineering. The National Institute of Standards and Technology has two divisions that provide bioengineering grants and the National Science Foundation's Division of Bioengineering and Environmental

Systems also provides grants and assistance for tissue and cell engineers.

The America COMPETES Act, passed by Congress in August of 2007, provides funding for emerging research to promote American dominance in developing scientific fields. Funding from the federal government was used to create the Technology Innovation Program in the National Institute of Standards and Technology, which in turn provides funding for bioengineering programs in addition to a variety of other research initiatives. The Technology Innovation Program received more than $65 million of the institute's 2009 budget of $813 million.

**Regulation and Legal Issues.** Each country maintains its own laws and systems for regulating biotechnology. As much of the technology involved in cell and tissue engineering is newly developed or emerging, regulatory agencies and government officials are still debating the best way to monitor and regulate biotechnology.

The European health ministers completed a set of regulatory guidelines in 2007 designed to increase patient access to emerging treatments from biotechnology and to provide a framework for determining the health risks of biomedical research and applications. The legislation was accompanied by increased European funding for researchers investigating the benefits of bioengineering in the medical sciences.

In the United States, biotechnology is one of the most heavily regulated fields, and most emerging technology falls under the auspices of the Food and Drug Administration. While some researchers believe that United States regulations are too stringent and unnecessarily delay distribution of new technologies and therapies, the United States continues to be a world leader in the field.

**Biotech Corporations.** Biotechnology is experiencing rapid growth because in addition to the research being conducted in universities and government-sponsored programs, for-profit companies are investing in the field. As of 2009, more than two hundred companies in the United States were engaging in biotechnology research. A number of profit-driven bioengineering firms centered in Europe and Japan also have begun competing in the global market.

Major American companies involved in tissue and cell engineering include medical industry leaders, such as Pfizer, Johnson & Johnson, and Novartis, and dozens of smaller corporations. Although government funding for bioengineering has increased in the twenty-first century, private and corporate investment, largely from companies with other medical investments including pharmaceuticals and medical equipment manufacturing, accounts for more than 90 percent of available funding in the field.

## Careers and Course Work

Students interested in cell and tissue engineering might start at the undergraduate level, working toward a degree in biology or biochemistry, with a focus on cellular biology. Students might also enter the bioengineering field with a background in engineering, though students will still need a significant background in biology and medical science.

After achieving an undergraduate education, students can progress in the field by pursuing graduate studies in cell biology, bioengineering, or related fields. Many professionals working in other disciplines, such as orthopedic medicine, dermatology, and cardiac surgery may also become involved with cell and tissue engineering during their careers. Graduate institutions are increasingly trying to introduce programs that focus on cell and tissue engineering. The University of Pittsburgh and Duke University are among the universities offering specializations in tissue and cell engineering for qualified graduate candidates.

Professionals seeking work in the cell and tissue engineering field can seek employment with nonprofit research institutions, such as those in many universities. Positions in universities are generally funded by a combination of public and private funding. Additionally, those interested in bioengineering careers can find employment within a large number of corporations. The more than two hundred companies in the United States involved in biotechnology employ chemists, mechanical engineers, physicians, and individuals trained specifically in bioengineering.

## Social Context and Future Prospects

Bioengineering is intended to improve daily life, both for those suffering from injury and illness and for the population at large. Cell and tissue engineers are focusing on ways to replace damaged tissues, providing, for instance, new skin where skin has been destroyed, and technology to supplement the function of essential organs. One of the ultimate goals of the industry is to create artificial organs that can fully

---

### Fascinating Facts About Cell and Tissue Engineering

- In 2008, scientists from a number of European countries reported on the first procedure to install a bioengineered trachea, constructed from synthetic materials and cultures produced from stem cells, in a woman with a failing respiratory system.

- In August of 2009, a team of Italian scientists announced they were working on an innovative method to replace damaged bone with substitute bone made from wood.

- The Russ Prize, given by the National Academy of Engineering since 1999, is considered the Nobel Prize for bioengineering. Past winners have come from both the engineering and the medical fields.

- Scientists at the Fraunhofer Institute in Germany are working on creating artificial human organs that can be used to replace animal subjects in clinical experiments, allowing scientists to achieve more accurate results and to avoid costly and controversial animal trials.

- The large number of soldiers who lost limbs while serving in Iraq has prompted the U.S. Department of Defense to invest in a University of Michigan program aimed at creating prosthetic hands that can transmit touch sensations to a patient's brain.

- In 2009, scientists in Germany revealed a plan to institute an automated process to produce synthetic skin. Automation would be a first necessary step toward producing sufficient quantities of skin to meet all needs.

---

and permanently replace damaged organs. Bioengineers are confident that in the future it will be possible to provide patients with a variety of organs including a heart, liver, or pancreas.

Although most cell and tissue engineers focus on combating physical illness and injury, bioengineering also has the potential to produce technology that will allow humans to improve their functional abilities. At some point, combinations of synthetic computer technology and biological components could be used to improve human visual capacity or to endow humans with more precise access to memory.

Humans are not the only targets for bioengineers, as other organisms may also be altered to improve their basic physiological functions. Take, for instance, a 2008 project from the Australian Center for Plant Functional Genomics in which researchers are attempting to bioengineer plants that can withstand higher levels of salt in the soil, a breakthrough that could turn into a major benefit for agriculture. Salt-resistant strains of important agricultural crops could grow where agriculture was previously impossible because of the soil's alkalinity.

As a distinct discipline, bioengineering is relatively new and scientists have only begun to investigate the potential applications and discoveries possible with further research. As the field has begun to expand, so too have opportunities for scientists, engineers, and physicians interested in exploring the future of medicine and science. The bioengineering field has already created billions in revenue and is still in a state of rapid growth. Universities, hospitals, and biomedical corporations are likely to increase their investment in these emerging technologies and techniques, creating a strong and growing industry for many years to come.

*Micah L. Issitt, B.S.*

**FURTHER READING**

Chien, Shu, Peter C. Y. Chen, and Y. C. Fung, eds. *An Introductory Text to Bioengineering.* Hackensack, N.J.: World Scientific Publishing, 2008. While definitely written with advanced science students in mind, this text is one of the most basic and yet comprehensive texts available as an introduction to all types of bioengineering.

De Gray, Aubrey, and Michael Rae. *Ending Aging: The Rejuvenation Breakthroughs That Could Reverse Aging in Our Lifetime.* New York: St. Martin's Griffin, 2008. An investigation of research programs in bioengineering, nutrition, and other fields of medicine that are aimed at prolonging life. Provides interesting coverage of organ transplantation and cellular manipulation.

Mataigne, Fen. *Medicine by Design: The Practice and Promise of Biomedical Engineering.* Baltimore: The Johns Hopkins University Press, 2006. An introduction to and investigation of bioengineering and the potential future of the field. Provides discussions of issues such as bioreactors and organ replacements.

Rose, Nickolas. *The Politics of Life Itself: Biomedicine, Power, and Subjectivity in the Twenty-first Century.* Princeton, N.J.: Princeton University Press, 2006. An introduction to the moral, ethical, and political issues that surround medical engineering, genetic manipulation, and bioengineering. Addresses several prominent fields in cell and tissue engineering.

Valentinuzzi, Max. *Understanding the Human Machine: A Primer for Bioengineering.* Hackensack, N.J.: World Scientific Publishing, 2004. An accessible reference designed to give students much of the biological knowledge needed to pursue studies in bioengineering. Also provides useful information about the nature, goals, and development of the bioengineering field.

Zenios, Stefanos, Josh Makower, and Paul Yock, eds. *Biodesign: The Process of Innovating Medical Technologies.* New York: Cambridge University Press, 2010. Covers the biomedical industry, with a particular focus on the process of creating and marketing medical technology. Also provides information about the future of biotechnology and bioengineered products.

**WEB SITES**
*Johns Hopkins University School of Medicine*
Institute for Cell Engineering
http://www.hopkins-ice.org

*Tissue Engineering International and Regenerative Medical Society*
http://www.termis.org

**See also:** Artificial Organs; Bioengineering; Biomechanical Engineering; Bionics and Biomedical Engineering; Biosynthetics; Genetic Engineering; Human Genetic Engineering; Prosthetics; Stem Cell Research and Technology.

# CERAMICS

## FIELDS OF STUDY

Calculus; chemistry; glass engineering; materials engineering; physics; statistics; thermodynamics.

## SUMMARY

Ceramics is a specialty field of materials engineering that includes traditional and advanced ceramics, which are inorganic, nonmetallic solids typically created at high temperatures. Ceramics form components of various products used in multiple industries, and new applications are constantly being developed. Examples of these components are rotors in jet engines, containers for storing nuclear and chemical waste, and telescope lenses.

## KEY TERMS AND CONCEPTS

- **Ceramic:** Inorganic, nonmetallic solid processed or used at high temperatures; made by combining metallic and nonmetallic elements.
- **Coke:** Solid, carbon-rich material derived from the destructive distillation of low-ash, low-sulfur bituminous coal.
- **Glazing:** Process of applying a layer or coating of a glassy substance to a ceramic object before firing it, thus fusing the coating to the object.
- **Kiln:** Thermally insulated chamber, or oven, in which a controlled temperature is produced and used to fire clay and other raw materials to form ceramics.

### DEFINITION AND BASIC PRINCIPLES

Ceramic engineering is the science and technology of creating objects from inorganic, nonmetallic materials. A specialty field of materials engineering, ceramic engineering involves the research and development of products such as space shuttle tiles and rocket nozzles, building materials, as well as ball bearings, glass, spark plugs, and fiber optics.

Ceramics can be crystalline in nature; however, in the broader definition of ceramics (which includes glass, enamel, glass ceramics, cement, and optical fibers) they can also be noncrystalline. The most distinguishing feature of ceramics is their ability to resist extremely high temperatures. This makes ceramics very useful for tasks where materials such as metals and polymers alone are unsuitable. For example, ceramics are used in the manufacture of disk brakes for high-performance cars (such as race cars) and for heavy vehicles (such as trucks, trains, and aircraft). These brakes are lighter and more durable and can withstand greater heat and speed than the conventional metal-disk brakes. Ceramics can also be used to increase the efficiency of turbine engines used to operate helicopters. These aircraft have a limited travel range and cannot carry a great deal of weight because of the stress these activities place on engines made of metallic alloys. However, turbine engines using ceramic parts and thermal barrier coatings are currently in development—they already show superior performance when compared with existing engines. The ceramic engine parts are from 30 to 50 percent lighter than their metallic counterparts, and the thermal coatings increase the engine operating temperatures to 1,650 degrees Celsius (C). These qualities will enable future helicopters to travel farther and carry more weight.

### BACKGROUND AND HISTORY

One of the oldest industries on Earth, ceramics date back to prehistoric times. The earliest known examples of ceramics, animal and clay figures that were fired in kilns, date back to 24,000 B.C.E. These ceramics were earthenware that had no glaze. Glazing was discovered by accident and the earliest known glazed items date back to 5000 B.C.E. Chinese potters studied glazing and first developed a consistent glazing technique. Glass was first produced around 1500 B.C.E. The development of synthetic materials with better resistance to very high temperatures in the 1500's enabled the creation of glass, cement, and ceramics on an industrial scale. The ceramics industry has grown in leaps and bounds since then.

Many notable innovators have contributed to the growth of advanced ceramics. In 1709, Abraham Darby, a British brass worker and key player in the Industrial Revolution, first developed a smelting process for producing pig iron using coke instead of wood as fuel. Coke is now widely used in the production of carbide ceramics. In 1888, Austrian chemist

Karl Bayer first separated alumina from bauxite ore. This method, known as the Bayer process, is still used to purify alumina. In 1893, Edward Goodrich Acheson, an American chemist, electronically fused carbon and clay to create carborundum, also known as synthetic silicon carbide, a highly effective abrasive.

Other innovators include brothers Pierre and Jacques Curie, French physicists who discovered piezoelectricity around 1880; French chemist Henri Moissan, who combined silicon carbide with tungsten carbide around the same time as Acheson; and German mathematician Karl Schröter, who in 1923 developed a liquid-phase sintering method to bond cobalt with the tungsten-carbide particles created by Moissan.

The need for high-performance materials during World War II helped accelerate ceramic science and engineering technologies. Development continued throughout the 1960's and 1970's, when new types of ceramics were created to facilitate advances in atomic energy, electronics, and space travel. This growth continues as new uses for ceramics are researched and developed.

## HOW IT WORKS

There are two main types of ceramics: traditional and advanced. Traditional ceramics are so called because the methods for producing them have been in existence for many years. The familiar methods of creating these ceramics—digging clay, molding the clay by hand, or using a potter's wheel, firing, and then decorating the object—have been around for centuries, and have only been improved and mechanized to meet increasing demand. Advanced ceramics, which cover the more recent developments in the field, focus on products that make full use of specific properties of ceramic or glass materials. For example, ferrites, a type of advanced ceramics, are very good conductors of electricity and are typically used in electrical transformers and superconductors. Zirconia, another type of advanced ceramics, is strong, tough, very resistant to wear and tear, and does not cause an adverse reaction when introduced to biological tissue. This makes it ideal for use in creating joint replacements in humans. It works particularly well in hip replacements, but it is also useful for knee, shoulder, and finger-joint replacements. The unique qualities of these, and other advanced ceramics, and the research into the variety of ways they can be applied, are what differentiate them from traditional ceramics.

There are seven basic steps to creating traditional ceramics. They are described in detail below.

**Raw Materials.** In this first stage, the raw materials are chosen for creating a ceramic product. The type of ceramic product to be created determines the type of raw materials required. Traditional ceramics use natural raw materials, such as clay, sand, quartz, and flint. Advanced ceramics require the use of chemically synthesized powders.

**Beneficiation.** Here, the raw materials are treated chemically or physically to make them easier to process.

**Batching and Mixing.** In this step, the parts of the ceramic product are weighed and combined to create a more chemically and physically uniform material to use in forming, the next step.

**Forming.** The mixed material is consolidated and molded to create a cohesive body of the determined shape and size. Forming produces a "green" part, which is soft, pliable and, if left at this stage, will lose its shape over time.

**Green Machining.** This step eliminates rough surfaces, smooths seams, and modifies the size and shape of the green part to prepare for sintering.

**Drying.** Here, the water or other binding agent is removed from the formed material. Drying is a carefully controlled process that should be done as quickly as possible. After drying, the product will be smaller than the green part. It is also very brittle and must be handled carefully.

**Firing or Sintering.** The dried parts now undergo a controlled heating process. The ceramic becomes denser during firing, as the spaces between the individual particles of the ceramic are reduced as they heat. It is during this stage that the ceramic product acquires its heat-resistant properties.

**Assembly.** This step occurs only when ceramic parts need to be combined with other parts to form a complete product. It does not apply to all ceramic products.

This is not a comprehensive list. More steps may be required depending on the type of ceramic product being made. For advanced ceramics production, this list of steps will either vary or expand. For example, an advanced ceramic product may need to have forming or additives processes completed in addition to the standard forming processes. It may also require a post-sintering process such as machining or annealing.

## APPLICATIONS AND PRODUCTS

**Traditional Ceramics.** Applications and products include whiteware, glass, structural clay items, cement, refractories, and abrasives. Whiteware, so named because of its white or off-white color, includes dinnerware (plates, mugs, and bowls), sanitary ware (bathroom sinks and toilets), floor and wall tiles, dental implants, and decorative ceramics (vases, figurines, and planters). Glass products include containers (bottles and jars), pressed and blown glass (wineglasses and crystal), flat glass (windows and mirrors), and glass fibers (home insulation). Structural clay products include bricks, sewer pipes, flooring, and wall and roofing tiles. Cement is used in the construction of concrete roads, buildings, dams, bridges, and sidewalks. Refractories are materials that retain their strength at high temperatures. They are used to line furnaces, kilns, incinerators, crucibles, and reactors. Abrasives include natural materials such as diamonds and garnets, and synthetic materials such as fused alumina and silicon carbide, which are used for precision cutting as well.

**Advanced Ceramics.** Advanced ceramics focus on specific chemical, biomedical, mechanical, or optical uses of ceramic or glass materials.

Advanced ceramics fully came into being within the last few decades (beginning in the 1960's), and research and development is ongoing. The field has produced a wide range of applications and products. In aerospace, ceramics are used in space shuttle tiles, aircraft instrumentation and control systems, missile nose cones, rocket nozzles, and thermal insulation. Automotive applications include spark plugs, brakes, clutches, filters, heaters, fuel pump rollers, and emission control devices. Biomedical uses for ceramics include replacement joints and teeth, artificial bones and heart valves, hearing aids, pacemakers, dental veneers, and orthodontics. Electronic devices that use ceramics include insulators, magnets, cathodes, antennae, capacitors, integrated circuit packages, and superconductors. Ceramics are used in the chemical and petrochemical industry for ceramic catalysts, catalyst supports, rotary seals, thermocouple protection tubes, and pumping tubes. Laser and fiber-optics applications for ceramics include glass optical fibers (used for very fast data transmission), laser materials, laser and fiber amplifiers, lenses, and switches. Environmental uses of ceramics include solar cells, nuclear fuel storage, solid oxide fuel cells, hot gas

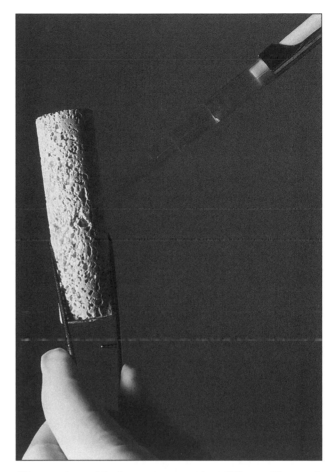

*Bioceramic used in bone reconstruction.* (Mauro Fermariello/Photo Researchers, Inc.)

filters (used in coal plants), and gas turbine components. Ceramic coatings include self-cleaning coatings for building materials, coatings for engine components, cutting tools, industrial wear parts, optical materials (such as lenses), and antireflection coatings.

Other products in the advanced ceramics segment include water-purification devices, particulate or gas filters, and glass-ceramic stove tops.

## IMPACT ON INDUSTRY

Ceramic engineering has come a long way from its humble beginnings in the 1800's. Various government agencies, universities, businesses, and the general public have contributed to, or benefited from, the innovations made in advanced ceramics. The North American market for advanced structural ceramics was worth $ 2.7 million in 2007 and is predicted to grow to $ 3.7 million by 2012. This represented a compound annual growth of 6 percent. In 2010, the global advanced ceramics market was predicted to grow to $ 56.4 million by 2015. These predictions are evidence that the field of advanced ceramics continues to grow even in the face of economic difficulties.

Japan is the largest regional market for advanced ceramics, and the market is expected to grow as it is driven by rapid developments in the information technology and electronics industries. The United States remains the largest market for military applications of advanced ceramics. The United Kingdom along with India, China, and Germany are making significant progress in research and development.

**Government and University Research.** Fraunhofer-Gesellschaft, a German research organization that is partially funded by the German government, has been making great strides in developing new advanced ceramic products. Recently, the company developed pump components that use a diamond-ceramic composite it invented. Scientists at the company also created digital projector lenses using flat arrays of glass microlenses as well as a credit card-size platform that uses magnetic nanoparticles to detect sepsis. Each of these products was a result of the joint efforts of the various institutes that fall under Fraunhofer's substantial umbrella.

The United States Department of Energy is a major investor in advanced ceramics research and development. The agency recently awarded funding for three U.S.-China Clean Energy Research Centers (CERCs). These CERCs, which will focus on clean vehicle and clean coal technologies, will receive $50 million in funding (governmental and nongovernmental) over a period of five years.

A joint Chinese research team from Tsinghua University and Peking University has developed a lightweight, durable floating sponge for use in cleaning up oil spills. The sponge is made of randomly oriented carbon nanotubes and attracts only oil. It expands to hold nearly 200 times its weight and 800 times the volume of the oil. It automatically moves toward higher concentrations of the oil and can be squeezed clean and reused dozens of times.

**Industry and Business.** Nearly all the major industries and business sectors rely to some degree on advanced ceramics. Corning, Alcoa, Boeing, and Motorola are just a few of the major corporations that use advanced ceramics in their products. DuPont recently opened a photovoltaic applications lab at its Chestnut Run facility in Wilmington, Delaware. The lab will support materials development for the fast-growing photovoltaic energy market. PolyPlus in Berkeley, California, is developing lightweight, high-energy, single-use batteries that can use the surrounding air as a cathode. The company is currently developing these batteries for the government, but it expects them to be on the market within a few years. The drive toward efficient, inexpensive, clean technology ensures continued advances in research and development in the field.

### CAREERS AND COURSE WORK

Ceramic scientists and engineers work in a variety of industries, as their skills are applicable in various contexts. These fields include aerospace, medicine, mining, refining, electronics, nuclear technology, telecommunications, transportation, and construction. A bachelor's degree is required for entrance into the above-mentioned fields. A master's degree in ceramic engineering qualifies the holder for managerial, consulting, sales, research, development, and administrative positions.

Pursuing a career in this field requires one have an aptitude for the sciences. Most colleges require that high school course work include four years of English, four years of math (at least one of which should be an advanced math course), at least three years of science (one of which should be a laboratory science), and at least two years of a foreign language.

Typical course work for a bachelor's degree in ceramics engineering includes calculus, physics, chemistry, statistics, materials engineering, and glass engineering. It also includes biology, mechanics, English composition, process design, and ceramic processing. There are fewer than ten universities in the United States that offer bachelor's degrees in ceramic engineering.

These include the Inamori School of Engineering at Alfred University and Missouri University of Science and Technology. Other universities, such as Iowa State University and Ohio State University, offer a bachelor's degree in materials engineering with a specialization in ceramics engineering. Additionally, some schools offer a combined-degree option: Undergraduate students can combine undergraduate and graduate course work to earn a bachelor's and a master's or doctorate in materials engineering with a concentration in ceramics simultaneously.

However, many American universities offer a bachelor's degree in materials engineering with a specialization in ceramics engineering.

An alternative to acquiring a bachelor's degree in ceramics engineering is to acquire a degree in a related field and then pursue a master's degree in ceramics engineering. Examples of related fields include biomedical engineering, chemical engineering, materials engineering, chemistry, physics, and mechanical engineering.

About ten universities in the United States offer master's degrees in ceramic engineering. As a result, admission into these programs is extremely competitive. Graduate students focus primarily on research and development, though they are required to take classes. Doctoral candidates also focus on research and development and after they have been awarded

their degree, they can choose to teach or continue working in research.

## SOCIAL CONTEXT AND FUTURE PROSPECTS

As mentioned in preceding sections, the advanced ceramics segment of the field, which is the primary focus of ceramic engineering, has plenty of room for growth. In 2008, there were about 24,350 materials engineers (including ceramic engineers) employed nationally; and a small number of ceramic engineers taught in colleges. The number of job openings is expected to exceed the number of engineers available.

Many of the industries in which ceramic engineers work—stone, clay, and glass products; primary metals; fabricated metal products; and transportation equipment industries—are expected to experience little employment growth through the year 2018. However, employment opportunities are expected to grow in service industries (research and testing, engineering and architectural). This is primarily because more firms, and by extension, more ceramic engineers, will be hired to develop improved materials for industrial customers.

*Ezinne Amaonwu, LL.B., M.A.P.W*

## FURTHER READING

Barsoum, M. W. *Fundamentals of Ceramics.* London: Institute of Physics, 2003. This text provides a detailed yet easy to understand overview of ceramics engineering. It is a good introductory and reference text, especially for readers with experience in other fields of science such as physics and chemistry, as it approaches ceramics engineering from these viewpoints.

Callister, William D., Jr., and David G. Rethwisch. *Materials Science and Engineering: An Introduction.* 8th ed. Hoboken, N.J.: John Wiley & Sons, 2010. This text provides an overview of materials engineering, with useful content on ceramics.

King, Alan G. *Ceramic Technology and Processing.* Norwich, N.Y.: Noyes, 2002. Published posthumously by the author's son, this text provides a technical, detailed description of every step in the advanced ceramics-production process. It focuses on implementing the production process in a laboratory, describes common problems associated with each step, and offers solutions for these problems.

Kingery, W. D., H. K. Bowen, and D. R. Uhlmann. *Introduction to Ceramics.* 2d ed. New York: John Wiley & Sons, 1976. A well-written guide to the basics of

---

### Fascinating Facts About Ceramics

- Advanced ceramics are the basis of the lightest, most durable body armor used by soldiers for small- to medium-caliber protection.

- Snowboards have become stronger and tougher, thanks to special composite materials that combine innovative glass laminates and carbon fiber materials.

- Prior to the 1700's, potters were criticized for digging holes in the roads to obtain more clay. The name of this offense is still used: pothole.

- Ceramics tend to be hard but brittle. To reduce the incidence of cracks, they are often applied as coatings on other materials that are resistant to cracking.

- Clay bricks are the only building product that will not burn, melt, dent, peel, warp, rot, rust, or succumb to termites. Buildings built with bricks are also more resistant to hurricanes.

- Radioactive glass microspheres are being used to treat patients with liver cancer. The microspheres, which are about one-third the diameter of a human hair, are made radioactive in a nuclear reactor. They are inserted into the artery supplying blood to the tumor using a catheter, and the radiation destroys the tumor and causes only minimal damage to the healthy tissue.

ceramics engineering, with an exhaustive index and questions to test the reader's retention.

Rahaman, M. N. *Ceramic Processing and Sintering.* 2d ed. New York: Marcel Dekker, 2003. Provides a clear description of all the steps in ceramics processing.

**WEB SITES**
*American Ceramic Society*
http://ceramics.org

*American Society for Testing and Materials*
http://www.astm.org

*ASM International*
http://www.asminternational.org

**See also:** Calculus.

# CHAOTIC SYSTEMS

## FIELDS OF STUDY

Mathematics; chemistry; physics; ecology; systems biology; engineering; theoretical physics; theoretical biology; quantum mechanics; fluid mechanics; astronomy; sociology; psychology; behavioral science.

## SUMMARY

Chaotic systems theory is a scientific field blending mathematics, statistics, and philosophy that was developed to study systems that are highly complex, unstable, and resistant to exact prediction. Chaotic systems include weather patterns, neural networks, some social systems, and a variety of chemical and quantum phenomena. The study of chaotic systems began in the nineteenth century and developed into a distinct field during the 1980's.

Chaotic systems analysis has allowed scientists to develop better prediction tools to evaluate evolutionary systems, weather patterns, neural function and development, and economic systems. Applications from the field include a variety of highly complex evaluation tools, new models from computer and electrical engineering, and a range of consumer products.

## KEY TERMS AND CONCEPTS

- **Aperiodic:** Of irregular occurrence, not following a regular cycle of behavior.
- **Attractor:** State to which a dynamic system will eventually gravitate—a state that is not dependent on the starting qualities of the system.
- **Butterfly Effect:** Theory that small changes in the initial state of a system can lead to pronounced changes that affect the entire system.
- **Complex System:** System in which the behavior of the entire system exhibits one or more properties that are not predictable, given a knowledge of the individual components and their behavior.
- **Dynamical System:** Complex system that changes over time according to a set of rules governing the transformation from one state to the next.
- **Nonlinear:** Of a system or phenomenon that does not follow an ordered, linear set of cause-and-effect relationships.

- **Randomness:** State marked by no specific or predictable pattern or having a quality in which all possible behaviors are equally likely to occur.
- **Static System:** System that is without movement or activity and in which all parts are organized in an ordered series of relationships.
- **System:** Group interrelated by independent units that together make up a whole.
- **Unstable:** State of a system marked by sensitivity to minor changes in a variety of variables that cause major changes in system behavior.

## DEFINITION AND BASIC PRINCIPLES

Chaotic systems analysis is a way of evaluating the behavior of complex systems. A system can be anything from a hurricane or a computer network to a single atom with its associated particles. Chaotic systems are systems that display complex nonlinear relationships between components and whose ultimate behavior is aperiodic and unstable.

Chaotic systems are not random but rather deterministic, meaning that they are governed by some overall equation or principle that determines the behavior of the system. Because chaotic systems are determined, it is theoretically possible to predict their behavior. However, because chaotic systems are unstable and have so many contributing factors, it is nearly impossible to predict the system's behavior. The ability to predict the long-term behavior of complex systems, such as the human body, the stock market, or weather patterns, has many potential applications and benefits for society.

The butterfly effect is a metaphor describing a system in which a minor change in the starting conditions can lead to major changes across the whole system. The beat of a butterfly's wing could therefore set in motion a chain of events that could change the universe in ways that have no seeming connection to a flying insect. This kind of sensitivity to initial conditions is one of the basic characteristics of chaotic systems.

## BACKGROUND AND HISTORY

French mathematician Henri Poincaré and Scottish theoretical physicist James Clerk Maxwell are considered two of the founders of chaos theory and complexity studies. Maxwell and Poincaré both worked on problems that illustrated the sensitivity of complex systems in the late nineteenth century. In 1960, meteorologist Edward Lorenz coined the term "butterfly effect" to describe the unstable nature of weather patterns.

By the late 1960's, theoreticians in other disciplines began to investigate the potential of chaotic dynamics. Ecologist Robert May was among the first to apply chaotic dynamics to population ecology models, to considerable success. Mathematician Benoit Mandelbrot also made a significant contribution in the mid-1970's with his discovery and investigation of fractal geometry.

In 1975, University of Maryland scientist James A. Yorke coined the term "chaos" for this new branch of systems theory. The first conference on chaos theory was held in 1977 in Italy, by which time the potential of chaotic dynamics had gained adherents and followers from many different areas of science.

Over the next two decades, chaos theory continued to gain respect within the scientific community and the first practical applications of chaotic systems analysis began to appear. In the twenty-first century, chaotic systems have become a respected and popular branch of mathematics and system analysis.

## How It Works

There are many kinds of chaotic systems. However, all chaotic systems share certain qualities: They are unstable, dynamic, and complex. Scientists studying complex systems generally focus on one of these qualities in detail. There are two ultimate goals for complex systems research: first, to predict the evolution of chaotic systems and second, to learn how to manipulate complex systems to achieve some desired result.

**Instability.** Chaotic systems are extremely sensitive to changes in their environment. Like the example of the butterfly, a seemingly insignificant change can become magnified into systemwide transformations. Because chaotic systems are unstable, they do not settle to equilibrium and can therefore continue developing and leading to unexpected changes. In the chaotic system of evolution, minor changes can lead to novel mutations within a species, which may eventually give rise to a new species. This kind of innovative transformation is what gives chaotic systems a reputation for being creative.

Scientists often study complexity and chaos by creating simulations of chaotic systems and studying the way the systems react when perturbed by minor stimuli. For example, scientists may create a computer model of a hurricane and alter small variables, such as temperature, wind speed and other factors, to study the ultimate effect on the entire storm system.

**Strange Attractors.** An attractor is a state toward which a system moves. The attractor is a point of equilibrium, where all the forces acting on a system have reached a balance and the system is in a state of rest.

Because of their instability, chaotic systems are less likely to reach a stable equilibrium and instead proceed through a series of states, which scientists sometimes call a dynamic equilibrium. In a dynamic equilibrium, the system is constantly moving toward some attractor, but as it moves toward the first attractor, forces begin building that create a second attractor. The system then begins to shift from the first to the second attractor, which in turn leads to the formation of a new attractor, and the system shifts again. The forces pulling the system from one attractor to the next never balance each other completely, so the system never reaches absolute rest but continues changing.

Visual models of complex systems and their strange attractors display what mathematicians call fractal geometry. Fractals are patterns that, like chaotic systems, are nonlinear and dynamic. Fractal geometry occurs throughout nature, including in the formation of ice crystals and the branching of the circulatory system in the human body. Scientists study the mathematics behind fractals and strange attractors to find patterns that can be applied to complex systems in nature. The study of fractals has yielded applications for medicine, economics, and psychology.

**Emergent Properties.** Chaotic systems also display emergent properties, which are system-wide behaviors that are not predictable from knowledge of the individual components. This occurs when simple behaviors among the components combine to create more complex behaviors. Common examples in nature include the behavior of ant colonies and other social insects.

Mathematic models of complex systems also yield emergent properties, indicating that this property is driven by the underlying principles that lead to the creation of complexity. Using these principals, scientists can create systems, such as computer networks, that also display emergent properties.

## APPLICATIONS AND PRODUCTS

**Medical Technology.** The human heart is a chaotic system, and the heartbeat is controlled by a set of nonlinear, complex variables. The heartbeat appears periodic only when examined through an imprecise measure such as a stethoscope. The actual signals that compose a heartbeat occur within a dynamic equilibrium.

Applying chaos dynamics to the study of the heart is allowing physicians to gain more accuracy in determining when a heart attack is imminent, by detecting minute changes in rhythm that signify the potential for the heart to veer away from its relative rhythm. When the rhythm fluctuates too far, physicians use a defibrillator to shock the heart back to its rhythm. A new defibrillator model developed by scientists in 2006 and 2007 uses a chaotic electric signal to more effectively force the heart back to its normal rhythmic pattern.

**Consumer Products.** In the mid-1990's, the Korean company Goldstar manufactured a chaotic dishwasher that used two spinning arms operated with an element of randomness to their pattern. The chaotic jet patterns were intended to provide greater cleaning power while using less energy.

Chaotic engineering has also been used in the manufacture of washing machines, kitchen mixers, and a variety of other simple machines. Although most chaotic appliances have been more novelty than innovation, the principles behind chaotic mechanics have become common in the engineering of computers and other electrical networks.

**Business and Industry.** The global financial market is a complex, chaotic system. When examined mathematically, fluctuations in the market appear aperiodic and have nonlinear qualities. Although the market seems to behave in random ways, many believe that applying methods created for the study of chaotic systems will allow economists to elucidate hidden developmental patterns. Knowledge of these patterns might help predict and control recessions and other major changes well before they occur.

The application of chaos theory to market analysis produced a subfield of economics known as fractal market analysis, wherein researchers conduct

---

## Fascinating Facts About Chaotic Systems

- Chaos theory gained popular attention after the 1993 film adaptation of Michael Crichton's novel *Jurassic Park* (1990). In the film, a scientist played by Jeff Goldblum evokes chaos theory to warn that complex systems are impossible to control.
- Fractal image compression uses the fractal property of self similarity to compress images.
- Chaotic systems can be found at both the largest and the smallest levels of the universe, from the evolution of galaxies to molecular activity.
- Fractal geometry is commonly found in the development of organisms. Among plants with fractal organization are broccoli, ferns, and many kinds of flowers.
- Some philosophers and neuroscientists have suggested that the creative properties of complex systems are what allow for free will in the human mind.
- The Korean company Daewoo marketed a washing machine known as the "bubble machine" in 1990, which the company claimed was the first consumer device to use chaos theory in its operation.

---

economic analyses by using models with fractal geometry. By looking for fractal patterns and assuming that the market, like other chaotic systems, is highly sensitive to small changes, economists have been able to build more accurate models of market evolution.

## IMPACT ON INDUSTRY

The fact that chaos theory applies to so many fields means that there are numerous funding opportunities available. Many countries offer some government funding for studies into chaos theory and related fields.

In the United States, the National Science Foundation offers a grant for researchers studying theoretical biology, which has been awarded to several research teams studying chaotic systems. The National Science Foundation, National Institute of Standards and Technology, National Institutes of Health, and the Department of Defense have also provided some funding for research into chaotic dynamics.

Any product created using chaotic principles may be subject to regulation governing the industry in question. For instance, the chaotic defibrillator is subject to regulation from the Food and Drug Administration, which has oversight over any new medical technology. Other equipment, such as the chaotic dishwasher, will be subject to consumer safety regulations at the federal and regional levels.

## CAREERS AND COURSE WORK

Those seeking careers in chaotic systems analysis might begin by studying mathematics or physics at a university. Those with backgrounds in other fields such as biology, ecology, economics, sociology, and computer science might also choose to focus on chaotic systems during their graduate training.

There are a number of graduate programs offering training in chaotic dynamics. For instance, the Center for Interdisciplinary Research on Complex Systems at Northeastern University in Boston, Massachusetts, offers programs training students in many types of complex system analyses. The Center for Nonlinear Phenomena and Complex Systems at the Université Libre de Bruxelles offers programs in thermodynamics and statistical mechanics.

Careers in chaotic systems span a range of fields from engineering and computer network design to theoretical physics. Trained researchers can choose to contribute to academic analyses and other pertinent laboratory work or focus on creating applications for immediate consumer use.

## SOCIAL CONTEXT AND FUTURE PROSPECTS

As chaotic systems analysis spread in the 1980's and 1990's, some scientists began theorizing that chaotic dynamics might be an essential part of the search for a grand unifying theory or theory of everything. The grand unifying theory is a concept that emerged in the early nineteenth century, when scientists began theorizing that there might be a single set of rules and patterns underpinning all phenomena in the universe.

The idea of a unified theory is controversial, but the search has attracted numerous theoreticians from mathematics, physics, theoretical biology, and philosophy. As research began to show that the basic principles of chaos theory could apply to a vast array of fields, some began theorizing that chaos theory was part of the emerging unifying theory.

Because many systems meet the basic requirements to be considered complex chaotic systems, the study of chaos theory and complexity has room to expand. Scientists and engineers have only begun to explore the practical applications of chaotic systems, and theoreticians are still attempting to evaluate and study the basic principles behind chaotic system behavior.

*Micah L. Issitt, B.S.*

## FURTHER READING

Ford, Kenneth W. *The Quantum World: Quantum Physics for Everyone.* Cambridge, Mass.: Harvard University Press, 2005. An exploration of quantum theory and astrophysics written for the general reader. Contains coverage of complexity and chaotic systems applied to quantum phenomena.

Gleick, James L. *Chaos: Making a New Science.* Rev. ed. New York: Penguin Books, 2008. Technical popular account of the development of chaos theory and complexity studies. Topics covered include applications of chaos systems analysis to ecology, mathematics, physics, and psychology.

Gribbin, John. *Deep Simplicity: Bringing Order to Chaos and Complexity.* New York: Random House, 2005. An examination of how chaos theory and related fields have changed scientific understanding of the universe. Provides many examples of complex systems found in nature and human culture.

Mandelbrot, Benoit, and Richard L. Hudson. *The Misbehavior of Markets: A Fractal View of Financial Turbulence.* New York: Basic Books, 2006. Provides a detailed examination of fractal patterns and chaotic systems analysis in the theory of financial markets. Provides examples of how chaotic analysis can be used in economics and other areas of human social behavior.

Stewart, Ian. *Does God Play Dice? The New Mathematics of Chaos.* 2d ed. 1997. Reprint. New York: Hyperion, 2005. An evaluation of the role that order and chaos play in the universe through popular explanations of mathematic problems. Includes accessible descriptions of complex mathematical ideas underpinning chaos theory.

Strogatz, Peter H. *Sync: How Order Emerges from Chaos in the Universe, Nature and Daily Life.* 2003. Reprint. New York: Hyperion Books, 2008. Strogatz provides an accessible account of many aspects of complex systems, including chaos theory, fractal organization, and strange attractors.

**WEB SITES**

*Society for Chaos Theory in Psychology and Life Sciences*
http://www.societyforchaostheory.org

*University of Maryland*
Chaos Group at Maryland
http://www-chaos.umd.edu

**See also:** Applied Mathematics; Applied Physics; Climate Modeling; Electrical Engineering; Fractal Geometry; Risk Analysis and Management.

# CHARGE-COUPLED DEVICES

## FIELDS OF STUDY

Physics; electrical engineering; electronics; optics; mathematics; semiconductor manufacturing; solid-state physics; chemistry; mechanical engineering; photography; computer science; computer programming; signal processing.

## SUMMARY

Originally conceptualized as a form of optical volatile memory for computers, charge-coupled devices were also seen to be useful for capturing light and images. The ability to capture images that can be read by a computer in digital form was profoundly useful, and charge-coupled devices quickly become known for their imaging abilities. Besides their utility in producing digital images that could be analyzed and manipulated by computers, charge-coupled devices also proved to be more sensitive to light than film. As technology advanced, charge-coupled devices became less expensive and more capable, and eventually they replaced film as the primary means of taking photographs.

## KEY TERMS AND CONCEPTS

- **Bias Frame:** Image frame of zero length exposure, designed to measure pixel-to-pixel variation across the charge-coupled device array.
- **Blooming:** Bleeding of charge from a full charge, often appearing as lines extending from overexposed pixels.
- **Dark Frame:** Image taken with the shutter closed, designed to measure noise and defects affecting images in the charge-coupled device.
- **Flat Field:** Image taken of uniform illumination across the charge-coupled device chip, designed to measure defects in the chip and optical system affecting images.
- **Photoelectric Effect:** Physical process whereby photons of light knock electrons off of atoms.
- **Pixel:** Individual charge-collecting region and ultimately a single element of a final image.
- **Quantum Efficiency:** Measure of the percentage of photons hitting a device that will produce electrons.
- **Radiation Noise:** Spurious charges produced in an image due to ionizing radiation, frequently produced by cosmic rays.
- **Readout Noise:** Electronic noise affecting image quality due to shifting and reading the charge in the pixels.

## DEFINITION AND BASIC PRINCIPLES

A charge-coupled device (CCD) is an electronic array of devices fabricated at the surface of a semiconductor chip. The typical operation of a CCD is to collect electrical charge in specific locations laid out as an array on the surface of the chip. This electrical charge is normally produced by light shining onto the chip, producing an electrical charge. Each collecting area is called a pixel and consists of an electrical potential well that traps charge carriers. The charge in one area of the array is read, and then the charges on other parts of the chip are shifted from potential well to potential well until all have been read. The brighter the light shining on a pixel, the more charge it will have. Thus, an image can be constructed from the data collected by each pixel. When the CCD is placed at the focal plane of an optical system, such as a lens, then the image on the chip will be the same as the image seen by the lens; thus CCD chips can be used as the heart of a camera system. Because the data is collected electronically in digital form for each pixel, the image from a CCD is inherently a digital image and can be processed by a computer. For this reason, CCDs (and digital cameras) have become more popular than film as a way to take photographs.

## BACKGROUND AND HISTORY

The charge-coupled device was invented based on an idea by Willard S. Boyle and George E. Smith of AT&T Bell Laboratories for volatile computer memory. They were seeking to develop an electrical analogue to magnetic bubble memory. Boyle and Smith postulated that electric charge could be used to store data in a matrix on a silicon chip. The charge could be moved from location to location within that array, shifting from one holding cell to another by applying appropriate voltages. The name "charge-coupled device" stemmed from this shifting of electrical charge around on the device. Initial research aimed to perfect the CCD as computer memory, but the inventors soon saw that it may be even more useful in converting light values into electrical charge that could be read. This property led to the development of the CCD as an imaging array. Subsequent developments made pixel sizes smaller, the arrays larger, and the price lower. The CCD became the detector at the heart of digital cameras, and their ease of use quickly made digital cameras popular. By the early 2000's, digital cameras and their CCDs had become more popular than film cameras.

## HOW IT WORKS

At the heart of a charge-coupled device is a semiconductor chip with electrical potential wells created by doping select areas in an array on the surface of the device. Doping of a semiconductor involves fabricating it with a select impurity that would tend to have either an extra electron or one too few electrons for the normal lattice structure of the semiconductor. Fabricating the CCD with select areas so doped would tend to make any electrical charge created at the surface of the semiconductor chip want to stay in the region of the potential well. By applying the proper electrical voltage between adjacent wells, however, charge is permitted to flow from one well to another. In this way, charge anywhere in the array can be moved from well to well to a location where it can be read.

**The Photoelectric Effect.** The key to using a charge-coupled device is to put charge on it. This is done by shining light onto the surface of the semiconductor chip. When photons of sufficient energy shine on a material, the absorbed light can knock electrons from atoms. This is known as the photoelectric effect, a physical phenomenon first observed by German physicist Heinrich Hertz in 1887 and explained by German physicist Albert Einstein in 1905. While the photoelectric electric effect works in many materials, in a semiconductor it produces a free electron and a missing electron, called a hole. The hole is free to move just like an electron. Normally, the electron and hole recombine (the electron going back to the atom it came from in the semiconductor lattice) after the electron-hole pair are created. However, if an electric field is present, the electron may go one way in the semiconductor and the hole another. Such an electric field is present in an operating CCD,

and so the electrons will congregate in regions of lower electrical potential (electrical potential wells), building up an electrical charge proportional to the intensity of the light shining on the chip. The potential wells are arranged in an array across the surface of the CCD. The individual wells serve as the individual picture elements in a digital image and are called pixels. A nearly identical technology to the CCD uses different materials and is called the complementary metal oxide semiconductor (CMOS).

**Shifting Charges.** Once charge is produced and captured on the CCD chip, it must be measured in order for the device to be of any use. Typically, charge is read only from one location on the CCD. After that charge is read, all of the charges in all the pixels on a row are shifted over to the adjacent pixel to allow the next charge to be read from the readout position. This process repeats until all charges in the row have been read. All columns of pixels are then shifted down one row, shifting the charges in the second row into the row that has just been read. The process of reading that row continues until all charges are read and all the columns shift down again, repopulating the empty row of charge with new charges. All the charges are shifted around until they have been read. If a pixel has too much charge (being overexposed), then the charge can bleed between adjacent pixels. This creates spurious surplus charge on charges throughout a column and appears as vertical lines running through the image. This image defect is known as blooming.

**Color Images.** Charge-coupled devices respond to light intensity, not color. Therefore, all images are inherently black and white. There are several techniques to get color images from CCDs. The simplest and most cost-effective method is to take pictures using different color filters and then to create a final image by adding those images in the colors of the filters. This is the technique usually used in astronomy. The disadvantage of this technique, though, is that it requires at least three images taken in succession. This does not permit "live" or action photography. A separate technique is to have filters constructed over each pixel on the CCD array. This permits one chip to take multiple images simultaneously. The disadvantage of this technique, besides the cost of constructing such an array of filters, is that each image uses only about one-third of the pixels on the chip, thus losing image quality and detail. Most color digital cameras use this technique. Though other strategies exist for making color images with CCDs, these two are the most common.

## APPLICATIONS AND PRODUCTS

CCDs and CMOSs are the detectors in virtually every digital imaging system. CCDs have also displaced the imaging tubes in television cameras.

**Cameras.** Because CCDs produce digital images, they can be directly viewed using computers, sent by e-mail, coded into Web pages, and stored in electronic media. Digital images can also be viewed almost immediately after taking them, unlike film, which must first be developed. This has made digital cameras very popular. In addition to providing nearly instant pictures, CCDs can be manufactured quite small, permitting cameras that are much smaller than used to exist. This has allowed cameras to be placed in cell phones, computers, tablet computers, and many other devices. Traditionally, the pixels in CCDs have been larger than the grains in film, so digital image quality has suffered. However, technological advantages have permitted CCD pixels to be made much smaller, and by 2010, most commercial CCD camera systems compared very favorably with film systems in terms of image quality, with high-end CCD cameras often performing better than most film cameras.

**Scientific Applications.** Some of the first applications of CCDs were in the scientific community. CCDs permit very accurate images, but these images also contain precise data about light intensity. This is particularly important in chemistry, where spectral lines can be studied using CCD detectors. It is also important in astronomy, where digital images can be studied and measured using computers. Astronomical observatories were among the first to adopt CCD imaging systems, and they are still leaders in working with companies to develop new more powerful and larger CCD arrays.

**Satellite Images.** Early satellite cameras used film that had to be dropped from orbit to be collected below. Soon, television cameras displaced film, but as soon as the first reliable charge-coupled devices capable of imaging became available, that technology was far better than any other. As of 2011, nearly all satellite imaging systems, both civilian and military, rely on a type of CCD technology.

## IMPACT ON INDUSTRY

**Government and University Research.** The digital nature of CCDs makes them of great use in scientific images, and since they are more sensitive than film, they can record dimmer light sources. This makes CCDs particularly useful for astronomy. By the 1990's, practically all professional astronomical research was done using CCDs. By 2010, nearly all amateur astronomers were using CCDs in astrophotography.

**Industry and Business.** CCDs are used in fax machines, scanners, and even many small copiers. The ubiquity of these devices changes business communications. Web cameras attached to computers make videos easy to produce and have led to an explosion of individual amateur video clips being posted online in such places as YouTube.

**Major Corporations.** The widespread use of CCDs in digital cameras has displaced film in most photographic activities. Many companies that used to make film have had to adjust their business model

---

### Fascinating Facts About Charge-Coupled Devices

- A charge-coupled device (CCD) is the imaging system at the heart of every digital camera, Web cam, and cell phone camera. CCDs are also found in scanners, fax machines, and many other places where digital images are needed.

- By 2005, digital-related sales revenue exceeded film-related sales revenue for Kodak, a company known for its high-quality film and film photography products.

- Astronomical and military applications of CCDs led the way before widespread commercial applications became available.

- In order to reduce thermal noise, astronomical charge-coupled devices are often cooled with cryogenic fluids, such as liquid nitrogen.

- The first experimental CCD imaging system, constructed in 1970 at AT&T Bell Laboratories, had only eight pixels.

- The first commercially produced CCD imaging system, made available by Fairchild Electronics in 1974, had only 10,000 pixels, arranged in a 100-by-100 array.

- Because charge-coupled devices are sensitive to infrared light, they can be used in some solid-state night-vision devices—and some commercial cameras and video cameras have night-vision capability.

to accommodate the change in customer activities. Companies that built their reputations on fine film products have had to shift their product lines to stay competitive. Most of these companies now sell more digital-related products than film-related products.

## CAREERS AND COURSE WORK

Charge-coupled devices have become the most common imaging system in use. Thus, any career that uses images will come into contact with CCDs. For most people using CCDs, there is no different training or course work than would be needed to use any camera.

Manufacturing CCDs or developing new CCD technology, however, requires specialized training. CCDs are semiconductor devices, so their manufacture requires a background in semiconductor-manufacturing technology. Both two-year and four-year degrees in semiconductor manufacturing technology exist, and these degrees require courses in electronics, semiconductor physics, semiconductor-manufacturing technology, mathematics, and related disciplines. Manufacturing CCDs is not much different from manufacturing any other semiconductor device.

Construction of equipment, such as cameras, to hold CCDs involves both optics and electronics. Manufacture of such equipment requires little detailed knowledge of these areas other than what is required for any other camera or electronic device. Most of the components are manufactured ready to be assembled. Some specialized electronics knowledge is needed in order to design the circuit boards to operate the CCDs, however.

Developing improved CCD technology or research into better CCDs or technologies to replace CCDs requires much more advanced training, generally a post-graduate degree in physics or electrical engineering. These degrees require extensive physics and mathematics courses, along with electrical engineering course work and courses in chemistry, materials science, electronics, and semiconductor technology. Most jobs related to research and development are in university or corporate laboratory settings.

## SOCIAL CONTEXT AND FUTURE PROSPECTS

CCDs were once very esoteric devices. Early digital cameras had CCDs with large pixels, producing pictures of inferior quality to even modest film cameras. However, CCDs quickly became less expensive and of

*Shown here is a charge-coupled device sensor used inside a monochrome digital video camera.* (GIPhotoStock/ Photo Researchers, Inc.)

higher quality and soon became ubiquitous. Most cameras sold by 2005 were digital cameras. Most cell phones now come equipped with cameras having CCD technology. Since most people own a cell phone, this puts a camera in the hands of almost everyone all the time. The number of photographs and videos being made has skyrocketed far beyond what has ever existed since the invention of the camera. Since these images are digital, they can easily be shared by e-mail or on Web pages. Social networking on the Internet has permitted these pictures to be more widely distributed than ever. There is no indication that this will change in the near future. The possibility exists that in the near future two-way video phone calls may become commonplace.

Though charge-coupled devices are far superior to film, other competing technologies that can do the same thing are being developed. A very similar technology (so similar that it is often grouped with CCD technology) is the complementary metal oxide semiconductor (CMOS). CMOS technology works the same as CCD technology for the end user, but the details of the chip operation and manufacture are different. However, with CCD technology becoming so commonplace in society, it is likely that other imaging technological developments will follow.

*Raymond D. Benge, Jr., B.S., M.S.*

## FURTHER READING

Berry, Richard, and James Burnell. *The Handbook of Astronomical Image Processing.* 2d ed. Richmond,

Va.: Willmann-Bell, 2005. Though the book discusses charge-coupled devices somewhat, it mainly focuses on how to use the images taken by such devices to produce pictures and do scientific research. Comes with a CD-ROM.

Holst, Gerald C., and Terrence S. Lomheim. *CMOS/CCD Sensors and Camera Systems.* 2d ed. Bellingham, Wash.: SPIE, 2011. A technical review of CCD and CMOS imaging systems and cameras with specifics useful for engineering applications and for professionals needing to know the specific capabilities of different systems.

Howell, Steve B. *Handbook of CCD Astronomy.* 2d ed. New York: Cambridge University Press, 2006. This excellent and easy to understand handbook covers how to use charge-coupled device cameras in astronomy. The book is written for the amateur astronomer but contains good information on how CCDs work.

Janesick, James R. *Scientific Charge-Coupled Devices.* Bellingham, Wash.: SPIE, 2001. A very technical, thorough overview of how charge-coupled devices work.

Jorden, P. R. "Review of CCD Technologies." *EAS Publications Series* 37 (2009): 239-245. A good overview of charge-coupled devices, particularly as they relate to astronomy and satellite imaging systems.

Nakamura, Junichi, ed. *Image Sensors and Signal Processing for Digital Still Cameras.* Boca Raton, Fla.: CRC Press, 2006. A thorough and somewhat technical overview of CCD and CMOS technologies. Includes a good history of digital photography and an outlook for future technological developments.

Williamson, Mark. "The Latent Imager." *Engineering & Technology* 4, no. 14 (August 8, 2009): 36-39. A very good history of the development of charge-coupled device imaging systems and some of their uses.

**WEB SITES**
*American Astronomical Society*
http://aas.org

*Institute of Electrical and Electronics Engineers*
http://www.ieee.org

*Professional Photographers of America*
http://www.ppa.com

**See also:** Computer Engineering; Computer Science; Electrical Engineering; Mechanical Engineering; Optics; Photography.

# CHEMICAL ENGINEERING

## FIELDS OF STUDY

Chemistry; mathematics; physics; electrical engineering; fluid dynamics; electronics; mechanical engineering; process control; material engineering; safety engineering; biology; communications; economics; critical path scheduling.

## SUMMARY

Chemical engineering, sometimes also called processing engineering, is the field of engineering that studies the conversion of raw chemicals into useful products by means of chemical transformation. Chemical engineering applies engineering concepts to design, construction, operation, and improvement of processes that create products from chemicals. For example, chemical engineering converts petroleum into products such as gasoline, lubricants, petrochemicals, solvents, plastics, processed food, electronic components, pharmaceuticals, agricultural chemicals, paints, and inks. Chemical engineering relies on all the technologies used in chemical and related industries, including distillation, chemical kinetics, mass transport and transfer, heat transfer, control instrumentation, and other unit operations, as well as economics and communications.

## KEY TERMS AND CONCEPTS

- **Activation Energy:** Amount of energy required to make a chemical reaction occur.
- **Azeotrope:** Mixture of two or more liquids in such a ratio that its composition cannot be changed by simple distillation; also known as constant-boiling mixture.
- **Catalyst:** Substance that changes the rate of a chemical reaction but is not consumed by it.
- **Chemical Decomposition:** Chemical reaction in which a chemical is decomposed into elements or smaller compounds; also known as analysis.
- **Chemical Reaction:** Interaction between chemicals that changes one or both of them in some way.
- **Distillation:** Process in which more volatile components are separated from materials that have a higher boiling point.

- **Intermediate:** Short-lived, unstable chemical in a chemical reaction.
- **Kinetics:** Study of chemical reaction rates and how these rates are affected by temperature, pressure, and catalysts.
- **Petroleum Fraction:** End product from refining petroleum.
- **Process Control:** Maintaining proper operations by controlling temperature, flows, levels, and other factors.
- **Product:** Result of a chemical reaction.
- **Reactant:** Substance initially involved in a chemical reaction and that is consumed in the process.
- **Reagent:** Substance added to a system to produce a chemical reaction; usually used to test for the presence of a specific substance.
- **Synthesis:** Chemical reaction that ends with the combination of two or more chemicals.
- **Unit Operations:** Individual steps that in combination with other steps are used to convert raw materials into finished products.

### DEFINITION AND BASIC PRINCIPLES

Chemical engineering is the discipline that studies chemical reactions used to manufacture useful products, which appear in almost everything people have and use. Chemical engineering also is involved with reduction and removal of waste, improvement of air and water quality, and production of new sources of energy. It is also the responsibility of chemical engineering to ensure the safety of all involved through design, training, and operating procedures.

The dividing line between chemistry and chemical engineering is hazy and there is much overlap. Primarily chemistry discovers and develops new reactions, new chemicals, and new analytical tools. It is the role of chemical engineering to take these discoveries and use them to evaluate the economic possibilities of a new product or a new process for making an existing product. Chemical engineers determine what processes or unit operation are needed to carry out the reactions required to produce, recover, refine, and store a particular chemical. In the case of a new process, chemical engineers consider what other products could be made using the process. They also examine what products newly discovered chemicals can be used to make.

The largest field of chemical engineering applications is the petroleum and petrochemical industries. Crude oil, if merely distilled, would yield less than 35 percent gasoline. Through reactions such a catalytic cracking and platforming, these yields have come to approach 90 percent. Other treatments produce other fuels, lubricants, and waxes. The lighter components of crude oil became the raw materials used to make most of the plastics, fibers, solvents, synthetic rubber, paper, antifreeze, pharmaceutical drugs, agricultural chemicals, and paints that are seen in people's daily lives.

### BACKGROUND AND HISTORY

Chemical engineering started in Germany with the availability of coal tar, a by-product of metallurgical coke. German chemists in the late nineteenth century used these tars to make aniline dyes and pharmaceuticals, most famously aspirin. Large-scale production meant that organic chemists needed help from engineers, usually mechanical, who designed and built the larger units needed. It became a tradition in Germany to pair a chemist with a mechanical engineer who would work together on this type of operation.

Americans first trained engineers in the chemical sciences to improve their understanding of chemical processes. In 1891, the Massachusetts Institute of Technology became the first school to offer a degree in chemical engineering. By 1904, the first handbook of chemical engineering, an encyclopedia of useful physical, chemical, and thermodynamic properties of chemicals and other information needed for design work, was published. A desire among professionals to share ideas, discoveries, and other useful information led to the founding in 1908 of the American Institute of Chemical Engineering.

World War I increased demand for military chemicals and the fuel needed for a mechanized war. Previously, chemicals had been developed in batches, but the petroleum industry converted to using continuous operations to raise the production rate. Soon the industry was developing systems to control production, first by empirical methods and then by scientific and mathematical techniques.

### HOW IT WORKS

The design, construction, and operation of chemical engineering projects are commonly divided into various unit operations. These unit operations, singly or in combination, require a basic knowledge and understanding of many scientific, mathematical, and economic principles.

Mathematics drives all aspects of chemical engineering. Calculations of material and energy balances are needed to deal with any operation in which chemical reactions are carried out. Kinetics, the study dealing with reaction rates, involves calculus, differential equations, and matrix algebra, which is needed to determine how chemical reactions proceed and what products are made and in what ratios. Control system design additionally requires the understanding of statistics and vector and non-linear system analysis. Computer mathematics including numerical analysis is also needed for control and other applications.

Chemistry, especially organic and physical, is the basis of all chemical processes. A full understanding of organic chemistry is essential in the fields of petroleum, petrochemicals, pharmaceuticals, and agricultural chemicals. A great deal of progress was made in the field of organic chemistry beginning in the 1880's. Physical chemistry is the foundation of understanding how materials behave with respect to motion and heat flow. The study of how gases and liquids are affected by heat, pressure, and flows is needed for the unit operations of mass transfer, heat transfer, and distillation.

Distillation, in which more volatile components are separated from less volatile materials, requires knowledge of individual physical and thermodynamic properties and how these interact with one another in mixtures. For example, materials with different boiling points sometimes form a constant-boiling mixture, or azeotrope, which cannot be further refined by simple distillation. A well-known example of this is ethyl alcohol and water. Although ethyl alcohol boils at 78 degrees Celsius as compared with 100 degree Celsius for water, a mixture of ethyl alcohol and water that contains 96 percent (by weight) or 190-proof ethyl alcohol is a constant-boiling mixture. Further concentration is not possible without extraordinary means.

Inorganic chemistry deals with noncarbon chemistry and is often considered basic as it is taught in high school and the first year of college. Many of the chemical industries do make inorganic chemicals such as ammonia, caustic, chlorine, oxygen, salts, cement, glass, sulfur, carbon black, pigments, fertilizers, and sulfuric, hydrochloric, and nitric acids. Catalysts are inorganic materials that influence organic reactions, and understanding them is very important. The effectiveness of any catalyst is determined by its chemical composition and physical properties such as surface area, pore size, and hardness.

Analytical chemistry was once strictly a batch process, in which a sample would be collected and taken to a laboratory for analysis. However, modern processes continually analyze material during various stages of manufacture, using chromatography, mass spectroscopy, color, index of refraction, and other techniques.

The disciplines of mechanical, electrical, and electronic engineering are needed by the chemical engineer to be able to consider process design, materials of construction, corrosion, and electrical systems for the motors and heaters. Electronic engineering is basic knowledge needed for control systems and computer uses.

Safety engineering is the study of all aspects of design and operation to find potential hazards to health and physical damage and how to correct them. This work begins at the start of any project. Not only must each and every part of a process be examined but also each chemical involved must be checked for hazards, either by itself or in combinations with other chemicals and materials in which it will come in contact. Safety engineering also is a part of operator training and the writing of operating procedures used for the proposed operation.

Communication is key to all progress. Chemical engineers must be able to interact with others; no idea—whether for a new process or product or an improvement in an existing operation—can be implemented unless others are convinced of its value and are willing to invest in it. Coherent and easily understood reports and presentations are as important as any other part of a project. Operating procedures must be reviewed with the operation personnel to ensure that they are understood and can be followed.

Economics is the driving force behind all design, construction, and operations. The chemical engineer must be conversant in finance, banking, accounting, and worldwide business practices. Cost estimates, operating balances to determine actual costs, market forces, and financing are a necessary part of the work of any chemical engineer.

## APPLICATIONS AND PRODUCTS

Chemical engineering is involved in every step in bringing a process from the laboratory to full-scale production. This involves determining methods to make the process continuous, safe, environmentally compatible, and economically sound. During these steps, chemical engineers determine the methods and procedures needed for a full-scale plant. Mathematical modeling is used to test various steps in the process, controls, waste treatment, environmental concerns, and economic feasibility.

**Acetonitrile Process.** The acetonitrile process demonstrates how these disciplines combine to produce a process. Acetonitrile is a chemical used as a solvent and an intermediate for agricultural chemicals. The chemistry of this process involves reacting acetic acid and ammonia to make acetonitrile and water. The reaction between these raw materials is carried out over a catalyst at 400 degrees Celsius. The reaction takes place in tubes loaded with a catalyst of phosphoric acid deposited on an alumina ceramic support, which allows the reaction to take place at high rates. The reaction tubes are located in a gas-fired furnace designed to provide even temperatures throughout the length of the tubes. The exiting gases are cooled and condensed by scrubbing with water. This mixture enters a train of distillation columns, which first removes a constant-boiling mixture of the acetonitrile along with some water. In another distillation column, an azeotroping agent is used to produce an overhead mixture that when condensed, produces two layers with all the water in one layer. The water layer is removed, and the other layer is recycled. The base material leaving the distillation column contains the water-free acetonitrile, which is then redistilled to produce the finished product, which is ready to package and ship.

Such a process involves chemical engineers in the design and assembly of all the equipment needed for the process: the furnace, reactor tubes, distillation columns, tanks, pumps, heat exchangers, and piping. The chemical engineers also create the controls, operating procedures, hazardous material data sheets, and startup and shut-down instructions, and provide operator training. Once the plant is running, the role of the chemical engineer becomes operating and improving the unit.

**Petroleum and Petrochemical Applications.** Chemical engineering is used in the petroleum and petrochemical industries. At first, all petroleum products

*Chemical plant storage tanks. The industrial-scale manufacture of chemicals requires specialized engineering to meet a wide range of storage requirements.* (Paul Rapson/Photo Researchers, Inc.)

were produced by simple batch distillations of crude oil. Chemical engineering developed continuous distillation processes that permitted marked increases in refinery production rates. Then high-temperature cracking methods permitted the conversion of high-boiling petroleum fractions (end products of refining) to useful products such as more gasoline. This was followed by the use of catalytic cracking that improved the gasoline output even more.

Distillation can take many forms in addition to simple atmospheric distillation. Chemical engineers can determine the need to use pressure distillation for the purification of components that are normally gases. Vacuum distillation may be useful if there are high-boiling components, which are sensitive to the elevated temperatures required for normal distillation. With the need for aviation gasoline and other high-octane fuels, chemical engineering developed such methods as platforming, which uses the addition of a platinum catalyst to speed up certain reactions, to convert lower-boiling petroleum components into high-octane additives.

Petrochemical industries convert surplus liquefiable gases into solvents, plastics, synthetic rubber, adhesives, coatings, paints, inks, intermediates for agricultural chemicals, food additives, and many more items.

**Inorganics.** Sulfuric acid is typical of a major inorganic product that requires the type of process improvement that is provided by chemical engineering. Sulfuric acid is a high-volume, low-profit-margin material that has been in production for more than a hundred years. The chemistry is well known. Sulfur is vaporized, mixed with air, and passed over a catalyst to make sulfur trioxide, which is then adsorbed into water to make sulfuric acid. Chemical engineers look for better catalysts and seek to improve the purity of raw materials and increase control over temperature, flows, safety, and environmental concerns.

**Biological Applications.** The ancient biological process of fermentation is best known for its role in producing alcoholic beverages and breads. Although many people do not realize that fermentation is a chemical reaction, it is much the same as other organic reactions. The biological component is a microorganism that acts as the catalyst. Biochemical engineering products created using fermentation include acetone and butyl acetate, chemicals that were in critical demand by the aviation industry in World War I.

Biochemical engineering also made antibiotics available on a large scale. In the 1920's, Sir Alexander Fleming discovered the antibiotic properties of penicillin, which was produced in laboratory flasks, a few grams at a time. During World War II, the need for penicillin increased, and chemical engineers developed a large-scale process for producing penicillin from corn. This process was adapted to produce other antibiotics, and chemical engineering processes were developed for the manufacture of many synthetic drugs. Genetic modifications are being developed that are expected to produce the drugs of the future.

**Fibers.** In 1905, the first synthetic fiber, reconstituted cellulose, commonly known as rayon, was developed. The second synthetic fiber, cellulose acetate, which was developed in 1924, is still manufactured in large amounts for use in cigarette filters. The first fully synthetic fiber was nylon, a commonly used polyamide, followed by polyesters. Chemically treated cotton, known as permanent press, is another product of chemical engineering. In addition to fibers, the textile industry uses lubricants, dyes, pigments, coatings, and inks, all derived by chemical processes.

**Plastics.** The first thermoplastic, a material that could be reheated and molded, were the cellulosics that are used to make large signs, toys, automobile parts, and other objects that can be easily fabricated. Acrylic polymers such as methyl methacrylate are formed into optically clear sheeting and used as a non-shattering replacement for plate glass and for eyeglass and camera lenses. Polyethylene, developed during World War II as a superior coating for the wiring needed in radar, is still used as an insulator. The most common use for polyethylene and polypropylene is as the thin film used in plastic bags. Molded items such as bottles, containers, kitchen items, and packaging are manufactured from these polymers as well as polyesters, nylon, polyvinyl chloride (PVC), polystyrene, Teflon, and polycarbonates. The manufacture of all these plastics was developed through chemical engineering and depends on it for operation and improvement.

**Refrigerants.** The first refrigerants were sulfur dioxide and ammonia. These substances were rather hazardous, so a new range of refrigerants was developed. These refrigerants, commonly known as Freon, are halogenated hydrocarbons that can be tailored to the needs of the application. Home air-conditioning, automotive air-conditioning, and industrial applications are examples of these uses.

**Nuclear Energy.** The nuclear energy industry requires many solvents and reaction agents for the separation and purification of the fuel used in the reactors. Special coatings and other materials used in the vicinity of intense radiation were developed by the chemical industry. Recovery of the spent nuclear fuel requires specially developed techniques, again requiring solvents and reactions.

**Coatings.** A typical example of corrosion is the rusting of iron. Corrosion causes the loss of equipment as well as physical and health hazards not only in the chemical industry but also in almost every aspect of modern life. Corrosion is avoided through metal alloys such as stainless steel and protective coatings. These coatings can be tailored to protective needs inside and outside of a product.

**Water Treatment.** Raw water, especially for industrial purposes, requires processing to remove suspended solids and soluble organics as well as inorganic ions such as sodium, calcium, iron, chlorides, sulfates and nitrates. Chemical engineering is used to design and manufacture the ion-exchange resins, coagulants, and adsorbents needed, as well as the procedures for use and regeneration of ion-removal systems.

**Waste Treatment.** Many chemical processes produce by-products or waste that may be hazardous to the environment and represent uncaptured value. The role of process improvement engineers is to first find ways to reduce waste and failing that, to develop methods to convert these wastes into nonhazardous materials that will not harm the environment.

**Agricultural Applications.** Agricultural chemicals such as insecticides, herbicides, fungicides, fertilizers, seed coatings, animal feed additives, and medicines such as hormones and antibiotics all are produced by chemical means. Other chemicals are used in the preparation of products for harvesting and transporting to market.

**Food Processing.** The manufacture of food stuffs such as dairy products, breakfast foods, soups, bread, canned goods, frozen foods, and other processed foods, as well as meats, fruits, and vegetables, use engineering operations such as heat transfer to heat or cool.

## IMPACT ON INDUSTRY

Chemical engineering has an impact on almost every aspect of people's lives. The chemical engineering profession is international in its scope. Chemical plants exist in almost every country in the world; however, almost all petrochemical industries are located in countries with or near petroleum refineries. The United States dominates this industry, followed by Europe and the Middle East. Industrial processes are developed throughout the world and are made available to others through licensing. Engineering research is carried out by government agencies, universities, and major industrial concerns.

**Government and University Research.** The U.S. government has many agencies and programs that promote the development of new energy products or the improvement of existing ones, the reduction of chemical by-products, and the recovery of energy and useful products from existing waste sites.

These agencies include the Bureau of Mines, the Environmental Protection Agency, the Department of Defense, the Department of Agriculture, and even the National Aeronautics and Space Administration. Government grants fund many academic research programs designed to develop new materials and processes that use less energy and produce fewer and less hazardous by-products.

One example of government-funded academic research is the recovery of liquid and gaseous fuels from coal. During World II, the Germans developed a process to make aviation gasoline and other fuels from low-grade coal. At the end of the war, the U.S. government confiscated the German data and gave it to Texas A&M University. In the 1970's, the Bureau of Mines and Texas A&M began converting this mass of literature into a program to design a research-scale plant to develop methods for the economic production of fuels from coal and oil shale. As naturally occurring petroleum becomes more expensive and harder to find, the recovery of fuel products from coal is likely to become practical and to lead to the creation of new industries.

Another area of government-funded academic research is the desalinization of sea water. Fresh water is becoming scarcer, and multiple studies are seeking to find ways to produce large quantities of fresh water.

**Industry and Business Sectors.** Chemical engineering affects almost every industry and business that manufactures, handles, transports, and sells chemicals. Chemical engineers are called on to help evaluate the practicality and earning potential of investment proposals in which the manufacture or use of chemicals is to be considered.

The major corporations in the chemical industry include DuPont, Union Carbide, the Dow Chemical Company, Celanese Chemicals, Monsanto, Eastman Chemical, ExxonMobil, Texaco, BP, and Royal Dutch Shell. Other major producers of chemicals include Procter & Gamble, Kraft Foods, Quaker Oats, Sherman-Williams, and PPG Pittsburgh Paints. Major pharmaceutical companies such as Pfizer and Ely Lily require biochemical engineering to produce the medicines they sell.

## CAREERS AND COURSE WORK

A bachelor's degree in chemical engineering takes four to five years of study. Course work includes a great deal of chemistry, including organic, physical,

## Fascinating Facts About Chemical Engineering

- Chaim Azreil Weizmann developed a process to produce acetone and butyl acetate, chemicals in critical demand during World War I. In return, the British government supported the establishment of a Jewish state, Israel, which named Weizmann its first president in 1948.
- The pure organic chemical produced in the greatest quantity per year is sucrose sugar. The process of converting raw sugarcane or beets into uniform-size sucrose crystals involves extraction, heat transfer, multieffect evaporation, adsorption, crystallization, filtration, and drying—not to mention mass transfer and instrumentation.
- One of the chemical by-products of paper manufacture is synthetic vanilla. One paper mill could produce the world's annual demand in one day.
- The first synthetic plastic, Bakelite, was developed to be a substitute for ivory in the manufacture of billiard balls.
- Aluminum, once more costly than gold or platinum, was so expensive that Russia minted aluminum coins to demonstrate its wealth and power. The Hall electrochemical engineering process for producing aluminum has lowered the cost of aluminum to where it has become a very low cost material.
- Latex paints were developed as a substitute for oil-based paints, not only to produce a water-based paint that would not contribute to air pollution but also to produce a drip-proof product.
- DuPont employee Roy Plunkett stored some tetrafluoroethylene gas in a container. The next day, he found it had polymerized, leaving a waxy substance called polytetrafluoroethylene, commonly known as Teflon.
- Water and the element gallium, a material used in semiconductor manufacture, are the only two significant substances known to expand when they are cooled and frozen.

and analytic chemistry; chemical engineering courses, and mechanical, electrical, and civil engineering courses. Advanced mathematics courses are also required. Courses in English, economics, history, and public speaking will also help further the career of a chemical engineer.

A bachelor's degree is generally sufficient for an entry-level position at most industrial companies. Many universities offer engineering degree programs in which students take classes one semester and work at a chemical plant the next semester, repeating this pattern until graduation. This enables students to pay for their education and to gain relevant work experience, as well as to make contacts within the industry.

A master's degree in chemical or computer engineering or a master's of business administration degree will help advance a career in chemical engineering. A doctorate is required for those seeking jobs in colleges and universities.

## SOCIAL CONTEXT AND FUTURE PROSPECTS

Many of the major issues that face society, particularly those concerning the supply of energy and water, the environment, and global warming, require immediate and continuing action from scientists such as chemists and chemical engineers.

**Energy.** The predominant sources of energy are coal, liquid petroleum, and natural gas. Coal use produces air pollutants such as sulfur dioxide, nitrogen oxides, mercury, and more carbon dioxide per British thermal unit of energy generated than any of the other energy sources. Liquid petroleum and natural gas also produce carbon dioxide and other air pollutants. Problems regarding the limited supply of all three sources of energy are causing worldwide economic and political disruptions. Chemical engineering may be able to provide economical answers in that it can create clean oil and gas from coal and oil shale, produce better biofuels from nonfood agricultural crops, and develop materials to make solar energy and hydrogen fuel cells practical.

**The Environment.** Both air and water pollution are affected by chemicals that are the result of manufacturing, handling, and disposing of materials such as solvents, insecticides, herbicides, and fertilizers. Chemical engineering has reduced factory emissions through the development of water-based coatings. In addition, new methods of converting solid wastes into usable fuels will help the environment and provide new fuel sources.

**Water Access.** The supply of fresh water for personal and agricultural use is already limited and will become more so as the world population grows. Two methods for recovering fresh water are distillation and reverse osmosis. To become practical, distillation systems must use better materials to prevent corrosion. Chemical engineers are likely to develop new alloys and ways to manufacture them, as well as better heat-recovery methods to reduce cost and lower carbon dioxide emissions. Although the long-range effect of increased concentrations of carbon dioxide on the climate may not be known for sure, chemical engineers are striving to develop methods to control and reduce these emissions.

Reverse osmosis uses membranes that allow water to pass through but not soluble salts. Chemical engineers are working on improved membrane materials that will allow higher pressures to improve efficiency and membrane life.

*Max Statman, B.S., M.S.*

## FURTHER READING

Dethloff, Henry C. *A Unit Operation: A History of Chemical Engineering at Texas A&M University.* College Station: Texas A&M University Press, 1988. Describes how chemical engineering helped create the petrochemical industry that dominates Gulf Coast industry.

Dobre, Tanase, and José G. Sanchez Marcano. *Chemical Engineering: Modeling, Simulation, and Similitude.* Weinheim, Germany: Wiley-VCH, 2007. Looks at how computer-aided modeling is used to develop, implement, and improve industrial processes. Covers the entire process, including mathematical modeling, results analysis, and performance evaluation.

Lide, David R., ed. *CRC Handbook of Chemistry and Physics: A Ready-Reference Book of Chemical and Physical Data.* 90th ed. Boca Raton, Fla.: CRC Press, 2009. A vital source of information for designing chemical processes, analyzing results, and estimating costs.

Perry, R. H., and D. W. Green, eds. *Perry's Chemical Engineers Handbook.* 8th ed. New York: McGraw-Hill, 2007. First published in 1934, this handbook provides information about the processes, operations, and equipment involved in chemical engineering, as well as chemical and physical data, and conversion factors. More than seven hundred illustrations.

Towler, Gavin P., and R. K. Sinnott. *Chemical Engineering Design.* Oxford, England: Butterworth-Heinemann, 2009. Examines how chemical engineers design chemical processes and discusses all the elements and factors involved.

## WEB SITES

*American Chemical Society*
http://portal.acs.org

*American Institute of Chemical Engineering*
http://www.aiche.org

*Society of Chemical Industry*
http://www.soci.org

**See also:** Biochemical Engineering; Coal Gasification; Coal Liquefaction; Computer-Aided Design and Manufacturing; Cracking; Distillation; Gasoline Processing and Production; Industrial Fermentation; Petroleum Extraction and Processing.

# CINEMATOGRAPHY

## FIELDS OF STUDY

Physics of light; acoustics; holography; photography; computer graphics; computer animation; mathematics; chemistry; sensitometry; mechanical engineering; electronic engineering; perception.

## SUMMARY

Cinematography is the science and practice of making motion pictures. It includes the technical processes behind all imaging formats, including film and analogue and digital video. Cinematography is concerned with the careful manipulation of technical and mechanical tools, such as lighting, exposure, lenses, filters, and special effects, to create a coherent visual expression of information, emotion or narrative. Besides the filmmaking and commercial industries, in which cinematography plays a central role, virtually any area where the production of a moving image is useful relies on the principles of this field of applied science. For example, cinematography has important applications in disciplines as diverse as military, education, scientific research, marketing, and medicine.

## KEY TERMS AND CONCEPTS

- **Cut:** Transition between one camera shot and another without any effect in between, such as a dissolve or a fade; also called "edit."
- **Film Stock:** Traditional photographic film on which a moving image is captured, consisting of a base painted with a silver-halide emulsion that reacts to light. Common film stock gauges, or widths, are 16 and 35 millimeters.
- **High-Definition Technology:** Method of capturing and displaying images at a much higher resolution than standard-definition footage.
- **Interlaced Scanning:** Technology in which each frame of a motion picture is composed by scanning two separate fields and displaying them simultaneously.
- **Progressive Scanning:** Technology in which each frame of a motion picture is processed as a complete, separate image.

- **Sampling:** Process of repeatedly measuring an analogue signal at regular intervals and using this data to produce a digital signal; sampling allows motion pictures shot on film to be converted to a digital format.
- **Sensitometry:** Measurement of the sensitivity of various materials, especially photographic film, to light.
- **Widescreen:** Imaging format in which the ratio of the width of the image frame to its height (aspect ratio) is greater than 4:3. Widescreen formats appear much wider than they are tall.

## DEFINITION AND BASIC PRINCIPLES

Cinematography is the science behind the techniques involved in creating a motion picture, as well as the practical application of those techniques. Basic cinematographic tools include film and video cameras; different lenses, filters, and film stock; the equipment involved with artificially lighting a set; the machinery used to mount and transport cameras and control their angles and movements; and a wide variety of computerized special effects that can be created or integrated into the motion picture in the postproduction stage. Unlike still photography, in which a complete product is composed of a single image, cinematography makes use of the relationships between quickly moving images to produce a narrative arc.

At heart, cinematography is based on an illusion caused by the interplay between technology and human perception. The typical motion-picture projection, whether the format is film, video, or digital, consists of twenty-four separate still frames per minute. These still frames are transformed into what appears to be a moving image through two related features of the human visual system known as persistence of vision and the phi phenomenon. Persistence of vision refers to the fact that when light hits the retina of the eye, the images it creates persist in the brain for a tiny fraction of a second longer than the physical stimulus itself. The phi phenomenon refers to the fact that when the eye is shown two separate images in rapid succession, the brain creates the appearance of seamless movement between the two frames. The combination of these two psychophysical characteristics makes it possible for people to ignore the tiny fractions of darkness that appear for a moment in between each still frame.

## BACKGROUND AND HISTORY

Motion-picture technology was born in the late nineteenth century, when a research team led by American inventor Thomas Alva Edison and engineers working in England and France all independently developed a means of photographing still images at a rate fast enough to capture movement and of projecting successive images at a rate fast enough to create the illusion of seamless motion. Among the first motion-picture projection technologies were Edison's Kinetograph and Kinetoscope.

Early motion pictures were recorded in black and white (though they were often hand colored) and had no sound. Early in the twentieth century, various experiments with sound recording and amplification equipment were carried out, including a technique for using wax phonograph discs to record sound and an electronic loudspeaker for amplifying it in theaters. The Vitaphone system, created by the Warner Bros. studio, used these developments to produce a short musical known as *The Jazz Singer* (1927). By the late 1920's, studios had moved to recording sound optically on a separate reel of film stock. For the sound track of a film to be properly synchronized with the visual image, the speed of motion picture projection was standardized at twenty-four frames per second. Color cinematography was first achieved through a system of filters that recorded information from the red, green, and blue spectrums of light separately onto a strip of black-and-white film—later, color film stock was developed in which three layers of emulsion on the film served the same purpose.

In the 1990's, digital cinematography, which captures moving images in a digital, rather than analogue, format, began to take off. The first major cinematic release shot on digital video was George Lucas's *Star Wars Episode II: The Attack of the Clones* (2002).

## HOW IT WORKS

**Cameras.** The body of a traditional motion-picture camera consists of a sturdy, lightproof housing, a system of motors that control the movement of film and mechanical components such as the shutter, and a viewing system through which the camera operator can monitor the footage being shot. Onto the body of the camera is affixed a lens, which uses one or more

*Invented by Thomas Alva Edison's Scottish employee, William Dickson, the Kinetoscope was the first device to show motion pictures.* (SSPL via Getty Images)

convex glass elements to gather and focus rays of light onto the film. Depending on the type of framing the cinematographer wishes to achieve, standard, wide-angle, telephoto, or zoom lenses may be used. To produce a smooth, not jerky, moving image, cameras are usually mounted on sturdy supports during filming. A basic dolly mount consists of a large, heavy tripod on a sturdy base with casters. A mount has various adjustment levers that allow the camera operator to raise, lower, and tilt the camera. To raise the camera to an elevated height, a crane is added to the dolly; for long panning shots, the entire contraption is often moved along specially built rails that are laid on the floor. Steadicam systems, in which a camera is harnessed to the operator's body, reduce the jerkiness caused by his or her movements during shooting and allow for smoother handheld operation of cameras.

**Film Cinematography.** To capture a rapid succession of images, film is inserted behind the camera lens and automatically unrolled by an electric motor from a supply reel, through a gate, and onto a take-up reel. As the film travels through the gate, a motorized shutter lifts up and down at regular intervals, usually twenty-four times per second. Every time the shutter lifts, the film is exposed to light. Film stock is composed of a base, traditionally made of celluloid but later usually made of polyester or Mylar, painted with a layer or layers of silver-halide emulsion. This emulsion reacts when exposed to light. (Color films have three layers of emulsion, each of which reacts with only one of the three primary colors of light: blue, green, or red.) Once the film has been shot, a chemical known as a developer is used to process the exposed areas of the film and produce a negative image. With black-and-white film, a negative shows dark areas of the image as light areas and vice versa. With color film, a negative shows complementary colors—for example, red areas appear green. Further chemical processes make a positive print out of these negatives. When a positive print is projected onto a screen, the image appears in its correct form. A type of film known as reversal film is also available. It has the advantage of producing a positive rather than a negative image at the end of the developing process.

**Video Cinematography.** Although film remains an important tool for many cinematographers and motion-picture producers, other imaging formats have become increasingly relevant. With analogue video imaging, light gathered by the camera lens is not captured by the emulsions on film stock but instead is converted into electric signals by tiny photosensitive diodes on a component known as a charge-coupled device (CCD). In some cameras, three separate CCDs are used, with each capturing information about one of the three primary colors. This electric signal is then recorded as a pattern on a strip of magnetic tape. Analogue video imaging is used to produce news broadcasts and other television broadcasts. Because it is very inexpensive, it is also a popular tool for at-home or amateur motion-picture production. Digital video imaging also makes use of CCDs. With digital video, however, the electric signal from a CCD is not converted into a magnetic pattern on tape. Instead, it is translated into a digital signal or binary code consisting solely of 1's and 0's. Digital cameras can record onto hard disks, digital video discs (DVDs), flash drives, or any other digital format.

## APPLICATIONS AND PRODUCTS

**Filmmaking.** With the advent of motion-picture imaging came an entirely new form of visual communication, one that—unlike photographs—was able to depict a series of events, and thereby express a narrative rather than a single moment in time. In addition, the illusion of movement and the fidelity with which the camera records images lends viewers of motion pictures an irresistible sense that the events they are watching not only are somehow real but also are occurring in the present rather than the past.

### Fascinating Facts About Cinematography

- The word cinematography is derived from two ancient Greek words that can be literally translated as "writing with motion"—the term distinguishes cinematography from photography, which comes from roots that mean "writing with light."

- An aerial shot is usually captured by a camera mounted on a helicopter. Before this became possible, filmmakers who wanted a shot from the air teamed camera men with stunt pilots to get the desired effect.

- In the 1939 film *The Wizard of Oz*, a scene in which the witch writes a message in the sky with her broom was filmed by a camera mounted beneath a glass tank. The "sky" was composed of a cloudy mixture of water and oil.

- The first motion-picture recordings were not shown on a theater screen but in a large machine the size of a cabinet. People would step into the machine and deposit a coin to watch the film.

- In 1903, projectionist Edwin S. Porter created the first motion picture that presented a narrative. Known as *The Great Train Robbery*, the 14-minute film contained the world's first motion-picture chase scene.

- When astronomers on the space shuttle *Atlantis* went up into space to repair the Hubble telescope in 2009, they carried a massive IMAX high-definition camera on board with which to shoot a film about the project. The camera required a mile of film to record 8 minutes of action.

- Before the development of computer-generated special effects, filmmakers used inventive techniques to create illusions. In the 1895 film *The Execution of Mary Queen of Scots*, the camera was stopped, the actress playing Mary ran off-screen and was replaced with a dummy, and the other actors froze in position. Then, the camera was restarted and the guillotine fell to "behead" Mary.

Arguably the most significant application of cinematography, then, at least in terms of its cultural impact, is the use of motion-picture imaging to create narrative, documentary, and newsreel films. Practicing cinematographers are able to apply the full range of cinematographic tools and techniques in different ways to achieve sophisticated effects. For instance, the use of a dolly mounted on tracks allows a camera operator to pan (turn) the camera in a horizontal plane. This is often used to produce the illusion that the camera's perspective is that of a character in the film. Although this application has largely been taken over by computer-generated imaging, cinematographic techniques also enable the creation of animated films. Animation is created by using a motion-picture camera to film a series of carefully constructed illustrations (or objects, such as clay puppets). When replayed, the still images appear to move.

**Education.** Cinematography has made it possible for students of all subjects to engage in learning through direct observation without ever leaving their classrooms. Techniques such as time-lapse photography, in which still frames that were captured at a very slow rate are replayed at a much faster speed, enable teachers to demonstrate incredibly drawn-out and subtle processes, such as the development of vegetation from tiny seed to fruiting tree, or the movement of the stars across the sky. In the same way, high-speed photography, in which still frames that were captured at a very fast rate are replayed at a slower speed, enable students to see clearly the steps involved in processes that happen in the blink of an eye, such as the formation of droplets as a bead of water hits the surface of a pond. Time-lapse photography is created with a standard camera attached to a device set to trigger its shutter at extended, regular intervals—such as an hour apart or more. The technology behind high-speed photography is more complex and requires careful control over the precise timing of the camera's shutter and the amount of light the film or CCD is exposed to during the time the shutter is open. A shutter, for instance, may be triggered by a sound associated with the event the cinematographer is trying to capture, such as the firing of a gun.

Another cinematographic technology that has transformed both education and entertainment is the use of three-dimensional (3D) imaging and projection systems such as IMAX 3D. With three-dimensional cinematography, two camera lenses are used to capture two separate streams of images, one corresponding to the view a human observer would receive through his or her left eye, and one corresponding to the view through his or her right eye. In the theater, these two images are them projected onto the screen simultaneously. With the use of a special pair of three-dimensional glasses, viewers are able to perceive the dual cinematic streams as a single three-dimensional image. Three-dimensional imaging and projection is widely used in science museums as a means of giving museum-goers a more vivid entryway into presentations about topics such as space exploration and underwater habitats.

**Medical Video Imaging.** A host of specialized cinematographic techniques have found useful homes in the realm of clinical diagnosis and surgery. For example, stereo endoscopes make it far easier for physicians to perform minimally invasive surgeries. A stereo endoscope is an instrument that can be inserted into a patient's body through a small incision or down a natural orifice such as the throat. It is used to transmit a three-dimensional video image of a patient's internal parts to the surgeon as he or she works. A stereo endoscope consists of a tube, often flexible, containing a dual lens system and a bundle of optical fibers. These fibers bring light from an external light source into the patient's body, then transmit the video image from the lens back out to a large screen in the operating room. When cinematographic techniques are combined with tools from communications technology, another application of real-time video imaging in medicine emerges—Web-based conferencing, which enables collaborations between clinicians who are physically distant from each other.

## IMPACT ON INDUSTRY

Apart from the multibillion-dollar filmmaking industry, which is the biggest global user of cinematographic technology, many other industry sectors rely on this field of applied science to carry out a significant element of their business operations. The advertising industry, for example, uses the tools and visual conventions of motion-picture imaging to generate what are essentially short narrative films—commercials. Commercials promote virtually every consumer product on the market, from baby food to prescription drugs. The music industry has been revolutionized by the advent of cinematography in the form of music videos. These brief, lavishly produced motion pictures help recording studios showcase their performers as flesh-and-blood personalities rather than faceless voices, helping drive additional sales.

Cinematography also has important applications in the security industry. Both private security firms and public agencies make use of closed-circuit television (CCTV) technology—a system in which the video imaging signals from one or more cameras are transmitted to a restricted set of monitors—to keep a continuous record of what takes place in businesses such as banks, convenience markets, and hospitals, and public areas with a high record of criminal activity.

**Major Corporations.** Much of the research into new cinematographic techniques takes place within the context of the motion-picture industry rather than in academic or government research laboratories. In the early twenty-first century, for example, the Eastman Kodak company introduced a new generation of color motion-picture films that—because of a change in the chemistry of the silver-halide emulsion used to produce these films—was twice as sensitive to light as previous films, enabling camera operators to shoot in conditions with lower light without compromising the quality of the footage produced. Besides Kodak, other large players in the field of cinematographic technology include Technicolor, Deluxe, and Fujifilm. The Arri Group and Panavision are among the largest suppliers of traditional film cameras to the motion-picture industry, and both groups have also entered the digital cinema market alongside camera-producing corporations such as Silicon Imaging and RED. The six most important corporations involved in filming and producing motion pictures in the United States are Paramount Pictures, Walt Disney Pictures/Touchstone Pictures, Warner Bros. Pictures, Twentieth Century Fox Film Corporation, Universal Pictures, and Columbia Pictures.

## CAREERS AND COURSE WORK

Within the motion-picture industry, the role of the cinematographer is more creative than technical. Cinematographers work with the director of a film to decide on how each shot will be composed. Students interested in more hands-on career options may consider becoming camera operators, lighting technicians, sound engineers, set electricians, or postproduction editors.

Anyone interested in a career as a cinematographer or camera operator or who plans on pursuing some other kind of technical occupation related to cinematography such as lighting, sound recording, and postproduction and editing, should acquire a strong body of knowledge about the physics of light. Besides courses in optics, it is important to take specialized classes in the mechanical and electrical engineering of film, video, and digital cameras, and to understand how each type of cinematographic equipment, including lenses, interacts with light to produce an image. Other important educational requirements include an understanding of the chemistry of film emulsions and development and the mathematics of exposure.

Finally, it would be appropriate for any student of cinematography to take a course in human visual perception through a psychology or physiology department. Understanding how the eye and brain work together to process light and visual information allows those who work in motion picture imaging to carefully manipulate those processes to achieve the desired emotional or visual effect. Many practicing cinematographers have graduated from a technical college rather than a traditional liberal arts university, or have pursued specialized training in motion-picture technology from a formal film school program. A degree in computer science or computer graphics would also be an appropriate starting point for someone wishing to work in special effects, animation, or video editing.

## SOCIAL CONTEXT AND FUTURE PROSPECTS

It is hard to overestimate the social impact that cinematography and its associated technologies have had. One important cultural role for motion-picture imaging has been expanding the horizons of viewers beyond the concerns of their own lives, their own communities, and even their own countries—enabling them to see directly into entirely different worlds. In this way, many films have served as powerful mechanisms for social and humanitarian change. In the 1980's, for example, when documentary cameras captured dramatic footage of starving Ethiopian men, women, and children and delivered it to the eyes of the Western world, aid money poured into the impoverished nation. The 1993 film *Philadelphia*, which told the story of a lawyer infected with the human immunodeficiency virus and his partner, highlighted the growing devastation wrought by the acquired immunodeficiency syndrome (AIDS) epidemic in the United States and put a human face on what was still considered by many Americans to be a shameful disease.

The use of motion-picture technology by nonprofessionals has also had a profound cultural impact. In the second half of the twentieth century, the development of less expensive cameras for making home movies (particularly home video cameras) enabled individuals across the world to create a tangible archive of their personal experiences and domestic milestones. Unlike a photograph, which is often posed, the home movie preserves not just formal, fleeting glimpses of the past but rather ceremonies, events, and candid interactions. In addition, the home movie transports the viewer straight back to the moment when the footage was recorded.

One of the most promising emerging technologies in cinematography is a technique known as the digital intermediate (DI) process. This is a means of scanning motion pictures recorded on film—which many advocates argue is of a far superior visual quality than digital video—at an extremely high resolution, thus converting it into a digital format. The DI process may enable the motion-picture industry to take advantage of the best of both film and digital technologies to produce imaging that is stunningly crisp and clear as well as far easier and cheaper to distribute.

*M. Lee, B.A., M.A.*

### FURTHER READING

Barclay, Steven. *The Motion Picture Image: From Film to Digital.* Boston: Focal Press, 2000. A narrative history of the development of cinematographic technology, as well as a practical overview of related technical issues. Contains a comprehensive index.

Brown, Blaine. *Cinematography: Theory and Practice.* Boston: Focal Press, 2002. A hands-on guide for the practicing cinematographer, heavily illustrated with photographs and diagrams. Includes a bibliography.

Meza, Philip E. *Coming Attractions? Hollywood, High Tech, and the Future of Entertainment.* Stanford, Calif.: Stanford Business Books, 2007. An examination of how technological developments in cinematography are acting to shape the U.S. entertainment market. Organized into a series of real-life business case studies.

Sawiki, Mark. *Filming the Fantastic: A Guide to Visual Effect Cinematography.* Boston: Focal Press, 2007. Includes hundreds of full-color photographs illustrating special-effects techniques.

Saxby, Graham. *The Science of Imaging: An Introduction.* Philadelphia: Institute of Physics Publishing, 2002. An overview of the fundamentals of physics and mathematics involved in creating television, holography, and other forms of imaging. Includes diagrams, glossaries, and numerous explanatory notes in the sidebars.

### WEB SITES

*Eastman Kodak Company*
The Essential Reference Guide for Filmmakers
http://motion.kodak.com/US/en/motion/Education/Publications/Essential_reference_guide/index.htm

*U.S. Department of Labor, Bureau of Labor Statistics*
Motion Picture and Video Industries
http://www.bls.gov/OCO/CG/CGS038.HTM

**See also:** Audio Engineering; Computer Graphics; Mirrors and Lenses; Optics; Photography; Television Technologies; Video Game Design and Programming.

# CIVIL ENGINEERING

## FIELDS OF STUDY

Mathematics; chemistry; physics; engineering mechanics; strength of materials; fluid mechanics; soil mechanics; hydrology; surveying; engineering graphics; environmental engineering; structural engineering; transportation engineering.

## SUMMARY

Civil engineering is the branch of engineering concerned with the design, construction, and maintenance of fixed structures and systems, such as large buildings, bridges, roads, and other transportation systems, and water supply and wastewater-treatment systems. Civil engineering is the second oldest field of engineering, with the term "civil" initially used to differentiate it from the oldest field of engineering, military engineering. The major subdisciplines within civil engineering are structural, transportation, and environmental engineering. Other possible areas of specialization within civil engineering are geotechnical, hydraulic, construction, and coastal engineering.

## KEY TERMS AND CONCEPTS

- **Abutment:** Part of a structure designed to withstand thrust, such as the end supports of a bridge or an arch.
- **Aqueduct:** Large pipe or conduit used to transport water a long distance.
- **Backfill:** Material used in refilling an excavated area.
- **Cofferdam:** Temporary structure built to keep water out of a construction zone in a river.
- **Design Storm:** Storm of specified return period and duration at a specified location, typically used for storm water management design.
- **Foundation:** Ground that is used to support a structure.
- **Freeboard:** Difference in height between the water level and the top of a tank, dam, or channel.
- **Girder:** Large horizontal structural member, supporting vertical loads.
- **Invert:** Curved, inside, bottom surface of a pipe.
- **Percolation Test:** Test to determine the drainage capability of soil.

- **Rapid Sand Filter:** System for water treatment by gravity filtration through a sand bed.
- **Reinforced Concrete:** Concrete that contains wire mesh or steel reinforcing rods to give it greater strength.
- **Sharp-Crested Weir:** Obstruction with a thin, sharp upper edge, used to measure flow rate in an open channel.
- **Tension Member:** Structural member that is subject to tensile stress.
- **Ultimate Bearing Capacity:** Theoretical maximum pressure that a soil can support from a load without failure

### DEFINITION AND BASIC PRINCIPLES

Civil engineering is a very broad field of engineering, encompassing subdisciplines ranging from structural engineering to environmental engineering, some of which have also become recognized as separate fields of engineering. For example, environmental engineering is included as an area of specialization within most civil engineering programs, many colleges offer separate environmental engineering degree programs.

Civil engineering, like engineering in general, is a profession with a practical orientation, having an emphasis on building things and making things work. Civil engineers use their knowledge of the physical sciences, mathematics, and engineering sciences, along with empirical engineering correlations to design, construct, manage, and maintain structures, transportation infrastructure, and environmental treatment equipment and facilities.

Empirical engineering correlations are important in civil engineering because useable theoretical equations are not available for all the necessary engineering calculations. These empirical correlations are equations, graphs, or nomographs, based on experimental measurements, that give relationships among variables of interest for a particular engineering application. For example the Manning equation gives an experimental relationship among the flow rate in an open channel, the slope of the channel, the depth of water, and the size, shape, and material of the bottom and sides of the channel. Rivers, irrigation ditches, and concrete channels

*An aerial view of Hoover Dam and the under-construction Hoover Dam bypass in 2009.* (Getty Images)

used to transport wastewater in a treatment plant are examples of open channels. Similar empirical relationships are used in transportation, structural, and other specialties within civil engineering.

## BACKGROUND AND HISTORY

Civil engineering is the second oldest field of engineering. The term "civil engineering" came into use in the mid-eighteenth century and initially referred to any practice of engineering by civilians for nonmilitary purposes. Before this time, most large-scale construction projects, such as roads and bridges, were done by military engineers. Early civil engineering projects were in areas such as water supply, roads, bridges, and other large structures, the same type of engineering work that exemplifies civil engineering in modern times.

Although the terminology did not yet exist, civil engineering projects were carried out in early times. Examples include the Egyptian pyramids (about 2700-2500 B.C.E.), well-known Greek structures such as the Parthenon (447-438 B.C.E.), the Great Wall of China (220 B.C.E.), and the many roads, bridges, dams, and aqueducts built throughout the Roman Empire.

Most of the existing fields of engineering split off from civil engineering or one of its offshoots, as new fields emerged. For example, with increased use of machines and mechanisms, the field of mechanical engineering emerged in the early nineteenth century.

## HOW IT WORKS

In addition to mathematics, chemistry, and physics, civil engineering makes extensive use of principles from several engineering science subjects: engineering mechanics (statics and strength of materials), soil mechanics, and fluid mechanics.

**Engineering Mechanics—Statics.** As implied by the term "statics," this area of engineering concerns objects that are not moving. The fundamental principle of statics is that any stationary object must be in static equilibrium. That is, any force on the object must be cancelled out by another force that is equal in magnitude and acting in the opposite direction. There can be no net force in any direction on a stationary object, because if there were, it would be moving in that direction. The object considered to be in static equilibrium could be an entire structure or it could be any part of a structure down to an individual member in a truss. Calculations for an object in static equilibrium are often done through the use of a free body diagram, that is a sketch of the object, showing all the forces external to that object that are acting on it. The principle then used for calculations is that the sum of all the horizontal forces acting on the object must be zero and the sum of all the vertical forces acting on the object must be zero. Working with the forces as vectors helps to find the horizontal and vertical components of forces that are acting on the object from some direction other than horizontal or vertical.

**Engineering Mechanics—Strength of Materials.** This subject is sometimes called mechanics of materials. Whereas statics works only with forces external to the body that is in equilibrium, strength of materials uses the same principles and also considers internal forces in a structural member. This is done to determine the required material properties to ensure that the member can withstand the internal stresses that will be placed on it.

**Soil Mechanics.** Knowledge of soil mechanics is needed to design the foundations for structures. Any structure resting on the Earth will be supported in some way by the soil beneath it. A properly designed foundation will provide adequate long-term support for the structure above it. Inadequate knowledge of soil mechanics or inadequate foundation design may lead to something such as the Leaning Tower of Pisa. Soil mechanics topics include physical properties of soil, compaction, distribution of stress within soil, and flow of water through soil.

**Fluid Mechanics.** Fundamental principles of physics are used for some fluid mechanics calculations. Examples are conservation of mass (called the continuity equation in fluid mechanics) and conservation of energy (also called the energy equation or the first law of thermodynamics). Some fluid mechanics applications, however, make use of empirical (experimental) equations or relationships. Calculations for

flow through pipes or flow in open channels, for example, use empirical constants and equations.

**Knowledge from Engineering Fields of Practice.** In addition to these engineering sciences, a civil engineer uses accumulated knowledge from the civil engineering areas of specialization. Some of the important fields of practice are hydrology, geotechnical engineering, structural engineering, transportation engineering, and environmental engineering. In each of these fields of practice, there are theoretical equations, empirical equations, graphs or nomographs, guidelines, and rules of thumb that civil engineers use for design and construction of projects related to structures, roads, storm water management, or wastewater-treatment projects, for example.

**Civil Engineering Tools.** Several tools available for civil engineers to use in practice are engineering graphics, computer-aided drafting (CAD), surveying, and geographic information systems (GIS). Engineering graphics (engineering drawing) has been a mainstay in civil engineering since its inception, for preparation of and interpretation of plans and drawings. Most of this work has come to be done using computer-aided drafting. Surveying is a tool that has also long been a part of civil engineering. From laying out roads to laying out a building foundation or measuring the slope of a river or of a sewer line, surveying is a useful tool for many of the civil engineering fields. Civil engineers often work with maps, and geographic information systems, a much newer tool than engineering graphics or surveying, make this type of work more efficient.

**Codes and Design Criteria.** Much of the work done by civil engineers is either directly or indirectly for the public. Therefore, in most of civil engineering fields, work is governed by codes or design criteria specified by some state, local, or federal agency. For example, federal, state, and local governments have building codes, state departments of transportation specify design criteria for roads and highways, and wastewater-treatment processes and sewers must meet federal, state, and local design criteria.

**APPLICATIONS AND PRODUCTS**

**Structural Engineering.** Civil engineers design, build, and maintain many and varied structures. These include bridges, towers, large buildings (skyscrapers), tunnels, and sports arenas. Some of the civil engineering areas of knowledge needed for structural engineering are soil mechanics/geotechnical engineering, foundation engineering, engineering mechanics (statics and dynamics), and strength of materials.

When the Brooklyn Bridge was built over the East River in New York City (1870-1883), its suspension span of 1,595 feet was the longest in the world. It remained the longest suspension bridge in North America until the Williamsburg Bridge was completed in New York City in 1903. The Brooklyn Bridge joined Brooklyn and Manhattan, and helped establish the New York City Metropolitan Area.

The Golden Gate Bridge, which crosses the mouth of San Francisco Bay with a main span of 4,200 feet, had nearly triple the central span of the Brooklyn Bridge. It was the world's longest suspension bridge from its date of completion in 1937 until 1964, when the Verrazano-Narrows Bridge opened in New York City with a central span that was 60 feet longer than that of the Golden Gate Bridge. The Humber Bridge, which crosses the Humber estuary in England and was completed in 1981, has a single suspended span of 4,625 feet and is the longest suspension bridge in the world.

One of the most well-known early towers illustrates the importance of good geotechnical engineering and foundation design. The Tower of Pisa, commonly known as the Leaning Tower of Pisa, in Italy started to lean to one side very noticeably, even during its construction (1173-1399). Its height of about 185 feet is not extremely tall in comparison with towers built later, but it was impressive when it was built. The reason for its extreme tilt (more than 5 meters off perpendicular) is that it was built on rather soft, sandy soil with a foundation that was not deep enough or spread out enough to support the structure. In spite of this, the Tower of Pisa has remained standing for more than six hundred years.

Another well-known tower, the Washington Monument, was completed in 1884. At 555 feet in height, it was the world's tallest tower until the Eiffel Tower, nearly 1,000 feet tall, was completed in 1889. The Washington Monument remains the world's tallest masonry structure. The Gateway Arch in St. Louis, Missouri, is the tallest monument in the United States, at 630 feet.

The 21-story Flatiron Building, which opened in New York City in 1903, was one of the first skyscrapers.

It is 285 feet tall and its most unusual feature is its triangular shape, which was well suited to the wedge-shaped piece of land on which it was built. The 102-floor Empire State Building, completed in 1931 in New York City with a height of 1,250 feet, outdid the Chrysler Building that was under construction at the same time by 204 feet, to earn the title of the world's tallest building at that time. The Sears Tower (now the Willis Tower) in Chicago is 1,450 feet tall and was the tallest building in the world when it was completed in 1974. Several taller buildings have been constructed since that time in Asia.

Some of the more interesting examples of tunnels go through mountains and under the sea. The Hoosac Tunnel, built from 1851 to 1874, connected New York State to New England with a 4.75-mile railway tunnel through the Hoosac Mountain in northwestern Massachusetts. It was the longest railroad tunnel in the United States for more than fifty years. Mount Blanc Tunnel, built from 1957 to 1965, is a 7.25-mile long highway tunnel under Mount Blanc in the Alps to connect Italy and France. The Channel Tunnel, one of the most publicized modern tunnel projects, is a rather dramatic and symbolic tunnel. It goes a distance of 31 miles beneath the English Channel to connect Dover, England, and Calais, France.

**Transportation Engineering.** Civil engineers also design, build, and maintain a wide variety of projects related to transportation, such as roads, railroads, and pipelines.

Many long, dramatic roads and highways have been built by civil engineers, ever since the Romans became the first builders of an extensive network of roads. The Appian Way is the most well known of the many long, straight roads built by the Romans. The Appian Way project was started in 312 B.C.E. by the Roman censor, Appius Claudius. By 244 B.C.E., it extended about 360 miles from Rome to the port of Brundisium in southeastern Italy. The Pan-American Highway, often billed as the world's longest road, connects North America and South America. The original Pan-American Highway ran from Texas to Argentina with a length of more than 15,500 miles. It has since been extended to go from Prudhoe Bay, Alaska, to the southern tip of South America, with a total length of nearly 30,000 miles. The U.S. Interstate Highway system has been the world's biggest earthmoving project. Started in

## Fascinating Facts About Civil Engineering

- The Johnstown flood in Pennsylvania, on May 31, 1889, which killed more than 2,200 people, was the result of the catastrophic failure of the South Fork Dam. The dam, built in 1852, held back Lake Conemaugh and was made of clay, boulders, and dirt. An improperly maintained spillway combined with heavy rains caused the collapse.

- The I-35W bridge over the Mississippi River in Minneapolis, Minnesota, collapsed during rush-hour traffic on August 1, 2007, causing 13 deaths and 145 injuries. The collapse was blamed on undersized gusset plates, an increase in the concrete surface load, and the weight of construction supplies and equipment on the bridge.

- On November 7, 1940, 42-mile-per-hour winds twisted the Tacoma Narrows Bridge and caused its collapse. The bridge, with a suspension span of 2,800 feet, had been completed just four months earlier. Steel girders meant to support the bridge were blocking the wind, causing it to sway and eventually collapse.

- Low-quality concrete and incorrectly placed rebar led to shear failure, collapsing the Highway 19 overpass in Laval, Quebec, on September 30, 2006.

- The first design for the Gateway Arch in St. Louis, Missouri, had a fatal flaw that made it unstable at the required height. The final design used 886 tons of stainless steel, making it a very expensive structure.

- On January 9, 1999, just three years after it was built, the Rainbow Bridge, a pedestrian bridge across the Qi River in Sichuan Province, collapsed, killing 40 people and injuring 14. Concrete used in the bridge was weak, parts of it were rusty, and parts had been improperly welded.

- On December 7, 1982, an antenna tower in Missouri City, Texas, collapsed, killing 2 riggers on the tower and 3 who were in an antenna section that was being lifted. U-bolts holding the antenna failed, and as it fell, it hit a guy wire on the tower, collapsing the tower. The engineers on the project declined to evaluate the rigger's plans.

- On February 26, 1972, coal slurry impoundment dam 3 of the Pittston Coal Company in Logan County, West Virginia, failed, four days after it passed inspection by a federal mine inspector. In the Buffalo Creek flood that resulted, 125 people were killed. The dam, which was above dams 1 and 2, had been built on coal slurry sediment rather than on bedrock.

1956 by the Federal Highway Act, it contains sixty-two highways covering a total distance of 42,795 miles. This massive highway construction project transformed the American system of highways and had major cultural impacts.

The building of the U.S. Transcontinental Railroad was a major engineering feat when the western portion of the 2,000-mile railroad across the United States was built in the 1860's. Logistics was a major part of the project, with the need to transport steel rails and wooden ties great distances. An even more formidable task was construction of the Trans-Siberian Railway, the world's longest railway. It was built from 1891 to 1904 and covers 5,900 miles across Russia, from Moscow in the west to Vladivostok in the east.

The Denver International Airport, which opened in 1993, was a very large civil engineering project. This airport covers more than double the area of all of Manhattan Island.

The first oil pipeline in the United States was a 5-mile-long, 2-inch-diameter pipe that carried 800 barrels of petroleum per day. Pipelines have become much larger and longer since then. The Trans-Alaska Pipeline, with 800 miles of 48-inch diameter pipe, can carry 2.14 million barrels per day. At the peak of construction, 20,000 people worked 12-hour days, seven days a week.

**Water Resources Engineering.** Another area of civil engineering practice is water resources engineering, with projects like canals, dams, dikes, and seawater barriers.

The oldest known canal, one that is still in operation, is the Grand Canal in China, which was constructed between 485 B.C.E. and 283 C.E. The length of the Grand Canal is more than 1,000 miles, although its route has varied because of several instances of rerouting, remodeling, and rebuilding over the years. The 363-mile-long Erie Canal was built from 1817 to 1825, across the state of New York from Albany to Buffalo, thus overcoming the Appalachian Mountains as a barrier to trade between the eastern United States and the newly opened western United

States. The economic impact of the Erie Canal was tremendous. It reduced the cost of shipping a ton of cargo between Buffalo and New York City from about $100 per ton (over the Appalachians) to $4 per ton (through the canal).

The Panama Canal, constructed from 1881 to 1914 to connect the Atlantic and Pacific oceans through the Isthmus of Panama, is only about 50 miles long, but its construction presented tremendous challenges because of the soil, the terrain, and the tropical illnesses that killed many workers. Upon its completion, however, the Panama Canal reduced the travel distance from New York City to San Francisco by about 9,000 miles.

When the Hoover Dam was built from 1931 to 1936 on the Colorado River at the Colorado-Arizona border, it was the world's largest dam, at a height of 726 feet and crest length of 1,224 feet. The technique of passing chilled water through pipes enclosed in the concrete to cool the newly poured concrete and speed its curing was developed for the construction of the Hoover Dam and is still in use. The Grand Coulee Dam, in the state of Washington, was the largest hydrolectric project in the world when it was built in the 1930's. It has an output of 10,080 megawatts. The Itaipu Dam, on the Parana River, along the border of Brazil and Paraguay, is also one of the largest hydroelectric dams in the world. It began operation in 1984 and is capable of producing 13,320 megawatts.

Dikes, dams and similar structures have been used for centuries around the world for protection against flooding. The largest sea barrier in the world is a 2-mile-long surge barrier in the Oosterschelde estuary of the Netherlands, constructed from 1958 to 1986. Called the Dutch Delta Plan, the purpose of this project was to reduce the danger of catastrophic flooding. The impetus that brought this project to fruition was a catastrophic flood in the area in 1953. A major part of the barrier design consists of sixty-five huge concrete piers, weighing in at 18,000 tons each. These piers support tremendous 400-ton steel gates to create the sea barrier. The lifting and placement of these huge concrete piers exceeded the capabilities of any existing cranes, so a special U-shaped ship was built and equipped with gantry cranes. The project used computers to help in guidance and placement of the piers. A stabilizing foundation used for the concrete piers consists of foundation mattresses made up of layers of sand, fine gravel, and coarse gravel. Each

foundation mattress is more than 1 foot thick and more than 650 feet by 140 feet, with a smaller mattress placed on top.

## IMPACT ON INDUSTRY

In view of its status as the second oldest engineering discipline and the essential nature of the type of work done by civil engineers, it seems reasonable that civil engineering is well established as an important field of engineering around the world. Civil engineering is the largest field of engineering in the United States. The U.S. Bureau of Labor Statistics estimates that 278,400 civil engineers were employed in the United States in 2008. The bureau also projected that civil engineering employment would grow at the rate of 24 percent rate until 2018, which is much faster than average for all occupations. Civil engineers are employed by a wide variety of government agencies, by universities for research and teaching, by consulting engineering firms, and by industry.

**Consulting Engineering Firms.** This is the largest sector of employment for civil engineers. There are many consulting engineering firms around the world, ranging in size from small firms with a few employees to very large firms with thousands of employees. In 2010, the *Engineering News Record* identified the top six U.S. design firms: AECOM Technology, Los Angeles; URS, San Francisco; Jacobs, Pasadena, California; Fluor, Irving, Texas; CH2M Hill, Englewood, Colorado; and Bechtel, San Francisco. Many consulting engineering firms have some electrical engineers and mechanical engineers, and some even specialize in those areas; however, a large proportion of engineering consulting firms are made up predominantly of civil engineers. About 60 percent of American civil engineers are employed by consulting engineering firms.

**Construction Firms.** Although some consulting engineering firms design and construct their own projects, some companies specialize in constructing projects designed by another firm. These companies also use civil engineers. About 8 percent of American civil engineers are employed in the nonresident building construction sector.

**Other Industries.** Some civil engineers are employed in industry, but less than 1 percent of American civil engineers are employed in an industry other than consulting firms and construction firms. The industry sectors that hire the most civil engineers are oil and gas extraction and pipeline companies.

**Government Agencies.** Civil engineers work for many federal, state, and local government agencies. For example, the U.S. Department of Transportation uses civil engineers to handle its many highway and other transportation projects. Many road or highway projects are handled at the state level, and each state department of transportation employs many civil engineers. The U.S. Corps of Engineers and the Department of the Interior's Bureau of Reclamation employ many civil engineers for their many water resources projects. Many cities and counties have one or more civil engineers as city or county engineers, and many have civil engineers in their public works departments. About 15 percent of American civil engineers are employed by state governments, about 13 percent by local government, and about 4 percent by the federal government.

**University Research and Teaching.** Because civil engineering is the largest field of engineering, almost every college of engineering has a civil engineering department, leading to a continuing demand for civil engineering faculty members to teach the next generation of civil engineers and to conduct sponsored research projects. This applies to universities not only in the United States but also around the world.

## CAREERS AND COURSE WORK

A bachelor's degree in civil engineering is the requirement for entry into this field. Registration as a professional engineer is required for many civil engineering positions. In the United States, a graduate from a bachelor's degree program accredited by the Accreditation Board for Engineering and Technology is eligible to take the Fundamentals of Engineering exam to become an engineer in training. After four years of professional experience under the supervision of a professional engineer, one is eligible to take the Professional Engineer exam to become a registered professional engineer.

A typical program of study for a bachelor's degree in civil engineering includes chemistry, calculus and differential equations, calculus-based physics, engineering graphics/AutoCAD, surveying, engineering mechanics, strength of materials, and perhaps engineering geology, as well as general education courses during the first two years. This is followed by fluid mechanics, hydrology or water resources, soil mechanics, engineering economics, and introductory courses for transportation engineering, structural engineering,

and environmental engineering, as well as civil engineering electives to allow specialization in one of the areas of civil engineering during the last two years.

A master's degree in civil engineering that provides additional advanced courses in one of the areas of specialization, an M.B.A., or engineering management master's degree complement a bachelor's of science degree and enable their holder to advance more rapidly. A master's of science degree would typically lead to more advanced technical positions, while an M.B.A. or engineering management degree would typically lead to management positions.

Anyone aspiring to a civil engineering faculty or research position must obtain a doctoral degree. In that case, to provide proper preparation for doctoral level study, any master's-level study should be in pursuit of a research-oriented master of science degree rather than a master's degree in engineering or a practice-oriented master of science degree.

## SOCIAL CONTEXT AND FUTURE PROSPECTS

Civil engineering projects typically involve basic infrastructure needs such as roads and highways, water supply, wastewater treatment, bridges, and public buildings. These projects may be new construction or repair, maintenance or upgrading of existing highways, structures, and treatment facilities. The buildup of such infrastructure since the beginning of the twentieth century has been extensive, leading to a continuing need for the repair, maintenance, and upgrading of existing structures. Also, governments tend to devote funding to infrastructure improvements to generate jobs and create economic activity during economic downturns. All of this leads to the projection for a continuing strong need for civil engineers.

*Harlan H. Bengtson, B.S., M.S., Ph.D.*

## FURTHER READING

Arteaga, Robert R. *The Building of the Arch.* 10th ed. St. Louis, Mo.: Jefferson National Parks Association, 2002. Describes how the Gateway Arch in St. Louis, Missouri, was built up from both sides and came together at the top. Contains excellent illustrations.

Davidson, Frank Paul, and Kathleen Lusk-Brooke, comps. *Building the World: An Encyclopedia of the Great Engineering Projects in History.* Westport, Conn.: Greenwood Press, 2006. Examines more than forty major engineering projects from the Roman aqueducts to the tunnel under the English Channel.

Hawkes, Nigel. *Structures: The Way Things Are Built.* 1990. Reprint. New York: Macmillan, 1993. Discusses many well-known civil engineering projects. Chapter 4 contains information about seventeen projects, including the Great Wall of China, the Panama Canal, and the Dutch Delta Plan. Contains illustrations and discussion of the effect of the projects.

National Geographic Society. *The Builders: Marvels of Engineering.* Washington, D.C.: National Geographic Society, 1992. Documents some of the most ambitious civil engineering projects, including roads, canals, bridges, railroads, skyscrapers, sports arenas, and exposition halls. Discussion and excellent illustrations are included for each project.

Weingardt, Richard G. *Engineering Legends: Great American Civil Engineers—Thirty-two Profiles of Inspiration and Achievement.* Reston, Va.: American Society of Civil Engineers, 2005. Looks at the lives of civil engineers who were environmental experts, transportation trendsetters, builders of bridges, structural trailblazers, and daring innovators.

**WEB SITES**
*American Institute of Architects*
http://www.aia.org

*American Society of Civil Engineers*
http://www.asce.org

*Institution of Civil Engineers*
http://www.ice.org.uk

*WBGH Educational Foundation and PBS*
http://www.pbs.org/wgbh/buildingbig/index.html

**See also:** Architecture; Bridge Design and Barodynamics; Hydraulic Engineering; Transportation Engineering.

# CLIMATE ENGINEERING

## FIELDS OF STUDY

Atmospheric chemistry; ecology; meteorology; plant biology; ecosystem management; marine systems and chemistry; geoengineering; environmental engineering; aeronautical engineering; naval architecture; engineering geology; applied geophysics; Earth system modeling.

## SUMMARY

Climate engineering (more commonly known as geoengineering) is a field of science that aims to deliberately control both micro (local) and macro (global) climates—by actions such as seeding clouds or shooting pollution particles into the upper atmosphere to reflect the Sun's rays—with the intention of reversing the effects of global warming.

## KEY TERMS AND CONCEPTS

- **Albedo:** Measure of the reflectivity of the Earth surface (that is, the proportion of solar radiation or energy that is reflected back into space).
- **Carbon Dioxide Removal (CDR):** Geoengineering techniques and technologies that remove carbon dioxide from the atmosphere in an attempt to combat climate change and global warming.
- **Climate Change:** Alterations over time in global temperatures and rainfall because of human-caused increases in greenhouse gases, such as carbon dioxide and methane.
- **Fossil Fuel:** Deposit of either solid (coal), liquid (oil), or gaseous (natural gas) hydrocarbon derived from the decomposition of organic plant and animal matter over many million of years within the Earth's crust.
- **Ocean Acidification:** Process in which the pH (acidity) level of the Earth's oceans becomes more acidic because of the increase in atmospheric carbon dioxide.
- **Phytoplankton:** Microscopic photosynthesizing aquatic plants that are responsible for absorbing carbon dioxide and producing more than half of the world's oxygen.

- **Radiative Forcing:** Measure of the effect that climatic factors have in modifying the energy balance (incoming and outgoing energy) of the Earth's atmosphere.
- **Solar Radiation Management (SRM):** Geoengineering techniques and technologies that reflect solar radiation from the Earth's atmosphere (or surface) back into space in an attempt to combat climate change and global warming.
- **Stratosphere:** The region of the Earth's atmosphere (approximate altitude of 10 to 50 kilometers) below the mesosphere and above the troposphere.

### DEFINITION AND BASIC PRINCIPLES

The term "geoengineering" comes from the Greek word "geo," meaning earth, and the word "engineering," a field of applied science that incorporates and uses data from scientific, technical, and mathematical sources in the invention and execution of specific structures, apparatuses, and systems.

Not to be confused with geotechnical engineering, which is related to the engineering behavior of earth materials, geoengineering is a field of science that aims to manipulate both micro (local) and macro (global) climates with the intention of reversing the effects of climate change and global warming. Although geoengineering, or climate engineering, can be undertaken on the local level, such as cloud seeding, most of the proposed technologies are based on a worldwide scale for the wholesale mediation of the global climate and the effects of climate change.

### BACKGROUND AND HISTORY

Human society and ancient cultures had long believed power over the weather to be the province of gods, but the advent of human-made climate control began with the concept of rainmaking (pluviculture) in the mid-nineteenth century. James Pollard Espy, an American meteorologist, was instrumental in developing the thermal theory of storms. Although such a discovery placed him in the annals of scientific history, he also became known (and disparaged) for his ideas regarding artificial rainmaking. According to Espy, the burning of huge areas of forest would create sufficient hot air and updraft to create clouds and precipitation.

In 1946, Vincent Schaefer, a laboratory technician at General Electric Research Laboratory in New York, generated a cloud of ice from water droplets that had been supercooled by dry ice. That same year, Bernard Vonnegut discovered that silver iodide smoke produced the same result. These processes came to be known as cloud seeding. Cloud seeding attempts to encourage precipitation to fall in arid or drought-stricken agricultural areas by scattering silver iodide in rain clouds. Traditional climate modification has generally been limited to cloud seeding programs on a regional or local level.

In the following years, particularly during the Cold War, researchers in countries including the United States and the Soviet Union investigated climate control for its potential as a weapon. The Soviets also looked at climate control to warm the frozen Siberian tundra. However, in the 1970's, as concern regarding the greenhouse effect began to be expressed within conventional scientific circles, the concept of climate engineering took on new global significance.

Cesare Marchetti, an Italian physicist, first coined the term "geoengineering" in 1977. The word initially described the specific process of carbon dioxide capture and storage in the ocean depths as a means of climate change abatement. Since then, however, the term "geoengineering" has been used to cover all engineering work performed to manipulate the global and local climate. In later years, many researchers have begun using the term "climate engineering," which is a more accurate way to describe this type of applied science.

## How It Works

Significant scientific evidence points to human activities, particularly the burning of fossil fuels, as playing a role in global climate change. An increase in the level of greenhouse gases, in particular carbon dioxide and methane, has been identified as the main culprit in relation to global climate change. The majority of scientists agree that the safest and best way to tackle climate change is through a reduction in fossil fuel consumption via the implementation of green technology, societal change, and industry regulation. However, although carbon reduction technology is already available and affordable, many scientists are increasingly concerned that carbon reduction

schemes will not be introduced in time to stop the possible and probable effects of climate change. As a result, interest in geoengineering technology as a means to provide rapid solutions to climate change has increased.

There are two main areas of geoengineering research—carbon dioxide removal (CDR) and solar radiation management (SRM). Techniques based on CDR propose to remove carbon dioxide from the atmosphere and store it, while SRM techniques seek to reflect solar radiation from the Earth's atmosphere (or surface) back into space. Despite having the same objective of combating climate change and reducing global temperatures, these two approaches differ significantly in regard to their implementations, time scales, temperature effects, and possible consequences. According to the Royal Society of London, CDR techniques "address the root cause of climate change by removing greenhouse gases from the atmosphere," whereas SRM techniques "attempt to offset effects of increased greenhouse gas concentrations by causing the Earth to absorb less solar radiation."

The CDR approach removes carbon dioxide from the atmosphere and sequesters it underground or in the ocean. Many consider this approach to be the more attractive of the two as it not only helps reduce global temperatures but also works to combat issues such as ocean acidification caused by escalating carbon dioxide levels. Conversely, SRM techniques have no effect on atmospheric carbon dioxide levels, instead using reflected sunlight to reduce global temperatures.

## Applications and Products

Traditionally, climate modification took the form of regional or local cloud seeding programs. In cloud seeding, a substance, usually silver iodide, is scattered in rain clouds to cause precipitation in arid or drought-stricken agricultural areas. By the start of the twenty-first century, however, deleterious global climate change and increasing levels of atmospheric carbon dioxide had pushed climate engineering to the forefront of science. Geoengineering technology, once considered as more fringe then functional, is being investigated as a possible weapon in the fight against global warming.

Despite the surge in research and interest, no longitudinal or large-scale geoengineering projects have been conducted. Geoengineering applications and products are generally regarded as highly speculative and environmentally untested, with significant ambiguity in regard to global and institutional regulation. Although climate engineering theories abound, only a limited few have captured global attention for their feasibility and applicability.

**Iron Fertilization of the Oceans.** The intentional introduction of iron into the upper layers of certain areas of the ocean to encourage phytoplankton blooms is a form of CDR. The concept relies on the fact that increasing certain nutrients—such as iron—in nutrient-poor areas stimulates phytoplankton growth. Carbon dioxide is absorbed from the surface of the ocean during the processes of photosynthesis; when the phytoplankton, marine animals, and plankton die and sink in the natural cycle, that carbon is removed from the atmosphere and sequestered in the ocean's depths.

**Scrubbers and Artificial Trees.** Both scrubbers and artificial trees aim to remove and store carbon dioxide from the Earth's atmosphere and assist in reducing the effect of climate change. Scrubbing towers involve the use of large wind turbines funneling air into specially designed structures, where the air reacts with a number of chemicals to form water and carbonate precipitates, essentially capturing the carbon. These carbon precipitates can then be stored. The use of artificial trees seeks the same result but by a different method. Large artificial trees or structures act as filters to capture and convert atmospheric carbon dioxide into a carbonate, which is then removed and stored.

**Biochar.** Biochar is a form of charcoal created from the pyrolysis (chemical decomposition by heating) of plant and animal waste. It captures and stores carbon by sequestering it in biomass. The biochar can be returned to the soil as fertilizer, as it helps the soil retain water and necessary nutrients, as well as to store carbon.

**Stratospheric Sulfur Aerosols.** Stratospheric sulfur aerosols are minute sulfur-rich particles that are found in the Earth's stratosphere and are often observed following significant volcanic activity (such as after the 1991 Mount Pinatubo eruption). The presence of these aerosols in the stratosphere results in a cooling effect. The SRM geoengineering technique of intentionally releasing sulfur aerosols into the stratosphere is based on the concept that they would produce a cooling or dimming effect by reflecting solar radiation. A workable delivery system has not yet been developed, but proposals include using high-altitude aircraft, balloons, and rockets.

**Orbital Mirrors and Space Sunshades.** The SRM technique of orbital mirrors and space sunshades entails the release of many billions (possibly trillions) of small reflective objects at a Lanrangian point in space to partially reflect solar radiation or impede it from entering the Earth's atmosphere. The theory is that the decrease in sunlight hitting the Earth's surface would help decrease average global temperatures.

**Marine Cloud Whitening.** Marine cloud whitening involves increasing the reflective properties of cloud cover so that solar radiation entering the Earth's atmosphere is reflected back into space. Proposed methods for achieving this include mounting large-scale mist-producing structures on seafaring vessels. The theory is that the spray of minute water droplets released by these structures would increase cloud cover and whitening, which would in turn increase sunlight reflection.

**Reflective Roofs.** Reflective or white roofs are often considered to be the most cost effective and easily implemented SRM method of reducing global temperatures. The concept relies on reflecting solar radiation back into space by using white materials (or paint) on the surface of building roofs.

## IMPACT ON INDUSTRY

Geoengineering has been experiencing a revolution. For the most part, many scientists and governments have not been interested in climate engineering because of questions about its feasibility or have been opposed to research and implementation of such technology. Because of the global consequences of climate change, however, many government agencies, universities, industries, corporations, and businesses worldwide are becoming interested in researching and working in this field of applied science.

**Government and University Research.** The governments of the United States and the United Kingdom have increasingly expressed interest in researching geoengineering technology and determining its applicability. Although the U.S. government has traditionally shied away from the climate engineering controversy, the Congress heard testimony related to geoengineering in November, 2009. This follows on the July, 2009, joint seminar, "Geoengineering: Challenges and Global Impacts," held by the United Kingdom's House of Commons, the Institute of Physics, the Royal Society of Chemistry, and the Royal Academy of Engineering. The seminar explored and discussed the possibility of applying geoengineering technology to help mitigate the effects of climate change.

Although a growing number of governments are studying the possible use of geoengineering technology to help alleviate the effects of climate change, many scientists and other experts hesitate to embrace and apply the technologies. They cite the unanticipated consequences of these yet untested theories and technologies.

An increasing number of universities are exploring the potential of climate engineering. Some

---

## Fascinating Facts About Climate Engineering

- Human activity, particularly the burning of fossil fuels such as oil and coal, is responsible for the release of some 30 billion metric tons of carbon dioxide into the Earth's atmosphere every year.
- Most climate engineering proposals fall into one of two fundamentally different approaches–the removal of carbon dioxide from the atmosphere or the reflection of solar radiation from Earth's atmosphere or surface back into space.
- Climate engineers claim they can cool the planet and reverse ice-sheet melting by mimicking a natural volcanic eruption through the release of sulfate aerosols into the Earth's stratosphere, where they will reflect sunlight back into space.
- Giant artificial trees, ocean fertilization, and trillions of space mirrors are some of the most popular proposals being considered and researched in the fight against global warming.
- Reflecting the Sun's rays back into the atmosphere could quickly affect global temperatures but would not have an impact on the levels of carbon dioxide in the atmosphere.
- Biochar, the end result of the burning of animal and agricultural waste, can help clean the air by storing carbon dioxide that would have been released into the atmosphere during decomposition of the waste, and it assists plants by aiding in the storage of carbon from photosynthesis.

---

of the more well-known institutions that conduct geoengineering research include the Department of Atmospheric Chemistry at the Max Planck Institute for Chemistry, the Carnegie Institution for Science's Department of Global Ecology at Stanford University, the Earth Institute at Columbia University, the Energy and Environmental Systems Group at the University of Calgary, and the Oxford Geoengineering Institute.

**Major Organizations.** Many world-renowned organizations have entered into the geoengineering debate. Although there is growing support for increasing research into such technology, many organizations are hesitant to fully endorse its actual implementation and use. Such organizations include the Intergovernmental Panel on Climate Change, which published information about climate engineering in its fourth assessment report, "Climate Change 2007," and the American Meteorological Society, which released a guarded geoengineering policy statement in 2009. One of the most comprehensive reports, *Geoengineering the Climate: Science, Governance, and Uncertainty* (2009), was issued by the Royal Society of London, the world's oldest scientific academy. The society detailed the potential applications and costs of geoengineering and stated that although some climate engineering schemes are utterly implausible or are being endorsed without due consideration, some are more realistic. Its report, the first by any national science academy, concludes that much greater research is required before the implementation of any such technology is undertaken.

### CAREERS AND COURSE WORK

Most commonly, students who wish to pursue careers in climate engineering begin by majoring in scientific fields such as atmospheric science, marine systems, and civil engineering. However, given the multitude of fields covered in geoengineering, a career in this applied science could follow many different paths, and students should have a solid understanding of subjects such as atmospheric chemistry, ecology, meteorology, plant biology, ecosystem management, marine systems and chemistry, and engineering. The majority of graduate programs in this area are open to students who have backgrounds in engineering, applied sciences, or closely related disciplines. University research covers many areas, and students who obtain a doctorate or master's degree in climate engineering can expect to have careers in geoengineering research and design, atmospheric sciences, aeronautical and nautical engineering, and environmental management consulting.

## SOCIAL CONTEXT AND FUTURE PROSPECTS

In the past, global geoengineering was regarded as more science fiction than fact. With greenhouse gas emissions continuing to increase from the burning of fossil fuel, however, the concept of deliberately engineering the Earth's climate is garnering interest and gaining credibility.

Engineering the climate could assist in lowering atmospheric carbon dioxide and reducing the impact of climate change. Although the majority of geoengineering scientists stress that such technology should be used only for emergency quick fixes or as a last resort, many are also stressing that research into such technology is imperative. The 2009 Royal Society report strongly advocated increased research and recommended the world's governments allocate some £100 million ($165 million), collectively, per year to examine geoengineering options. Given the probable economic and environmental costs associated with climate change, many researchers have claimed that implementing geoengineering technology may be cheaper and more viable than doing nothing. Many researchers are also quick to state, however, that while dollar costs may be affordable, the possible environmental and economic costs if the technology fails to work or has unexpected deleterious effects may be immeasurable.

Despite the move into more mainstream science, climate engineering is still controversial and full of significant technological, social, ethical, legal, diplomatic, and safety challenges. The concept of purposely modifying the climate to correct climate change caused by humans has been labeled as, at best, ironic and, at worst, catastrophic. Significant concern has been raised about encouraging geoengineering research and technology. For example, many conservation organizations are concerned that access to such technology not only will promote the "if we build it, they will use it" mentality but also will lessen the resolve of governments and people to tackle climate change by reducing people's ecological footprint and consumption of fossil fuels.

In addition, geoengineering technology may ignite tensions between nations The ethical ramifications of climate engineering are unclear, and significant confusion and uncertainty exists regarding who should implement and control the global thermostat. If a country implements technology to fix one area and inadvertently adversely alters the climatic patterns in another country, who pays for the mishap? The consequences of climate manipulation on a global scale will almost certainly be unequal across nations. Such concerns stress the importance of conducting further research and of using caution regarding any technological advance in geoengineering.

*Christine Watts, Ph.D., B.App.Sc., B.Sc.*

## FURTHER READING

Flannery, Tim F. *The Weather Makers: How Man Is Changing the Climate and What It Means for Life on Earth.* New York: Grove Press, 2006. A comprehensive look at how human activity has influenced the global climate and its impact on life on the Earth.

Keith, David. "Geoengineering the Climate: History and Prospect." *Annual Review of Energy and the Environment* 25 (November, 2000): 245-284. Provides an interesting review of the history of climate control technology and future directions.

Launder, Brian, and J. Michael T. Thompson, eds. *Geoengineering Climate Change: Environmental Necessity or Pandora's Box?* New York: Cambridge University Press, 2010. Presents a comprehensive examination of the problems of climate change and the potential of different geoengineering technologies.

The Royal Society. *Geoengineering the Climate: Science, Governance, and Uncertainty.* London: The Royal Society, 2009. Twelve experts in the fields of science, economics, law, and social science conducted this study of the main techniques of climate engineering, focusing on how well they might work and examining possible consequences.

## WEB SITES

*American Geophysical Union*
Geoengineering the Climate
http://www.agu.org/sci_pol/positions/geoengineering.shtml

*Intergovernmental Panel on Climate Change*
http://www.ipcc.ch

**See also:** Air-Quality Monitoring; Atmospheric Sciences; Biofuels and Synthetic Fuels; Climate Modeling; Climatology; Environmental Engineering; Meteorology.

# CLIMATE MODELING

## FIELDS OF STUDY

Physics; environmental science; oceanography; meteorology; climatology; atmospheric physics; earth sciences; computer science; computer programming; advanced mathematics; biology; sociology; chemistry; game theory; statistics; agriculture; forestry; anthropology.

## SUMMARY

The goal of climate models is to provide insight into the Earth's climate system and the interactions among the atmosphere, oceans, and landmasses. Computer climate modeling provides the most effective means of predicting possible future changes and the potential effect such changes might have on societies and ecosystems. Shifting multiple variables and the unpredictability of climate model components have produced findings that fuel intense scientific and political debate. Climate models will continue to be refined as computing power increases, allowing models to be run with more integrative components and using wider, more refined space and time scales.

## KEY TERMS AND CONCEPTS

- **Climate:** Regional weather conditions averaged over an extended time period.
- **Climate Modification:** Changes to the climate due to either natural processes or human activities.
- **Climate System:** System consisting of the atmosphere, hydrosphere, cryospher, land surface, and biosphere and their interactions.
- **Cryosphere:** Portion of Earth covered by sea ice, mountain glaciers, snow cover, ice sheets, or permafrost.
- **Feedback:** Process changing the relationship between a forcing agent acting on climate and the climate's response to that agent.
- **General Circulation Model:** Model simulating the state of the entire atmosphere, ocean, or both at chosen locations over a discrete, selected incremental time scale.
- **Global Warming:** Event characterized by measurable increases in annual mean surface temperature of the Earth, most noticeably occurring since about 1860.
- **Greenhouse Gases:** Atmospheric gases that absorb infrared radiation; the most important are carbon dioxide, methane, water vapor, and chlorofluorocarbons.
- **Infrared Absorbers:** Molecules such as carbon dioxide and water vapor that absorb electromagnetic radiation at infrared wavelengths.
- **Infrared Radiation:** Electromagnetic radiation with wavelengths between microwaves and visible red light.
- **Microclimate:** Local climate regime distinguished by physical characteristics and processes associated with local geography.
- **Radiative Cooling:** Process by which the Earth's surface cools by emitting longwave radiation.
- **Regime:** Preferred or dominant state of the climate system.

## DEFINITION AND BASIC PRINCIPLES

All climate models are mathematical models derived from differential equations defining the principles of physics, chemistry, and fluid dynamics driving the observable processes of the Earth's climate. Climate models can be highly complex three-dimensional computer simulations using multiple variables and tens of thousands of differential equations requiring trillions of computations or fairly simple two-dimensional projections with a single equation defining a sole observable process. In all climate models, each additional physical process incorporated into the model increases the level of complexity and escalates the mathematical parameters needed to define the process's potential effects.

## BACKGROUND AND HISTORY

The ability to predict weather and climate patterns could have a major impact on the health and well-being of societies. The possibility to warn of impending harm, prepare for changes, estimate outcomes, and chose more opportune times for essential human activities has been a dream of cultures for thousands of years. Although observational records of climate phenomena have been kept and interpreted for generations, the advent of computer models and advanced procedures for gathering and assimilating global data allows for quantitative assessments of the complex interactions between climate processes.

Climate models are governed by the laws of physics, which produce climate system components affecting the atmosphere, cryosphere, oceans, and land. No climate model would exist without an understanding of Newtonian mechanics and the fundamental laws of thermodynamics. By the late eighteenth century, scientific understanding of these physical laws allowed for numerical calculations to define observable climate processes. By the early twentieth century, scientists understood atmospheric phenomena as the product of preceding phenomena defined by physical laws. If one has accurate observable data of the atmosphere at a particular time and understands the physical laws under which atmospheric phenomena take place, it is possible to predict the outcome of future atmospheric changes. Mathematical equations were subsequently developed to define atmospheric motions and simulate observable processes and features of climate system components.

By the late 1940's, electronic computing provided a means by which greater numbers of equations could be assigned to data and calculated to define observable climate phenomena. Throughout the 1950's, predictive climate models were refined through changes to equations, resulting in modifications to models of atmospheric circulation. During the 1960's, similar equations were used to define ocean, land, and ice processes and their role in climate. Once this was done, the development of coupled models began. Coupled models involve combining observational processes to achieve more realistic outcomes. For each new process, however, multiple observations must be added and defined by additional equations. As the quality of observational data increases, the complexity of the climate model grows. Each observation of sea ice movement, cloud density, atmospheric chemistry, land cover, water temperature, precipitation, and human activities can be defined mathematically and added to a climate model's program. By defining the multiple variables of climate processes and adding these to model programs, scientists became able to produce predictive simulations of climate ranging from small regional models to global general circulation models.

## HOW IT WORKS

The purpose of all climate models is to simulate, over a given time period, regional or global patterns of climatic processes: wind, temperature, precipitation, ocean currents, sea levels, and sea ice. By imposing changes to the physical process within a model simulation—such as altering the amount of solar input, carbon dioxide, or ice cover—predictions can be made concerning possible future climate scenarios. Although models constructed from purely observational evidence result in less accurate predictions, the use of computers to process numerical models representing multiple levels of climate system feedbacks improves the accuracy of predictions for multiple components of the climate model.

When formulating a truly practical climate model, all components of Earth's climate system must be taken into consideration. The model must represent the atmosphere, the oceans, ice and snow cover on both land and sea, and landmasses, including their biomass. The interactions between these climate components occur in multiple ways, at differing intensities, and over varied time scales: Practical climate models must be able to reflect these dynamics.

The key to understanding climate and climatic change is energy balance. Multiple feedback processes occur within Earth's climate system, and these interactions constantly fluctuate to maintain an efficient energy balance. Positive feedback processes amplify variations to the climate system; negative feedbacks dampen them. Examples of feedbacks include radiation feedback from clouds and water vapor; changes to reflective power from ice, snow, and deserts; and changes in ocean temperatures. These and many other forms of feedback must be considered when preparing a climate model. Because the physics of energy transfer between climate system processes is so varied, resolution scale becomes a major problem in preparing data for modeling. Flux compensations must be made in models to account for differences in energy transfers among atmosphere, oceans, and land. Heat, momentum, water densities, size of phenomena such as currents or eddies, energy radiation, rainfall, wind, humidity, barometric pressure, atmospheric chemistry, and natural phenomena such as forest fires and volcanic eruptions all must be considered.

The purpose of preparing climate models is to answer the fundamental questions of what can be

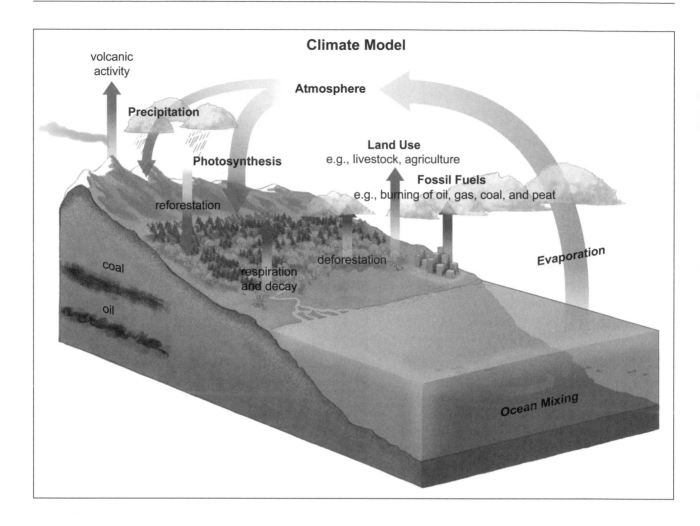

reliably predicted about climate and at what time scales these predictions can be made. The difficulty with all climate modeling is that the computers used must be powerful enough to deal with the complexity of observable climate system data; that the data used must be as free as possible of errors in interpretation; and that all observational data have limitations and these limit interpretive predictability. These factors are the limitations confining all existing climate-modeling capabilities and set the boundary ranges for predicting climate phenomena in space and time.

Variability of climate over space and time scales is a key component of Earth's history. Throughout Earth's history, climate has naturally fluctuated, often to extremes in comparison to present-day climatic conditions. Being able to distinguish between

naturally occurring and human-induced climate changes is an important aspect of climate modeling. The implications of climate models suggesting that human behaviors are altering the atmosphere—and in turn disrupting the planet's energy balance—drives vigorous debates and conflict.

## APPLICATIONS AND PRODUCTS

All climate models are simulations that make predictions about climate processes. Climate change skeptics and deniers make that case that model predictions are merely simulations and not "real" data produced by "real" science. Without climate models, however, there would be no climate data. No climate observation, no satellite imagery or observation, no meteorological data, no atmospheric sampling, no remote sensing, and no chemical analysis exists without passing through a series of computer data models. Almost everything known about Earth's climatic processes exists because of models. At present, almost all knowledge and understanding of climate change comes from three kinds of computer models: simulation models of weather and climate; reanalysis models re-creating climate history from historical weather data, including worldwide weather stations and remote data platforms; and data models combining and adjusting measurements from many different sources, including ice cores, gas analysis, tree rings, observation data, and archaeological and geological data. Because the amount of weather and climate data available are so vast and diverse and from so many different sources and at differing time scales, it can be managed and organized only by computer analysis. The result of modern computer-simulated climate models has been to create a stable and reliable basis for making predictions about climate change and the byproducts of that change.

## IMPACT ON INDUSTRY

Understanding how climate works and having the ability to predict possible changes to climate over certain regions and at specific time scales can have substantial impacts on the social, economic, and cultural well-being of any nation. More accurate prediction of cycles of increased precipitation or drought, rises in temperature, or shifts in oceanic currents allows people to determine more opportune times to plant and harvest crops, to change patterns of land use, to predict marine harvests, and to better plan other human activities. The ability to provide general predictions about these economically important aspects of society help control product costs and plan for social needs. High-quality climate models can also provide warnings that specific human behaviors may be affecting the environment and, as a result, people's quality of life. Climate models reflecting processes of deforestation, desertification, industrial insertion of greenhouse gases into the atmosphere, and ozone depletion all help establish grounds to alter human behaviors to increase survivability. Because of these factors and how they may affect the well-being and security of nations, governments throughout the world have created programs, earmarked research funds, and mandated agencies to study and model climate. A number of high-profile international treaties, most notably the Kyoto Protocol, have been initiated to address the observable and predicted effects of climate change.

Most major universities have climate change programs or research centers. A number of host government climate research centers and many universities have formed research consortiums to engage in regional climate studies and to share resources and supercomputing time. The result of high-profile interest in climate modeling related to issues of climate change provides large sums of research monies to qualified researchers and research facilities.

Although climate models are certainly helpful in predicting marine harvests, crop futures, energy needs, potential droughts, food shortages, and water resources, the biggest effect on industry from climate modeling is the concern they have generated regarding certain human behaviors. The use of environmentally detrimental technologies and the release of large amounts of greenhouse gases such as carbon dioxide, methane, sulfur dioxide, and chlorofluorocarbons into the atmosphere has begun to significantly alter Earth's future livability. Individuals do not wish to hear that driving their automobiles or heating their homes is damaging the planet. Companies do not want to be told that their methods of doing business are producing more harm than good and that changing these methods will require large financial investments. Additionally, governments do not wish to alter their strategies for national growth because their attempts to modernize are environmentally detrimental.

Since the mid-nineteenth century, many of humankind's economic, technological, and cultural advancements have been achieved through the exploitation of fossil fuels and other resources. Certain methods of forest harvesting have been shown to promote desertification, disrupt water resources, change the amount of radiation from the sun reflected in a region, and alter regional climate patterns. Because trees absorb carbon dioxide, the loss of large tracts of forest lessens Earth's ability to remove and recycle carbon dioxide from the atmosphere. Burning petroleum and coal results in the discharge of greenhouse gases and the release of harmful particulates into the atmosphere. Climate models examining historical and observational climate data and human behaviors have demonstrated a direct link between the use of fossil fuels and changes in the atmosphere. Climate observations and models show that increased greenhouse gas levels have produced real, noticeable, historically unprecedented, human-induced changes in the atmosphere and to climate patterns. In addition, the insertion of industrial chemicals—such as chlorofluorocarbons—into the atmosphere is known to deplete ozone and have a serious effect on both the climate and human health.

**Laws and Regulations.** In attempts to control industrial and commercial release of greenhouse gases and other toxic chemicals into the atmosphere, governments have imposed regulations on consumers and companies. Laws and regulations have been enacted that mandate limits on automobile exhaust, industrial air pollution, aerosol propellants, power plant emissions, field burning, biofuels, and certain agricultural practices. The difficulty with these attempts to alter industrial, personal, and governmental behaviors is that their reach is limited. Although one nation, state, or individual seeks to reduce harmful emissions, not all nations, states, or individuals share the same goals. Earth's atmosphere is an open system of circulation without human-imposed boundaries; therefore, pollutants created in one region affect all others.

**Financial Impact.** The impact of climate modeling on industry is often financial, decreasing corporations' profits. As climate models become more accurate in their predictions, certain industrial practices are likely to be found clearly responsible for some negative atmospheric effects. Altering business practices often requires large financial investments in new technologies; these investments cut into short-term profits. Although such investments are likely to result in long-term growth and allow the business to survive, unquestionably, they negatively affect short-term profits. For many nations and all nonprofit industries, the bottom line is of utmost importance. Changing the way one does business, retooling, reequipping, changing priorities, changing business models, and altering national and institutional identities is unthinkable for many. Some nations and many businesses have reacted to regulations and attempts to formulate new laws restricting atmospheric pollution by denying the existence of climate change. They seek to maintain the status quo for as long as possible.

### CAREERS AND COURSE WORK

Careers in climate modeling are limited. Most positions involving climate modeling are in academic institutions, government agencies, and private research organizations. The majority of climate modelers are employed within university settings, which usually

---

## Fascinating Facts About Climate Modeling

- Some of the earliest known long-term observations leading to predictive climate models are those of ancient native peoples of South and Central America who noted regional climate changes related to what are now known as El Niño and La Niña ocean currents.
- In the early 1960's, Warren M. Washington and Akira Kasahara, scientists with the National Center for Atmospheric Research, developed a computer model of atmospheric circulation. Data were input to a CDC computer using punch cards and seven-channel digital magnetic tape, and data were output through two line printers, a card punch, a photographic plotter, and standard magnetic tape.
- In December, 2009, the Copenhagen United Nations Climate Conference was the most-searched topic on the Google Internet search engine. In December, 2010, "climate change" was one of the top phrases searched daily on the Internet.
- In 2010, the National Center for Atmospheric Research released the Community Earth System Model, which creates computer simulations of Earth's past, present, and future climates. Experiments using this model will be part of the Intergovernmental Panel on Climate Change's 2013-1014 assessments.
- The complexity of climate models requires that they be run on supercomputers. The fastest supercomputer in the United States as of November, 2010, was the Cray XT5, at the U.S. Department of Energy's Oak Ridge Leadership Computing Facility. It can perform at 1.75 petaflops (quadrillions of calculations per second).
- Pioneering climate modeler Warren M. Washington was awarded the National Medal of Science by President Barack Obama on November 17, 2010. Washington was also a member of the Intergovernmental Panel on Climate Change, which received the 2007 Nobel Peace Prize.

means that their time must be divided between active research and teaching responsibilities. In some university consortiums, such as the University Corporation for Atmospheric Research, climate modelers work in teams. Climate modelers working for government agencies focus on predictive applications to help fulfill their governmental branch mandates. At the international level, a number of climate modelers work under contract or are funded by the United Nations and may do research for any of a number of United Nations agencies or advisory boards. Climate modelers at the federal level may find work with the U.S. Geological Survey, the Department of Agriculture, all branches of the military and intelligence communities, the National Oceanic and Atmospheric Agency, the National Aeronautics and Space Administration, the National Center for Atmospheric Research, and the Department of Energy. A limited number of climate modeling positions are available in private research organizations that offer services to businesses, lobbying organizations, agricultural interests, commodities traders, maritime shipping and fishing concerns, and politicians.

Students interested in careers involving climate modeling need to take classes in chemistry, physics, atmospheric sciences, geosciences, meteorology, oceanography, biology, mathematics, statistics, computer science, and environmental science, and obtain a bachelor of science degree. Almost all careers in climate modeling require a graduate-level education. Although obtaining a master's degree may allow for an entry-level position or journeyman status within certain agencies, for nearly all climate-modeling opportunities in academia, civil service, or private research, a doctorate is a necessity. In graduate school, studies in advanced mathematics and computer programming combined with intensely focused studies in climatology, oceanography, and physics are the norm. Individual areas of research interest require students to narrow their courses to reflect the direction of their research and make course and seminar selections accordingly.

As interest in climate change issues continues to grow, the need for qualified climate modeling scientists will increase. The rate at which observable changes to the climate begin to reflect predictions made by climate models will most likely raise the value and significance of trained climatologists and their predictive modeling skills.

## SOCIAL CONTEXT AND FUTURE PROSPECTS

Using climate models to accurately predict future climate trends is the ultimate goal. If climate models can help define the difference between natural climate fluctuations and human-induced climate change, positive human activities can be developed to counter the impact of environmentally unsustainable behaviors. Accurate predictive climate models can also increase economic outcomes by indicating more opportune times for planting, harvesting, and fishing, and by predicting droughts and temperature shifts. Climate shifts are known to be associated with pandemic outbreaks of disease: Climate models may allow people to prepare in advance for disease-formulating conditions.

Existing climate modeling is limited by computational power and the nonlinear nature of climate system phenomena. As computing technology advances and observational data of climate processes over longer time scales becomes available, flux compensations will be more accurately defined and the accuracy of future climate models will increase. Public acceptance of climate model predictions will remain complicated as long as politics and economics, rather than observational facts, are allowed to drive the climate change debate.

*Randall L. Milstein, B.S., M.S., Ph.D.*

## FURTHER READING

Edwards, Paul N. *A Vast Machine: Computer Models, Climate Data, and the Politics of Global Warming.* Cambridge, Mass.: MIT Press, 2010. Tells the history of how scientists learned to understand the atmosphere.

Kiehl, J. T., and V. Ramanathan. *Frontiers of Climate Modeling.* New York: Cambridge University Press, 2006. A good general overview of climate modeling, with an emphasis on how greenhouse gases are altering the climate system.

McGuffie, Kendall, and Ann Henderson-Sellers. *A Climate Modelling Primer.* West Sussex, England: John Wiley & Sons, 2005. Explains the basis and mechanisms of existing physical-observation-based climate models.

Mote, Philip, and Alan O'Neill, eds. *Numerical Modeling of the Global Atmosphere in the Climate System.* Boston: Kluwer Academic, 2000. Meant for those actively creating climate models, this work explains the uses of numerical constraints to define climate processes.

Robinson, Walter A. *Modeling Dynamic Climate Systems.* New York: Springer, 2001. A basic book for understanding climate modeling that includes a good description of how climate systems function and interact with each other and vary over time and space.

Trenberth, Kevin E., ed. *Climate System Modeling.* New York: Cambridge University Press, 2010. A comprehensive textbook covering the most important topics for developing a climate model.

Washington, Warren M., and Claire L. Parkinson. *An Introduction to Three-Dimensional Climate Modeling.* Sausalito, Calif.: University Science Books, 2005. An introduction to the use of three-dimensional climate models. Includes a history of climate modeling.

**WEB SITES**

*American Meteorological Society*
http://www.ametsoc.org

*Intergovernmental Panel on Climate Change*
http://www.ipcc.ch

*National Center for Atmospheric Research*
http://ncar.ucar.edu

*National Oceanic and Atmospheric Administration*
National Climatic Data Center
http://www.Ncdc.noaa.gov

*National Weather Association*
http://www.nwas.org

*University Corporation for Atmospheric Research*
http://www2.ucar.edu

*U.S. Geological Survey*
Climate and Land Use Change
http://www.usgs.gov/climate_landuse

*World Meteorological Organization*
Climate
http://www.wmo.int/pages/themes/climate/index_en.php

**See also:** Air-Quality Monitoring; Atmospheric Sciences; Climate Engineering; Climatology; Industrial Pollution Control; Meteorology.

# CLIMATOLOGY

## FIELDS OF STUDY

Atmospheric sciences; meteorology; physical geography; climate change; climate classification; climate zones; tree-ring analysis; climate modeling; bioclimatology; climate comfort indices; hydroclimatology.

## SUMMARY

Climatology deals with the science of climate, which includes the huge variety of weather events. These events change at periods of time that range from months to millennia. Climate has such a profound influence on all forms of life, including human life, that people have made numerous attempts to predict future climatic conditions. These attempts resulted in research efforts to try to understand future changes in the climate as a consequence of anthropogenic and naturally caused activity.

## KEY TERMS AND CONCEPTS

- **Climate:** Average weather for a particular area.
- **Climograph:** Line graph that shows monthly mean temperatures and precipitation.
- **Coriolis Effect:** Rghtward (Northern Hemisphere) and leftward (Southern Hemisphere) deflection of air due to the Earth's rotation.
- **Energy Balance:** Balance between energy received from the Sun (shortwave electromagnetic radiation) and energy returned from the Earth (longwave electromagnetic radiation).
- **Front:** Boundary between air masses that differ in temperature, moisture, and pressure.
- **Greenhouse Effect:** Process in which longwave radiation is trapped in the atmosphere and then radiated back to the Earth's surface.
- **Occluded Front:** Front that occurs when a cold front overtakes a warm front and forces the air upward.
- **Semipermanent Low-Pressure Centers:** Semipermanent patterns of low pressure that occur in northern waters off the Alaskan Aleutian Islands and Iceland.
- **Sublimation:** Change from ice in a frozen solid state to water vapor in a gaseous state without going through a liquid state.
- **Troposphere:** Lowest level of the atmosphere that contains water vapor that can condense into clouds.

## DEFINITION AND BASIC PRINCIPLES

"Weather" pertains to atmospheric conditions that constantly change, hourly and daily. In contrast, "climate" refers to the long-term composite of weather conditions at a particular location, such as a city or a state. Climate at a location is based on daily mean conditions that have been aggregated over periods of time that range from months and years to decades and centuries. Both weather and climate involve measurements of the same conditions: air temperature, water vapor in the air (humidity), atmospheric pressure, wind direction and speed, cloud types and extent, and the amount and kind of precipitation.

Estimates of ancient climates, going back several thousand years or more, are produced in various ways. For example, the vast amount of groundwater discovered in southern Libya indicates that during some period in the past, that part of the Sahara Desert was much wetter. In ancient Egypt, Nilometers, stone markers built along the banks of the Nile, were used to gauge the height of the river from year to year. They are similar to the staff gauges that are used by the U.S. Geological Survey to indicate stream or canal elevation. The height of the Nilometer reflects the extent of precipitation and associated runoff in the headwaters of the Nile in east central Africa.

## BACKGROUND AND HISTORY

Measurements of precipitation were being made and recorded in India during the fourth century B.C.E. Precipitation records were kept in Palestine about 100 C.E., Korea in the 1440's, and in England during the late seventeenth century. Galileo invented the thermometer in the early 1600's. Physicist Daniel Fahrenheit created a measuring scale for a liquid-in-glass thermometer in 1714, and Swedish astronomer Anders Celsius developed the centigrade scale in 1742. Italian physicist Evangelista Torricelli, who worked with Galileo, invented the barometer in 1643.

The first attempt to explain the circulation of the atmosphere around the Earth was made by English astronomer Edmond Halley, who published a paper charting the trade winds in 1686. In 1735, English meteorologist George Hadley further explained the movement of the trade winds, describing what became known as a Hadley cell, and in 1831, Gustave-Gaspard Coriolis developed equations to describe the movement of air on a rotating Earth. In 1856, American meteorologist William Ferrel developed a theory that described the mid-latitude atmospheric circulation cell (Ferrel cell). In 1860, Dutch meteorologist Christophorus Buys Ballot demonstrated the relationship between pressure, wind speed, and direction (which became known as Buys Ballot's law).

The first map of average annual isotherms (lines connecting points having the same temperature) for the northern hemisphere was created by German naturalist Alexander von Humboldt in 1817. In 1848, German meteorologist Heinrich Wilhelm Dove created a world map of monthly mean temperatures. In 1845, German geographer Heinrich Berghaus prepared a global map of precipitation. In 1882, the first world map of precipitation using mean annual isohyets (lines connecting points having the same precipitation) appeared.

## How It Works

**Earth's Global Energy Balance.** The Earth's elliptical orbit about the Sun ranges from 91.5 million miles at perihelion (closest to the Sun) on January 3, to 94.5 million miles at aphelion (furthest from the Sun) on July 4, averaging 93 million miles. The Earth intercepts about two-billionth of the total energy output of the Sun. Upon reaching the Earth, a portion of the incoming radiation is reflected back into space, while another portion is absorbed by the atmosphere, land, or oceans. Over time, the incoming shortwave solar radiation is balanced by a return to outer space of longwave radiation.

The Earth's atmosphere extends to an estimated height of about 6,000 miles. Most of it is made up of nitrogen (78 percent by volume) and oxygen (about 21 percent). Of the remaining 1 percent, carbon dioxide ($CO_2$) accounts for about 0.0385 percent of the atmosphere. This is a minute amount, but carbon dioxide can absorb both incoming shortwave radiation from the sun and outgoing longwave radiation from the Earth. The measured increase in carbon dioxide since the early 1900's is a major cause for concern as it is a very good absorber of heat radiation, which adds to the greenhouse effect.

**Air Temperature.** Air temperature is a fundamental constituent of climatic variation on the Earth. The amount of solar energy that the Earth receives is governed by the latitude (from the equator to the poles) and the season. The amount of solar energy reaching low-latitude locations is greater than that reaching higher-latitude sites closer to the poles. Another factor pertaining to air temperature is the fivefold difference between the specific heat of water (1.0) and dry land (0.2). Accordingly, areas near the water have more moderate temperatures on an annual basis than inland continental locations, which have much greater seasonal differences.

Anthropogenic (human-induced) changes in land cover in addition to aerosols and cloud changes can result in some degree of global cooling, but this is much less than the combined effect of greenhouse gases in global warming. The gases include carbon dioxide from the burning of fossil fuels (coal, oil, and natural gas), which has been increasing since the second half of the twentieth century. Other gases such as methane ($CH_4$), chlorofluorocarbons (CFCs), ozone ($O_3$), and nitrous oxide ($NO_3$) also create additional warming effects.

Air temperature is measured at 5 feet above the ground surface and generally includes the maximum and minimum observation for a twenty-four-hour period. The average of the maximum and minimum temperature is the mean daily temperature for that particular location.

**Earth's Available Water.** Water is a tasteless, transparent, and odorless compound that is essential to all biological, chemical, and physical processes. Almost all the water on the Earth is in the oceans, seas, and bays (96.5 percent) and another 1.74 percent is frozen in ice caps and glaciers. Accordingly, 98.24 percent of the total amount of water on this planet is either frozen or too salty and must be thawed or desalinated. About 0.76 percent of the world's water is fresh (not saline) groundwater, but a large portion of this is found at depths too great to be reached by drilling. Freshwater lakes make up 0.007 percent, and atmospheric water is about 0.001 percent of the total. The combined average flows of all the streams on Earth—from tiny brooks to the mighty Amazon River—account for 0.0002 percent of the total.

**Air Masses.** The lowest layer of the atmosphere is the troposphere, which varies in height from 10 miles at the equator and lower latitudes to 4 miles at

## Fascinating Facts About Climatology

- Mean sea levels for the world have increased an average of 0.07 inches per year from 1904 to 2003, an amount that is much larger than the average rate in the past.
- Serbian astronomer Milutin Milankovitch developed the astronomical hypothesis in 1938 by observing that glacial and interglacial periods were related to insolation variations caused by small cycles in the Earth's axial rotation and orbit about the Sun.
- The larger input of solar energy received at and near the equator creates a very large circulation path known as the Hadley cell. The air converges at a narrow zone known as the intertropical convergence zone, which seasonally varies from 15 degrees south in northern Australia to 25 degrees north in northern India, representing a latitudinal shift of 40 degrees.
- The coldest temperatures in the world were recorded at the Russian weather station at Vostok, Antarctica, at 78 degrees south: –127 degrees Fahrenheit in 1958 and –128.5 degrees Fahrenheit on July 21, 1983, at an ice sheet elevation of 11,220 feet.
- There is a fivefold difference between the specific heat of water (1.0) and dry land (0.2). This results in a one- to two-month lag in average maximum and minimum temperatures after the summer and winter solstices.
- Atlantic storm names can be repeated after six years unless the hurricane was particularly severe. The names of noteworthy storms–such as Camille (1969), Hugo (1989), Andrew (1992), Floyd (1999), and Katrina (2005)–are not used again, so as to avoid confusion in later years.

the poles. Different types of air masses within the troposphere can be delineated on the basis of their similarity in temperature, moisture, and to a certain extent, air pressure. Air masses develop over continental and maritime locations that strongly determine their physical characteristics. For example, an air mass starting in a cold, dry interior portion of a continent would develop thermal, moisture, and pressure characteristics that would be substantially different from those of an air mass that developed over water. Atmospheric dynamics also allow air masses to modify their characteristics as they move from land to water and vice versa.

Air mass and weather front terminology were developed in Norway during World War I. The Norwegian meteorologists were unable to get weather reports from the Atlantic theater of operations; consequently, they developed a dense network of weather stations that led to impressive advances in atmospheric modeling.

**Greenhouse Effect.** Selected gases in the lower parts of the atmosphere trap heat and radiate some of that heat back to Earth. If there was no natural greenhouse effect, the Earth's overall average temperature would be close to 0 degrees Fahrenheit rather than 57 degrees Fahrenheit.

The burning of coal, oil, and gas makes carbon dioxide the major greenhouse gas. Carbon dioxide accounts for nearly half of the total amount of heat-producing gases in the atmosphere. In mid-eighteenth century Great Britain, before the Industrial Revolution, the estimated level of carbon dioxide was about 280 parts per million (ppm). Estimates for the natural range of carbon dioxide for the past 650,000 years are 180-300 ppm. All of these values are less than the October, 2010, estimate of 387 ppm. Since 2000, atmospheric carbon dioxide has been increasing at a rate of 1.9 ppm per year. The radiative effect of carbon dioxide accounts for about one-half of all the factors that affect global warming. Estimates of carbon dioxide values at the end of the twenty-first century range from 490 to 1,260 ppm.

The second most important greenhouse gas is methane ($CH_4$), which accounts for about 14 percent of all of the global warming factors. This gas originates from the natural decay of organic matter in wetlands, but anthropogenic activity in the form of rice paddies, manure from farm animals, the decay of bacteria in sewage and landfills, and biomass burning (both natural and human-induced) doubles the amount produced.

Chlorofluorocarbons (CFCs) absorb longwave energy (warming effect) but also have the ability to destroy stratospheric ozone (cooling effect). The warming radiative effect is three times greater than the cooling effect. CFCs account for about 10 percent of all of the global warming factors. Tropospheric ozone ($O_3$) from air pollution and nitrous oxide ($N_2O$) from motor vehicle exhaust and bacterial emissions from nitrogen fertilizers account for about

10 percent and 5 percent, respectively, of all the global warming factors.

Several human actions lead to a cooling of the Earth's climate. For example, the burning of fossil fuels results in the release of tropospheric aerosols, which acts to scatter incoming solar radiation back into space, thereby lowering the amount of solar energy that can reach the Earth's surface. These aerosols also lead to the development of low and bright clouds that are quite effective in reflecting solar radiation back into space.

### APPLICATIONS AND PRODUCTS

Climatology involves the measurement and recording of many physical characteristics of the Earth. Therefore, numerous instruments and methods have been devised to perform these tasks and obtain accurate measurements.

**Measuring Temperature.** At first glance, it would appear that obtaining air temperatures would be relatively simple. After all, thermometers have been around since 1714 (Fahrenheit scale) and 1742 (Celsius scale). However, accurate temperature measurements require a white (high-reflectivity) instrument shelter with louvered sides for ventilation, placed where it will not receive direct sunlight. The standard height for the thermometer is 5 feet above the ground.

**Remote-Sensing Techniques.** Oceans cover about 71 percent of the Earth's surface, which means that large portions of the world do not have weather stations and places where precipitation can be measured with standard rain gauges. To provide more information about precipitation in the equatorial and tropical parts of the world, the National Aeronautics and Space Administration (NASA) and the Japanese Aerospace Exploration Agency began a program called the Tropical Rainfall Monitoring Mission (TRMM) in 1997. The TRMM satellite monitors the area of the world between 35 degrees north and 35 degrees south latitude. The goal of the study is to obtain information about the extent of precipitation, its intensity, and length of occurrence. The major instruments on the satellite include radar to detect rainfall, a passive microwave imager that can acquire data about precipitation intensity and the extent of water vapor, and a scanner that can examine objects in the visible and infrared portions of the electromagnetic spectrum. The goal of collecting this data is to obtain the necessary climatological information about atmospheric circulation in this portion of the Earth so as to develop better mathematical models for determining large-scale energy movement and precipitation.

**Geostationary Satellites.** Geostationary orbiting earth satellites (GOES) enable researchers to view images of the planet from what appears to be a fixed position. To achieve this, these satellites circle the globe at a speed that is in step with the Earth's rotation. This means that the satellite, at an altitude of 22,200 miles, will make one complete revolution in the same twenty-four hours and direction that the Earth is turning above the Equator. At this height, the satellite is in a position to view nearly half the planet at any time. On-board instruments can be activated to look for special weather conditions such as hurricanes, flash floods, and tornadoes. The instruments can also be used to make estimates of precipitation during storm events.

**Rain Gauges.** The accurate measurement of precipitation is not as simple as it may seem. Collecting rainfall and measuring it is complicated by the possibility of debris, dead insects, leaves, and animal intrusions occurring. Standards were established, although the various national climatological offices use more than fifty types of rain gauges. The location of the gauge, its height above the ground, the possibility for splash and evaporation, its distance from trees, and turbulence all affect the results. Accordingly, all gauge records are really estimates. Precipitation estimates are also affected by the number of gauges per unit area. The number of gauges in a sample area of 3,860 square miles for Britain, the United States, and Canada is 245, 10, and 3, respectively. Although the records are reported to the nearest 0.01 inch, discrepancies occur in the official records. It is important to have a sufficiently dense network of rain gauges in urban areas. Some experts think that 5 to 10 gauges per 100 square miles is necessary to obtain an accurate measure of rainfall.

**Doppler Radar.** Doppler radar was first used in England in 1953 to pick up the movement of small storms. The basic principle behind Doppler radar is that the back-scattered radiation frequency detected at a certain location changes over time as the target, such as a storm, moves. The mode of operation requires a transmitter that is used to send short but powerful microwave pulses. When a foreign object

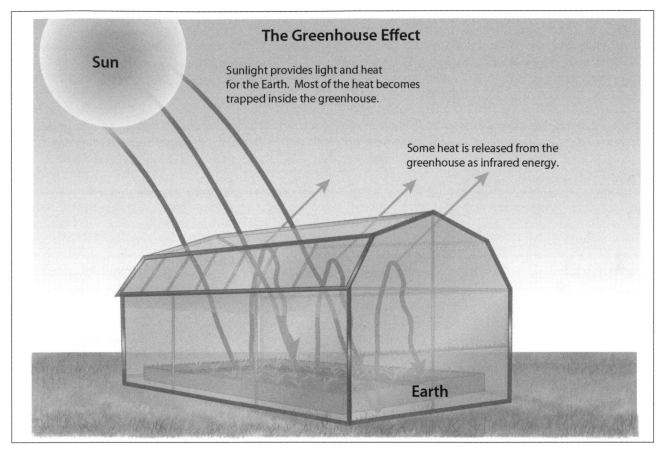

The greenhouse effect refers to the process in which longwave radiation is trapped in the atmosphere and then radiated back to the Earth's surface.

(or target) is intercepted, some of the outgoing energy is returned to the transmitter, where a receiver can pick up the signal. An image (or echo) from the target can then be enlarged and shown on a screen. The target's distance is revealed by the time between transmission and return. The radar screen can indicate not only where the precipitation is taking place but also its intensity by the amount of the echo's brightness. Doppler radar has developed into a very useful device for determining the location of storms and the intensity of the precipitation and for obtaining good estimates of the total amount of precipitation.

## IMPACT ON INDUSTRY

**Global Perspective.** The World Meteorological Organization, headquartered in Geneva, Switzerland, was established to encourage weather station networks that would facilitate the acquisition of climatic

data. In 2007, the organization decided to expand the global observing system and other related observing systems, including the global ocean observing system, global terrestrial observing system, and the global climate observing system. Data are being collected by 10,000 manned and automatic surface weather stations, 1,000 upper-air stations, more than 7,000 ships, 100 moored and 1,000 floating buoys that can drift with the currents, several hundred radars, and more than 3,000 commercial airplanes, which record key aspects of the atmosphere, land, and ocean surfaces on a daily basis.

**Government Research.** About 183 countries have meteorological departments. Although many are small, the larger countries have well-established organizations. In the United States, meteorology and climatology are handled by the National Oceanic and Atmospheric Administration. The agency's National

Climatic Data Center has records that date back to 1880 and provide invaluable information about previous periods for the United States and the rest of the world. For example, sea ice in the Arctic Ocean typically reaches its maximum extent in March. The coverage at the end of March, 2010, was 5.8 million square miles. This was the seventeenth consecutive March with below-average coverage. The center issues many specialized climate data publications as well as monthly temperature and precipitation summaries for all fifty states.

**University Research.** In the United States, forty-eight states (except Rhode Island and Tennessee) have either a state climatologist or someone who has comparable responsibility. Most of the state climatologists are connected with state universities, particularly the colleges that started as land-grant institutions.

The number of cooperative weather stations established to take daily readings of temperature and precipitation in each of these states has varied in number since the late nineteenth century.

**Industry and Business Sectors.** The number, size, and capability of private consulting firms has increased over the years. Some of the first of these firms were frost-warning services that served the citrus and vegetable growers in Arizona, Florida, and California. These private companies expanded considerably as better forecasting and warning techniques were developed. For example, AccuWeather.com has seven global forecast models and fourteen regional forecast models for the United States and North America and also prepares a daily weather report for *The New York Times.*

## CAREERS AND COURSE WORK

Although many consider meteorologists to be people who forecast weather, the better title for such a person is atmospheric scientist. For example, climatologists focus on climate change, and environmental meteorologists are interested in air quality. Broadcast meteorologists work for television stations. The largest number of jobs in the field are with the National Weather Service, which employs about one-third of the atmospheric scientists who work for the federal government.

Meteorologists are predicted to have above-average employment growth in the 2010's. Employment at the National Weather Station generally requires a bachelor's degree in meteorology, or at least twenty-four credits in meteorology courses along with college physics and physical science classes. Anyone who wants to work in applied research and development needs a master's degree. Research positions require a doctorate.

The median annual average salary in 2008 was $81,290. Entry-level positions at the National Weather Service earn about $35,000, and those in the highest 10 percent bracket earn more than $127,000.

## SOCIAL CONTEXT AND FUTURE PROSPECTS

Climate change may be caused by both natural internal and external processes in the Earth-Sun system and human-induced changes in land use and the atmosphere. The United Nations Framework Convention on Climate Change states that the term "climate change" should refer to anthropogenic changes that affect the composition of the atmosphere as distinguished from natural causes, which should be referred to "climate variability." An example of natural climate variability is the global cooling of about 0.5 degrees Fahrenheit in 1992-1993 that was related to the 1991 Mount Pinatubo volcanic eruption in the Philippines. The 15 million to 20 million tons of sulfuric acid aerosols that were released into the stratosphere reflected incoming radiation from the sun and created a cooling effect. Many experts suggest that climate change is caused by human activity, as evidenced by the above-normal temperatures in the 2000's. Based on a variety of techniques that estimate temperatures in previous centuries, the year 2005 was the warmest in the last thousand years.

Numerous observations strongly suggest a continuing warming trend. Snow and ice have retreated from areas such as Mount Kilimanjaro, which at 19,340 feet is the highest mountain in Africa, and glaciated areas in Switzerland. In the Special Report on Emission Scenarios (2000), the Intergovernmental Panel on Climate Change examined the broad spectrum of possible concentrations of greenhouse gases by examining the growth of population and industry along with the efficiency of energy use. The panel estimated future trends using computer climate models. For example, it estimated that the global temperature would increase 35.2-39.2 degrees Fahrenheit by the year 2100.

Given the effect that climate change will have on humanity, many agencies and organizations will be doing research in the area, and climatologists are likely to be needed by a variety of governmental and private entities.

*Robert M. Hordon, B.A., M.A., Ph.D.*

## FURTHER READING

Coley, David A. *Energy and Climate Change: Creating a Sustainable Future.* Hoboken, N.J.: John Wiley & Sons, 2008. A detailed review of energy topics and their relationship to climate change and energy technologies.

Gautier, Catherine. *Oil, Water, and Climate: An Introduction.* New York: Cambridge University Press, 2008. A good discussion of the impact of fossil fuel burning on climate change.

Lutgens, Frederick K., and Edward J. Tarbuck. *The Atmosphere: An Introduction to Meteorology.* 9th ed. Upper Saddle River, N.J.: Prentice Hall, 2004. A useful and standard text that is written with considerable clarity.

Strahler, Alan. *Introducing Physical Geography.* 5th ed. Hoboken, N.J.: John Wiley & Sons, 2011. An excellent text covering weather and climate with superlative illustrations, clear maps, and lucid discussions.

Wolfson, Richard. *Energy, Environment, and Climate.* New York: W. W. Norton, 2008. Provides an extensive discussion of the relationship between energy and climate change.

## WEB SITES

*American Association of State Climatologists*
http://www.stateclimate.org

*American Meteorological Society*
http://www.ametsoc.org

*Intergovernmental Panel on Climate Change*
http://www.ipcc.ch

*International Association of Meteorology and Atmospheric Sciences*
http://www.iamas.org

*National Oceanic and Atmospheric Administration*
National Climatic Data Center
http://www.Ncdc.noaa.gov

*National Weather Association*
http://www.nwas.org

*U.S. Geological Survey*
Climate and Land Use Change
http://www.usgs.gov/climate_landuse

*World Climate Research Programme*
http://www.wcrp-climate.org

*World Meteorological Organization*
http://www.wmo.int/pages/themes/climate/index_en.php

**See also:** Atmospheric Sciences; Barometry; Climate Engineering; Climate Modeling; Measurement and Units; Meteorology; Remote Sensing; Temperature Measurement.

# CLINICAL ENGINEERING

## FIELDS OF STUDY

Biomedical engineering; medicine; computer engineering; electrical engineering; mechanical engineering; systems engineering; biomechanics; mathematics; physiology; measurement; quality control; strategic planning; systems analysis.

## SUMMARY

Clinical engineering is one of many subfields of biomedical engineering. In this subspecialty, practitioners support, monitor, and advance patient care by applying engineering and managerial skills to health care technology and delivery. A clinical engineer can function as a bridge between modern medicine and modern technology, interpreting and implementing sophisticated technology and complicated equipment to improve and enhance patient care and delivery of that care. Possibly the most important characteristic of a clinical engineer is the ability to identify and solve problems, translating solutions into usable clinical care to improve patient outcomes.

## KEY TERMS AND CONCEPTS

- **Biomechanics:** Study of the mechanics of a living body and how the forces exerted on it, including those provided as part of clinical care, affect it and its function.
- **Biomedical Engineering:** Application of engineering techniques and principles to the understanding of biological systems, the development of therapeutic technologies and devices, and solutions to medical problems experienced by patients.
- **Clinical Outcome:** End outcome of patient's treatment; includes any problems experienced by delivery of health care.
- **Health Care Technology:** Equipment or device that interacts with the human body to deliver clinical care or diagnose a condition.
- **Human Factors:** Human-machine interface and the psychological, social, physical, biological, and safety factors involved in a system that interfaces with a human body.
- **Strategic Planning:** Process whereby an organization defines the direction it plans to take and allocates time, money, and people to accomplishing this goal.
- **Systems Analysis:** Method of analyzing a sequence of activities or operations to determine which are necessary and how they can best and most effectively be accomplished.

## DEFINITION AND BASIC PRINCIPLES

Clinical engineering is a field involving many aspects of biomedical engineering. It is a combination of technology and medicine that focuses on the practical side of implementing sophisticated medical technology into clinical care of patients. A clinical engineer's job involves monitoring the interaction of humans and medical equipment to ensure that the machines or equipment are providing service to a patient in the most comfortable or efficient way.

As part of their job responsibilities, clinical engineers may train and supervise other equipment technicians who handle biomedical equipment in a hospital-type setting. In a manufacturing setting, a clinical engineer may advise manufacturers of medical devices on how to improve the design of such equipment to function better in a clinical situation. The field of clinical engineering tends toward redesign rather than new design, because it focuses on improving patient care.

Clinical engineers can be confused with biomedical equipment technicians (BMETs). The basic difference between these two jobs is that BMETs are usually responsible for service and repair of any medical equipment when it fails to function properly. Clinical engineers may supervise and, possibly, even perform these functions depending on the size of the hospital in which the clinical engineer is employed. Typically, however, the clinical engineer is more involved in developing equipment or suggesting changes to existing equipment to improve patient care and delivery of health care services.

## BACKGROUND AND HISTORY

Medicine and engineering have been partners for many years, nearly from the beginning of both fields. In the modern age, this relationship possibly started in the early eighteenth century with English physiologist Stephen Hales and his work, which led to the invention of a ventilator and the discovery of blood pressure. The first meeting focusing on the collaboration between medicine and engineering is thought to have been held in 1948 by the Alliance for Engineering in Medicine and Biology. The term "clinical engineer" was first used in 1969 in a paper published by cardiologist Cesar Caceres, who is generally credited with coining the term.

The first formal accreditation process for clinical engineers was started in the early 1970's by the Association for the Advancement of Medical Instrumentation (AAMI). This body formed the International Certification Commission for Clinical Engineering and Biomedical Technology (ICC) to provide an avenue for clinical engineers to be formally certified. Another body, the American Board of Clinical Engineering, started a similar program, but this version of the program was based in academic institutions offering graduate degrees in clinical engineering. This program was dissolved in 1979, and those who had been certified under its program were absorbed into the ICC. However, in the early 1980's, this body had certified only 350 clinical engineers, and in the late 1990's it suspended its program.

In 2002, the American College of Clinical Engineering started a clinical engineering certification program sponsored by its Healthcare Technology Foundation. This body awarded certification to several individuals who had certified previously under other programs and has continued awarding certifications since then.

The field of clinical engineering as a subfield of bioengineering has had a relatively rocky history. The delineation between the fields has always been hazy and continues to be so. Many of those who function as clinical engineers have a background in biomedical engineering or one of its other subdisciplines.

The Healthcare Technology Foundation offers an Excellence in Clinical Engineering Leadership Award (ExCEL), which identifies clinical engineering professionals who demonstrate leadership in best practices in the management and advancement of health care technology. AAMI also has a Clinical Engineering Management Committee, which follows the career development of clinical engineers and has opportunities for career development and recognition. The *Journal of Clinical Engineering* is an industry publication that provides articles about career development and innovations in the field. There are also many journals covering the biomedical engineering field that include items of interest to clinical engineers.

## HOW IT WORKS

Clinical engineers focus on the interaction of medical technology and the human body. They may perform systems analysis to understand how this interaction takes place and suggest or implement changes to make sure the equipment is providing the necessary service to a particular patient. For example, different types of equipment have certain sizes of tubing that enter a patient's body through a variety of methods. A clinical engineer may examine how this tubing enters the body of a particular patient and make suggestions that may include ways to incorporate larger or smaller tubing to ensure that the device is functioning correctly and providing the patient with the best service. If the clinical engineer sees that a different type of delivery, such as different size tubing or any other equipment modification, makes a functional difference to many patients, he or she may be involved in decisions to modify equipment or suggest a different type of equipment or another manufacturer who would help provide consistently better outcomes for patients.

A clinical engineer may also be involved in implants or prostheses and ensuring that these devices are working correctly and providing the patient with the best possible outcome. He or she may make suggestions as to device modifications that may help the patient. Engineers may also be involved with manufacturers who make these types of equipment to help incorporate changes.

## APPLICATIONS AND PRODUCTS

Clinical engineers are necessary for the safe operation of the diagnostic and therapeutic equipment that is necessary for the operation of a health care facility to deliver top-notch care to its patients.

**Evaluate.** Clinical engineers evaluate equipment before it is purchased to ensure it meets standards of current patient care and safety. They may make recommendations on the equipment a health care system should purchase to provide patients with cutting-edge care. They may manage or coordinate service contracts or purchase negotiations and participate in strategic planning or systems analysis to assure that the recommended equipment is the best possible solution at the best possible price.

**Install and Test.** When new equipment is purchased, clinical engineers may be involved in the installation or in supervising the technicians who actually install the equipment. They make sure that the equipment meets the requirements of any regulatory agency and provide or supervise inspection, installation, and preventive or corrective maintenance.

**Repair and Maintain.** Clinical engineering departments oversee or coordinate the maintenance and continued functionality of technological medical equipment. This extends beyond the mere functionality of the equipment itself and into the realm of warranting that the equipment is functioning at the best level for patient care and comfort and suggesting any changes that would lead to improvement.

**Improve.** Clinical engineers are on the front lines of patient care and are charged with ensuring that technological advances improve that care. As such, they may make adjustments to equipment or even suggest changes to device makers to make sure that patient's needs are met. If incidents occur, clinical engineers investigate the cause and correct and improve the situation. They provide continuous quality assurance and control.

## IMPACT ON INDUSTRY

The field of clinical engineering has had a wide impact on the delivery of health care. As health care delivery becomes more and more technologically advanced, the impact of this field will only increase. The United States and Europe have possibly the best and most comprehensive programs for those interested in biomedical engineering in general and clinical engineering specifically. The Food and Drug Administration, World Health Organization, Department of Veterans Affairs, and U.S. military organizations all have interest in and programs for biomedical and clinical engineers. The Indian Health Service, the Federal Health Program for American Indians and Alaska Natives, established a clinical engineering program in 1973 to service all Indian Health Service facilities. All these types of facilities have recognized the importance of clinical engineering and have jobs available for clinical engineers.

In addition to government agencies, many universities have clinical engineering departments that

### Fascinating Facts About Clinical Engineering

- Clinical engineers are not relegated to the hospital basement alongside an ancient boiler. They are in the forefront of health care technology, using evidence-based approaches to implement new technology that is beneficial to patients.
- There is no such thing as a "typical day" for a clinical engineer; in the fast-paced health care environment, technology that is critical to patient care can go haywire at any time, creating emergency situations that rival those of emergency medical technicians.
- Health innovation groups, a relatively new type of health-system group, perform analysis, conduct post-market trials, recommend new technologies, and plan implementation projects for large health care organizations. A clinical engineer is a perfect fit for this emerging field.
- Clinical engineers tend to focus on delivery of medical technology to the patient. Though this means they often help make small, incremental changes that focus on patient care, they are also in a position to see and suggest or act on major changes in health care delivery that could revolutionize the biomedical engineering field.
- When the field of clinical engineering began to develop, hospitals expected to hire many new clinical engineers, possibly as many as one clinical engineer for every 250 beds in a hospital. However, as the field progressed, not as many hospitals hired clinical engineers as was first thought. However, many clinical engineers work in research and technology development.
- To become certified as a clinical engineer, one must have an accredited engineering degree, obtain relevant clinical experience, and pass a written exam that includes 150 multiple-choice questions (with a three-hour time limit) and an oral exam. If one is already licensed as a registered professional engineer (PE), these requirements may be altered.

feed into health care centers. Some of these include Duke University, University of Kentucky, University of Connecticut, University of Arkansas, California State University at Long Beach, Wayne State University, and University of Toronto.

Worldwide, India continues to offer a number of clinical engineering programs at a university level. Australia also has recently developed programs to enhance the education of clinical engineers.

## CAREERS AND COURSE WORK

A clinical engineer generally completes at least a bachelor's degree in a field of engineering then completes specific training related to clinical engineering. As this field is still evolving, there are not many programs that specifically relate to clinical engineering, so often clinical engineers have a background in biomedical engineering with further training in human factors engineering or a similar field.

After course work is completed, clinical engineers often complete an internship in a teaching hospital, which gives him or her a practical background in hospital functions. He or she may also become certified by the American College of Clinical Engineering, which involves written and oral tests to demonstrate specific knowledge and a portfolio review to determine applicable experience.

A clinical engineer may then begin a career in nearly any aspect of the medical field, including academia, design, or research, but usually this type of engineer puts his or her skills to work in a practical clinical setting, where he or she assesses, manages, and solves the problems that occur when complicated, technologically advanced medical equipment and the human body interact.

In a hospital setting, a clinical engineer may be the technological manager of medical equipment and related systems. This type of job requires knowledge of budget and finance, service agreements, data-processing systems, planning, assessing the effectiveness and efficacy of new equipment, quality control, incident investigation, training, and hospital operations in general. An executive-level job for a clinical engineer might be a chief technology officer position.

Another possible career for a clinical engineer involves research and development of new medical equipment. He or she may work in clinical trials during the development of new equipment to help evaluate and improve products or may even work in a medical-equipment company to develop and design equipment.

A clinical engineer may be asked to consult in a variety of situations, including acting as an expert witness or serve on governmental commissions overseeing the rapidly changing world of medical technology.

## SOCIAL CONTEXT AND FUTURE PROSPECTS

When the term "clinical engineering" was first coined in the late 1960's, expectations were high as to the number of clinical engineers that would be needed in the upcoming years in the medical field. However, not nearly as many clinical engineering positions were created as expected. This lack of positions was partly due to the confusion as to what clinical engineering actually is and how it differs from biomedical engineering in general and biomedical equipment technicians specifically. These fields have been muddled and interchanged since their beginnings.

As health care technology becomes more and more complex and complicated, a clinical engineer can act as a translator between information systems, medical equipment, and the patient. Clinical engineers bring creativity, curiosity, and analytical, communication, and problem-solving skills to the medical field. They may serve on health care system-wide committees that relate to process and performance improvement, quality control and inspection, new technology purchases, and patient safety.

As evidence-based medicine becomes implemented into health care services, the convergence of information technology and medical technology becomes more and more important to control health care costs and improve clinical outcomes for patients. Clinical engineers are perfectly positioned to bridge the gap between developing and implementing emerging technologies in the intricate web of interconnected machinery and equipment.

The American College of Clinical Engineering is a major professional organization that promotes clinical engineering as a career of the future. It has established a code of ethics, advanced workshops, and certification programs to pursue recognition of the field. This organization has committees and task forces to educate and improve the field of clinical engineering as it relates to different pressing issues such as quality control and medical errors. This committee and its members work internationally to improve knowledge of and education in this field.

*Marianne M. Madsen, M.S.*

**FURTHER READING**

Carr, Joseph J., and John M. Brown. *Introduction to Biomedical Equipment Technology.* 4th ed. Upper Saddle River, N.J.: Prentice Hall, 2001. Discusses quality improvement and continuous quality assurance for those who use, monitor, or modify medical equipment.

Chan, Anthony Y. K. *Biomedical Device Technology: Principles and Design.* Springfield, Ill.: Charles C. Thomas, 2008. Discusses basic medical equipment and the principles behind their use and operation.

David, Yadin, et al. *Clinical Engineering.* Boca Raton, Fla.: CRC Press, 2003. Essays about the discipline of clinical engineering from leading practitioners in the field; includes graphics, tables, figures, formulas, definitions of terms, and comprehensive index.

Dyro, Joseph F. *Clinical Engineering Handbook.* Burlington, Mass.: Academic Press, 2004. Comprehensive guide to clinical engineering, including history and worldwide practice in the field, ethics, and future trends.

Enderle, John, Susan Blanchard, and Joseph Bronzino. *Introduction to Biomedical Engineering.* 2d ed. Burlington, Mass.: Academic Press, 2005. A reference book for the field of biomedical engineering with practical examples.

McGill, Jennifer. "Clinical Engineering." In *Career Development in Bioengineering and Biotechnology*, edited by Guruprasad Madhavan, Barbara Oakley, and Luis Kun. New York: Springer Science + Business, 2008. Overview of clinical engineering as a profession including background, required course work, possible careers, and first-person stories about being a clinical engineer.

Montaigne, Fen. *Medicine by Design: The Practice and Promise of Biomedical Engineering.* Baltimore: The Johns Hopkins University Press, 2006. Nontechnical introduction to the field in which each chapter focuses on a different subspecialty of the field; includes stories and case studies.

Saltzman, W. Mark. *Biomedical Engineering: Bridging Medicine and Technology.* New York: Cambridge University Press, 2009. Basic overview of the field of biomedical engineering including discussion of range of fields available; includes further reading, suggested Web sites, and appendixes.

**WEB SITES**

*American College of Clinical Engineering*
http://www.accenet.org

*American Society for Healthcare Engineering*
http://www.ashe.org

*Biomedical Engineering Society*
http://www.bmes.org/aws/BMES/pt/sp/home_page

*Healthcare Technology Foundation*
http://thehtf.org

**See also:** Biomechanics; Bionics and Biomedical Engineering; Computer Engineering; Electrical Engineering; Mechanical Engineering.

# CLONING

## FIELDS OF STUDY

Biology; genetics; biotechnology; animal husbandry; aquaculture; veterinary medicine; developmental biology; embryology; biochemistry; theriogenology; agriculture; horticulture; botany; cell biology; conservation biology; viticulture; enology; medicine; pharmacology; microbiology; pomology; toxicology; zoology.

## SUMMARY

Cloning is any type of biological reproduction that produces offspring that are genetically identical to their parents. Cloning occurs naturally, since many organisms routinely reproduce through natural cloning processes. Artificial cloning technologies include molecular cloning, which reproduces large quantities of discrete segments of DNA; reproductive cloning, which uses assisted reproductive technologies to produce animals that share the same desirable genetic characteristics as another living or previously existing organism; and therapeutic cloning, which uses the same techniques as reproductive cloning but instead derives useful cell lines from cloned embryos.

## KEY TERMS AND CONCEPTS

- **Clone:** Organism whose genetic information is identical to the donor organism from which it was created, or a macromolecule that is an exact replicate of another macromolecule.
- **Embryonic Stem Cell:** Stem cell made from the inner cell mass of very young mammalian embryos.
- **Enucleation:** Microsurgical technique that removes nuclei from cells.
- **Genome:** Sum total of the DNA stored in the cells of an organism.
- **Parthenogenesis:** Biological process whereby an egg initiates embryonic development without having first undergone fertilization.
- **Pharming:** Use of genetic engineering to express cloned genes that encode useful pharmaceutical products in host animals or plants.
- **Pluripotent:** Ability of a cell to differentiate into any fetal or adult cell type.

- **Restriction Endonuclease:** Special enzyme that cuts DNA at a specific sequence motif.
- **Somatic Cell Nuclear Transfer:** Implantation of nuclei from somatic (body) cells into an egg to make a cloned embryo.
- **Transgenic Organism:** Biological entity that has had a foreign gene inserted into its genome. Commercially available transgenic organisms are often called genetically modified organisms (GMOs).

## DEFINITION AND BASIC PRINCIPLES

Cloning is a means of producing biological organisms, cells, or DNA molecules that are genetically identical to their progenitors. There are natural forms of cloning and three main types of artificial cloning: molecular, reproductive, and therapeutic cloning.

Natural mechanisms of cloning occur in organisms such as bacteria that simply split or fragment into identical copies of themselves. In other organisms, reproductive cells, or gametes, undergo a process called parthenogenesis, in which they initiate development without the benefit of fertilization. Cloning is uncommon in mammals, but rarely, early mammalian embryos undergo a form of cloning called twinning, in which the embryo splits into two embryos, which develop into genetically identical twins.

Molecular cloning, also known as recombinant DNA technology or DNA cloning, involves the transfer of an isolated fragment of DNA from an organism of interest to a host cell that replicates it. Such isolated DNA fragments are known as cloned DNA or genes.

Reproductive cloning uses assisted reproductive technologies to generate animals with the same nuclear genome as another animal. The particular procedure used during reproductive cloning is called somatic cell nuclear transfer ( SCNT). Cloned embryos are gestated in the womb of a surrogate mother until they come to term. Cloned organisms are not genetically modified organisms but are simply produced through a type of assisted reproduction.

Therapeutic cloning uses the same procedures as reproductive cloning; however, instead of transferring the cloned embryo into the womb of a surrogate mother, the embryo is further manipulated in the laboratory to make cell cultures of embryonic cells for basic or clinical research.

## BACKGROUND AND HISTORY

Sea urchins were the first animal cloned in the laboratory. In 1894, Hans Dreisch isolated sea urchin embryo cells and watched them develop into small, separate larvae. In 1902, Hans Spemann used the same procedure, embryo splitting, to isolate cells from salamander embryos, which also developed into identical adult salamanders. In 1903, U.S. Department of Agriculture employee Herbert Webber coined the word "clon" for asexually produced cells or organisms, which later evolved into "clone." This term comes from the Greek *klon*, which means "trunk" or "branch." Horticulturists have used this term for more than a century, since an entire new plant can grow from a cutting, resulting in a plant that is genetically identical to the plant from which the cutting was taken.

In 1928, Spemann cloned salamanders by transferring the nucleus, the subcellular compartment that houses the chromosomes, from one salamander embryo into the egg of another. Since Spemann's seminal experiments, scientists have adapted nuclear transfer technology to clone other organisms. In 1952, frogs were cloned, and in 1963, the Chinese embryologist Tong Dizhou cloned a carp to produce the first cloned fish. During the 1980's and 1990's, sheep, cows, and mice were cloned. However, all these animals were cloned by using nuclei from embryos. In 1996, Ian Wilmut and his team at the Roslin Institute in Edinburgh, Scotland, cloned a sheep from an adult cell, demonstrating that adult cells could serve as the source of genetic material for animal clones. This technological feat was followed by the cloning of goats, mules, gaurs (an endangered species), horses, pigs, mouflons (a wild sheep), mice, rats, dogs, cats, water buffalos, camels, rabbits, deer, wolves, and African wildcats, and even embryos from nonhuman primates and humans.

## HOW IT WORKS

**Molecular Cloning.** To clone a gene, the DNA of the model organism is selectively fragmented by enzymes called restriction endonucleases (REs) and inserted into another piece of DNA called a cloning vector. Cloning vectors are either small circles of DNA called plasmids, bacterial viruses, or bacterial or yeast artificial chromosomes. They ferry the DNA fragments from the genome of the model organism into a host cell (either a bacterium or yeast). This population of host cells collectively carries the entire genome of the model organism in small fragments, and is called a gene library.

To isolate a gene from a gene library requires a probe, which is a fragment of DNA or RNA of any length that has a sequence that is complementary to the sequence of the gene that is to be isolated. Probes can be made synthetically or can come from the genes of closely related organisms. By screening the gene library with the probe, the gene of interest is cloned, which simply means to isolate it from all the other sequences found in the genome of the model organism.

Alternatively, scientists can synthesize small strands of DNA called primers, whose sequences are complementary to different locations in the gene. These primers can be used to specifically amplify the gene from the library by means of a polymerase chain reaction (PCR). A polymerase chain reaction makes large quantities of the gene of interest from a very small amount of starting material, and the amplified DNA can also be cloned into a cloning vector or analyzed directly.

**Reproductive Cloning.** To clone an animal, mature eggs are isolated from females of the animal species that is to be cloned. The egg is enucleated by piercing it with a microscopically narrow (0.0002-inch-wide) glass tube that is used to vacuum out the egg nucleus. The enucleated egg is fused with a cell from the body of the animal to be cloned and activated with either chemicals or an electric current. This procedure is called somatic cell nuclear transplantation (SCNT).

After activation, the egg divides and grows like a newly formed embryo. However, if the animal is a mammal, the embryo can survive only for a limited period of time before it must implant into the inner layer of the mother's womb. Therefore, a surrogate female from the same species of the animal to be cloned, or a closely related species, is made pseudopregnant by feeding her hormones, and the embryo is released into her receptive womb, where it implants. Barring any technical or biological mishap, the cloned embryo will develop, and the process will result in a live birth.

**Therapeutic Cloning.** To make embryonic cell cultures, cloned embryos are made by means of somatic cell nuclear transplantation. They are then either disassembled in the laboratory and used to establish embryonic cell cultures or gestated in a surrogate mother to the fetal stage, at which time the fetus is aborted, and cells from the fetus are used to establish fetal cell cultures.

By culturing specific cells from cloned embryos, scientists can make embryonic stem cell (ESC) cultures. During mammalian development, two distinct cell populations form after the first few days of embryonic development. The trophoblast, or the flattened, outer layer of cells, will eventually form the placenta and its associated structures. The inner cell mass (ICM) is the round, inner clump of cells that develop to form the embryo proper and a few structures associated with the placenta. If ICM cells are isolated and cultured on feeder cells, a layer of nondividing skin cells that secrete a cocktail of growth-promoting chemicals, the ICM cells will grow and spread over the surface of the culture dish. Such a culture is an embryonic stem cell culture, and these cells are pluripotent, which means that they can differentiate into any cell type in the adult body.

## APPLICATIONS AND PRODUCTS

**Molecular Cloning.** Organisms that express cloned genes make many useful pharmaceuticals such as human insulin, growth hormone, clotting factors, fertility drugs, and vaccines. Cloned genes are also used to genetically screen individuals for genetic diseases. Pharmacologists even use cloned genes for pharmacogenetics, which screens patients for the presence of gene variants that can profoundly affect the efficacy and toxicity of particular drugs. This allows clinicians to tailor treatment to the exact genetic makeup of the patient to maximize treatment efficacy and minimize side effects. Such a strategy is called personalized medicine. Cloned genes are also used in gene therapy, which delivers cloned genes into the bodies of patients who suffer from genetic diseases in an attempt to cure them. Patients with cancer and inherited deficiencies of the immune system, blindness, and blood-based defects have been treated with gene therapy protocols.

In agriculture, the introduction of cloned genes into plants that are used as food crops has generated transgenic crops. These crops display several advantageous traits: reduced dependence on agrochemical applications (for example, Bt-corn and herbicide-resistant crops), increased nutritional value (for example, Golden Rice), increased resistance to environmental stresses, and reduced spoilage (for example, the Flavr Savr tomato).

**Reproductive Cloning.** When farmers identify food animals with desirable traits, they typically breed those animals as much as possible to improve the genetic quality of their herds and flocks. However, such prize animals inevitably die. Propagating these animals by reproductive cloning and mating them to as many animals as possible preserves the exceptional genetic content of a prize animal and allows it to produce far more offspring. This significantly raises the genetic quality of the flock or herd, and commercial dissemination of such cloned animals to other farmers raises the overall genetic quality of food animals. Reproductive cloning also eliminates the need for artificial insemination, which is often expensive and inconvenient.

Cloning effectively maintains high-quality animal stocks. Reproductive cloning of only the healthiest and most productive animals increases their numbers and improves the gene pool (sum total of genetic diversity) and overall health of food animals. This results in safer and healthier food and reduces the use of growth hormones, antibiotics, and other chemicals in the raising of animals.

In the field of conservation biology, the numbers of endangered species are often increased by captive breeding programs. However, not all endangered species can effectively breed in captivity. Reproductive cloning can aid in the preservation of those organisms that do not reproduce in captivity. Cloning can also resurrect genetic material from dead animals and potentially expand the gene pool of endangered species. In 2001, scientists at the University of Teramo, Italy, cloned the European mouflon, an endangered sheep, from cells sampled from a dead animal. When combined with other reproductive technologies, cloning can help save endangered species.

Cloned animals also serve as excellent research models. Because each cloned animal is genetically identical, experiments on cloned animals are devoid of differences caused by heterogeneous genetic backgrounds. Genetic manipulation of cloned animals allows researchers to modify genes of interest and more completely analyze their contribution to development and disease. Modifying particular genes of cloned animals also generates model systems for particular genetic diseases. Cloned, transgenic mice and cloned knockout mice, which have had a specific gene inactivated, are examples of the vast usefulness of such model systems.

## Fascinating Facts About Cloning

- Scientists at Advanced Cell Technology used fetal heart muscle cells from cloned cow fetuses to reverse the effects of heart attacks in adult cows.

- The first cloned cat, CC (CopyCat), made at Texas A & M University in 2001, has a completely different personality than the donor cat. Even though CC is genetically identical to her donor, she is shy and timid whereas the donor cat is outgoing and playful.

- In 2008, BioArts International held an essay contest that invited people to argue why their dog should be cloned. The winner was Trakr, a German Shepherd police dog, who discovered the last survivor of the September 11, 2001, terrorist attacks on the World Trade Center in New York City.

- By cloning vaccines into plants, scientists have made edible vaccines against digestive diseases such as cholera, the Norwalk virus, some food poisonings, and enterotoxigenic *Escherichia coli*. These vaccines are not injected but rather eaten.

- Ingo Potrykus and Peter Beyer invented Golden Rice in the 1990's. This genetically engineered strain of rice produces beta-carotene, a precursor for vitamin A biosynthesis, which is not found in normal rice in appreciable quantities. Children who live in countries where rice is the main food staple are at higher risk for vitamin A deficiency, and Golden Rice was developed to help prevent this deficiency. Subsequent development has increased the nutritional value of Golden Rice even further. Even though the makers of Golden Rice want to give it to farmers completely free of charge, opposition to genetically modified organisms has prevented it from ever being cultivated for food.

Of enormous interest is modifying the genomes of cloned animals so that they can produce clinically and pharmaceutically significant products. By genetically modifying pigs, it is possible to make cloned pigs that contain organs that are fit for transplantation into humans (xenotransplantation). Also, producing antibodies, clotting factors, or even vaccines in the blood or milk of farm animals provides a means to mass-produce potentially expensive pharmaceutical agents at a fraction of the normal cost. This process is called pharming.

**Therapeutic Cloning.** Therapeutic cloning has tremendous potential for numerous clinical applications.

Embryonic stem cells (ESCs) made from therapeutic cloning procedures are pluripotent. Therefore, injured, diseased, or failing tissues or organs could potentially be replaced by tissues or organs manufactured from embryonic stem cells in the laboratory or fetal cells from cloned fetuses. Furthermore, embryonic stem cells made from cloned embryos, or any tissues or organs fashioned from these cells, would not be regarded by the patient's body as foreign. Experiments in laboratory animals have shown that such scenarios are possible. Therapeutic cloning, coupled with embryonic stem cells technology, could christen a new era of regenerative medicine.

Embryonic stem cells from cloned embryos have toxicological applications. Toxicologists typically use laboratory animals or cultured cells to gauge the biological effects of natural or industrially produced molecules on human beings. Unfortunately, laboratory animals show limited utility as a model for human toxicology, and cultured cells do not represent the response of an organ or tissue to foreign molecules. Furthermore, neither of these model systems can assess the individual responses people will have to such molecules, because the genetic variation between individual humans causes differential responses to drugs, toxins, or environmental pollutants. However, cultured embryonic stem cells from cloned embryos can test the biological effects of drugs or environmental pollutants on cells made from a specific person. In addition, because these cells can be differentiated into various tissues and even organs, they can be used to evaluate the individual and tissue-specific responses people might have to particular drugs or pollutants.

### IMPACT ON INDUSTRY

Biotechnology companies that use cloning technology in the United States, Europe, Canada, and Australia reported a combined net profit of $3.7 billion in 2009. The United States has been the leader in cloning research, but there are many high-quality laboratories that study cloning technology in the United Kingdom, continental Europe, South Korea, Australia, China, Canada, Iran, and Israel.

**Governmental Regulatory Agencies.** The U.S. governmental agencies that regulate cloning are the Food and Drug Administration (FDA), Environmental Protection Agency (EPA), and the Department of Agriculture (USDA). The FDA regulates any foods made by genetically modified organisms. This agency concerns itself with only the safety of foods and not the manner in which they are made. The EPA has regulatory authority over all pest-resistant plants to ensure that genetically engineered crops do not adversely affect the environment. Field testing of genetically modified organisms is overseen by a division of the USDA, the Animal and Plant Health Inspection Service.

**Government and University Research.** The largest funder of cloning research is the National Institutes of Health (NIH). Other governmental funding agencies include the National Science Foundation and the USDA. The NIH not only funds the research of other laboratories but also houses many of its own laboratories, some of which use investigate cloning technologies.

Most of the cloning on university campuses is basic research. Many universities house cloning research centers on their campuses. In other cases, the cloning centers are extensions of state universities. The Roslin Institute, for example, where Dolly the cloned sheep was made, is an extension of the University of Edinburgh. Some universities have even formed partnerships with biotechnology companies that allow the company to work on university property in exchange for funds and increased collaboration between the company and the university.

**Industry and Business.** Biotechnology companies from all over the world participate in cloning research. Many of these companies have even formed associations. Ausbiotech represents Australian biotechnology companies, the European Federation of Biotechnology represents institutions from European and non-European countries, and BIOTECanada represents more than 250 Canadian biotechnology companies. These trade associations represent the interests of biotechnology to governing bodies.

Pharmaceuticals made by transgenic organisms that express cloned genes constitute the largest proportion of products developed and manufactured by biotechnology companies. These products include diabetes treatments, vaccines, cytokines (special proteins that signal to white blood cells), and other medicines. The demand for new medicines drives research and development in this area, and the understanding of the entire human genome is ushering in many previously unknown medical treatment strategies.

Agricultural biotechnology companies focus largely on developing new crops with improved characteristics. Some of their work is focused on making crops that can grow in underdeveloped countries. For example, Monsanto has begun field trials of a genetically engineered cassava plant that is virus resistant, less poisonous, and much more nutritious than its native counterpart. Some 800 million people globally rely on cassava as their main food staple, but viral infections, poor processing that tends to generate poisonous cyanides, and a lack of nutritional content tend to limit the food potential of cassava.

A few animal cloning industries market techniques for cloning pets. Genetic Savings and Clone is one such company. A related company, Viagen, which is part of Exeter Life Sciences, offers commercial cloning services for farm animals.

Human cloning industries are working toward therapeutic cloning strategies. Advanced Cell Technologies (ACT) works on human cloning. Based in Worchester, Massachusetts, ACT seeks to produce patient-specific stem cells from cloned embryos and cloned fetuses that can cure degenerative diseases without the risk of rejection by the immune system.

## CAREERS AND COURSE WORK

Students who wish to pursue a career in cloning should possess a foundational grasp of biology and chemistry. Of cardinal importance is advanced course work in cell, molecular, and developmental biology. Many entry-level jobs exist in academic and industrial laboratories for those with a bachelor's degree in biology, biochemistry, or chemistry. Such jobs are usually laboratory technician positions, and good laboratory skills are required for such work. For those who wish to work as a leader of a research group, a Ph.D. in either cell or developmental biology, a D.V.M., or M.D. degree is required.

Cloning work requires highly skilled technicians who are very dexterous and can look through microscopes for a long period of time while performing extremely fine manipulations. Such techniques often require many weeks of practice to perfect, and a patient, forbearing personality greatly helps people in the cloning field. Because cloning experiments are also very labor intensive, a collaborative mind-set is also very helpful.

Many scientists are involved in cloning work in industry, and several biotechnology companies have divisions that investigate cloning technology. Because such companies normally attempt to clone organisms for profit, cloning divisions of biotechnology companies usually examine improving the efficiency and cost-effectiveness of cloning procedures to standardize the manufacture of clones.

A large percentage of cloning scientists work in academic laboratories, where they devote their time to more basic research questions. Scientists in academic institutions must split their time between teaching and research. Other scientists who work for government laboratories such as the NIH can work on cloning, safety testing of cloned products, and other aspects of cloning without teaching responsibilities.

## SOCIAL CONTEXT AND FUTURE PROSPECTS

Despite the reservations of some people, cloning is a part of everyday life. Many of the foods Americans consume contain some genetically engineered products. Physicians prescribe medicines, give vaccines, and apply other biological products made by genetically engineered microorganisms on a quotidian basis. Given the inroads molecular cloning has already made into people's lives, it is unlikely that people would suffer any revulsion from eating meat from cloned cattle or sheep or having their lives saved by the transplantation of an organ that came from a cloned pig. People would also probably not protest seeing cloned versions of endangered species at their local zoos.

Nevertheless, many people have raised concerns over cloning technologies. First, conservation biologists have suggested that cloning endangered species does not address the habitat destruction and environmental degradation that pushed these species to near extinction in the first place. Second, cloning only makes one species and does not re-create an ecosystem. For example, cloning cannot recapitulate a coral reef or an old growth forest. Thus, it is the wrong solution for the problem.

Genetically modified organisms have become the focal point of concern for several environmental activism groups. Such groups oppose GMOs because they believe that the cloned genes inserted into them can spread to other species and cause severe environmental disruption and that genetically engineered foods have not been sufficiently tested and are potentially dangerous to human health.

The most contentious aspect of cloning technologies is human genetic engineering and reproductive cloning. Transhumanists are some of the most energetic proponents of human cloning and genetic enhancement. As a movement, Transhumanism regards infirmity, disease, aging, and death as undesirable and unnecessary and views science and technology as the means to defeat human limitations. Transhumanists' main argument for human cloning is that reproductive freedoms extend to everyone, and therefore, every human being has an inherent right to clone himself or herself.

Opponents of human cloning object to the manufacturing of human beings. Cloned children are made to be identical to someone else and therefore will always live in the shadow of the original person and never be completely the person they choose to be. These unreasonable expectations can psychologically damage them and violate their human dignity and individuality. Cloning would also alter the concept of human nature and therefore undermine the very foundation of liberal democracy.

In the future, the argument over cloning will not dissipate, but cloning research will certainly advance and provide more and more examples of the utility of this remarkable technology.

*Michael A. Buratovich, B.S., M.A., Ph.D.*

## FURTHER READING

Alexander, Brian. *Rapture: A Raucous Tour of Cloning, Transhumanism, and the New Era of Immortality.* New York: Basic Books, 2004. A reporter examines the fringe groups that support human cloning and genetic enhancement and finds people who want to defeat the effect of entropy and live forever.

Fukuyama, Francis. *Our Posthuman Future: Consequences of the Biotechnology Revolution.* New York: Picador, 2003. A historian's admonition of the consequences of the biotechnology revolution and its potential to abolish human rights and erode the foundations of liberal democracy.

Mitchell, C. Ben, et al. *Biotechnology and the Human Good.* Washington D.C.: Georgetown University Press, 2007. A distinctly Christian assessment of the application of biotechnology to humans that remains optimistic but cautious and concerned.

Shanks, Pete. *Human Genetic Engineering: A Guide for Activists, Skeptics, and the Very Perplexed.* New York: Nation Books, 2005. A helpful explication of the science behind cloning, coupled with stern warnings against it, by a noted social activist.

Silver, Lee. *Challenging Nature: The Clash Between Biotechnology and Spirituality.* New York: Harper Perennial, 2006. A Princeton stem cell scientist explains the science behind biotechnology and stem cells. He offers some rather harsh critiques of more conservative thinkers who do not agree with his optimistic views of genetic enhancement and embryonic stem cells.

_____. *Remaking Eden: How Genetic Engineering and Cloning Will Transform the American Family.* New York: Harper Perennial, 2007. A very readable introduction to the science of cloning and genetic engineering by a noted mammalian embryologist, who believes that humans should be cloned and that people should welcome the profound changes that it will invoke within human societies.

Wilmut, Ian, Keith Campbell, and Colin Trudge. *The Second Creation: Dolly and the Age of Biological Control.* New York: Farrar, Straus and Giroux, 2000. The two researchers who made Dolly team up with a noted British science writer to give a personal but rigorous explanation and thoughtful examination of cloning. Contains a helpful glossary of terms.

## WEB SITES

*Human Cloning Foundation*
http://www.humancloning.org

*MedlinePlus*
Cloning
http://www.nlm.nih.gov/medlineplus/cloning.html

*National Human Genome Research Institute*
Cloning
http://www.genome.gov/25020028

**See also:** Agricultural Science; Animal Breeding and Husbandry; Bioengineering; DNA Sequencing; Genetically Modified Food Production; Genetically Modified Organisms; Genetic Engineering; Plant Breeding and Propagation; Reproductive Science and Engineering; Stem Cell Research and Technology.

# COAL GASIFICATION

## FIELDS OF STUDY

Process engineering; chemical engineering; mechanical engineering; electrical engineering; systems engineering; mining engineering; chemistry; physics; earth science; statistics.

## SUMMARY

Coal gasification is the chemical and physical process of converting coal into coal gas, a type of synthesis gas (syngas) that is composed of varying amounts of carbon monoxide and hydrogen gas. The syngas is subsequently used as fuel for power generation or as feedstock in chemical processes such as the production of synthetic fuels and production of fertilizers.

## KEY TERMS AND CONCEPTS

- **Biomass:** Renewable energy source derived from a living or recently living organism such as wood, alcohol, or oil.
- **Coal:** Combustible, sedimentary rock that is primarily composed of carbon and smaller amounts of sulfur, hydrogen, oxygen, and nitrogen and used as the largest source of energy for the generation of electricity worldwide and as one of the main feedstocks to create steel.
- **Coal Gas:** Type of synthesis gas composed of a mixture of carbon monoxide and hydrogen gas.
- **Feedstock:** Raw material that is used for processing or manufacturing.
- **Fischer-Tropsch Gasification Process:** Catalyzed chemical reaction in which carbon monoxide and hydrogen are converted into liquid hydrocarbons of various forms, typically used to produce synthetic petroleum substitute for use as lubrication oil or as fuel. Named for German chemists Hans Fischer and Franz Tropsch.
- **Flue Gas:** Exhaust gas that exits into the atmosphere from a device such as a boiler, furnace, or steam generator through a pipe-like channel called a flue.
- **Fuel Cell:** Electrochemical cell that uses an external source fuel to generate an electric current.

- **Gasification:** Process of converting carbonaceous material such as coal, petroleum coke, refuse-derived fuel (RDF), or biomass into gas that can be used as an energy source.
- **Integrated Gasification Combined Cycle (IGCC):** High-efficiency gasification system that uses two types of turbines, combustion and steam.
- **Synthesis Gas:** Gas mixture that contains carbon monoxide and hydrogen; also known as syngas.
- **Synthetic Fuel:** Liquid fuel created from carbon-rich materials such as coal, natural gas, oil shale, biomass, or industrial waste.
- **Underground Coal Gasification (UCG):** Industrial method of converting coal that is underground, or in situ, within natural coal seams into a combustible gas to be used for heating, power generation, or other industrial uses; also called in situ coal gasification (ISCG).

## DEFINITION AND BASIC PRINCIPLES

Coal gasification is a method of converting coal into a combustible gas that can be used for heating, power generation, and manufacture of hydrogen, synthetic natural gas, or diesel fuel. Coal gasification plants are in operation throughout the world. Unlike conventional coal-fired power plants, gasification involves a thermochemical process that breaks down coal into its basic chemical constituents. This is accomplished in modern-day gasifiers by reacting coal with a mixture of steam and air or oxygen under high temperature and pressure. The end product is a gaseous mixture of carbon monoxide, hydrogen, and other gas compounds. Many experts predict that coal gasification will lead the way to a clean-energy future: Whereas burning coal can contribute to global warming by increasing the concentration of carbon dioxide in the atmosphere, coal gasification produces lean gas because pollutants or impurities such as sulfur and mercury are removed in the system.

## BACKGROUND AND HISTORY

Coal gasification dates back to about 1780, when it was first developed, but the technology and its applications have evolved significantly over the years. When coal gasification was first implemented, the carbon monoxide that was produced was used as a source of energy for municipal lighting and heating because industrial-scale production of natural gas was not yet available. This coal gas was referred to as blue gas, producer gas, water gas, town gas, or fuel gas. Natural gas became widely available in the 1940's and by the 1950's, coal gas was nearly replaced with natural gas because it burned cleaner, had a greater heating value, and was safer to use. Later, when the price of alternative fuels such as oil and natural gas were low, interest in coal gasification fell further; however, renewed interest has arisen for coal gasification solutions because of the high cost of oil and gas, and energy security and environmental considerations.

*Coal gasification plant in Beulah, North Dakota.* (David R. Frazier/Photo Researchers, Inc.)

### HOW IT WORKS

**Coal Selection and Preparation.** The first step in successful coal gasification is the careful selection, analysis, and preparation of coal. Coal is first analyzed for its percentage of sulfur, fixed carbon, oxygen, ash, and other volatile content. In general, use of a coal feedstock with low sulfur, low moisture, high fixed-carbon content, and low ash content will result in low oxygen consumption, a high volume of syngas, and a small volume of waste-product generation. Some adjustments to gasification systems can be made to accommodate for different coal qualities.

**Coal Gasifier.** Central to coal gasification is the gasifier. A gasifier converts hydrocarbon feedstock into gaseous components by providing heat under pressure along with steam. A gasifier, unlike a combustor, relies on the careful regulation of the quantity of air or oxygen that is permitted to enter the reaction so that only a small portion of the fuel, the coal, burns completely. At a high temperature (about 900 degrees Celsius) and high pressure, the oxygen and water in the gasifier partially oxidizes the coal. During this stage, called the steam-forming reaction, rather than burning, the coal feedstock is converted into a syngas mixture of carbon dioxide, carbon monoxide, molecular hydrogen, and water in the form of vapor. To get the syngas out of the gasifier, the syngas is cooled to room temperature by using exhausts and filters that remove solid particles.

**Types of Gasifiers.** There are three main types of gasification technologies: entrained flow, moving bed, and fluid bed. Of the three, the entrained-flow gasifier possesses the greatest efficiency. Because it operates at a high temperature, nearly 99 percent of

the coal is converted into high-purity syngas, as the majority of the tar and oil contained in the coal is destroyed with the high heat. The one drawback, however, is that the entrained-flow gasifier has a high oxygen demand, and this is easily exacerbated through use of a coal feedstock with high ash content. The moving-bed gasifier has the lowest oxygen demand. In this system, coal moves slowly in a downward fashion and is gasified by a counter-current blast. The low operational temperature tends to be inhibitory to the reaction rate, resulting in lower syngas purity and volume generation. The fluid-bed gasifier, named so because of the fluidlike manner in which coal particles behave in the gasifier, facilitates good interaction between the coal feedstock and oxygen without requiring a membrane, leading to lower overall cost to implement and operate. The fluid-bed gasifier is the best option for low-rank (low-quality) coals; however, overall, it possesses the lowest conversion rate of carbon, resulting in a low-purity syngas.

**Integrated Gasification Combined Cycle.** The technology that combines coal gasification and the subsequent burning of the gas is called integrated gasification combined cycle (IGCC). IGCC combines a coal gasifier with a gas turbine and a steam turbine to produce electric power. The hydrogen-rich syngas from the gasifier is purified to remove acidic compounds and particulates. It then enters the first turbine (the gas turbine) to generate electricity, and the waste heat from the gas turbine works to power the second turbine (the steam turbine), which in turn produces additional electricity. Because steps are taken to remove the majority of acidic compounds and particulates, the resulting combustion-exhaust gas, or flue gas, that leaves the gas turbines has minimal effect on the environment. The combined use of a gas turbine and steam turbine, which both work to produce electricity, makes IGCC the preferred technology in this carbon-constrained world. It is more energy-efficient and generates less carbon dioxide per ton of coal used compared with conventional coal-fired power plants.

## APPLICATIONS AND PRODUCTS

**Underground Coal Gasification.** Coal gasification can be applied to coal in situ— in underground coal seams. Gasification of coal contained underground is called underground coal gasification (UCG). It is particularly well suited for technologically or economically unmineable coal deposits. The traditional method of extracting coal is through the excavation of open pits to expose coal-containing seams or through the excavation of underground mines. The design, construction, equipment, and labor to build such mines are a costly undertaking. UCG-candidate coal seams possess one or more features that render it unsuitable, from a technical and economic standpoint, for conventional coal mining: faulting, volcanic intrusions, complex depositional and tectonic features, and environmental constraints.

The UCG process uses injection and production wells drilled from the surface to access the underground coal, and the coal is not mined to the surface. A horizontal connection underground between the injector and extractor is made normally by hydrofracturing, which is a process that uses high-pressure water to break up the rock or coal. Through the injection well, oxidants (such as water and air or a water and oxygen mixture) are sent down into the coal seam. As in the case of conventional coal gasification, the coal is heated to temperatures that would normally cause the coal to burn, but through careful control of the oxidant flow, the coal is separated into syngas. The product syngas is drawn out through the second well.

**Gasification By-Products.** Some by-products of coal gasification have commercial value, so they are isolated and set aside to be sold or used on-site for industrial use. Mineral components in the coal that are unable to gasify leave the gasifier or fall to the bottom of the gasifier as an inert, glasslike material,

known as slag, which can be used in cement or road construction.

Mercury that is isolated by passing the syngas through a bed of charcoal has no commercial value, but the final cleaning step that follows in the acid gas removal units handles sulfur impurities, which are converted into valuable by-products. Sulfur impurities are converted to hydrogen sulfide ($H_2S$) and carbonyl sulfide (COS), which are isolated in the form of elemental sulfur or sulfuric acid, which can be used for industrial processes such as incorporation into fertilizer.

Nitrogen, in the form of ammonia, is also extracted out of the product gas stream for industrial use. Other usable by-products include tar and phenols.

## IMPACT ON INDUSTRY

**Impact on Seaborne Coal Trade.** There is significant merit to the deployment of coal gasification technology, particularly for emerging countries such as China, India, and Africa, as low plant development and operational costs are involved. As the price for thermal coal used for conventional coal-fired plants and the price for metallurgical coal used for making steel (to build rail systems to transport coal) continues its upward trend in trading price, more countries will see that embracing implementation of UCG for their coal resources on hand is the most economical choice. This may lead to a relative reduction in the volume of the seaborne coal trade market.

**Alternative Feedstocks.** In addition to the research to advance coal gasification plant efficiency and environmental performance, the application of other feedstocks such as municipal waste and biomass are currently being explored. If a significant volume of conventionally unrecyclable material such as municipal waste or solid human waste can be converted into useful energy, there will be further development in the area of isolating certain materials.

**Enhanced Oil Recovery.** Another area that could flourish because of advancement in coal gasification is oil extraction. An optional addition to coal gasification is the isolation of carbon dioxide ($CO_2$) and subsequent storage underground or in depleted oil wells. This process is called $CO_2$ sequestration or carbon capture. The collected carbon dioxide may be transported and used for other industrial processes such as enhanced oil recovery

(EOR), which is an oil-extraction method that uses carbon dioxide to help extraction of oil that is difficult to recover by conventional means. When $CO_2$ is injected in underground oil fields, it increases the pressure and enhances the recovery of oil

remaining underground. In EOR, once the oil field is depleted, the oil field is sealed to trap $CO_2$ permanently underground, instead of allowing it to release into the atmosphere. Coal gasification systems can also be designed subsequently to route the product gas to another processing device called a Fischer-Tropsch reactor. Fischer-Tropsch reactors are used to produce alkanes, which are hydrocarbons often added to natural gas, gasoline, and diesel fuel.

### Fascinating Facts About Coal Gasification

- The benefits of coal gasification are based on the capability to cleanse or capture as much as 99 percent of the impurities from coal-derived gases that have been associated with global warming.
- Gasification plants typically consume 30 percent less water than conventional coal-fired power plants. Gasification plants can also be designed so that water is never discharged from the system but instead circulates to be used over and over again.
- The fuel efficiency of a coal gasification power plant can exceed 50 percent. On the other hand, in a conventional coal combustion plant, only one-third of the energy value of coal is converted into usable energy. The remainder is lost as heat and is usually not recovered.
- Integrated gasification combined cycle (IGCC) may help people realize a clean-energy future. Hydrogen is rapidly gaining popularity as a clean-burning fuel source for vehicles, and IGCC is an economic solution to generating hydrogen.
- Syngas produced from coal gasification can be used for power generation in IGCC or for making fertilizers, methanol, synthetic natural gas, hydrogen, or carbon dioxide. Alternatively, syngas put through the Fischer-Tropsch process can produce diesel fuel, kerosene, jet fuel, gasoline, detergents, waxes, and lubricants.
- Underground coal gasification is not yet commercial. There are several pilot projects in North America.
- Coal is used to generate about 50 percent of the electricity in North America.
- Each person in North America uses just under four tons of coal per year.
- Gasifiers can use a dry or a wet coal feed. A dry coal feed requires about 25 percent less oxygen.
- Gasification is thought to become the driving force for the development and widespread acceptance of hydrogen-powered automobiles and fuel cells.

### CAREERS AND COURSE WORK

Courses in advanced engineering such as process engineering, chemical engineering, petroleum engineering, civil engineering, mechanical engineering, electrical engineering, as well as mathematics and physics will form the foundational requirements for students interested in pursuing careers as gasification or process engineers. Software programs specific to engineering such as CAD and AutoCAD are also essential areas of mastery. Earning a bachelor of science or bachelor of applied science degree in any of the aforementioned fields would prepare a student for graduate studies in a similar field. A professional engineer (PE) license, earned usually through fulfillment of work in a field of engineering for a prescribed number of years and successful completion of an engineering ethics examination and a comprehensive engineering examination, will facilitate career advancement. A master's degree or doctorate would equip students to pursue advanced career opportunities in industry. To obtain an upper management or executive role in industry, one would likely additionally be required to hold a management or finance degree, such as a master of business administration (M.B.A.). An advanced degree is typically not a requirement for a technician or administrator position in this field. A coal gasification engineer or manager in the field would work closely with various other engineering specialties to build, monitor, and maintain coal gasification plants. They may work with vendors and customers for procurement of equipment, feedstock, or other industrial materials and supplies.

### SOCIAL CONTEXT AND FUTURE PROSPECTS

Fuel flexibility is important in an increasingly carbon-constrained world. Future technology conceivably could accommodate economical use of a wide variety of feedstocks such as municipal waste, biomass, and recycled materials. Existing coal gasification technologies perform best on high rank, costly coal or petroleum refinery products but are inefficient and expensive to operate when using poorer-quality coal, despite being available worldwide in abundance. Advancements to bring down the cost of construction and operation of gasification plants will be important to ensure widespread use of the technology. One way to bring down the operational cost of gasification plants is to reduce the cost of oxygen used in the gasification process. Oxygen is made by an expensive cryogenic process. Research involving the use of ceramic membranes is demonstrating promising results for the separation of oxygen from air at higher temperatures. Innovations in membrane technology may enhance the utility of coal gasifiers. The development of inexpensive membranes that can readily separate hydrogen from syngas would be useful. Economical sequestration of hydrogen will help drive the use and advancement of hydrogen fuel cells and hydrogen-powered vehicles.

*Rena Christina Tabata, B.Sc., M.Sc.*

**FURTHER READING**

Asplund, Richard W. *Profiting from Clean Energy: A Complete Guide to Trading Green in Solar, Wind, Ethanol, Fuel Cell, Carbon Credit Industries, and More.* Hoboken, N.J.: John Wiley & Sons, 2008. This guidebook provides summaries on clean energy topics and informs on how to invest financially in the clean-energy market.

Bell, David A., Brian F. Towler, and Maohong Fan. *Coal Gasification and Its Applications.* Burlington, Mass.: William Andrew, 2011. Includes chapters on the fundamentals of gasification, gasification technologies, gas cleaning processes, gasification kinetics, conversion of syngas to electricity, and process economics.

De Souza-Santos, Marcio L. *Solid Fuels Combustion and Gasification: Modeling, Simulation, and Equipment Operations.* 2d ed. Boca Raton, Fla.: CRC Press, 2010. The book covers operational features of equipment dealing with combustion and gasification of solid fuels, such as coal and biomass, presents basic concepts of solid and gas combustion mechanism, introduces fundamental approaches to formulate mathematical models for gasification systems, and case studies demonstrating computer simulation.

Fehl, Pamela. *Green Careers: Energy.* New York: Ferguson, 2010. The handbook provides comprehensive coverage on the range of green careers both traditional and new age, including those that focus on conservation of resources and the development and application of alternative energy sources.

Girard, James. *Principles of Environmental Chemistry.* 2d ed. Sudbury, Mass.: Jones and Bartlett, 2010. This textbook covers the principles of environmental chemistry and is suitable for students who have a basic fluency of general chemistry.

Schaeffer, Peter. *Commodity Modeling and Pricing: Methods for Analyzing Resource Market Behavior.* Hoboken, N.J.: John Wiley & Sons, 2008. This textbook introduces readers to the economic analysis and modeling of commodity markets, including those pertaining to coal and clean-energy trade.

Taylor, Allan T., and James Robert Parish. *Career Opportunities in the Energy Industry.* New York: Ferguson, 2008. Includes career profiles of a variety of technological specialties including petroleum engineering, energy processing, and mining.

Williams, A., et al. *Combustion and Gasification of Coal.* New York: Taylor & Francis, 2000. Provides an overview of coal combustion and gasification with special attention to the properties of coal, combustion mechanism of coal, combustion in fluidized beds, the gasification process, industrial applications, and environmental impact.

**WEB SITES**

*Coal Gasification News*
http://www.coalgasificationnews.com

*Underground Coal Gasification Association*
http://www.ucgp.com

*United States Department of Energy*
Fossil Energy: Coal Gasification R&D
http://www.fossil.energy.gov

*World Coal Association*
http://www.worldcoal.org

**See also:** Chemical Engineering; Electrical Engineering; Engineering; Mechanical Engineering.

# COAL LIQUEFACTION

## FIELDS OF STUDY

Energy; chemistry; environmental science; political economy; physics.

## SUMMARY

Coal liquefaction is a catalytic process that converts different varieties of coal into synthetic petroleum by reacting coal with hydrogen gas at high temperature and under high pressure. The resultant synoil product has the potential to contribute significantly to the U.S. petroleum supply and reduce the need for the United States and other coal-producing countries to rely on imported oil.

## KEY TERMS AND CONCEPTS

- **Anthracite Coal:** Coal composed of 90 percent carbon and prized for coal-fired electrification because of its low level of impurities.
- **Bituminous Coal:** Most common form of coal, which has anthracite's high heating value but usually contains a high level of sulfur.
- **Commercialization:** Idea of developing only those technologies that can compete in the marketplace and leaving that development to the private sector, which can profit from those technologies.
- **Direct Liquefaction:** Process in which coal is subjected to hydrogenation under high pressure and at a high temperature, thereby converting it directly into a synthetic liquid fuel; also known as the Bergius process.
- **Fossil Fuel:** General term applied to any fuel created by the fossilization of plant and animal matter dating back millions of years; these fuels are coal, petroleum, and natural gas.
- **Hydrogenation:** The chemical process resulting from the addition of hydrogen to an element at a very high temperature or in the presence of a catalyst.
- **Indirect Liquefaction:** Process for converting coal into oil or synfuel by first gasifying it; also know as the Fischer-Tropsch method.
- **Pyrolysis:** Process of chemically decomposing organic materials by incineration in an oxygen-free environment.

## DEFINITION AND BASIC PRINCIPLES

Coal liquefaction involves converting bituminous or (more rarely) anthracite coal into a liquid that can be refined into the end product. The focus is on transportation fuels to lower the direct and indirect costs that oil-importing states incur in purchasing large volumes of petroleum from abroad.

The chemical processes being used in the twenty-first century favor the indirect Fischer-Tropsch method of coal liquefaction. In this process, coal is initially subjected to very high heat to create a charred substance that can be combined with carbon dioxide and steam to produce a synthesis gas composed of hydrogen and carbon monoxide. The gas is then chemically subjected to a metallic catalyst, which transforms it into a synthetic crude oil. The resultant synthetic oil can then be refined into the desired fuel.

Whatever the method, the amount of coal necessary to produce oil is about the same: Nearly three-quarters of a ton of coal is required to produce slightly over one barrel of oil. Therefore, large-scale production of oil from coal would be a massive undertaking.

## BACKGROUND AND HISTORY

Neither the idea nor the practice of producing oil from coal is new. As early as 1819, Charles Macintosh distilled naphtha from coal for the purpose of waterproofing textiles. However, the major breakthroughs in coal liquefaction did not occur until the period between the early 1900's, when the two processes long used as the starting points for producing oil from coal—the Bergius and Fischer-Tropsch methods—were developed.

Since then, political considerations have trumped economics in pushing countries into developing oil from coal on only two occasions: when Germany experienced an oil shortage during World War II and when South Africa feared the imposition of an oil embargo against its white minority government during apartheid. However, the high cost of imported oil or concern with its inadequate or insecure supply have sparked interest in coal liquefaction on four separate occasions. In 1928, Standard Oil of New Jersey and I. G. Farben joined forces to pursue a significant liquefaction project based on the Bergius process, but their venture was short-lived. No sooner was the

Standard Oil-Farben deal made than the market conditions that had brought it about disappeared. The Great Depression dampened the Western demand for oil, and the discovery of rich oil fields in Texas in 1930 meant an increase in the supply of cheap crude oil in the United States.

During World War II and shortly after, concerns over the security of the oil supply and fears that the supply would not be adequate to meet the rapid postwar expansion drove a short period of government-sponsored research into creating oil from coal in the United States. However, soon after two pilot plants were established to test the Bergius and Fischer-Tropsch processes in the postwar period, imported oil from the vast, newly developed fields of the Middle East began to arrive in such abundance that an oil glut appeared likely. In 1949, the National Petroleum Council began questioning the commercial feasibility of coal-based oil, and four years later, citing the changed economic climate, the administration of President Dwight D. Eisenhower ordered the government-owned research facilities closed.

During the next twenty years, private industry continued to conduct coal liquefaction experiments, and major petroleum corporations acquired ownership of several coal companies with coal liquefaction pilot plants. Nevertheless, with the cost of oil remaining low throughout this period, research projects tended to remain very small in scale. It was not until the oil crises of the 1970's, which drove the price of imported oil from under $3 per barrel in mid-1973 to over $35 per barrel in 1979, that a third round of interest in a commercial synthetic fuels industry began. This time, major research and development programs emerged across five continents, and both public agencies and private companies initiated multibillion-dollar projects. The outcome, nonetheless, was again the same. Within a decade of the 1973 oil crisis, the rising price of oil had generated a massive recession in the oil-importing world. Demand for oil dropped precipitously, and when the price of oil began to drop, Organization of the Petroleum Exporting Countries (OPEC) members began trying to undersell one another to gain a wider share of the shrinking market, further accelerating the downward spiral in the price of oil. By then, President Ronald Reagan had drained the Synthetic Fuel Corporation created by the U.S. Congress in 1980 of the $88 billion allocated to it for the development of an American coal liquefaction

and gasification industry, and around the world, others were following his lead.

Except for a momentary spike in the price of oil when Iraq invaded Kuwait in 1990, which generated fears that the supply of oil would be disrupted, low oil prices remained the norm from the mid-1980's until the U.S. invasion of Iraq in 2003. The growing demand for oil in China and India had already pushed the price toward $30 per barrel, but that figure was too low for investors to revisit the synthetic fuels option. However, in 2003, when the United States invaded Iraq, the gap between OPEC's production capacity and global demand for oil was closing. By the end of 2003, imported petroleum had reached more than $30 per barrel; two years later, it had doubled to $60 per barrel. Costs eventually peaked at more than $147 per barrel in July, 2008, before the high price of oil dampened demand and oil prices began to drop sharply. By then, interest in coal liquefaction as a commercial alternative to OPEC oil had been resuscitated, and several pilot projects based on the Fischer-Tropsch process were under way around the world.

## HOW IT WORKS

Although there are variations in terms of catalytic agents, the degree of heat, and the level of pressure employed, there are three broad methods for liquefying coal into oil: hydrogenation, catalytic synthesis, and pyrolysis.

**Hydrogenation.** The first hydrogenation process to develop was the Bergius process, which produces heavy crude oil directly from coal by adding hydrogen to unprocessed coal and recycled heavy oil. Hydrogenation is the central element in the Begrius process; however, unlike the Fischer-Tropsch approach, which also uses hydrogenation, the Bergius system employs a combination of various catalysts (typically tungsten, tin, or nickel oleate) to convert the carbon directly, under very high pressure and at much higher temperatures (from 700 to more than 925 degrees Fahrenheit), into oil liquids capable of being refined into industrial and heating oil and transportation fuels.

Developed in Germany by Friedrich Bergius in 1913 and sold to the German chemical conglomerate I. G. Ferben in 1924, this process was the more popular means of coal liquefaction in Germany from 1925 until the end of World War II. It was resurrected there again during the oil crises of the 1970's with the creation of a demonstration plant in Bottrop, a coal-mining center in the Ruhr region; however, that plant closed in the early 1990's.

**Catalytic Synthesis.** The liquefaction method more commonly used in the twenty-first century was also developed in Germany by Franz Fischer and Hans Tropsch in 1923. This indirect method produces oil from coal by first gasifying the hydrocarbons through a combination of complex chemical reactions involving catalytic synthesis. At the heart of the process, typically at temperatures ranging from 300 to 575 degrees Fahrenheit, carbon monoxide is mixed with hydrogen in a series of catalytic chemical reactions.

As of 2011, the Republic of South Africa's synthetic fuels industry and all confirmed liquefaction projects under way were employing some variant of this process and for understandable reasons. In a world in which liquid fuels and gas are important sources of energy, the Fischer-Tropsch process has the advantage of producing both petroleum liquids and a high volume of natural gas liquids and ethane.

**Pyrolysis.** In the pyrolysis process for converting coal into oil, high heat is applied to coal in an oxygen-free environment, decomposing it into oil and coal tar, then the oil is subjected to hydrogenation to remove sulfur and other extraneous content before it is processed into fuel sources. Substantially fewer demonstration efforts have been devoted to this process because the amount of oil produced per ton of coal is less than that yielded by the Fischer-Tropsch or Bergius method, and no major liquefaction operation at even the pilot plant level has been based on it.

## APPLICATIONS AND PRODUCTS

Because the Earth's coal reserves are substantially greater than its oil reserves, a general consensus exists within the energy industry that a liquefied coal industry will eventually emerge around the globe. That day, however, has yet to come. Hence, any discussion of the products that can emerge from such an industry must necessarily be divided into two parts: the likely applications of a future liquefied coal industry and what has occurred within the framework of the Fischer-Tropsch method in the Republic of South Africa, the one country where a significant oil-from-coal industry exists.

The development of a liquefied coal industry in the oil-importing world is likely. The United States,

---

### Fascinating Facts About Coal Liquefaction

- For six years during World War II, Nazi Germany used a network of twenty-five synthetic fuel plants, manned mostly by slave labor, that, at peak production, converted coal into the 124,000 barrels of oil per day that Germany used to power most of its war machine before the U.S. Army Air Force closed those I. G. Farben plants in 1944.

- After World War II, the United States employed German scientists to explore the Fischer-Tropsch method as a possible way of creating fuel in the United States. The project was code-named Operation Paperclip, but it collapsed when the discovery of oil in Kuwait and new pools in Venezuela ended postwar fears of an oil shortage.

- Representative Carl Perkins from Kentucky's coal-mining district, the same congressman who encouraged the federal government to explore coal liquefaction in 1946, also spearheaded the quest for synthetic fuels in the 1970's that led to the creation of the Synthetic Fuels Corporation.

- In campaigning during his failed 1980 run for the Senate from oil-producing Oklahoma, Edward Noble–who was appointed by President Ronald Reagan to head the Synthetic Fuels Corporation—pledged that if elected, he would shut down the corporation.

- In 2005, the U.S. Air Force launched a coal-to-fuel project designed to produce aviation fuel but abandoned it in 2009, when the drop in oil prices made it unlikely that the fuel could be produced at competitive prices.

- Audi has competed in the Le Mans races with R15 TDI cars powered by synthetic diesel fuel produced using the Fischer-Tropsch method.

---

for example, throughout the first decade of the twenty-first century, was persistently importing 60 percent or more of its petroleum needs. Mid-decade testimony before the U.S. Congress placed the cost of these imports—a direct transfer of wealth from the United States to petroleum-exporting states—at a minimum of $320 billion per year, with oil in the $80 per barrel range. Indirect costs, measured in terms of lost jobs, inflationary erosion of purchasing power, and other consequences of this transfer of wealth, were estimated at an additional $800 billion per year.

Moreover, because more than half of the petroleum used in the United States goes into the transportation sector, it is assumed that the majority of any synthetic oil produced in the United States or elsewhere in the oil-importing world will be used in that sector to achieve the target goal of reducing the amount of oil imported as much as possible.

As for South Africa, liquefied coal has figured into its overall energy equation since 1955, when the South Africa Coal Oil and Gas Corporation (SASOL) began liquefying cheap South African coal mined by inexpensive local labor into state-subsidized gasoline. That first Fischer-Tropsch-based plant was subsequently devoted entirely to producing gas from coal, but in the interim, SASOL opened two additional coal-into-oil plants that together produce nearly 200,000 barrels of oil per day. SASOL has more than half a century of experience in converting coal to liquids, and in addition to transportation fuel, it produces a variety of chemical solvents and waxes using the Fischer-Tropsch method. However, even the increase in petroleum prices after 2003 did not encourage it to expand its liquefaction operations, and South Africa still imports about two-thirds of its petroleum needs.

### IMPACT ON INDUSTRY

The U.S. coal industry is marred by collaboration between mine owners and workers in evading safety standards that, if implemented, could close mines and eliminate jobs because of their cost. Most coal goes to the production of electricity, and the cleaner coal deposits of the West, largely harvested by deep-shovel strip mining, is preferred for environmental reasons over the deep-shaft coal of the East and Midwest. A vibrant coal liquefaction industry, whatever the process used, would revive the economically depressed coal-mining areas in the East and Midwest. Consequently, it is not hard to find enthusiastic supporters for a coal liquefaction industry among those members of Congress who represent these regions, as occurred in 1979-1980 when Congress created the quasi-public Synthetic Fuels Corporation. Japan, Germany, Britain, Australia, and China's coal regions would likewise profit from a major commitment to an coal liquefaction industry.

### CAREERS AND COURSE WORK

Careers involving the development of alternative energy sources will almost certainly expand in the twenty-first century as the demand for oil inevitably continues to rise even if Western countries are able to reduce the number of internal combustion engines on their highways. With that demand, a long-term rise in the price of oil is also probable. Consequently, scientists trained in the applied fields of chemistry, earth sciences, geology, environmental science, and physics, with specialization in alternative fuel development, can expect to be in demand. Whether those specializing in coal liquefaction technologies will be swept up in that demand and find more employment opportunities, however, is more debatable. The case for pursuing coal liquefaction has altered little in a century.

On the plus side, coal remains far more abundantly available than petroleum and is found in significant deposits on every continent. Moreover, new liquefaction processes based on the Fischer-Tropsch method have reportedly reduced the cost of coal liquefaction to levels that should be competitive at the likely cost of future OPEC oil imports. On the negative side, the déjà vu quality of these arguments has prevented investors from diving heavily into coal liquefaction projects as has the developed world's swing toward more environmentally clean energy sources. By its nature, coal is a dirtier fossil fuel than oil or natural gas, and increased coal production for and processing in large volumes in synthetic oil industries carries a risk of greater air pollution and groundwater contamination. On balance, then, in at least the near future, those inclined to pursue courses in chemistry and thermal physics for future careers involving alternative energy fuels would probably be better off specializing in coal gasification or the production of oil from tar sands than coal liquefaction.

### SOCIAL CONTEXT AND FUTURE PROSPECTS

The economic assumptions driving early twenty-first century interest in coal liquefaction are that coal can be converted into oil for about $40 per barrel and that it is highly unlikely that OPEC oil will ever again drop below that figure because the production costs have become higher for some OPEC members. To this may be added a third assumption, that Western societies will understand that the social costs of importing so much oil are greater than the environmental risks involved in reducing those imports.

Moving to a large-scale synthetic oil industry, however, is not blocked by just the uncertainties surrounding the likelihood of greater environmental restrictions being placed on the industry. In addition, some experts fear that given the low production costs of OPEC producers in the Arabian peninsula (where Saudi Arabia can still produce oil at $2 per barrel), these oil producers might drop the price of oil to the point where synthetic fuels cannot compete, and coal liquefaction investors will lose—as they did in the 1980's—billions of dollars. Therefore, although several synthetic fuel projects were begun or announced between 2004 and 2010, including a Shell project in Malaysia and Shell and Marathon Oil projects in Qatar, the size of those ventures has generally been that of pilot plants. As a global commercial undertaking, coal liquefaction remains a technology in search of its time.

*Joseph R. Rudolph, Jr., Ph.D.*

## FURTHER READING

Bartis, James T., Frank Camm, and David S. Ortiz. *Producing Liquid Fuels from Coal: Prospects and Policy Issues.* Santa Monica, Calif.: RAND Corporation, 2008. Good description of the issues, written by a cautious advocate of oil-from-coal projects.

Crow, Michael, et al. *Synthetic Fuel Technology Development in the United States: A Retrospective Assessment.* Westport, Conn.: Greenwood Press, 1988. Examines the economic and political factors affecting synthetic fuel projects during the Carter and Reagan administrations.

International Energy Agency. *Coal Liquefaction: A Technology Review.* Paris: Organization for Cooperation and Development, 1982. An excellent, easy-to-understand explanation of coal liquefaction technologies and the oil-from-coal projects being pursued in Europe in the early 1980's.

Speight, J. G. *Synthetic Fuels Handbook: Properties, Process, and Performance.* New York: McGraw-Hill, 2008. An easy-to-follow explanation of the costs and benefits of various synthetic fuel technologies, including coal liquefaction.

Toman, Michael, et al. *Unconventional Fossil-based Fuels: Economic and Environmental Tradeoffs.* Santa Monica Calif.: RAND Corporation, 2008. A rather upbeat view of the potential cost competitiveness of oil derived from coal and oil sands, combined with a hard look at the economic benefits versus environmental costs of developing these sources.

Yanarella, Ernest J., and William C. Green, eds. *The Unfulfilled Promise of Synthetic Fuels: Technological Failure, Policy Immobilism, or Commercial Illusion.* New York: Greenwood Press, 1987. A one-stop reader on the politics and technology of synthetic fuels at the time of the last big push to develop liquefied coal in the United States and abroad.

Yeng, Chi-Jen. *Belief-Based Energy Technology in the United States: A Comparative Study of Nuclear Power and Synthetic Fuel Policies.* Amherst, N.Y.: Cambria Press, 2009. An excellent study of the importance of political clout as well as technological feasibility in explaining energy choices in the American political process.

## WEB SITES

*Coal Utilization Research Center*
Clean Coal 101
http://www.coal.org/clean_coal_101/index.asp

*U.S. Department of Energy*
National Energy Technology Laboratory, Clean Coal Demonstrations
http://www.netl.doe.gov/technologies/coalpower/cctc/cctdp/bibliography/misc/bibm_cl.html

**See also:** Chemical Engineering; Coal Gasification; Gasoline Processing and Production.

# COASTAL ENGINEERING

## FIELDS OF STUDY

Marine/coastal science; marine biology; civil engineering; oceanography; meteorology; hydrodynamics; geomorphology and soil mechanics; numerical and statistical analysis; structural mechanics; fluid mechanics; electronics; geology; chemistry.

## SUMMARY

Coastal engineering is a branch of civil engineering that aims to solve coastal zone problems such as shoreline erosion and the destruction caused by tsunamis through the development, construction, and preservation of mechanisms and structures. Coastal engineering also entails understanding the theories and processes of wave actions and forces, ocean currents, and wave-structure interactions. Given that coastal areas are often highly populated and especially vulnerable to human impact, coastal engineering attempts to blend conservation and management of the world's coastal zones with the development requirements of humans.

## KEY TERMS AND CONCEPTS

- **Accretion:** Buildup of sediment (usually sand or shingle) through natural fluid flow processes on a beach or coastal area.
- **Estuary:** Semi-enclosed body of water with a mouth to the ocean; contains a mixture of fresh water from inland and salt water from the sea.
- **Hard Coastal Engineering:** Use of conventional tools and hard structures, including groins, seawalls, and breakwaters, to mitigate coastal erosion and flooding.
- **Land Reclamation:** Process of creating dry land by taking an area of land below sea level, pushing back the water, and preventing the water from returning with structures such as dykes, or introducing sediment and material into the sea, raising the level of the seabed.
- **Littoral (or Longshore) Drift/Transportation:** Movement of material or sediment (usually sand or shingle) along the coastline, caused by waves and currents.

- **Longshore Current:** Current formed or driven by waves breaking obliquely (diagonally) to the shore; flows in the same direction as the wave.
- **Shingle:** Coarse gravel, such as beach pebbles.
- **Soft Coastal Engineering:** Use of long-term, naturalistic tools, including beach nourishment and artificial dunes, to manage coastal areas.
- **Tsunami:** Very large and potentially damaging oceanic wave caused by underwater seismic activity, such as an earthquake or volcanic eruption.

## DEFINITION AND BASIC PRINCIPLES

Coastal engineering is a branch of civil engineering involved in the development, design, construction, and preservation of structures in coastal zone areas. Its primary function is to monitor and control shoreline erosion, design and develop harbors and transport channels, and protect low-lying areas from tidal flooding and tsunamis. To achieve such objectives, coastal engineers must have a strong understanding of the sciences of engineering, oceanography, meteorology, hydrodynamics, geomorphology, and geology. They must also have a strong understanding of the theories and processes of wave motion and action, wave-structure interaction, wave-force forecasting, and ocean current prediction.

Coastal areas are socially, economically, and ecologically crucial. They are home to a significant proportion of the world's human population, serve as important breeding grounds for many animal species, and are important for tourism, aquaculture, fishing, transport, and trade. These multiple uses of a relatively small area have led to rapid environmental degradation and have created conflict and the need to achieve a sustainable balance.

## BACKGROUND AND HISTORY

Although the specialty of coastal engineering emerged only in the latter half of the twentieth century, coastal engineering and management have been practiced for many centuries. Human beings have long lived in coastal zones and, for just as long, have attempted to control and manage the effects of tidal flooding and wave action and to engage in land reclamation. People have developed ports to make transportation and trading easier for thousands of years. The first river port—port of A-ur on the Nile—was built before 3000 B.C.E. and the first coastal port—the port on Pharos near Alexandria—about 2000 B.C.E. For centuries, many countries wanting to protect strategic coastal areas used structures such as embankments as sea defenses.

Early engineering projects mainly involved hard coastal engineering, which is the design and building of docks, ports, defenses, and walkways. Although coastal engineering has long been practiced, the actual coastal processes and their driving forces were little understood. Therefore, formal engineering development required much more comprehensive knowledge of the ocean and the processes and principles involved.

Traditionally, engineering projects in coastal zones were the province of civil and military engineers. However, because of the growing need for specific research in coastal engineering, the first International Conference on Coastal Engineering was held in 1950 in California, establishing the branch of applied science and introducing the terminology. The conference proceedings stated that coastal engineering was not a novel science or unconnected to civil engineering but rather was a branch of civil engineering that was strongly reliant on and influenced by sciences such as oceanography, meteorology, electronics, and fluid and structural mechanics.

By the 1960's, significant inroads had been made into understanding coastal processes and this, coupled with advances in modeling, highlighted the importance of incorporating such processes into design and construction. By the 1970's, the increase in understanding coastal processes led to the introduction of soft coastal engineering, which included beach nourishment and artificial dunes. In the following decades, significant progress was made in the development of coastal engineering techniques, such as modeling techniques for deep-water wave prediction, and the understanding of processes, including wave-structure interaction and coastal sediment erosion.

Essentially, coastal engineering has transformed from a science that simply responds to anthropocentric demands (docks, ports, and harbors) to one that aims to balance human requirements and environmental protection, with a particular focus on mitigating both natural and human-induced coastal erosion.

## HOW IT WORKS

It is the job of a coastal engineer to recognize the characteristics and natural processes of the shoreline environment and to apply fundamental engineering theories and philosophies to solving ecological problems. To be successful, a coastal engineer must have a strong understanding of other sciences such as oceanography, meteorology, geomorphology, structural and fluid mechanics, and hydrodynamics. In particular, this means clarifying and using theories of wave action, wave-structure interaction, wave-force forecasting, ocean current prediction, and beach profile modification.

Coastlines across the world are the natural division between the terrestrial landscape and the seas and oceans. The geological composition of these areas is distinctive, as are the natural and anthropomorphic processes that affect them. Coastal engineering techniques must take into account the morphological development of coasts, such as erosion and accretion, and relate this to shore protection through engineering approaches and structures.

The action of waves on the shoreline forms one of the fundamental areas of coastal engineering research. Waves contain a very large amount of energy formed by the action of wind over vast tracts of open water, the gravitational attraction of the Sun and Moon, and occasionally seismic activity (as with tsunamis). Although this energy is collected over very large areas, it is released along relatively small areas of coastline. This release of energy, in the form of breaking waves, has a strong influence on the currents, sediment, and geological structures of coastal areas. The shoreline or beaches, which are frequently composed of sediment such as sand, shingle, or gravel, are constantly battered and reshaped by the actions of wind and water.

Erosion is a natural process, and the movement of sediment is achieved by wave uprush, which transports sand onshore, and backwash, which transports sand offshore. The natural processes of erosion in coastal areas are very complex and involve flow fields created through not only the action of breaking waves but also erratic turbulent sediment transport in the water column and a shifting shoreline. Worldwide research aims to develop predictive models of this erosion process. Over time and under normal conditions, erosion is also a relatively cyclical process. Sediment can be carried offshore during one season and dumped back onshore during the next season, as well as moved obliquely along the shore over time. Erosion processes can, however, be significantly disrupted by human activity and structures or can be considered undesirable because of human requirements and desired aesthetics. Coastal engineering structures can be, in fact, both detrimental and beneficial in regard to coast erosion, and most coastal engineers believe that only an integrated and holistic approach to planning and design can generate long-term sustainability.

Within this energetic and process-driven natural boundary between land and sea, people have constructed coastal engineering structures such as ports and harbors, which have had both positive and negative effects on the environment. The design of such structures must predict wave dynamics and their impact on individual coastal zones and beach environments. Since the emergence of coastal engineering science, much greater emphasis has been placed on understanding coastal processes, such as wave action, and designing and developing policies and structures to protect coastal areas from erosion.

Although the study of coastal erosion is a fundamental concept for coastal engineering, coastal engineers must also recognize and study coastal protection measures and applications such as hard and soft shoreline protection structures, understand the effects of these structures on the morphology of the coastal areas, and develop effective coastal zone management plans and policies.

## APPLICATIONS AND PRODUCTS

Coastal engineering structures and activities vary significantly from country to country. They depend on social aspects, such as the country's history and development and its people's relationship to the ocean, and environmental and structural aspects, such as the nature of the ocean along the country's coast, geomorphological and geological conditions, the ecosystem, climate, extent of tectonic activity, currents, and wave action.

Generally speaking, however, there are two types of coastal engineering applications: hard and soft coastal engineering. Although hard stabilization techniques have been used for many years and are considered to be appropriate under certain circumstances, they can also be expensive and, in trying to rectify human impact, can actually disrupt and

destroy natural shoreline processes and habitats, intertidal areas, and wetlands. In some cases, they can even increase erosion. As coastal engineering has evolved and become more sophisticated, engineers have moved away from constructing hard structures with little understanding of their impact on the environment and toward incorporating soft engineering and minimizing the ecological impact.

**Hard Structures and Applications.** Groins are solid structures, usually constructed from wood, concrete, or rocks, running perpendicular from the foreshore into the water (under normal wave levels). They aim to disrupt water flow and reduce the transportation of sediment from longshore drift. They trap sediment to extend or preserve the beach area on the up-drift side and reduce erosion on the down-drift side. The size and length of these structures is usually determined based on specific local conditions, including wave dynamics, beach slope, and environmental factors. Seawalls are very large, rigid, and usually vertical structures constructed from concrete. They are found at the transition between the low-lying beach and the higher mainland and run parallel to the shoreline. The main function of a seawall is to preserve the shoreline by preventing additional shoreline erosion and recession during direct wave-energy impact and flooding. Seawalls are, however, not effective in preventing longshore erosion.

Revetments are a concrete, rock, or stone "veneer," or facing, constructed on a beach slope with the aim of preventing erosion caused by wave action and storm surges. Unlike seawalls, which can assist in flood prevention, revetments are not usually constructed for this function. Revetments are always built as sloping structures, and although they may sometimes be completely solid and rigid, they are often built with interlocking slabs or stone and designed to be permeable. This permeability tends to enhance the strength of the revetment and the absorption of wave energy while reducing erosion and wave run-up. Revetments can also be made from sand-filled bags, interlocking tires, concrete-filled bags, and wire-mesh stone-filled gabions (sunken cylinders filled with earth or stones). As revetments tend to be passive structures, their application is often limited to areas that are already protected in some other way or by some other engineering structure.

Breakwaters are fixed or floating structures built parallel or at an angle to the shoreline. The main function of a breakwater is to protect the shore and activities along the coastline by reducing the impact of wave energy. The ability and impact of a breakwater depend on whether it is submerged or floating, its distance from the shoreline, its length and orientation, and if it is solid or segmented. The four main types of floating breakwaters are box, pontoon, mat, and tethered float; the three main types of detached breakwaters are offshore, coastal, and beach.

**Soft Structures and Applications.** Shore nourishment has become one of the most common soft coastal engineering applications. The three main types are backshore, beach, and shoreface nourishment. As the name suggests, nourishment is the action of artificially adding sand to the backshore (upper part of the beach), beach, or shoreface (usually the seaside of the bar) in an attempt to modify the effects of erosion. Although nourishment replaces sand in an eroded area and is considered to be a rather natural form of coastal engineering, it does not address the causes or processes of erosion nor does it reduce the impact of wave energy.

Sand dune stabilization is a relatively basic and common form of soft coastal engineering. It involves the planting of vegetation to stabilize and protect sand dunes along the coastline. The creation of artificial dunes is also considered to be an effective form of soft coast engineering protection, particularly in conjunction with shore nourishment.

Beach drainage, or beach dewatering, is a system of shore protection based on the concept of physically draining water from a beach. Drainpipes are installed and buried below the beach and parallel to the shoreline. These pipes collect seawater and transport it to a pumping station, where it is collected and either returned to the ocean or used. Beach drainage helps reduce erosion by lowering the watertable in the uprush zone and reducing the force of the backwash by increasing the volume of water that seeps into the beach.

## IMPACT ON INDUSTRY

The field of coastal engineering is being developed and researched by many countries and organizations all over the world. A definite shift from hard to soft engineering has occurred, and in some areas, hard engineering has been opposed and even actively prohibited. Many organizations and government agencies have become actively in-

volved in coastal engineering research, particularly in holistic and soft approaches to coastal zone management.

**Major Organizations.** Because of the global interest in coastal management and the rather ubiquitous nature of coastal areas, many international organizations are interested in and are actively conducting research and contributing to the field of coastal engineering. Many countries have specific organizations that represent the specialist interests of coastal engineers, such as the National Committee on Coastal and Ocean Engineering (NCCOE), found within Engineers Australia. The NCCOE states that its priorities are to improve management of the coastal zone through understanding of the coastal environment, to develop strategies for hazards and risk in the coastal zone, and to establish a national coastal and near-ocean data program integrating advanced technologies such as satellite, airborne, and shore-based remote sensing.

**Government and University Research.** Many world-wide governments have a stake in the management of coastal areas. Numerous governments have increased their interest in monitoring and managing the land-sea interface through such methods as engineering projects, involving themselves not only in the economics of a project but also in its environmental aspects. Coastal engineering is important to the continued survival and sustainability of the world's shorelines. Although governments are increasingly aware of the importance of coastal engineering, many universities and organizations state that the funding of academic research in the field is still inadequate and affects the ability of researchers to conduct effective and extensive studies. Programs in coastal engineering are offered at many universities, including the University of California, Berkeley, the University of Florida, Oregon State University, the University of Queensland, Tokyo University, the University of Nottingham, and the Technical University of Denmark.

In many cases, no single federal department or agency is solely responsible for funding and managing coastal engineering research and education. In the United States, academic funding can come from many sources, including the National Science Foundation, the Sea Grant program of the National Oceanic and Atmospheric Administration (NOAA), the Office of Naval Research, and the U.S. Army Corps of Engineers. The NOAA is one of the world's largest governmental agencies involved in ocean management. The NOAA's Office of Ocean and Coastal Resource Management (OCRM) provides management and guidance for national, state, and territory coastal programs and research. The six divisions of the OCRM implement research and engineering techniques for shoreline protection. The Coastal Engineering Research Center, which was established in 1963 and is part of the U.S. Army Corps of Engineers' Coastal and Hydraulics Laboratory, is also at the forefront of coastal engineering research both nationally and internationally.

## CAREERS AND COURSE WORK

Many universities offer undergraduate and graduate degrees in coastal engineering. Most commonly, students who wish to pursue a career in coastal engineering will study engineering or marine science as an undergraduate. By graduation, students should have a solid understanding of coastal engineering concepts, theories, processes, and practices, such as wave action, wave transformation, statistical and numerical analysis, tides and currents, beach dynamics and coastal structures, and the impact of natural and anthropogenic factors on the coastal environment. In addition, as coastal engineers need to provide the scientific base data used for coastal zone management, they are likely to require knowledge and experience in the use of modeling.

Students who study coastal engineering can enter such careers as consulting engineers, project managers, environmental consultants, and construction contractors and such fields as construction management, coastal and oceanographic engineering, water resource management, and hydrologic engineering. They may find employment in the private sector, with nongovernmental organizations, with specialized government organizations and agencies, or in universities undertaking teaching and research.

## SOCIAL CONTEXT AND FUTURE PROSPECTS

The social, economic, and ecological consequences of coastal erosion are significant. Almost 60 percent of the world's population lives within 100 kilometers of a coast. Coastal areas not only provide significant economic benefits—fish and other maritime products, a means of transportation, and easier access to trading partners—but also offer numerous recreational opportunities. However, the economic

---

### Fascinating Facts About Coastal Engineering

- Almost 650 million people living in low-lying coastal areas are at risk from the rising sea levels and associated erosion that are likely to be caused by climate change.
- Neskowin, a coastal town in northern Oregon that sits at 20 feet above sea level, began experiencing severe erosion of its beaches in the 1990's because of the rising sea level and high waves. Some relief has been provided by riprap (a stack of boulders) on the beach, but the future is uncertain.
- In 2008, a 45-foot-tall sheet pile ring and a rock apron were placed around the base of Morris Island Lighthouse to prevent further erosion. The lighthouse, built in 1875, was originally on high ground one-half mile from the Atlantic Ocean but has come to be located 2,000 feet offshore in 10 feet of water.
- The barrier island of Grand Isle, Louisiana, has experienced severe erosion problems since the 1980's. Numerous attempts to stop the erosion have had mixed results. In 2010, an artificial oyster reef was installed on the island's backside in an effort to end erosion.

---

and aesthetic appeal of coastal areas has created population pressure that has greatly increased the need for functional coastal structures, such as ports, wharfs, and jetties, and infrastructure, such as roads and sewerage facilities, which increase erosion and pollution. The erosion caused by human activities and structures can change beach dynamics and profiles, which in turn can have ecological effects, such as the loss of animal breeding grounds, and social effects, such as making the area unsafe for swimming and other recreational activities.

Research and climate change modeling has indicated that sea levels may rise significantly in the twenty-first century. This, coupled with an associated increase in coastal storm frequency and strength, will produce more floods and greater erosion, creating serious ecological repercussions for the coastal zones of the world, which are often densely populated and popular areas for tourism. The construction of such engineering structures as ports, harbors, recreational facilities, and resorts are considered necessary for continued economic and social development. Such engineering has, however, a very significant ecological impact on the coastal zone areas, particularly an increase in erosion. One of the primary goals in coastal engineering is to rectify erosion issues.

The difficulty for the future of coastal engineering is that some of the solutions to such environmental or aesthetic issues can be ineffective or ecologically unsound, thereby exacerbating the problem. As such, the future of coastal engineering requires integrated coastal zone management (ICZM), a holistic and integrated strategy and approach incorporating hard and soft engineering for the management of all aspects of coastal areas. It incorporates advances in modeling that are helping develop a framework for predicting coastal erosion hazards and processes. Many coastal engineers are advocating an integrated approach, stating that the future ecological and economic sustainability of the world's fragile and vulnerable coastal areas depends on its adoption. Such a strategy is optimal, as it considers such issues as climate change, the rise in sea level, navigational needs, the impact on plants and animals, and aesthetic considerations.

*Christine Watts, Ph.D., B.App.Sc., B.Sc.*

## FURTHER READING

Dean, Robert G., and Robert A. Dalrymple. *Coastal Processes with Engineering Applications.* 2002. Reprint. New York: Cambridge University Press, 2004. Provides comprehensive information on coastal processes and management of shoreline erosion as well as an overview of the topic, the hydrodynamics of the coastal zone, and coastal responses to erosion and engineering applications.

Kamphuis, J. William. *Introduction to Coastal Engineering and Management.* 2d ed. London: World Scientific Publishing, 2009. Aimed toward undergraduate and graduate students, it discusses traditional and contemporary issues, methods, and practices in coastal engineering design, as well as environmental issues and climate change.

Kraus, Nicholas, ed. *History and Heritage of Coastal Engineering: A Collection of Papers on the History of Coastal Engineering in Countries Hosting the International Coastal Engineering Conference 1950-1996.* New York: American Society of Civil Engineers, 1996. Documents the history of coastal engineering in fifteen countries, including the United States, Canada, Australia, France, Denmark, Germany, South Africa, and Great Britain.

Reeve, Dominic, Andrew Chadwick, and Christopher Fleming. *Coastal Engineering: Processes, Theory and Design Practice.* New York: Spon Press, 2004. A comprehensive examination of coastal processes, morphology, design, and the effect of engineering structures for coastal defense, including modeling techniques and case studies.

Shibayama, Tomoya. *Coastal Processes: Concepts in Coastal Engineering and Their Application to Multifarious Environment.* Hackensack, N.J.: World Scientific Publishing, 2009. Designed for coastal engineering graduate students. Describes wave-induced problems, coastal processes, sediment transport, and coastal disasters.

Sorensen, Robert. *Basic Coastal Engineering.* 3d ed. New York: Springer, 2006. Provides students of coastal engineering with the basics on coastal processes and wave mechanics, theoretical and applied hydromechanics, and engineering design and examination.

Young, Kim, ed. *Handbook of Coastal and Ocean Engineering.* Hackensack, N.J.: World Scientific Publishing, 2010. Presents a thorough compilation of topics, such as wave and water fluctuations, coastal and offshore structures, sediment and erosion processes, modeling and environmental issues from more than seventy international authorities and experts on coastal engineering.

## WEB SITES

*National Oceanic and Atmospheric Administration*
Ocean and Coastal Resource Management
http://coastalmanagement.noaa.gov

*U.S. Army Corps of Engineers*
Coastal and Hydraulics Laboratory
http://chl.erdc.usace.army.mil

**See also:** Civil Engineering; Climate Modeling; Earthquake Engineering; Erosion Control; Flood-Control Technology; Fluid Dynamics; Ocean and Tidal Energy Technologies; Oceanography; Water-Pollution Control.

# COATING TECHNOLOGY

## FIELDS OF STUDY

Materials engineering; chemical engineering; chemistry; physics; high-efficiency devices; corrosion protection; thermal protection; resource management; optics; biotechnology; nanotechnology.

## SUMMARY

A coating is a thin layer or film of a substance spread over a surface for protection or decoration. Coatings can significantly improve the performance of technology. Applications of coatings are far ranging, from corrosion protection of metals in vehicles to thermal protection in jet engine turbine blades. Functional coatings are also used to generate electricity in fuel cells and photovoltaic cells and can reduce thermal emissions from buildings through windows. There are numerous methods of applying these coatings, and each satisfies a different demand.

## KEY TERMS AND CONCEPTS

- **Capacitor:** Storage device for electric charge.
- **Cataphoretic Dip Coating:** Electrophoretic process that coats substrates such as cars in a complete, adhesive, protective film.
- **Cathodic Protection:** Method by which an electrochemically inferior material is applied to a metal substrate, preventing the corrosion of the substrate.
- **Chemical Vapor Deposition:** Variety of deposition methods to create thin films by chemical reactions on substrates.
- **Colloid:** Small particles dispersed within a continuous medium in a manner that prevents them from agglomerating, settling, or being filtered.
- **Corrosion:** Chemical or electrochemical reaction of a material with its surroundings, leading to the disintegration of the material and the formation of new, often undesired materials, such as rust.
- **Creep:** Form of permanent plastic deformation in metals at elevated temperatures.
- **Dip Coating:** Method by which a substrate is fully immersed in a liquid either to chemically alter the substrate surface or to deposit a coating.

- **Doping:** Method by which tiny trace amounts of one element are mixed with another element, thereby significantly changing the properties of the resulting compound.
- **Electrophoresis:** Migration of charged colloidal particles or molecules through a conductive solution under the influence of an applied electric field.
- **Galvanization:** Method of applying a sacrificial anode such as zinc on the surface of a different metal to protect the latter from corrosion.
- **High Velocity Oxy-Fuel Process (HVOF):** Mixture of a combustible gas (such as propane) and oxygen ignited inside a combustion chamber to melt a coating material before it is accelerated via an inert carrier gas at high speeds toward a surface, where the hot material forms a coating.
- **Hydroxyapatite:** Naturally occurring mineral form of calcium apatite with the chemical formula $Ca_{10}(PO_4)_6(OH)_2$.
- **Line-Of-Sight Process:** Coating that is applied only on the surfaces that face the emission source of the coating material.
- **Non-Line-Of-Sight Process:** Coating that is applied equally to all surfaces of a material, even inside convoluted shapes.
- **Physical Vapor Deposition:** Methods of using low pressure to deposit thin films by the condensation or deposition of volatilized materials.
- **Proton Exchange Membrane Fuel Cell:** Polymer-based electrochemical device that converts various fuels and air into water, heat, and electric energy at low operating temperatures (70 to 140 degrees Celsius).
- **Refraction:** Turning or bending of any wave, such as light or sound, when it passes from one material into a distinctly different material.
- **Slurry:** Thick suspension of solids in a liquid that may separate and can be filtered.
- **Sodium Hypophosphite:** Catalytic agent in nickel plating that reduces dissolved metal ions and deposits them as metal atoms on a substrate surface; chemical composition is $NaPO_2H_2\text{-}H_2O$.
- **Solid Oxide Fuel Cell:** Ceramic electrochemical device that converts various fuels and air into water, heat, and electric energy at high operating temperatures (600 to 1,000 degrees Celsius).

- **Spallation:** Delamination or de-bonding of one section of a material from another.
- **Stent:** Metal cylinder that can be moved inside a clogged blood vessel, where it is permanently inflated and remains in place, allowing blood to circulate once more.
- **Thermal Barrier Coating:** Porous coating applied to turbine blades, reducing the temperature between the combustion chamber and the metal of the blade.
- **Thermal Spraying:** A variety of methods involving elevated temperatures to deposit materials on a surface by first melting or vaporizing them and then accelerating them toward the substrate.

## DEFINITION AND BASIC PRINCIPLES

Coating technologies have been applied to the surface of materials for centuries. Corrosion of steels can cause structural degradation in buildings and vehicles. When exposed to the atmosphere, metals can corrode, forming ceramic materials such as oxides, which are much more brittle, and can lead to the mechanical failure of infrastructure. Coatings can significantly alter the performance of the substrates to which they are applied. Coating the surfaces of metals can prevent or at least delay the onset of detrimental corrosion, but modern coating technologies can do much more than that. Some examples of coatings

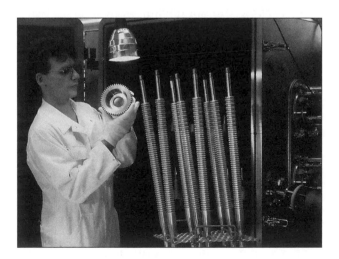

*Technician inspecting cogs for use in cars, after they have been coated with a thin film using a physical vapor deposition process.* (Rosenfeld Images Ltd/Photo Researchers, Inc.)

on polymers and glasses include antireflective and ultraviolet protection on sunglasses as well as thermochromic windows, which reduce heat influx into buildings in summer and prevent heat emissions in winter. Electrochemically active thin coatings can act as photovoltaic cells, super capacitors, fuel cells, and as electrodes of batteries. Thin films of the right chemical composition and microstructure typically increase the efficiency of the device in which they are used. Patterned corrosion-protection coatings applied to semiconductors in photolithography prevent chemical etching and determine the size, shape, and functionality of the semiconductor device. Magnetic coatings are used as data-storage devices. Ceramic-glaze coatings create colors on pottery pieces. Biomimetic coatings can prevent surgical implants from being attacked by the human immune system. Coating technologies can be physical, involving some type of melting or volatilization of a source material followed by deposition and solidification on a surface, or the solution or suspension of a coating material in a solvent, followed by the evaporation of the solvent after the deposition. Alternatively, the coating method can involve a chemical reaction that binds the coating to the substrate or chemically alters the substrate surface.

## BACKGROUND AND HISTORY

Early humans developed pottery methods that still exists. By heating specific ceramic particles deposited from slurry on the surface of prefered clay, the pottery develops a dense glaze. This technology is one of the earliest surface-coating technologies. Paint, which was initially developed for artwork, has also been used during the last centuries as a protective coating.

Infrastructural corrosion of metallic structures in the United States alone amounts to an estimated $276 billion, or 4 percent of the United States gross domestic product, in damages each year. Coatings have been applied to infrastructure and machinery for hundreds of years in order to prolong the useful lifetime of the equipment. Early coatings included the application of paint, which often contained metal-oxide materials. Paints usually involve particles suspended in a colloidal solution. Once the paint is applied to a surface, the solvent evaporates over time, leaving behind a coating. These coatings often did not stick well to their substrates and were often permeable to the environment, leading to their flaking off. In the early 1820's, the English chemist Sir Humphry Davy discovered that combining steels with easily corroding materials such as zinc results in significantly reduced corrosion rates—at the cost of corroding the sacrificial zinc.

Modern coating technologies have been developed as a result of the need to have devices with higher efficiencies and longer lifetimes. Although vehicles produced until the 1990's still exhibit an affinity to form rust, modern vehicles are produced with improved corrosion-protection coatings and rarely rust.

## How It Works

Coatings add an additional thin film between the surface of a material and the environment with which the material comes in contact. They can be applied as a design feature or with specific material properties in mind.

To apply a coating, the correct method of deposition has to be selected. Vehicle corrosion is addressed by applying several functional coatings, usually in a rapid, high-quality process called dip coating, in which the entire car components are immersed in an cataphoretic bath, which ensures complete, dense, and continuous coatings of all surfaces inside the structure. Further coatings may include color pigmentation and dyes, as well as a clear surface finish. Coatings can also be applied to much smaller substrates than vehicle parts. Small pigments such as mica and silica, with a size of a few micrometers, can change color if various oxide layers are applied by slurry casting to their surfaces, since the refraction of light is changed because of these surface oxide layers. For example, the chemical company Merck manufactures such coatings under the name Ronasphere for cosmetics and vehicles.

Whereas corrosion-protection coatings are supposed to be dense and not swell from water intake, other coatings must intentionally be porous. For example, the gas turbine engines, such as those in airplanes, can be made more efficient by increasing the temperature inside the combustion chamber. However, the turbine blades are rotating at a high velocity and high temperature, and consequently the metal surface of the blades may oxidize, and some deformation may occur because of creep. To prevent these detrimental effects, a 0.4-millimeter thin, porous layer of yttrium-stabilized zirconia (a type of

beach sand) is applied to the surface of the turbine blades. This is usually performed in an open environment using a hot deposition gun that is capable of melting even ceramic zirconia, which has a melting point above 2,750 degrees Celsius. Heat is delivered by plasma spraying and high velocity oxy fuel, in which the ceramic material passes at high velocities through a flame to impact in liquid form and solidify on the surface of the turbine blades. The resultant coating is very porous, which is quite advantageous. Different materials expand at varying rates when heated, and dense ceramic coatings may develop such high stresses on the surface of metals, that they spall when exposed to varying temperatures. In porous coatings, on the other hand, these stresses are dispersed due to the porosity, and the coatings remain on the surface of the turbine blades. All these thermal-spraying methods are line-of-sight methods and cannot consequently be used if intricate three-dimensional structures that cannot be evenly positioned into the incoming material stream are to be coated evenly.

Similar hot spraying methods can also be used to create multiple graded functional layers in electrochemical devices such as batteries, fuel cells, solar cells, or capacitors. For example, the National Research Council Canada has developed a reactive spray deposition method, in which metals are dissolved in an organic solvent, vaporized at elevated temperatures, and sprayed through a short flame on various substrate surfaces as a deposition straight from the gas phase. This method allows very good control of the microstructure of the functional layers.

For high-quality thin film coatings, physical vapor deposition methods have been developed that volatilize materials under extreme high vacuum conditions from high-purity emission sources referred to as targets. The growth rates of coating layers resulting from such a deposition are slow, and it may take several hours to form a layer with a thickness of even one micrometer. However, the layers are very clean, pure, dense, and thin. Even something as simple as a potato chip bag can have a thin metallic coating on the inside to improve the quality of the sealed products.

Chemical vapor deposition is another method to deposit well-adhered thin layers. Here, a chemical reaction occurs on the usually heated surfaces of the substrates, and the desired coating forms. Some of these deposition methods involving chemical reaction may include harsh chemicals such as hydrofluoric acid.

Biomedical coatings can have several functions. First and foremost they must prevent the human body from rejecting an implant. Furthermore, implants have coatings that either reduce the growth of scar tissue in soft tissues or aid in tissue growth inside bones, facilitating a solid attachment between bones

---

## Fascinating Facts About Coating Technology

- Thermal barrier coatings on gas turbine blades are only about 100 micrometers thick, but they can reduce the temperature from the combustion chamber by 200 degrees Celsius so that the metal underneath does not degrade.

- Copper and zinc have been shown to be toxic to some aquatic animals and plants that would typically befoul parts of vessels in contact with the water. However, when in contact with steel, zinc may act as a sacrificial anode and protect the steel from corroding, thereby corroding much faster itself. With less zinc present, the vessels become more sustainable to befouling.

- Metallic coatings inside some chips bags aid in keeping the inert atmosphere from the packaging inside the bag and prevent the chips from touching the polymer of the bag directly.

- Some deposition methods involve high-temperature plasmas that can exceed 16,000 degrees Celsius. These methods can melt any material, even ceramics, very rapidly, to form a coating.

- A coating of hydroxyapatite fewer than 10 micrometers on the surface of heart stents can significantly reduce inflammation and reduce the risk of biofilm formation inside the blood vessel.

- Thinner electrolytes in electrochemical devices mean shorter pathways that the charge carriers have to travel. Consequently, significantly improved batteries, fuel cells, solar cells and super capacitors result from the use of thin coatings.

- Most clothing has been coated multiple times—in color dyes, dye leach protection, insect repellants, ultraviolet radiation protection for the fibers and the wearer, stain repellant nonstick coatings, and antistatic coatings. It is, therefore, always advantageous to wash clothes before wearing them for the first time.

---

and the implant and consequently a longer lifetime of the implant. In heart surgery, for example, very thin (less than 0.01 millimeter), drug-laced hydroxyapatite coatings on stents can be used to create devices that will not become covered in human tissue, thus reducing the risk of difficult follow-up operations and replacements.

### APPLICATIONS AND PRODUCTS

**Corrosion Protection.** One of the largest application areas of coatings is corrosion protection. As a result of corrosion, fatal accidents can occur. For example, vessels that are under pressure or carry corrosive and abrasive liquids may rupture and corroding bridges may collapse. In 2010, a small oil pipeline in Wyoming ruptured because of corrosion, resulting in a spill of 85,000 gallons of crude oil. To prevent the corrosion of infrastructure, protective coatings are applied to the metal surfaces. In the production of steel, alloying elements such as nickel or chromium can be added to the liquid metals. Once exposed to oxygen, they form stable, protective oxide-layer coatings, which are very thin and invisible to the naked eye. This kind of metal is consequently called stainless steel. Although stainless steel is common, the alloying elements are more expensive than the iron-containing raw materials, and they are less often used in structural applications or in rebars within reinforced concrete. The most commonly produced stainless steel is called type 304; it contains 18 percent chromium and 8 percent nickel.

Once a type of steel has been produced, the metal surface can be protected by coatings that are applied externally. A method that was developed by French engineer Stanislas Sorel in 1837 is called galvanization. It involves dipping a metal part into a hot bath of zinc solution, leading to the formation of a visible zinc coating on the surface. Although steels are protected by this method, they degrade fast in corrosive environments—salted roads in winter, the ocean. An improved method to protect steels from corrosion results by cataphoretically coating the substrates. The entire substrate is hereby immersed in a conductive aqueous solution, and a dense, protective coating precipitates on the substrate surface due to an applied electric field. The German engineering firm Dürr works with many different industries worldwide using this coating method.

Surfaces can also be coated without involving electricity. Electroless nickel plating, for example, involves pretreating the surface of any material, including nonconductive materials, with a catalyst such as sodium hypophosphite. This treated surface is then immersed in a heated nickel-phosphorous or nickel-boron solution. The metal ions from the solution are reduced to metal in contact with the catalyst and form a dense alloy layer on the treated surface.

**Biomedical Coatings.** Blood vessel stems are often-used implants. Early implants were plagued by inflammation around the implant and a resulting growth of tissue around the implant as a defense mechanism of the body to isolate the implant from the body. Companies such as MIV Therapeutics use the application of thin ceramic hydroxyapatite coatings to facilitate a better uptake of the implant in the body. These coatings are porous and can also be used to release anti-inflammatory drugs into the wound, locally and long term, without the need to medicate the entire patient with high doses of potentially dangerous drugs. In places where cellular growth is desired, such as in bone scaffolding and artificial joints, the outsides of the material in contact with the bone are coated with porous materials, such as stainless steel or titanium foams or beads that match the three-dimensional structure of the bone. It has been shown that these surfaces are overgrown and integrated into the bone much more easily than smooth metallic surfaces and constitute a significant improvement in implant lifetime. Because this is a large global market, there are a significant amount of international companies involved. Johnson & Johnson operates a subsidiary in North America for these implants called DePuy Orthopaedics. Other North American manufacturers include Stryker Orthopedics, Wright Medical Technologies, and Biomet.

**Nonstick Coatings.** Other coatings that are applied in artificial blood vessels and also in large engineering pipes carrying liquids are nonstick surfaces such as polytetrafluoroethylene, which is marketed under the trademark Teflon. Blood clotting inside the artificial blood vessel is prevented by the use of such a functional polymer coating. In engineering pipelines, liquid flow is slowed because of the friction of the liquid on the walls of the vessel. Nonstick coatings can significantly cut down the friction in the pipelines, reducing the power required to transport liquids, while simultaneously providing a chemical barrier coating between the liquid and the inside of the pipe. In addition to polymer nonstick coatings such as Teflon, the insides of pipes can be coated with thin layers of ceramic glass. Manufacturers such as the Swedish company Trelleborg provide coating and sheathing solutions for transoceanic pipelines that can carry a large variety of liquids, including oil products. Of course, one of the more well-known applications for nonstick coatings is cookware. Nonstick coatings reduce the likelihood of heated materials sticking to the inside of a metallic pan, since it does not chemically react with other materials. For this type of coating to stick to the inside of the pan, the metal is prepared with groves or porosity generated by sandblasting and coated with a porous primer.

**Optical Coatings.** Most ceramic glasses permit infrared radiation but will block ultraviolet radiation from the sun. As a consequence, in summer, the insides of buildings are heated, and energy-intensive air-conditioning is required to cool the building. Some ceramics such as vanadium oxide that has been mixed with small amounts of tungsten oxide can block infrared radiation, effectively preventing solar heat radiation from reaching the insides of buildings. Applied to the surface of glass, they automatically insulate the building from heat, silently and efficiently. However, this particular material is even smarter. Infrared radiation is blocked only above 29 degrees Celsius. In winter, the infrared radiation can pass into the building and help in heating. Polymers, on the other hand, seldom block ultraviolet radiation. Most eyeglasses and sunglasses are manufactured using polymers, which are coated with materials that block ultraviolet radiation to protect the eyes. Similarly, polymers and polymer fabrics can be protected from degradation due to ultraviolet radiation by applying thin coatings of zinc and titanium oxides. Other transparent coating materials, such as indium tin oxide, can be applied to glass surfaces to make them conductive, as is done in solar cells, where the transparent layer facing the sun acts as one of the electrodes. Schott and Carl Zeiss are Europe-based companies that manufacture high-quality specialty glasses with any number of functional coatings for the global market.

**Magnetic Coatings.** Videotapes, cassettes, floppy discs, and zip drives are all now technologically obsolete, but they have one thing in common: They used magnetic materials to store data. In the case of magnetic tapes, a polymer film was coated with magnetic material, for example, iron oxide. The chemical company BASF used to be one of the largest providers of magnetic tapes. Modern computer hard disks, which are still produced in large numbers, typically use cobalt-based alloys to store the data, as they allow for a faster and safer magnetization.

## IMPACT ON INDUSTRY

**Government Research.** Because coatings are involved in most technology applications, major research is produced in academia, industry, and government facilities. In Canada, for example, the National Research Council includes many major research facilities involved with coating technologies. The Institute for Fuel Cell Innovation in Vancouver has developed reactive spray-gas phase sublimation methods to deposit functional layers of fuel cells. In the United States, research facilities such as the Pacific Northwest Laboratories in Richland, Washington, are heavily involved in coatings research. Scientists there developed protective aluminide coatings for steel. Instead of forming a separate layer on the surface of the metal, the atoms of the coating diffuse into the underlying steel, thereby making it stronger and creating a protective coating that cannot chip or scratch. In Europe, one example of coatings research is the development of dye-sensitized solar cells at the Max Planck Institute for Physics in Munich, where researchers apply several thin layers onto a glass or polymer substrate to form a solar cell. Dense fluorine-doped tin oxide and titanium oxide layers are coated with titanium dioxide spheres that have, in turn, been coated with organic dyes to increase the efficiency of the solar cell. These are just a few examples of the broad research in coatings worldwide.

**University Research.** The Shanghai Museum contains a collection of very rare ancient Chinese pottery from the Neolithic times to the end of the Qing Dynasty. It shows that advanced ceramic coatings mimicking gold, wood, or gemstones in color and refraction have been known to humanity for several millennia.

Also situated in Shanghai are the Shanghai University School of Materials Science and Engineering and the Shanghai Institute of Ceramics at the Chinese Academy of Sciences. Both of these academic research facilities develop methods for ceramic-coating technologies. For example, researchers have developed nanometer-size tin additives for gas products that can significantly reduce the friction inside motors and pistons, thereby improving the efficiency and longevity of the movable parts of a motor. Other research groups in Shanghai have discovered novel applications for simple rust. They have created nanometer-size hollow spheres of iron oxide that can assist in wastewater treatment. Another example for the global scope of coating applications are silica nanosprings developed at the Washington State University. Because of their unique small size, these springs can be used to store hydrogen atoms for later use by clean-energy devices such as fuel cells.

**Industry and Business.** The largest global revenues in coatings come from industries involved in paints, often including corrosion protection. Although only about 2 percent ($2 billion) of the total financial turnover of the French petrochemical giant Total Fina Elf is produced by its subsidiary paint manufacturer SigmaKalon, the figures still give an idea of the large scale of the international paint coating market. Another large global company in the coating business is the Netherlands-based AkzoNobel, with coatings-related sales volumes of almost $7 billion. Large paint manufacturers in the United States include Pittsburgh-based PPG Industries, Cleveland-based Sherwin-Williams Corporation, Minneapolis-based Valspar Corporation, and the Wilmington, Delaware-based DuPont Coatings and Color Technologies Group. The annual sales volume is estimated at between $4 billion and $5.3 billion. The BASF Coatings and Glasurit research facilities in Münster, Germany; ICI Paints in Berkshire, England; and the Japanese Nippon Paint and Kansai Paint in Osaka generate gross sales of about $9 billion annually to the global paint market.

## CAREERS AND COURSE WORK

Although it may seem trivial to pick up a paint roller to add color to a wall, the technological developments necessary to produce even a simple, stable paint are massive and have been ongoing for a long time. Detailed knowledge of the material properties, the chemistry of the solvents, and the fundamental physics of the optical, mechanical, thermal, and magnetic properties of coatings are essential to develop novel, innovative products for the global market.

Career advancement in the global coatings technology typically requires an undergraduate degree, ideally with many industrial internships for practical experience.

## Social Context and Future Prospects

Vehicles manufactured as late as the 1980's were prone to corrosion, reducing not only the optical appearance of a vehicle, but also impairing the mechanical stability and safety. Since then, coating technologies have significantly advanced. All surfaces of a car body can now be coated, no matter what shape, and they are less permeable to air and absorb significantly less water. Coatings adhere better to surfaces and are dense and have higher impact and scratch resistance, and as a result vehicles manufactured in the early twenty-first century corrode far less frequently. This development also applies to aeronautic and marine applications. The coated devices, such as ships, have to be re-coated less frequently, reducing downtime and preventing fatal material failures while the device is in service. The same applies to biomedical devices. If, for example, a hip implant with a modern coating has to be renewed inside a patient every fifteen years instead of every five years, the result is a significant improvement in life quality for the patient.

Coating technologies are far from perfect, which is evidenced by the huge global research effort in the development of all types of coatings. Modern automotive coatings can typically involve more than 200 chemicals, and the exact physical and chemical interaction between all these chemicals can be evaluated only in small steps. Likewise, coating technology methods are constantly evolving. Consequently, there is a huge potential in coatings research and development with applications in almost every piece of technology in use as of 2011.

*Lars Rose, M.Sc., Ph.D.*

## Further Reading

Birkmire, Robert. "Thin-Film Solar Cells and Modules." In *Solar Cells and Their Applications*, by Larry Partain and Lewis Fraas. 2d ed. Hoboken, N.J.: John Wiley & Sons, 2010. Birkmire, of the University of Delaware, has provided an excellent, detailed description of thin film solar cells.

Kumar, Challa, ed. *Nanostructured Thin Films and Surfaces*. Weinheim, Germany: Wiley-VCH, 2010. This assembly of research topics provides an overview of thin film coatings and nanoscale materials with a focus on their uses in the life sciences.

Lakhtakia, Akhlesh, and Russell Messier. *Sculptured Thin Films: Nanoengineered Morphology and Optics*. Bellingham, Wash.: SPIE, 2005. This very scientific book contains some good examples of how the morphology (the shape, size, and orientation) of thin surface films can significantly alter the properties of the material.

Lüth, Hans. *Solid Surfaces, Interfaces, and Thin Films*. 4th ed. Berlin: Springer-Verlag, 2001. Provides very detailed drawings and examples of different coating methods and the devices used to produce them.

Smeets, Stefan, Egbert Boerrigter, and Stephan Peeters. "UV Coatings for Plastics." *European Coatings Journal* 6 (2004): 42-48. Cytec (formerly Surface Specialties UCB) researchers' account of how to coat plastic surfaces with the help of ultraviolet radiation.

Weldon, Dwight G. *Failure Analysis of Paints and Coatings*. Rev. ed. John Wiley & Sons, 2009. There is a plethora of literature available on paints and paint technologies, but this book takes a unique look at how and why these coatings fail in service and how research can address these issues.

## Web Sites

*American Coatings Association*
http://www.paint.org

*Chemical Coaters Association International*
http://www.ccaiweb.com

*Society for Protective Coatings*
http://www.sspc.org

**See also:** Chemical Engineering; Nanotechnology; Optics.

# COMMUNICATION

## FIELDS OF STUDY

Interpersonal communication; conflict management; group dynamics; intercultural communication; linguistics; nonverbal communication; semiotics; communication technology; negotiation and mediation; organizational behavior; public speaking; speech pathology; rhetorical theory; diction; mass communication; advertising; broadcasting and telecommunications; journalism; media ethics; public relations.

## SUMMARY

Communication is the complex, continuous, two-way process of sending and receiving information in the form of messages. Communication engages all the senses and involves speech, writing, and myriad nonverbal methods of data exchange. It is a vital component in the everyday existence of all living creatures. An interdisciplinary, multidimensional field that incorporates language, linguistic structure, symbols, interpretation, and meaning in both personal and professional life, communication is an essential part of any division of scientific study.

## KEY TERMS AND CONCEPTS

- **Etymology:** Study of the origins of words and their changes in meaning and form over time.
- **Feedback:** Response of a recipient to a message received.
- **Haptics:** Study of how touch and bodily contact affects communication.
- **Kinesics:** Study of how nonverbal forms of communication such as posture, gestures, and facial expressions affect meaning.
- **Paralinguistics:** Study of the nonverbal elements that individualize speech to help convey meaning, including intonation, pitch, word stress, volume, and tempo.
- **Phonetics:** Study of how speech is physically produced and received.
- **Proxemics:** Study of the use of personal space in face-to-face communication.
- **Semantics:** Interpretation of meaning in language.
- **Semiotics:** Study of how signs, symbols, and signifiers create meaning.
- **Syntax:** Study of the grammatical rules of a language that aid effective communication.

## DEFINITION AND BASIC PRINCIPLES

Communication is the science and art of transmitting information. Communication science is objective, involving research, data, methodologies, and technological approaches to procedures of information exchange. Communication art is subjective, concentrating on the aesthetics and effectiveness with which messages are composed, sent, received, interpreted, understood, and acted on. Art and science are equally important in understanding how communication is supposed to work, why it succeeds or fails, and how best to use that knowledge to improve the dissemination of information. Communication is the oldest, broadest, most complicated, and most versatile of all scientific disciplines, with attitude-changing, life-influencing applications in every field of human endeavor.

The process of communication requires three basic components: a sender, a message, and a receiver. The sender encodes information into a message to send to the recipient for decoding. The message usually has an overt purpose, based on the sender's desire: to inform, educate, persuade, or entertain. Some messages can also be covert: to attract attention, make connections, gain support, or sell something.

In theory, the communication ideal is to match sender intent to receiver interpretation of a message closely enough to generate a favorable response. In practice, communication is often unsuccessful in achieving such objectives, considering the quirky, complicated nature of human beings and the variety and intricacy of languages used to transmit messages.

There are two fundamental forms of communication: verbal and nonverbal. Verbal communication includes interpersonal or electronically transmitted conversations and chats, lectures, audiovisual mass media such as television and radio, and similar forms of oral speech. Verbal communication also includes written communication: books, letters, magazines, signs, Web sites, and e-mails. Nonverbal communication incorporates an infinite range of facial expressions, gestures and sign language, body language, and paralanguage. In face-to-face interactions, nonverbal messages frequently convey more meaning than verbal content.

## BACKGROUND AND HISTORY

Human communication has been a vital part of evolution. Protohumans communicated through gestures before developing the ability to speak, perhaps as far back as 150,000 years ago. An opposable thumb allowed early humans to make tools and to leave long-lasting marks, such as cave paintings. As communities formed, spoken language evolved. Verbal language spawned written language; pictographs appeared 10,000 years ago. Cuneiform originated several thousand years later in Mesopotamia about the time hieroglyphics were born in Egypt. By 3000 B.C.E., writing had developed independently in China, India, Crete, and Central America. Written symbols became alphabets, which were organized into words and structured linguistic systems.

From the dawn of civilization, humans have reached out to one another, devising ingenious solutions to expand range, enhance exactitude, and ensure the permanence of messages. A library was begun in Greece in the sixth century B.C.E., and paper was invented soon afterward. By 1455, Johannes Gutenberg had devised a printing press with movable type and printed his first communication: the Gutenberg Bible. Each innovation represented a leap forward in the spread of information.

Communication has greatly accelerated and expanded during the last two hundred years as it became a scientific study. The nineteenth century witnessed the development of inventions such as railroads, the telegraph, postal systems, the typewriter and phonograph, automobiles, and telephones. The twentieth century ushered in movies, airplanes, radio and television, audio and video recorders, communication satellites, photocopiers, and facsimile machines. Since the 1980's personal computers, cellular phones, fiber optics, the Internet, and a variety of mobile, handheld devices have widened and sped human integration into the information age.

## HOW IT WORKS

Though necessary to interaction throughout human history, communication is still an imperfect science. As in the past, modern senders have vastly different abilities in composing comprehensible messages. Contemporary receivers, for a variety of reasons, may have an equally difficult time interpreting messages.

**Channeling.** How a message is sent—the medium of transmission, called a channel in communications science—is key to the communication process. An oral message too faint to be heard, a written message too garbled to be understood, or a missed nonverbal signal can interrupt, divert, delay, or derail the passage of information from one source to another. In the modern world, there are dozens of methods of sending messages: telephone, text messaging, ground mail, mass media, e-mail or social networking via the Internet, and face-to-face contact are most common. Deciding the best means of information transmission depends on a number of factors, including the purpose of the message, the amount and profundity of data to be sent, the audience for the message, and the desired outcome. All media have distinct advantages and disadvantages. Each type of communication requires a particular skill set from both sender and receiver.

A vital consideration is the relationship between sender and receiver: They must have something in common. A message written in Chinese and sent to someone who reads only English is ineffective. Likewise, shouting at the hearing impaired or sending text messages to newborn infants are counterproductive activities. Even among individuals with a great degree of similarity in language and culture there can be gaping discrepancies between sender intent and receiver interpretation. Ultimately, communication is a result-oriented discipline. Whatever the purpose of a message, a response of some kind from the recipient (feedback) is expected. If there is no reply, or an inappropriate response is generated, the process is incomplete and communication fails.

**Context.** Communication does not occur in a vacuum. Many factors determine and define how the process of information transfer will unfold and the likelihood of successful transmission, reception, and response to a message.

Psychological context, for example, entails the emotional makeup of individuals who originate and receive information: the desires, needs, values, and personalities of the participants, which may be in harmony or at variance. Environmental context deals with the physical setting of an interaction: weather, surroundings, noise level, or other elements with the potential to impair communication. Situational context is specific to the relationship between the participants in communication: Senders tailor messages differently to receivers who are friends, relatives, coworkers, or strangers. Linguistic context concerns the relationship among words, sentences, and paragraphs used throughout a speech or written work that help to clarify meaning. Interpretation of any part of a message is relative to the preceding and following parts of the message.

Particularly relevant to effective communication in the ever-widening global community is cultural context. Every culture has particular, ingrained rules governing verbal and nonverbal behavior among its members. What might be acceptable within one cultural group—such as extended eye contact, the use of profanity, or frequent touching during conversation—might be offensive to a different cultural group. High-context cultures are homogeneous communities with long-established, well-defined traditions that help preserve cultural values; such cultures are found throughout Asia, Africa, the Middle East, and South America. Low-context cultures, like the United States, are heterogeneous, a blend of many traditions that has produced a less rigid, broader-based, more open-ended set of behavioral rules. Though adhering to certain patterns learned from diverse national and ethnic heritages, Americans as a group are more geared toward individual values. When high- and low-context cultures collide in communication, such differences in attitude can wreak havoc on the implied intent of a message and the interpretation of meaning.

**Communication Barriers.** There are universal emotions (such as fear, surprise, happiness, sadness, and anger), common to all peoples in all cultures at all times, that can serve as building blocks to understanding. However, numerous obstacles interfere with the clear, unambiguous transmission of messages at one end, and the full grasp of meaning at the other end. Some of the more prevalent physical, social, and psychological impediments to be overcome include racial or ethnic prejudice, human ego, noise (in transmission equipment or surroundings), and distractions. Gender issues are also a primary concern. For a number of reasons, men and women think, speak, and act quite differently in interpersonal relations. There are generational issues, too:

Children, parents, and grandparents can speak different languages. It is the task of communication to identify and find methods of circumventing or accommodating such information blockers.

**APPLICATIONS AND PRODUCTS**

**Linguistics.** The scientific study of the structure and function of spoken, written, and nonverbal languages, linguistics has many academic, educational, social, and professional applications. A foundation in linguistics is essential in understanding how words are formed and fitted together to create meaning and establish connections among people who are simultaneously senders and receivers of billions of messages over a lifetime. The stronger a linguistic base, the easier it is to discriminate among the plethora of messages received daily from disparate sources, to interpret meaning and respond accordingly, to prioritize, and to bring personal order to a disordered world of information overload.

Linguistics incorporates numerous threads, each worthy of close examination in its own right—word origins, vocabulary, grammar, phonetics, symbols, gestures, dialects, colloquialisms, or slang—that contribute to the rich tapestry of communication. Linguistics can be approached from several basic, broadly overlapping directions.

Structural linguistics deals with how languages are built and ordered into rules for communication. This subdiscipline, a basis for scientific research, clinical studies, sociological explorations, or scholarly pursuits, focuses on how speech is physically produced. Structural linguistics involves acoustic features, the comparison of different languages, grammatical concepts, sentence construction, vocabulary function, and other analytical aspects of language.

Historical linguistics concentrates on the development of written, oral, and nonverbal languages over time. An appropriate field for historians, educators, and comparative linguists, the discipline is concerned with why, where, when, and how words change in pronunciation, spelling, usage, and meaning.

Geolinguistics, a newer discipline with sociological, political, and professional implications in the modern global marketplace, focuses on living languages. The study is primarily concerned with historical and contemporary causes and effects of language distribution and interaction as a means to improve international and intercultural communication.

**Interpersonal Communication.** Interpersonal communication is that which takes place between two or more people. An understanding of the process is useful in grasping how message transfer, receipt, interpretation, and response succeed or fail in the real world.

A broad discipline relevant to all applied science—since it involves everyday relationships and interactions among people—interpersonal communication serves as the basis for a considerable amount of scientific and popular study. Thousands of books and articles are published annually on various aspects of the subject: how to talk to a spouse, how to deal with a child, how to succeed in business, how to communicate with pets.

Interpersonal communication can be subdivided by specific fields of study or professional emphasis.

- Conflict management and problem solving are communication-based specialties invaluable for students of family therapy, sociology, psychology, psychiatry, law, criminology, business management, and political science.
- Gender, racial, sexual, intergenerational, and cultural communication issues are the concerns of many disciplines, particularly sociologists, psychologists, intercultural specialists, and international business students.
- Health communication, a subdiscipline that involves translation of jargon to plain language and the development of empathy, is of particular relevance to medical and psychology students, and to intercultural therapists and clinicians.
- Organizational communication, including departmental interaction, group dynamics, and behavior, is indispensable to the successful function of businesses, professional associations, government, and the military.

**Mass Communication.** This is the study of the dissemination of information through various media (newspapers, magazines, billboards, books, radio, television, Web sites, blogs, and a host of other methods) with the potential to influence large audiences. Modern mass media is the culmination of centuries of technology enhancing the power and glory of communication. For better or worse, messages sent and received are the manifestation of qualities and possibilities that have elevated humans to a predominant position on earth. As languages

---

### Fascinating Facts About Communication

- Humans the world over typically express basic emotions–happiness, sadness, surprise, anger, fear, disgust, and embarrassment–through virtually identical facial expressions, though such emotions can be hidden or controlled with training.
- Throughout his or her lifetime, the average American will spend an entire year (365 days × 24 hours per day = 8,760 hours) talking on the telephone.
- By first grade, average children understand about 10,000 words; by fifth grade, they understand about 40,000 words.
- Radio took thirty-eight years to accumulate 50 million listeners; television took thirteen years to acquire 50 million viewers and the Internet only five years for 50 million users.
- Humans typically speak at a rate of 100 to 175 words per minute but are capable of listening intently to information transmitted at 600 to 800 words per minute.
- More than a dozen languages are spoken by at least 100 million people. Mandarin Chinese is the widest-spoken language, with about 900 million speakers. English, in third place behind Mandarin and Spanish, is the first language of nearly 500 million people.

---

have been defined and refined, as methods of transmission have improved, the range of communication has expanded. The distribution of data, once confined to the limitations of the human voice, through a series of quantum leaps is now a worldwide phenomenon. There are multiple outlets (Internet, satellite television, international phone lines) available to send messages, solicit responses, and record aftereffects on a global scale. Mass communication offers numerous opportunities for specialization.

Advertising, marketing, and public relations represent the persuasive power of communication. The modern world is a global marketplace. There are billions of potential customers for every conceivable product, service, or cause, and there are thousands of businesses and agencies whose task it is to create consumer demand. Marketing is sales strategy: planning, researching, testing, budgeting, setting objectives, and measuring results. Advertising is message strategy: what to say; how, when, where, and how often to say it to achieve desired aesthetic and marketing goals. Public relations is image strategy: the manipulation of words and pictures that establish and preserve public perception of corporations, institutions, organizations, governmental entities, or individuals.

Performance or public communication often involves a particular motivation: to generate immediate response from an audience. Motivational speakers, lecturers, debaters, politicians, actors, and other public performers all have the common goal of eliciting instant emotion in listeners or viewers.

Telecommunications refers to electronic means of sending information, primarily via radio or television broadcast, but the field also incorporates broadband, mobile wireless, information technology, networks, cable, satellite, unified communication, and emergency communication. A burgeoning, far-ranging industry, telecommunications offers global possibilities in business management, on-air and on-screen performance, research, marketing, journalism, programming, advertising, editing, sales, information science, and technology.

Creative communication concerns the composition of messages for a variety of informational purposes and is a component found to some degree throughout all facets of mass media. Writers of all kinds, graphic artists, photographers, designers, illustrators, and critics have the ability to influence behavior through the use of words or pictures.

Related fields of study fall under the umbrella of mass communication. Demographics, the collection of data that quantify an intended audience according to economic, political, ethnic, religious, professional, or educational factors, is an important consideration for many segments of mass media. There are also numerous legal, ethical, environmental, political, and regulatory issues to be dealt with that require specialized training.

## IMPACT ON INDUSTRY

A few facts demonstrate the scope and trend of contemporary communication, a field that continues to grow exponentially in all directions. More than a million new books were released in 2009, three-quarters of which were self-published; digital book sales are expected to increase 400 percent by 2015. In 2010, the leading one hundred worldwide corporations—led by Proctor & Gamble's $8.68 billion expenditures—spent an average of $1 billion apiece in a $900 billion advertising industry. In 2011, the telecommunications industry will comprise $4 trillion, about 3 percent of the entire global economy. Internet advertising, expected to top $105 billion, will soon surpass newspaper advertising. Though the United States leads the way in online spending, with Japan in second place, China has supplanted Germany for third place, with Central and Eastern Europe, the Middle East, and Africa all fast-growing regions for such spending. Some 1.2 billion people, more than 20 percent of Earth's inhabitants, were connected to the Internet in 2010, and the penetration of the medium (about 75 percent in the United States) to the farthest corners of the world is expected to continue apace throughout the century. There is a constant influx of new individuals and groups establishing a presence through Web sites, blogs, and social networks. Mind-boggling numbers of messages about a bewildering array of subjects are sent and received daily, making effective communication more important than ever.

**Academics.** Though rhetoric and oratory have been studied since the days of Aristotle, communication science as a formal discipline dates from the 1940's, when research facilities were established at Columbia University, the University of Chicago, and the University of Illinois, Urbana-Champaign. In the twenty-first century, virtually all universities and colleges worldwide offer communication studies, most have degree programs, and many offer postgraduate work in various areas of concentration. The Annenberg School for Communication at the University of Pennsylvania, for example, has a long tradition in examining the effects of mass media on culture. Rensselaer Polytechnic Institute specializes in technical communications. New York University has particular interest in exploring media ecology, the study of media environments. At many institutions, original research in a wide variety of communication-related studies is encouraged and supported via grants and scholarships.

**Government.** The U.S. Government Printing Office (GPO) was established 150 years ago to keep track of mountains of information generated in the course of governing behavior across a sprawling, complicated society. The agency physically or electronically prints and disseminates—to officials, libraries, and the public— Supreme Court documents, the text of all congressional proceedings and legislation, presidential and executive department communications, and the records or bulletins of many independent agencies. The Federal Communications Commission (FCC), founded in 1934 and segmented to deal with various issues, regulates wired and wireless interstate and international transmissions via radio, television, satellite, cable, and fiber optics. Likewise, such bureaus as the Federal Trade Commission (FTC) oversee the commerce generated from correspondence between businesses and citizens. Millions of dollars in research funding is available at federal and state government levels across a spectrum of concerns related to communication. Finally, many government agencies, bureaus, and administrations—from the Census Bureau to the National Aeronautics and Space Administration to the U.S. Geological Survey and the White House—disperse aggregated data, images, and other information to the tax-paying public.

**Corporations.** While universities study causes and effects and governments regulate legal form and use, the business world presents numerous opportunities for practical application of communication principles. A broad-based, interdisciplinary field, communication allows many points of corporate entry, such as technology, creative disciplines, public relations, management, or sales. A computer maven can find a niche in many modern industries; prowess in journalism can be adapted to other forms of writing; performance skills can translate from television to film to stage; artistic talent can be suited to graphic design or commercial illustration; an individual who can successfully sell one product can easily learn to sell another product. Corporations reward innovation.

Many professional organizations exist to provide detailed, updated data on issues that affect both for-profit and not-for-profit industries. The National Communication Association, the International Communication Association, and a United Nations agency, the International Telecommunications Union, are leaders in the exchange of ideas, knowledge, research, technology, and regulatory information pertinent to communication.

## CAREERS AND COURSE WORK

Communication is an all-purpose discipline that offers a multitude of career paths to personally and professionally rewarding occupations in a booming, always relevant field. Core courses—English language and grammar, sociology, psychology, history and popular culture, speech and business—provide a firm foundation on which a successful specialization can be built in any of several broad, compatible areas always in demand.

In academics, communication offers careers in teaching, counseling, research, and technology. Students should plan to pursue degrees past the bachelor's level, adding courses in education, communication theory, media ethics and history, information technology, and telecommunications in the undergraduate program.

Many creative niches are available in corporate communications, journalism, consumer advertising, public relations, freelance writing and art, and the media. Verbal artists need course work in rhetoric, literature, composition, linguistics, persuasion, and technical writing. Visual artists require graphic design, illustration, semiotics, and computer-aided design. On-screen or on-air performers would benefit from courses in organization, diction, public speaking, nonverbal communication, and mass media.

There are dozens of jobs and professions (positions including editor, agent, producer, director, publicist, and journalist) in media, government, politics, corporations, nonprofits, and other organizations for those with skills and credentials in communication and its various subfields. Courses in interpersonal and intercultural communication, business management, group dynamics, interviewing, problem solving, negotiation, and motivation are particularly useful.

Communication also has many applications in the social and health sciences. These include family therapy, psychiatry, marriage counseling, speech pathology, gender and sexuality services, legal providers, ethical concerns, international relations, and intercultural, intergenerational, and interspecies specialties that often require advanced degrees and original research. Course work can include concentrated studies in sociology and psychology, cultural history and geolinguistics, law, and foreign languages.

## SOCIAL CONTEXT AND FUTURE PROSPECTS

From the beginning of civilization, communication has been the glue that binds together all elements of human society. In the modern global community—with billions of information exchanges passing rapidly among members of a vast, diverse, largely receptive audience eager to connect—competence in oral and written skills is mandatory for personal and professional success. More than ever, messages must be concise, unambiguous, accurate, and compelling to cut through clutter streaming from dozens of sources.

Several issues are of particular concern in contemporary communication studies. Publishing is changing from print to electronic. There is high interest across the multitrillion-dollar telecommunications industry in green technology that reduces air and noise pollution. A new field, unified communications, which promotes system interoperability, is expected to reach worldwide revenues of $14.5 billion by 2015. With fresh markets of information exchangers emerging around the world, there is great demand for intercultural expertise.

All disciplines are subject to economic climate. If employment slumps in one field, such as journalism, skill sets can often be easily and profitably applied to a related field, such as advertising or public relations. People and management and verbal skills translate well across industries, geography, and time.

One of the greatest challenges facing communication in the age of the Internet is the educated assessment and consumption of the vast array of information available to receivers at all levels. The need for courses and instruction in critical thinking, always important, has risen exponentially as various electronic sources of information have proliferated—especially when those sources are not well understood, their authoritativeness is open to question, and their existence is ephemeral.

For the foreseeable future, there will always be news to report, information to supply, products and services to sell, causes to promote, politicians to elect, legislation to be enacted, and government actions to document and disseminate. There will always be ethical debates about what can be done versus what should be done. There will always be a place for those who advance the art and science of communication through performance, education, therapy, research, or innovation. And there will always be a need for people who can consistently connect successfully with fellow humans through the evocative use of words and images.

*Jack Ewing*

## FURTHER READING

Baran, Stanley J., and Dennis K. Davis. *Mass Communication Theory: Foundations, Ferment, and Future*. 6th ed. Boston: Wadsworth, 2011. This textbook examines the field of mass communication, exploring theories and providing examples that aid in understanding the roles and ethics of various media.

Belch, George, and Michael Belch. *Advertising and Promotion: An Integrated Marketing Communications Perspective*. 9th ed. New York: McGraw-Hill/Irwin, 2008. This work spotlights the variety of strategies and methods used to build relationships between advertisers and customers in highly competitive, in-demand communication specialties.

Knapp, Mark L., and Judith A. Hall. *Nonverbal Communication in Human Interaction*. 7th ed. Boston: Wadsworth, 2009. A thorough examination of the theory, research, and psychology behind one of the most significant aspects of interpersonal communication.

Seiler, William J., and Melissa L. Beall. *Communication: Making Connections*. 8th ed. Needham Heights, Mass.: Allyn & Bacon, 2010. This book concentrates on the theory and practice of speech communication as applied to public discourse, interpersonal relationships, and group dynamics in a variety of settings.

Tomasello, Michael. *Origins of Human Communication*. Cambridge, Mass.: MIT Press, 2008. An award-winning study of how human communication evolved from gestures, founded on a fundamental need to cooperate for survival.

Varner, Iris, and Linda Beamer. *Intercultural Communication in the Global Workplace*. 5th ed. New York: McGraw-Hill/Irwin, 2010. This work addresses the impact of the Internet in particular on communication, especially as it affects relationships between businesses and governments in a shrinking world.

## WEB SITES

*International Communication Association*
http://www.icahdq.org

*International Telecommunications Union*
http://www.itu.int/en/pages/default.aspx

*Linguistic Society of America*
http://www.lsadc.org

*National Communication Association*
http://www.natcom.org

**See also:** Anthropology; Grammatology; Information Technology; Telecommunications; Telephone Technology and Networks; Wireless Technologies and Communication.

# COMMUNICATIONS SATELLITE TECHNOLOGY

## FIELDS OF STUDY

Avionics; computer engineering; computer information systems; computer network technology; computer numeric control technology; computer science; electrical engineering; electronics engineering; mechatronics; physics; telecommunications technology; television broadcast technology.

## SUMMARY

Communications satellite technology has evolved from its first applications in the 1950's to become a part of most people's daily lives and thereby producing billions of dollars in yearly sales. Communications satellites were initially used to help relay television and radio signals to remote areas of the world and to aid navigation. Weather forecasts routinely make use of images transmitted from communications satellites. Telephone transmissions over long distances, including fax, cellular phones, pagers, and wireless technology, are all examples of the increasingly large impact that communications satellite technology continues to have on daily, routine communications.

## KEY TERMS AND CONCEPTS

- **Baseband:** Transmission method for communications signals that uses the entire range of frequencies (bandwidth) available. It differs from broadband transmission, which is divided into different frequency ranges to allow multiple signals to travel simultaneously.
- **Bit:** Binary digit, of which there are only two, the numbers zero and one. Computer technology is based on the binary number system because early computers contained switches that could be only on or off.
- **Browser:** Software program that is used to view Web pages on the Internet.
- **Downlinking:** Transmitting data from a communications satellite or spacecraft to a receiver on Earth.
- **Downloading:** Process of accessing information on the Internet and then allowing a browser to make a copy to save on a personal computer.
- **Gravity:** Force of attraction between two objects, expressed as a function of their masses and the distance between them. Typically, the Earth's gravity is the most important consideration for satellites, and it has the constant value of 9.8 meters per second squared ($m/s^2$).
- **Hyperlinks:** Clickable pointers to online content in Web pages other than the page one is reading.
- **Internet Service Provider (ISP):** Organization that provides access to the Internet for a fee.
- **Ionosphere:** Part of the upper atmosphere that is ionized because of radiation from the sun, and therefore it affects the propagation of the radio waves within the electromagnetic spectrum.
- **Kilobit:** Quantity equal to 1,000 bits.
- **Orbit:** Curved path that an object travels in space because of gravity.
- **Period:** Time required to complete one revolution around the Earth.
- **Satellite:** Object that travels around the Earth.
- **Transponder:** The electronic component of a communications satellite that automatically amplifies and retransmits signals that are received.
- **Uplinking:** Transmitting data from a station on Earth up to a communications satellite or spacecraft.

### DEFINITION AND BASIC PRINCIPLES

Sputnik 1, launched on October 4, 1957, by the Soviet Union, was the first artificial satellite. It used radio transmission to collect data regarding the distribution of radio signals within the ionosphere in order to measure density in the atmosphere. In addition to space satellites, the most common artificial satellites are the satellites used for communication, weather, navigation, and research. These artificial satellites travel around the Earth because of human action, and they depend on computer systems to function. A rocket is used to launch these artificial satellites so that they will have enough speed to be accelerated into the most common types of circular orbits, which require speeds of about 27,000 kilometers per hour. Some satellites, especially those that are to be used at locations far removed from the Earth's equator, require elliptical-shaped orbits instead, and their acceleration speeds are 30,000 kilometers per hour. If a launching rocket applies too much energy

*AT&T satellite station.* (Michael P. Gadomski/Photo Researchers, Inc.)

to an artificial satellite, the satellite may acquire enough energy to reach its escape velocity of 40,000 kilometers per hour and break free from the Earth's gravity. It is important that the satellite be able to maintain a constant high speed. If the speed is too low, gravity may cause the satellite to fall back down to the Earth's surface. There are also natural satellites that travel without human intervention, such as the Moon.

## BACKGROUND AND HISTORY

In 1945, science fiction writer Arthur C. Clarke first described the concept of satellites being used for the mass distribution of television programs in his article "Extra-Terrestrial Relays," published in *Wireless World*. John Pierce, who worked at Bell Telephone Laboratories, further expanded on the idea of using satellites to repeat and relay television channels, radio signals, and telephone calls in his article "Orbital Radio Relays," published in the April, 1955, issue of *Jet Propulsion*. The first transatlantic telephone cable was opened by AT&T in 1956. The first transatlantic call was made in 1927, but it traveled via radio waves. The cable vastly improved the signal quality. The Soviet Union launched Sputnik 1, the first satellite, in 1957, which began the Space Race between the Soviet Union and the United States. The Communications Satellite Act of 1962 was passed by the United States Congress to regulate and assist the developing communications satellite industry. The first American television satellite transmission was made on July 10, 1962, five years into the

Space Race, with the National Space and Aeronautics Administration's (NASA) launch of the world's first communications satellite, AT&T's Telstar.

The many new communications satellites followed, with names such as Relay, Syncom, Early Bird, Anik F2, Westar, Satcom, and Marisat. Since the 1970's, communications satellites have allowed remote parts of the world to receive television and radio, primarily for entertainment purposes. Technology advances have continued to evolve and now use these satellites to facilitate mobile phone communication and high-speed Internet applications.

## HOW IT WORKS

Communications satellites orbit the Earth and use microwave radio relay technology to facilitate communication for television, radio, mobile phones, weather forecasting, and navigation applications by receiving signals within the six-gigahertz (GHz) frequency range and then relaying these signals at frequencies within the four-GHz range. Generally there are two components required for a communications satellite. One is the satellite itself, sometimes called the space segment, which consists of the satellite and its telemetry controls, the fuel system, and the transponder. The other key component is the ground station, which transmits baseband signals to the satellite via uplinking and receives signals from the satellite via downlinking.

These communications satellites are suspended around the Earth in different types of orbits, depending on the communication requirements.

**Geostationary Orbits.** Geostationary orbits are most often used for communications and weather satellites because this type of orbit has permanent latitude at zero degrees, which is above the Earth's equator, and only longitudinal values vary. The result is that satellites within this type of orbital can use a fixed antenna that is pointed toward one location in the sky. Observers on the ground view these types of satellites as motionless because their orbit exactly matches the Earth's rotational period. Numerically, this movement equates to an orbital velocity of 1.91 miles per second, or a period of 23.9 hours. Because this type of orbit was first publicized by the science fiction writer Arthur C. Clarke in the 1940's, it is sometimes called a Clarke orbit. Systems that use geostationary satellites to provide images for meteorological applications include the Indian National Satellite System (INSAT), the European Organisation for the Exploitation of Meteorological Satellites' (EUMETSAT) Meteosat, and the United States' Geostationary Operational Environmental Satellites (GOES). These geostationary meteorological satellites provide the images for daily weather forecasts.

**Molniya Orbits.** Molniya orbits have been important primarily in the Soviet Union because they require less energy to maintain in the area's high latitudes. These high latitudes cause low grazing angles, which indicate angles of incidence for a beam of electromagnetic energy as it approaches the surface of the Earth. The angle of incidence specifically measures the deviation of this approach of energy from a straight line. As a result, geostationary satellites would orbit too low to the Earth's surface and their signals would have significant interference. Because of Russia's high latitudes, Molniya orbits are more energy efficient than geostationary orbits. The word Molniya comes from the Russian word for "lightning," and these orbits have a period of twelve hours, instead of the twenty-four hours characteristic of geostationary orbits. Molniya orbits have a large amount of incline, with an angle of incidence of about 63 degrees.

**Low Earth Orbits.** Low earth orbit (LEO) refers to a satellite orbiting between 140 and 970 kilometers above the Earth's surface. The periods are short, only about ninety minutes, which means that several of them are necessary to provide the type of uninterrupted communication characteristic of geostationary orbits, which have twenty-four-hour periods. Although a larger number of low earth orbits are needed, they have lower launching costs and require less energy for signal transmission because of how close to the Earth they orbit.

## APPLICATIONS AND PRODUCTS

**DISH Network and Direct Broadcast Satellites.** DISH Network is a type of direct broadcast satellite (DBS) network that communicates using small dishes that have a diameter of only 18 to 24 inches to provide access to television channels. This DBS service is available in several countries through many commercial direct-to-home providers, including DIRECTV in the United States, Freesat and Sky Digital in the United Kingdom, and Bell TV in Canada. These satellites transmit using the upper portion of the microwave Kμ band, which has a range of between 10.95 and 14.5 GHz. This range is divided based on the

---

### Fascinating Facts About Communications Satellite Technology

- In 2011, there are more than 2,000 communications satellites orbiting the Earth.
- There are at least nineteen Orbiting Satellite Carrying Amateur Radio (OSCAR). These artificial communication satellites have been launched by individuals.
- The Communications Satellite Act was passed in 1962 with the intention of making the United States a worldwide leader in communications satellite technology so that it could encourage peaceful relations with other nations. This act also established the Communications Satellite Corporation (Comsat), with headquarters in Washington, D.C.
- In 2003, Optus and Defence C1 carried sixteen antennas to provide eighteen satellite beams across Australia and New Zealand, making it one of the most advanced communications satellites ever launched.
- Sputnik, the first artificial satellite, was only 58 centimeters in diameter.
- The first known ideas for communications satellites were actually made public not by scientists or engineers but by the writers Arthur C. Clarke and Edward Everett Hale.

---

geographic regions requiring transmissions. Law enforcement also uses the frequencies of the electromagnetic spectrum to detect traffic-speed violators.

**Fixed Service Satellites.** Besides the DBS services, the other type of communication satellite is called a fixed service satellite (FSS), which is useful for cable television channel reception, distance learning applications for universities, videoconferencing applications for businesses, and local television stations for live shots during the news broadcasts. Fixed service satellites use the lower frequencies of the Kμ bands and the C band for transmission. All of these frequencies are within the microwave region of the electromagnetic spectrum. The frequency range for the C band is about 4 to 8 GHz, and generally the C band functions better when moisture is present, making it especially useful for weather communication.

**Intercontinental Telephone Service.** Traditional landline telephone calls are relayed to an Earth station via the public switched telephone network (PSTN). Calls are then forwarded to a geostationary satellite to allow intercontinental phone communication. Fiber-optic technology is decreasing the dependence on satellites for this type of communication.

**Iridium Satellite Phones.** Iridium is the world's largest mobile satellite communications company. Satellites are useful for mobile phones when regular mobile phones have poor reception. These phones depend only on open sky for access to an orbiting satellite, making them very useful for ships on the open ocean for navigational purposes. Iridium manufactures several types of satellite phones, including the Iridium 9555 and 9505A, and models that have water resistance, e-mail, and USB data ports. Although it has the largest market share for satellite phones in Australia, Iridium does face competition from two other satellite phone companies: Immarsat and Globalstar.

**Satellite Trucks and Portable Satellites.** Trucks equipped with electrical generators to provide the power for an attached satellite have found applications for mobile transmission of news, especially after natural disasters. Some of these portable satellites use the C-band frequency for the transmission of information via the uplink process, which requires rather large antennas, whereas other portable satellites were developed in the 1980's to use the $K_u$ band for transmission of information.

**Global Positioning System (GPS).** GPS makes use of communications satellite technology for navigational purposes. The GPS was first developed by the government for military applications but has become widely used in civilian applications in products such as cars and mobile phones.

## IMPACT ON INDUSTRY

**Cable Television and Satellite Radio.** Ted Turner led the way in the 1980's with the application of communications satellite technology for distributing cable television news and entertainment channels (CNN, TBS, and TNT, to name a few). Since the 1980's, access to additional cable television stations has continued to grow and has evolved to include access to radio stations as well, such as SiriusXM radio. Communications satellite technology is transforming the radio industry by allowing listeners to continue to listen to the radio stations of their choice no matter where they are. Specifically, many new cars incorporate the Sirius S50 Satellite Radio Receiver and MP3/

WMA player. However, the overall effect of communications satellites on the radio industry has not been as significant as it has on the television industry.

**Internet Access and Cisco.** In addition to providing access to additional television stations, the satellite dish can be used for Internet access as well. Downloading data occurs at speeds faster than 513 kilobits per second and uploading is faster than 33 kilobits, which is more than ten times faster than the dial-up modems that use the plain old telephone service. Cisco has been a leader in developing the tools to connect the world via the Internet for more than twenty years. High-speed connections to the Internet have been replacing the slower dial-up modems that connect via telephone lines. High-speed connections use fiber distributed data interface (FDDI), which is composed of two concentric rings of fiber-optic cable that allow it to transmit over a longer distance. This is expensive, so Cisco developed internet routing in space (IRIS), which will have a huge impact on the industry as it places routers on the communications satellites that are in orbit. These routers will provide more effective transmission of data, video, and voice without requiring any transmission hubs on the ground at all. IRIS technology will radically transform Internet connections in remote, rural areas. An Internet server connects to the antenna of a small satellite dish about the same size of the dishes used to enhance television reception. This connection speed can be as fast as 1.5 megabits per second and is a viable alternative to wireless networks. Wireless networks have been transforming the way that people communicate on a daily basis, by using radio and infrared signals for a variety of handheld devices (iPhones, iPods, iPads). Huge economic growth potential exists for Cisco and other companies that can effectively incorporate more reliable satellite technology, such as IRIS.

**EchoStar and High-Speed Internet Access.** Further indication of the large economic effect of communications satellites is from the February, 2011, report that EchoStar Corporation is going to pay more than $1 billion to purchase Hughes Communications. Hughes Communications uses satellites to provide high-speed, broadband access to the Internet. EchoStar, located in Englewood, Colorado, is the major provider of satellites and fiber networking technology to Dish Network Corporation (the second largest television provider in the United States), as well as additional military, commercial, and government customers that involve various data, video, and voice communications. The combined impact of merging Hughes Communications and Echo-Star will be to continue the expansion of communications satellite technology.

## Careers and Course Work

Careers working with communications satellite technology can be found primarily in the radio, television, and mobile phone industries. Specifically, these careers involve working with the wired telecommunications services that often include direct-to-home satellite television distributors as well as the newer wireless telecommunications carriers that provide mobile telephone, Internet, satellite radio, and navigational services. Government organizations also need employees who are trained in working with communications satellite technology for weather forecasting and other environmental applications as well as communication of data between public-safety officials.

The highest salaries are earned by those with a bachelor's degree in avionics technology, computer engineering, computer science, computer information systems, electrical engineering, electronics engineering, physics, or telecommunications technology, although a degree in television broadcast technology can also lead to lucrative career after obtaining several years of on-the-job-training. The work environments for those with these types of degrees are primarily office and technology, with more than 14 percent of the workers working more than 40 hours per week. Although the telecommunications and communications technology industries are expected to continue to grow faster than many other industries, the actual job growth is expected to be less than other high-growth industries because of computer optimization. Those without a bachelor's degree are the most at-risk, as their jobs also can involve lifting and climbing around electrical wires outdoors in a variety of weather and locations. The National Coalition for Telecommunications Education and Learning (NACTEL), the Communications Workers of America (CWA), and the Society of Cable Telecommunications Engineers (SCTE) are sources of detailed career information for anyone interested in communications satellite technology.

## Social Context and Future Prospects

Advances in satellite technology have accompanied the rapid evolution of computer technology to such an extent that some experts describe this media revolution as an actual convergence of all media (television, motion pictures, printed news, Internet, and mobile phone communications). In 1979, Nicholas Negroponte of the Massachusetts Institute of Technology began giving lectures describing this future convergence of all forms of media. As of the twenty-first century, this convergence seems to be nearly complete. Television shows can be viewed on the Internet, as can news from cable television news stations such as CNN, Fox, and MSNBC. Hyperlinks provide digital connections between information that can be accessed from almost anywhere in the world instantly because of communication satellite technology. The result is that there is a twenty-four-hour news cycle, and the effects are sometimes positive but can also be negative if the wrong information is broadcast. The instantaneous transmission of political and social unrest by communications satellite technology can lead to further actions, as shown by the 2011 protests in Egypt, Iran, Yemen, Libya, and Bahrain.

*Jeanne L. Kuhler, B.S., M.S., Ph.D.*

## Further Reading

Baran, Stanley J., and Dennis K. Davis. *Mass Communication Theory: Foundations, Ferment, and Future.* 5th ed. Boston: Wadsworth Cengage Learning, 2008. Provides a detailed historical discussion of the communications technologies and their impacts on society.

Bucy, Erik P. *Living in the Information Age: A New Media Reader.* 2d ed. Belmont, Calif.: Wadsworth Thomson, 2005. Describes the societal implications of the evolution of media technology and the convergence of communications media made possible by new technologies.

Giancoli, Douglas C. *Physics for Scientists and Engineers with Modern Physics.* 4th ed. Upper Saddle River, N.J.: Pearson Education, 2008. This introductory physics textbook provides mathematical information regarding satellites and their orbits.

Grant, August E., and Jennifer Meadows. *Communication Technology Update and Fundamentals.* 12th ed. Burlington, Mass.: Focal Press, 2010. This introductory textbook provides technical information regarding communications satellites in terms of historical development, detailed applications, and existing uses.

Hesmondhalgh, David. *The Cultural Industries.* 2d ed. Thousand Oaks, Calif.: Sage, 2007. Describes the worldwide cultural effects of communications technologies in the past and present.

Mattelart, Armand. *Networking the World: 1794-2000*. Translated by Liz Carey-Libbrecht and James A. Cohen. Minneapolis: University of Minnesota Press, 2000. Describes the impact of communications satellites and other technologies on political, economic, and cultural phenomena, including nationalism, liberalism, and universalism.

Parks, Lisa, and Shanti Kumar, eds. *Planet TV:, A Global Television Reader*. New York: New York University Press, 2003. This textbook focuses on the application of communication satellites to television programming and discusses its historical evolution as well as its impact on the societies of various nations, especially India.

**WEB SITES**
*Communications Workers of America*
http://www.cwa-union.org

*National Coalition for Telecommunications Education and Learning*
http://www.nactel.org

*Society of Cable Telecommunications Engineers*
http://www.scte.org

**See also:** Avionics and Aircraft Instrumentation; Computer Engineering; Computer Networks; Telecommunications.

# COMPUTED TOMOGRAPHY

## FIELDS OF STUDY

Physics; physiology; medical physics; radiology; mathematics; computer science; electrical engineering; biomedical engineering; electronics; health care; medicine.

## SUMMARY

Computed tomography (CT) is an imaging modality that relies on the use of X rays and computer algorithms to provide high-quality image data. CT scanners are an integral part of medical health care in the developed world. Physicians rely on CT scanners to acquire important anatomical information on patients. CT images provide a three-dimensional view of the body that is both qualitative and quantitative in nature. CT scans are often used in the diagnosis of cancer, stroke, bone disorders, lung disease, atherosclerosis, heart problems, inflammation, and a range of other diseases and physical ailments, such as a herniated disc and digestive problems.

## KEY TERMS AND CONCEPTS

- **Attenuation Coefficient:** Physical quantity that characterizes the response of a specific material to incoming radiation and the associated reduction in intensity.
- **Back Projection:** Process of reversing the acquired projection measurements from the detector to reconstruct an image set.
- **Ionization:** Process through which an atom loses one or more of its electrons or acquires an electron. The resulting atom is ionized while the radiation that caused the ionization is termed ionizing radiation.
- **Nondestructive Evaluation (NDE):** Process of testing the integrity of a material according to specified standards; usually done using a form of X-ray imaging. In contrast, destructive testing is the process of placing the test material under extensive forces (hence destructive) to evaluate the integrity and strength of the material.
- **Photon:** Elementary particle that carries electromagnetic radiation from all wavelengths, including those of visible light and X rays. Photons travel at the speed of light and interact with matter in different ways depending on their characteristic frequency.
- **Picture Archiving And Computer System (PACS):** Computer-based system of managing radiological information, including patient-specific data and images. PACS is used in all imaging modalities to facilitate the transfer, retrieval, and presentation of images.
- **Tomography:** Method of X-ray photography in which a single section or plane is imaged while eliminating other sections or planes. When this process is done through a computer, it is known as computed tomography.
- **X Ray:** Form of electromagnetic radiation that is characterized by its ionizing potential and short wavelength, ranging from 0.01 to 10 nanometers.

### DEFINITION AND BASIC PRINCIPLES

Computed tomography (CT; also known as computer-aided tomography, or CAT) is an imaging modality that uses X rays and computational and mathematical processes to generate detailed image data of a scanned subject. X rays are generated through an evacuated tube containing a cathode, an anode, and a target material. The high voltage traveling through the tube accelerates electrons from the cathode toward the anode. This is very similar to a light bulb, with the addition of a target that generates the X rays and directs them perpendicularly to the tube. As electrons interact with the target material, small packets of energy, called photons, are produced. The photons have energies ranging from 50 to 120 kilovolts, which is characteristic of X-ray photons. The interaction of X-ray photons with a person's body produces a planar image with varying contrast depending on the density of the tissue being imaged. Bone has a relatively high density and is more readily absorbed by X rays, resulting in a bright image on the X-ray film. Less dense tissues, such as lungs, do not absorb X rays as readily, and as a result, the image produced is only slightly exposed and therefore dark.

In computed tomography, X rays are directed toward the subject in a rotational manner, which generates orthogonal, two-dimensional images. The X-ray tube and the X-ray detector are placed on a rotating gantry, which allows the X rays to be detected at every possible gantry angle. The resulting two-dimensional images are processed through computer algorithms, and a three-dimensional image of the subject is constructed. Because computed tomography relies on the use of ionizing radiation, it has an associated risk of inducing cancer in the patient. The radiation dose obtained from CT procedures varies considerably depending on the size of the patient and the type of imaging performed.

### BACKGROUND AND HISTORY

X rays were first discovered by Wilhelm Conrad Röntgen in November, 1895, during his experiments with cathode-ray tubes. He noticed that when an electrical discharge passed through these tubes, a certain kind of light was emitted that could pass through solid objects. Over the course of his experimentations with this light, Röntgen began to refer to it as an "X ray," as *x* is the mathematical term for an unknown. By the early 1900's, medical use of X rays was widespread in society. They were also used to entertain people by providing them with photographs of their bodies or other items. In 1901, Röntgen was awarded the first Nobel Prize in Physics for his discovery of X rays. During World War II, X rays were frequently used on injured soldiers to locate bullets or bone fractures. The use of X rays in medicine increased drastically by the mid-twentieth century.

The development of computer technology in the 1970's made it possible to invent computed tomography. In 1972, British engineer Godfrey Hounsfield and South African physicist Allan Cormack, who was working at Tufts University in Massachusetts, independently developed computed tomography. Both scientists were awarded the 1979 Nobel Prize in Physiology or Medicine for their discovery.

### HOW IT WORKS

The first generation of CT scanners was built by Electric and Musical Industries (EMI) in 1971. The CT scanner consisted of a narrow X-ray beam (pencil beam) and a single detector. The X-ray tube moved linearly across the patient and subsequently rotated to acquire data at the next gantry angle. The process of data acquisition was lengthy, taking several minutes for a single CT slice. By the third generation of CT scanners, numerous detectors were placed on an arc across from the X-ray source. The X-ray beam in

these scanners is wide (fan beam) and is covered by the entire area of detectors. At any single X-ray emission, the entire subject is in the field of view of the detectors, and therefore linear movement of the X-ray source is eliminated. The X-ray tube and detectors remain stationary, while the entire apparatus rotates about the patient, resulting in a drastic reduction in scan times. Most medical CT scanners in the world are of the third-generation type.

**Mode of Operation.** The process of CT image acquisition begins with the X-ray beam. X-ray photons are generated when high-energy electrons bombard a target material, such as tungsten, that is placed in the X-ray tube. At the atomic level, electrons interact with the atoms of the target material through two processes to generate X rays. One mode of interaction is when the incoming electron knocks another electron from its orbital in the target atom. Another electron from within the atom fills the vacancy, and as a result, an X-ray photon is emitted. Another mode of interaction occurs when an incoming electron interacts with the nucleus of the target atom. The electron is scattered by the strong electric field in the nucleus of the target atom, and as a result, an X-ray photon is emitted. Both modes of interaction are very inefficient, resulting in a considerable amount of energy that is dissipated as heat. Cooling methods need to be considered in the design of X-ray machines to prevent overheating in the X-ray tube. The resulting X-ray beam has a continuous energy spectrum, ranging from low-energy photons to the highest energy photons, which corresponds with the X-ray tube potential. However, since low-energy photons increase the dose to the body and do not contribute to image quality, they are filtered from the X-ray beam.

Once a useful filtered X-ray beam is generated, the beam is directed toward the subject, while an image receptor (film or detector) is placed in the beam direction past the subject to collect the X-ray signal and provide an image of the subject. As X rays interact with the body's tissues, the X-ray beam is attenuated by different degrees, depending on the density of the material. High-density materials, such as bone, attenuate the beam drastically and result in a brighter X-ray image. Low-density materials, such as lungs, cause minimal attenuation of the X-ray beam and appear dark on the X-ray image because most of the X rays strike the detector.

**Image Acquisition.** In CT, X-ray images of the subject are taken from many angles and reconstructed into a three-dimensional image that provides an excellent view of the scanned subject. At each angle, X-ray detectors measure the X-ray beam intensities, which are characteristic of the attenuation coefficients of the material through which the X-ray beam passes. Generating an image from the acquired detector measurements involves determining the attenuation coefficients of each pixel within the image matrix and using mathematical algorithms to reconstruct the raw image data into cross-section CT image data.

### APPLICATIONS AND PRODUCTS

The power of computed tomography to provide detailed visual and quantitative information on the object being scanned has made it useful in many fields and suitable for numerous applications. Aside from disease diagnosis, CT is also commonly used as a real-time guide for surgeons to accurately locate their target within the human body. It is also used in the manufacturing industry for nondestructive evaluation of manufactured products.

**Disease Diagnosis.** The most common use for CT is in radiological clinics, where it is used as an initial procedure to evaluate specific patients' complaints or symptoms, thereby providing information for a diagnosis, and to assess surgical or treatment options. Radiologists, medical professionals specialized in reading and analyzing patient CT data, look for foreign bodies such as stones, cancers, and fluid-filled cavities that are revealed by the images. Radiologists can also analyze CT images for size and volume of body organs and detect abnormalities that suggest diseases and conditions involving changes in tissue density or size, such as pancreatitis, bowel disease, aneurysms, blood clots, infections, tuberculosis, narrowing of blood vessels, damaged organs, and osteoporosis.

In addition to disease diagnosis, CT has been used by private radiological clinics to provide full-body scans to symptom-free people who desire to obtain a CT image of their bodies to ascertain their health and to detect any conditions or abnormalities that might indicate a developing problem. However, the use of CT imaging for screenings in the absence of symptoms is controversial because the X-ray radiation used in CT has an associated risk of cancer. The dose of radiation deposited by a CT scan is between fifty and two hundred times the dose deposited by a

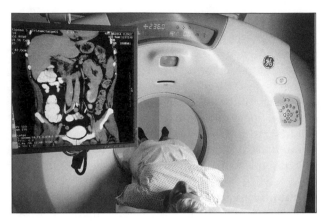

*Patient undergoing a CT scan.* (Mauro Fermariello/ Photo Researchers, Inc.)

conventional X-ray image. Although the association between CT imaging and cancer induction is not well established, its casual use remains an area of considerable debate.

**Guided Biopsy.** A biopsy is a time-consuming, sometimes inaccurate, and invasive procedure. Traditionally, doctors obtained a biopsy (sample) of the tissue of interest by inserting a needle into a patient at the approximate location of the target tissue. However, real-time CT imaging allows doctors to observe the location of the biopsy needle within the patient. Therefore, doctors can obtain a more accurate tissue biopsy in a relatively short time and without using invasive procedures.

**CT Microscopy.** The resolution of clinical CT scanners is limited by practical scan times for the patients and the size of X-ray detectors used. In the case of small animals, higher resolutions can be obtained by use of smaller detectors and longer scan time. The field of microCT, or CT microscopy, has rapidly developed in the early twenty-first century as a way of studying disease pathology in animal models of human disease. Numerous disorders can be modeled in small animals, such as rats and mice, to obtain a better understanding of the disease biology or assess the efficacy of emerging treatments or drugs. Traditionally, animal studies would involve killing the animal at a specific time point and processing the tissue for viewing under the microscope. However, the development of CT microscopy has allowed scientists and researchers to investigate disease pathology and treatment efficacy at very high resolutions while the animal is alive, reaching one-fifth of the resolution of a light microscope.

**Nondestructive Evaluation.** Computed tomography has gained wide use in numerous manufacturing industries for nondestructive evaluation of composite materials. Nondestructive evaluation is used to inspect specimens and ensure the integrity of manufactured products, either through sampling or through continuous evaluation of each product. CT requirements for nondestructive testing differ from those for medical imaging. For medical imaging, scan times have to be short, exposure to radiation has to be minimal, and patient comfort throughout the procedure must be taken into consideration. For nondestructive evaluation, patient comfort and exposure to radiation are no longer important issues. However, keeping scan times short is advantageous, especially for large-scale industries. Furthermore, the X-ray energy for scanning industrial samples can vary significantly from the energy used for patient imaging, since patient composition is primarily water and industrial samples can have a wide range of compositions and associated densities. Engineers have custom-designed CT scanners for specific materials, including plastics, metals, wood, fibers, glass, soil, concrete, rocks, and various composites. The capability of CT to provide excellent qualitative image data and accurate quantitative data on the density of the specimen has made it a powerful tool for nondestructive evaluation. CT is used in the aerospace industry to ensure the integrity of various airplane components and in the automotive industry to evaluate the structure of wheels and tires. In addition to industrial applications, CT is commonly used in research centers to further their imaging and analytical power.

## IMPACT ON INDUSTRY

CT has revolutionized many industries by providing detailed information on specific physical parameters of materials along with high-quality three-dimensional images. CT has been successfully incorporated into the workflow of many companies—including aerospace and aviation, plastic, casting, electronic, medical device, pharmaceutical, and dentistry industries—and research laboratories, museums, and archaeological centers. The continued development of CT scanners will allow for improved spatial resolution, reduced scan times, and customized designs.

**Academic Research.** Computed tomography has found many scientific applications in academic and research centers across North America and Western Europe. Many engineering departments have dedicated research centers that incorporate CT for research into nondestructive testing.

Medical research into the application of CT for robotic surgery and biopsy is becoming increasingly popular. Furthermore, the manufacturing of compact CT scanners used for animal imaging has allowed the incorporation of CT into basic science laboratories for investigating disease mechanisms and progression and for evaluating new drugs. Using CT microscopy, researchers can track the progression of disease and the effects of treatment in animals while they are alive, rather than killing the animals to obtain the necessary tissues for analysis with conventional light microscopy.

**Major Corporations.** Some of the major corporations that have been involved in the manufacture and development of CT scanners are GE Healthcare, Hitachi Medical Corporation, Philips Medical Systems, Siemens Medical Systems, Shimadzu Medical Systems, and Toshiba Medical Systems.

## CAREERS AND COURSE WORK

CT is an imaging modality that exists primarily in the health care sector and is available in every major hospital and in most small hospitals in the developed world. CT is also widely distributed in hospitals in the developing world. Experts in the operation and maintenance of CT scanners who live in developed nations are often recruited to work in less-developed nations.

Those who wish to pursue careers in CT can take many paths. A degree in physics or engineering with a specialization in biomedical or electrical engineering can provide a solid grounding in the electronics and hardware of CT. Engineers can find work in hospitals or in industries that design and manufacture CT scanners. Degrees in mathematics and computer science provide the necessary background for working with image reconstruction algorithms and advancing software-related operation of CT scanners. Computer programmers can find work in the research and development sector of CT industries.

A graduate degree in medical physics provides theoretical and practical experience in medical imaging, from data acquisition to image reconstruction,

## Fascinating Facts About Computed Tomography

- Computed tomography (CT) is one of the milestone discoveries of the twentieth century that are based on X-ray imaging.
- CT provides three-dimensional and sectional views of the human body. CT images of a patient's body can be correlated to a cross section of the human body in real life. The ability of CT to see inside the body has revolutionized modern medicine.
- CT can produce images of tissues in small animals, such as mice and rats, that approximate the resolution of a microscope. CT microscopy has become an integral tool for understanding disease pathology and evaluation of treatment efficacy in animal models of disease.
- The automotive industry uses CT scanners to evaluate the integrity of parts, such as rims and tires, without destroying them.
- CT produces not only high-quality images but also quantitative data on the density of the scanned material.
- In the United States, more than 6,000 CT scanners are being used to create more than 62 million CT scans per year.

interpretation, and troubleshooting. Medical physicists often work in imaging facilities and hospitals, where they usually supervise the personnel operating CT scanners. A degree in medicine with specialization in radiology provides theoretical and practical experience in understanding human anatomy, pathology, and physiology; in interpreting CT images; and in diagnosing specific disease conditions. Technical colleges can provide education in the operation of CT scanners. Graduates with a technical degree in CT imaging often work as technologists in hospitals, where they are responsible for patient scheduling and CT operation. CT is invaluable in the medical sector, and career prospects have been good.

### SOCIAL CONTEXT AND FUTURE PROSPECTS

The number of CT scanners and scans being performed have risen considerably since the 1980's. By the early twenty-first century, there were more than 6,000 scanners in the United States, and more than

62 million CT scans were being performed every year. The rapid and wide acceptance of CT scanners in health care institutions has sparked controversy in the media and among health care practitioners regarding the radiation doses being delivered through the scans. Risk of cancer induction rises with increased exposure to radiation, and this risk has to be carefully weighed against the benefits of a CT scan. The issue of cancer induction is more alarming when CT procedures are performed on young children or infants. Studies have recommended that CT scanning of children should not be performed using the same protocols as used for adults because the children are generally more sensitive to radiation than adults. More strict federal and state regulations are being instituted to better control the use of CT in health care.

*Ayman Oweida, B.Sc., M.Sc.*

### FURTHER READING

Aichinger, H., et al. *Radiation Exposure and Image Quality in X-ray Diagnostic Radiology: Physical Principles and Clinical Applications.* New York: Springer, 2003. Provides the reader with a strong foundation in radiation physics, especially in relation to X-ray medical imaging. It also describes many of the concepts of X-ray production, as well as its interaction with matter.

Bossi, R. H., K. D. Friddell, and A. R. Lowrey. "Computed Tomography." In *Non-destructive Testing of Fibre-reinforced Plastics Composites,* edited by John Summerscales. New York: Elsevier Science, 1990. Covers various applications of CT in relation to composite material. Provides a basic review of various CT concepts important to nondestructive testing of materials, as well as various illustrations and data on CT measurements in fiber-reinforced plastics.

Brenner, David, and Eric Hall. "Computed Tomography: An Increasing Source of Radiation Exposure." *New England Journal of Medicine* 357, no. 22 (2007): 2277-2284. An analytical overview of the use of CT and the risk of cancer induction from radiation exposure. Illustrations and statistical data on various CT procedures.

Hsieh, Jiang. *Computed Tomography: Principles, Design, Artifacts, and Recent Advances.* 2d ed. Bellingham, Wash.: International Society for Optical Engineering, 2009. A comprehensive overview of the fundamentals of CT in medical imaging. Reviews concepts in physics and mathematics that are necessary for a detailed understanding of CT operation and development and image analysis.

Otani, Jun, and Yuzo Obara, eds. *X-ray CT for Geomaterials: Soils, Concrete, Rocks.* Rotterdam, the Netherlands: A. A. Balkema, 2004. This collection of lectures and papers presented at a workshop on X-ray CT for geomaterials in 2003 is a broad and technical overview of the numerous applications of CT in imaging composite materials. Abundant diagrams and illustrations of the various applications.

**WEB SITES**
*MedlinePlus*
CT Scan
http://www.nlm.nih.gov/medlineplus/ency/article/003330.htm

*U.S. Food and Drug Administration*
What Is Computed Tomography?
http://www.fda.gov/Radiation-EmittingProducts/RadiationEmittingProductsandProcedures/MedicalImaging/MedicalX-Rays/ucm115318.htm

**See also:** Magnetic Resonance Imaging; Radiology and Medical Imaging; Ultrasonic Imaging.

# COMPUTER-AIDED DESIGN AND MANUFACTURING

## FIELDS OF STUDY

Three-dimensional modeling; aerospace engineering; architecture; civil engineering; computer animation; computer engineering; electrical engineering; graphic design; industrial design; landscaping; mechanical engineering; plant design; product design; ray tracing; textiles; texture mapping; virtual engineering.

## SUMMARY

Computer-aided design and manufacturing is a method by which graphic artists, architects, and engineers design models for new products or structures within a computer's "virtual space." It is alternatively referred to by a host of other terms related to specific fields: computer-aided drafting (CAD), computer-aided manufacturing (CAM), computer-aided design and drafting (CADD), computer-aided drafting/computer-aided manufacturing (CAD/CAM), and computer-aided drafting/numerical control (CAD/NC). The generic term computer-aided technologies is, perhaps, the most useful in describing the computer technology as a whole. Used by professionals in many different disciplines, computer-aided technology allows designers in traditional fields, such as architecture, to render and edit two-dimensional (2-D) and three-dimensional (3-D) structures. Computer-aided technology has also revolutionized once highly specialized fields, such as video game design and digital animation, into broad-spectrum fields that have applications ranging from the film industry to ergonomics.

## KEY TERMS AND CONCEPTS

- **Assembly Modeling:** Design that describes how parts and subassemblies can be put together into a single unit.
- **Critical Path Method:** Spreadsheet or activity diagram that schedules the time frame for each stage of production.
- **Gannt Chart:** Graphical representation (perhaps in bar chart or line format) that expresses a project schedule.

- **Kinematics:** Study of the movement of objects.
- **Numeric Control (NC):** Control of a machine's activity by the use of commands recorded in alphanumeric format.
- **Parametric Design:** Ability of a computer system to translate graphics (computer-drawn images) into mathematical calculations and calculations back into graphics.
- **Part Libraries:** Catalog or database of standard drawings of small parts or symbols used frequently in CAD designs.
- **Product Lifecycle Management (PLM):** Process describing the entire life cycle of a product, from concept to disposal.
- **Rapid Prototyping:** Function of computer-aided manufacturing that constructs physical objects by layer fabrication (for example, fabricating parts by melting layers of metal powder).
- **Vector Graphics:** Use of mathematically derived points, lines, and curves (or polygons based on the combination of lines and curves) to create computer graphic images.

### DEFINITION AND BASIC PRINCIPLES

Computer-aided design and manufacturing is the combination of two different computer-aided technologies: computer-aided design (CAD) and computer-aided manufacturing (CAM). Although the two are related by their mutual reliance on task-specific computer software and hardware, they differ in how involved an engineer or architect must be in the creative process. A CAD program is a three-dimensional modeling software package that enables a user to create and modify architectural or engineering diagrams on a computer. CAD software allows a designer to edit a project proposal or produce a model of the product in the virtual "space" of the computer screen. CAD simply adds a technological layer to what is still, essentially, a user-directed activity. CAM, on the other hand, is a related computer-aided technology that connects a CAD system to laboratory or machine shop tools. There are many industrial applications for computer-directed manufacturing. One example of CAD/CAM might be when an automotive designer creates a three-dimensional model of a proposed design using a CAD program, verifies the physical details of the design in a computer-aided engineering (CAE) program, and then constructs a physical model via CAD or CAE interface with CAM hardware (using industrial lasers to burn a particular design to exact specifications out of a block of paper). In this case, the automotive designer is highly involved with the CAD program but allows the CAD or CAE software to direct the activity of the CAM hardware.

### BACKGROUND AND HISTORY

CAD/CAM is, essentially, a modernization of the age-old process of developing an idea for a structure or a product by drawing sketches and sculpting models. Intrigued by the idea that machines could assist designers in the production of mathematically exact diagrams, Ivan Sutherland, a doctoral student at the Massachusetts Institute of Technology, proposed the development of a graphical user interface (GUI) in his 1963 dissertation "Sketchpad: A Man-Machine Graphical Communication System." GUIs allow users to direct computer actions via manipulation of images, rather than by text-based commands and sometimes requires the use of an input device (like a mouse or a light pen) to interact with the GUI images, also known as icons. Although it was never commercially developed, Sketchpad was innovative in that a user could create digital lines of a screen and determine the constraints and parameters of the lines via buttons located along the side of the screen. In succeeding decades, the CAD software package employed the use of a mouse, light pen, or touch screen to input line and texture data through a GUI, which was then manipulated by computer commands into a virtual, or "soft," model.

In the twenty-first century, CAM is more or less the mechanical output aspect of CAD work, but in the early days, CAM was developed independently to speed up industrial production on assembly lines. Patrick J. Hanratty, regarded as the father of CAD/CAM, developed the first industrial CAM using numerical control (NC) in 1957. Numerical control—the automation of machining equipment by the use of numerical data stored on tape or punched on cards—had been in existence since the 1940's, as it was cheaper to program controls into an assembly machine than to employ an operator. Hanratty's PRONTO, a computer language intended to replace the punched-tape system (similar to punch cards in that the automaton motors reacted to the position of

holes punched into a strip of tape) with analogue and digital commands, was the first commercial attempt to modernize numerical control and improve tolerances (limiting the range of variation between the specified dimensions of multiple machined objects).

### HOW IT WORKS

Traditionally, product and structural designers created their designs on a drawing board with pencils, compasses, and T squares. The entire design process was a labor-intensive art that demanded not just the ability to visualize a proposed product or structure in exact detail, but also demanded designers to possess extensive artistic abilities to render what, up until that point, had been only imagined. Changes to a design demanded either by a client's specific needs or by the limitations of the materials needed for construction frequently required the designer to "go back to the drawing board" and create an entirely new series of sketches. CAD/CAM was intended to speed up the design and model-making processes dramatically.

**Computer-Aided Design (CAD).** CAD programs function in the same way that many other types of computer software function: One loads the software into a computer containing an appropriate amount of memory, appropriate video card, and appropriate operating system to handle the mathematical calculations required by the three-dimensional modeling process. There are quite literally hundreds of different CAD programs for different applications available for trial use or purchase. Some programs are primarily two-dimensional drawing applications, while others allow three-dimensional drawing, shading, and rendering. Some are intended for use in construction management, incorporating project schedules, or for engineering, focused on structural design. There are a range of CAD programs of varying complexity for computer systems of different capabilities or differing operating systems.

**Computer-Aided Manufacturing (CAM).** CAM, on the other hand, tends to be a series of automated mechanical devices that interface with a CAD program installed in a computer system—typically, but not exclusively, a mainframe computer. The typical progress of a CAD/CAM system from design to soft model might involve an automotive designer using a light pen on a CAD system to sketch out the basic outlines of suggested improvements for an existing vehicle. Once the car's structure is defined, the designer can

## Fascinating Facts About Computer-Aided Design and Manufacturing

- The earliest commercial uses of CAD, in the late 1960's, were in the automobile and aerospace industries. Ford, McDonnell-Douglas and Lockheed Martin all created proprietary CAD software programs.
- Will Wright, who designed *The Sims*, initially intended his engrossing people simulator to be a home creation and design tool.
- *Second Life*, another virtual life simulator, created by Linden Lab, had an active online player base of around 500,000 in 2007.
- *World of Warcraft*, developed by Blizzard Entertainment, had an active online player base of 10 million in 2008.
- Computer-aided tissue engineering is an example of the growing field of biotechnology. This specialty uses computers to assist in the development of artificial scaffolds for the growth of human tissues, such as ear and nose cartilage replacement, and the development of tissue models for artificial replacement of lost tissue.
- Fashion design is another industry where CAD is employed. The patternmakers who bring to fruition the designers' creations use CAD to determine the most efficient cuts of fabrics and to alter the pattern sizes easily.

then command the CAD program to overlay a surface, or "skin," on top of the skeletal frame. Surface rendering is a handy tool within many CAD programs to play with the use of color or texture in a design. When the designer's conceptual drawings are complete, they are typically loaded into a computer-aided engineering program to ascertain the structural integrity and adherence to the requirements of the materials from which the product will ultimately be constructed. If the product design is successful, it is then transferred to a computer-aided manufacturing program, where a product model, machined out of soft model board, can be constructed to demonstrate the product's features to the client more explicitly.

Advertisements for CAD programs tend to emphasize how the software will enhance a designer's imagination and creativity. They stress the many different artistic tools that a designer can easily

manipulate to create a virtual product that is as close as possible to the idealized conceptual design that previously existed only in the designer's imagination. On the other hand, CAM advertisements stress a connection of the CAM hardware to real-world applications and a variety of alterable structural materials ranging from blocks of paper to machinable steel. The two processes have fused over the succeeding decades partly because of the complementary nature of their intended applications and partly because of clients' need to have a model that engages their own imaginations, but certain aspects of CAD (those technological elements most appropriate to the creation of virtual, rather than actual, worlds) may eventually allow the software package to reduce the need for soft models in terms of product development.

**CAD in Video Games and Film Animation.** CAD work intended for video games and film animation studios, unlike many other types of designs, may intentionally and necessarily violate one basic rule of engineering—that a design's form should never take precedence over a design's function. Industrial engineers also stress the consideration of ergonomics (the requirements of the human body are considered when a design for a product is developed) in the design process. In the case of the virtual engineer, however, the visual appeal of a video game or animation element may have the greater say in just how a design is packaged for a client. *Snow White and the Seven Dwarfs* was an animated film originally drawn between 1934 and 1937. Since animation was drawn by hand in those days—literally on drawing boards—with live models for the animated characters, the studio's founder, Walt Disney, looked for ways to increase the animated character's visual appeal. One technique he developed was to draw human characters (in this case, Snow White herself) with a larger-than-standard size head, ostensibly to make her appear to have the proportions of a child and, thus, to be more appealing to film viewers. In order that the artists might draw their distorted heroine more easily, the actress playing Snow White, Marjorie Bell, was asked to wear a football helmet while acting out sections of the story's plot. Unfortunately for Disney, Bell complained so much about the helmet that it had to be removed after only a few minutes. Ergonomics took precedence over design. Later animation styles, influenced by designers' use

*A designer using a CAD package to check the design of a piece of electrical equipment.* (Geoff Tompkinson/Photo Researchers, Inc.)

of CAD animation software, were able to be even more exaggerated without the need for a live model in an uncomfortable outfit. For example, the character of Jessica Rabbit in Amblin Entertainment's *Who Framed Roger Rabbit?* (1988) has a face and body that were designed exclusively for visual appeal, rather than to conform to external world reality. Form dominated function.

## APPLICATIONS AND PRODUCTS

CAD and CAD/CAM have an extensive range and variety of applications.

**Medicine.** In the medical field, for example, Abaqus 6.10-EF finite element analysis (FEA) allows designers to envision new developments in established medical devices (improving artificial heart valves, for example) or the creation of new medical devices. CAD software, ideally, allows a user to simulate real-world performance.

**Aerospace.** Femap and NEi Nastran are software applications used for aerospace engineering. It can simulate the tensile strength and load capacity of a variety of structural materials (aluminum, etc.) as well as the effect of weather on performance and durability.

**Mechanical Engineering.** Mechanical engineering firms sometimes use Siemens PLM Software NX suite for CAM or CAE functions because they require

their software to create and revise structural models quickly.

Some forms of CAD have engineering applications that can anticipate a design's structural flaws to an even greater extent than an actual model might. One example of this application is SIMULIA, one of the simulation divisions of Dassault Systèmes that creates virtual versions of the traditional automobile safety tester—the crash-test dummy. The company's program of choice is Abaqus FEA, but it has altered the base software to use BioRID (Biofidelic Rear Impact Dummy) and WorldSID (World Side Impact Dummy) as well as other altered models for a wider variety of weight and height combinations. In the real world, crash-test dummies are rather complex mechanical devices, but their virtual simulacra have been declared as effective in anticipating structural failure as the original models. Because the nature of CAD is to make duplicating a successful element of a design quickly and easily, SIMULIA has been able to reduce the time spent on validating its simulations from four weeks to four days.

**Animation.** CAD programs have caused the rapid expansion and development of the animation industry—both in computer games and in film animation. Models of living creatures, both animal and human, are generated through the use of three-dimensional CAD software such as Massive (which was used by the graphic artists creating the battle sequences in Peter Jackson's *The Lord of the Rings* trilogy) and Autodesk's Maya (considered an industry standard for animated character designs). Autodesk's 3ds Max is one of the forerunner programs in the video game industry, since it allows a great deal of flexibility in rendering surfaces. Certain video games, such as Will Wright's *Sims* series, as well as any number of massively multiplayer online role-playing games (MMORPGs) such as *World of Warcraft*, *Second Life*, and *EverQuest*, not only have CAD functions buried in the design of the game itself but accept custom content designed by players as a fundamental part of individualization of a character or role. A player must design a living character to represent him or her in the video game including clothing, objects carried, and weapons used, and the process of creation is easily as enjoyable as the actual game play.

*Avatar*, released by Twentieth Century Fox in 2009, was envisioned by director James Cameron as being a sweeping epic in very much the same vein as George Lucas's *Star Wars* trilogy (1977, 1980, and 1983). *Avatar* was originally scheduled to start filming in 1997, after Cameron completed *Titanic*, but Cameron felt the computer-aided technology was not ready to create the detailed feature film he envisioned. Given that the film was also supposed to take advantage of newly developed three-dimensional stereoscopic film techniques, Cameron started work on the computer-aided technology aspects of his film only when he was sure the technology was up to the task—2006.

Even with an extensive three-dimensional software animation package at his command, the task of creating and populating an entire alien world was a daunting prospect. Most of the action of *Avatar* takes place on a distant planet, Pandora, which is intrinsically toxic to human beings. To get around this difficulty and successfully mine Pandora for a much-needed mineral (called, amusingly, Unobtainium), humans must use genetically created compatible bodies and interact with the native population, the Na'vi. Both the native creatures, who are ten feet tall with blue skin, and the hybrid human-Na'vi "avatar" bodies were a challenge for animators to render realistically even using computer-aided technology. The full IMAX edition was released on December 18, 2009, having ultimately cost its creators somewhere between $280 million and $310 million. Fortunately for Cameron, the film ended up rewarding its creators with gross revenues of more than $2 billion. Having created the highest-grossing film of all time, which also won Academy Awards for art direction, cinematography, and visual effects, Cameron was able to negotiate successfully with Fox for two planned sequels (echoing the original *Star Wars* trilogy).

## IMPACT ON INDUSTRY

There have been two schools of thought regarding the impact of CAD/CAM on industry. On one hand, the cost of producing new ideas for products or structures has dropped dramatically because of the quick processing facility of the computer processor and video card. Designers are no longer required to anticipate every angle of a new building or aircraft—it is possible to input just enough data for the computer to interpolate the views from other perspectives. CAD/CAM also saves money in the fabrication of models because it is able to control the cutting of angles and lines to .0001 inch—something that could not reliably be produced by a human hand. The ease of correcting flaws in a CAD-generated design allows for multiple drafts and more intensive collaboration between designer and client. As CAD programs dropped in price in the early twenty-first century, even small design companies could afford to upgrade to CAD, increasing the competition (and lowering the cost of a service) in a given market.

The other school of thought, mostly espoused by designers in the traditional fields of architecture and engineering, is the difficulty of combining the sometimes foreign concepts of three-dimensional modeling with traditional, human methods of expressing traditional design concepts. For example, some architects dislike CAD because the computer system tends to be based on three-dimensional solids modeling rather than traditional architectural terminology. Architecture evolved to use special terminology that reflects a particular understanding of buildings. When an architect designs a house using hand-drawn techniques and conceives of the "eaves" and "hip" of the roof, he or she must, because of the terminology used, see the roof as a series of peaks and valleys that visually describes spaces as well as surfaces. On the other hand, when a designer uses an older-generation CAD program for architectural purposes, he or she may be completely free from the strictures of the construction materials and the weight of what has already been built. As a consequence, the designer may conceive of the house as a cube and a pyramid that intersect at a particular height—visualizations of solid models that, while they may function adequately in the virtual world of the software, alienate the real-world sensibilities of the traditional drawing-board architect. The complaints occasionally raised by engineers and architects have caused some to question CAD's feasibility, at least at the initial creation stage, for the development of designs that rely on as deep an understanding of engineering as art.

To a greater or lesser degree, depending on the application, computer-aided technology has lowered the costs and increased the efficiency of designers at the drawing board—corrections can be made easily and quickly to product designs, and this can save an engineering or product-design firm time and money in the design and production stages. For some fields, however, the changes caused by computer-aided technology are immense and groundbreaking—as in the field of film visual effects. *Star Wars*, writer-director

George Lucas's groundbreaking six-film science fiction classic, is a rare example of how computer-aided technology operates as a medium for the benefits of uninhibited imagination. The original film, released in 1977, was intentionally left incomplete because Lucas believed that computer-aided technology, then in its infancy, would eventually catch up so he could effectively realize the vision he had for his magnum opus. Several scenes were intentionally filmed without any attempt at visual effects, including one of actor Harrison Ford, in the role of Han Solo, trying to win over the confidence of the slug-like crime boss Jabba the Hutt (in the original shot, just a heavyset man in a tunic). Lucas retained these scenes, kept out of the 1977 version, and waited for the visual effects division of his Lucasfilm, Industrial Light & Magic, to integrate the new computer-animation technology he felt would change the film industry.

By 1997, Lucas felt that computer-aided technology had advanced enough to make the designed alterations to his original work. The unaltered scenes were extensively altered using CAD. The scene of Han Solo and Jabba the Hutt was changed so dramatically that the original actor standing in the place of the future Jabba the Hutt cannot be seen under the CAD rendering. Apparently, Lucas has not yet declared his film epic to be complete. In 2010, he announced that all six *Star Wars* films would be remade with the computer-animation software used for *Avatar*.

## CAREERS AND COURSE WORK

Careers in computer-aided design and manufacturing tend to be affected most obviously by the intended purpose of the designer's work. There could be significant differences in the course work required of a medical-device designer and that required of a film-animation specialist. Aside from course work in rapid visualization, CAD operation, graphic illustration, digital type and image manipulation, digital photography, new media, and three-dimensional modeling, those who wish to work with computer-aided technology should consider supplemental course work in the field most relevant to their area of design.

## SOCIAL CONTEXT AND FUTURE PROSPECTS

Computer-aided technologies, in general, are expected to keep developing and evolving. The possibilities of virtual reality are continuing to evolve as the need for structural and materials modeling increases.

One common description of the future of computer-aided design and manufacturing is the term "exponential productivity." In other words, so much productive time in the fields of design has been wasted by the constant need to create by hand what can be better (and more quickly) produced by a computer. An architect or engineer might spend hours carefully working out the exact proportions of a given structure, more hours using a straight edge and compass to determine the exact measurements needed by the design, and even more time to clean up mistakes. On the other hand, computers can help that same architect draw straight lines, evaluate numerical formulas for proportions, and edit mistakes in a matter of minutes. When one considers the added benefits of a compatible computer-aided manufacturing program, one can further see the time saved in the quick and efficient production of a soft model.

*Julia M. Meyers, M.A., Ph.D.*

## FURTHER READING

Duggal, Vijay, Sella Rush, and Al Zoli, eds. *CADD Primer: A General Guide to Computer-Aided Design and Drafting*. Elmhurst, N.Y.: Mailmax Publishing, 2000. An excellent introduction to CADD that is applicable to most CAD software programs.

Kerlow, Issac. *The Art of 3D Computer Animation and Effects*. 4th ed. Hoboken, N.J.: John Wiley & Sons, 2009. Written by a former Walt Disney production executive, this brings the traditional principles of animation into the computer age. The step-by-step guidelines to creating 3-D animation are written to work with any computer platform.

Lee, Kunwoo. *Principles of CAD/CAM/CAE Systems*. Reading, Mass.: Addison-Wesley, 1999. A good theoretical introduction to all computer-aided technologies for students of any level; also included is a case study that illustrates the product-development process.

Park, John Edgar. *Understanding 3D Animation Using Maya*. New York: Springer, 2005. An overview of computer animation using Autodesk's Maya, with hands-on tutorials and practical projects at the end of each chapter.

Schell, Jesse. *The Art of Game Design: A Book of Lenses*. Burlington, Mass.: Morgan Kaufmann, 2008. The author is the former chair of the International Game Developers Association, and his book is meant to inspire would-be designers and challenge them to look at their work through various "lenses": mathematics, music, design, film, among others.

Schodeck, Daniel, et al. *Digital Design and Manufacturing: CAD/CAM Applications in Architecture and Design.* Hoboken, N.J.: John Wiley & Sons, 2005. Comprehensive view of industrial design with case studies illustrating the work of architects Frank Gehry, Bernhard Franken, and Rafael Viñoly, among others.

**WEB SITES**

*Association for Computer Aided Design in Architecture (ACADIA)*
http://www.acadia.org

*Association for Computing Machinery's Special Interest Group on Computer Graphics and Interactive Techniques (ACM SIGGRAPH)*
http://www.siggraph.org

*Association of Professional Model Makers*
http://www.modelmakers.org

*International Game Developer's Association*
http://www.igda.org

**See also:** Civil Engineering; Computer Engineering; Computer Graphics; Human-Computer Interaction; Mechanical Engineering; Software Engineering; Video Game Design and Programming; Virtual Reality.

01028221
595 set